The Enzymes

VOLUME XXII

CO- AND POSTTRANSLATIONAL PROTEOLYSIS OF PROTEINS

Third Edition

THE ENZYMES

Edited by

Ross E. Dalbey

*Department of Chemistry
and Newman and Wolfrom
Laboratory of Chemistry
Ohio State University
Columbus, Ohio*

David S. Sigman

*Department of Biological Chemistry
and Molecular Biology Institute
University of California
School of Medicine
Los Angeles, California*

Volume XXII

CO- AND POSTTRANSLATIONAL
PROTEOLYSIS OF PROTEINS

THIRD EDITION

ACADEMIC PRESS
San Diego   London   Boston
New York   Sydney   Tokyo   Toronto

Academic Press
A Harcourt Science and Technology Company
525 B Street, Suite 1900, San Diego, California 92101-4495, USA
http://www.academicpress.com

Academic Press
Harcourt Place, 32 Jamestown Road, London NW1 7BY, UK
http://www.academicpress.com

Library of Congress Catalog Card Number: 00-107123

International Standard Book Number: 0-12-122723-5

PRINTED IN THE UNITED STATES OF AMERICA
01 02 03 04 05 06   SB   9 8 7 6 5 4 3 2 1

Contents

Section I
Signal Peptide Processing

1. The Eubacterial Lipoprotein-Specific (Type II) Signal Peptidase

HAROLD TJALSMA, GEESKE ZANEN, SIERD BRON, AND
JAN MAARTEN VAN DIJL

2. Bacterial Type I Signal Peptidases

JOSEPH L. CARLOS, MARK PAETZEL, PHILIP A. KLENOTIC,
NATALIE C. J. STRYNADKA, AND ROSS E. DALBEY

3. Structure and Function of the Endoplasmic Reticulum Signal Peptidase Complex

NEIL GREEN, HONG FANG, STEPHEN MILES, AND MARK O. LIVELY

4. Mitochondrial Processing Peptidase/Mitochondrial Intermediate Peptidase

JIRI ADAMEC, FRANTISEK KALOUSEK, AND GRAZIA ISAYA

5. Chloroplast and Mitochondrial Type I Signal Peptidases

CHRISTOPHER J. HOWE AND KEVIN A. FLOYD

6. Type IV Prepilin Leader Peptidases

MARK S. STROM AND STEPHEN LORY

Section II
Proprocessing

7. The Prohormone Convertases and Precursor Processing in Protein Biosynthesis

Donald F. Steiner

8. Furin

Sean S. Molloy and Gary Thomas

9. Cellular Limited Proteolysis of Precursor Proteins and Peptides

Nabil G. Seidah

10. Yeast Kex2 Protease

NATHAN C. ROCKWELL AND ROBERT S. FULLER

11. The Enzymology of PC1 and PC2

A. CAMERON, E. V. APLETALINA, AND I. LINDBERG

Section III
Other Proteases That Cleave Proteins

12. Self-Processing of Subunits of the Proteasome

ERIKA SEEMÜLLER, PETER ZWICKL,
AND WOLFGANG BAUMEISTER

13. Tsp and Related Tail-Specific Proteases

KENNETH C. KEILER AND ROBERT T. SAUER

14. Co- and Posttranslational Processing: The Removal of Methionine

RALPH A. BRADSHAW, CHRISTOPHER J. HOPE, ELIZABETH YI, AND KENNETH W. WALKER

15. Carboxypeptidases E and D

LLOYD D. FRICKER

Preface

Finding over 100 posttranslational modifications to proteins is remarkable and underscores the importance of these processes for the structure–function of proteins within living cells. Many of these modifications require a protease for the generation of the biological active mature proteins.

The key incentive for developing this volume was to concentrate on proteases involved in some of these maturation steps, while emphasizing the processing of proteins with signal peptides and propeptides. Proteases remove the N-terminal extension peptides from exported, secreted, and organellar proteins and are also involved in the maturation of prohormones and other precursor proteins. Other proteases are involved in C-terminal and amino-terminal maturation. Rarely are these types of proteases included in one volume. Increasing attention was paid to this field recently when Gunter Blobel received the Nobel Prize in Medicine for his work in protein targeting and his proposal that proteins have intrinsic targeting sequences that target them to their correct location in the cell.

Another incentive for this work was that many of the scientists studying the group of proteases on signal peptide processing were unaware of other work on propeptide processing because the work is often in nonoverlapping fields and, with few exceptions, scientists do not present their findings at the same meetings. The inclusion in one volume of the proteases involved both in signal peptide processing and propeptide processing is to emphasize their common importance in proteolysis in the export and secretion pathway.

The first section of this volume concentrates on the type I signal peptidases that process exported proteins, namely, in bacteria, the endoplasmic reticulum, and chloroplast. This section also includes chapters on signal peptidases involved in processing lipoprotein precursor proteins, type IV prepilin precursors, or in processing of mitochondrial preproteins. The second section emphasizes the proteases involved in preprocessing, including the prohormone convertases Furin, Kex2, and others that are responsible for precursor processing in the secretion pathway. The third and last section focuses on other proteases, including tail-specific proteases, methionine aminopeptidases, and carboxypeptidases E and D.

I would like to thank the contributors for their work and for making this project enjoyable.

Ross E. Dalbey

Section I

Signal Peptide Processing

1

The Eubacterial Lipoprotein-Specific (Type II) Signal Peptidase

HAROLD TJALSMA* • GEESKE ZANEN[†] • SIERD BRON* •
JAN MAARTEN VAN DIJL[†]

Department of Genetics
Groningen Biomolecular Sciences and Biotechnology Institute
9750 AA Haren, The Netherlands

[†] *Department of Pharmaceutical Biology*
University of Groningen
9713 AV Groningen, The Netherlands

THE ENZYMES, Vol. XXII

I. General Introduction

One of the most commonly used eubacterial sorting (retention) signals for proteins exported from the cytoplasm is an amino-terminal
lipid-modified Cys residue (*1, 2*). In gram-positive eubacteria, lipid-modified
proteins (lipoproteins) are retained in the cytoplasmic membrane. In
gram-negative eubacteria, these proteins are retained in the cytoplasmic
or the outer membrane; retention in the cytoplasmic membrane depends on
the presence of an additional sorting signal at the +2 position, relative to
the amino-terminal Cys residue of the mature lipoprotein [*3–7*]. Even the
organism with the smallest known genome, *Mycoplasma genitalium,* makes
use of lipid modification to retain proteins in the cytoplasmic membrane (*8*).
The number of putative lipoprotein-encoding genes per eubacterial genome
seems to range from approximately 18 in *M. genitalium* to approximately 94
in *Escherichia coli* and 114 in *Bacillus subtilis* (*7, 9*). Thus, lipoproteins appear
to represent about 1–7% of the proteome of eubacteria (see Table I). Bacterial lipoproteins are involved in a variety of processes, such as the uptake
of nutrients, resistance to antibiotics, protein secretion, sporulation, germi-
nation, cell wall biogenesis, and bacterial targeting to different substrates,
bacteria, and host tissues (*10*). Furthermore, surface-exposed lipoproteins
have been implicated as important mediators of the inflammatory response in human hosts during infections of gram-positive and gram-negative
pathogenic eubacteria, including *Borrelia, Treponema,* and *Mycoplasma*
species (*11–15*).

In gram-negative eubacteria, the outer membrane confines numerous proteins to the periplasm. In gram-positive eubacteria and mycoplasmas, which
lack an outer membrane, one of the roles of lipid modification of exported
proteins is to prevent their diffusion into the environment, as these proteins remain anchored to the cytoplasmic membrane by a lipid moiety. This
may explain why *B. subtilis* and *Mycoplasma pneumoniae* contain relatively
more putative lipoproteins than *E. coli* (see Table I) and why, for example, 32
lipoproteins of *B. subtilis* are homologs of periplasmic high-affinity substrate
binding proteins of gram-negative eubacteria (*9*).

Lipoprotein precursors are directed into the general (Sec) pathway for
protein secretion by their signal peptides, which show structural characteristics similar to those of the signal peptides of secretory proteins (*1, 18*):
a positively charged amino terminus (N-domain), a hydrophobic core region (H-domain), and a carboxyl-terminal region (C-domain) containing
the cleavage site for signal peptidase (SPase). The major difference between
signal peptides of lipoproteins and secretory proteins is the presence of a
well-conserved "lipobox" of four residues in lipoprotein signal peptides.
The invariable carboxyl-terminal Cys residue of this lipobox has to be lipid
modified in order to constitute the recognition and cleavage site for the

TABLE I

Predictions of Lipoproteins, Lipoprotein Modification, and Sorting Enzymes[a]

	Organism	Genome size (genes)	Predicted lipoproteins	%	Lgt	SPase II	Lnt	LolA	LolB
G⁻	*Escherichia coli*	4300	94	2.1	+	+	+	+	+
G⁻	*Haemophilus influenzae*	1740	49	2.8	+	+	+	+	+
G⁻	*Treponema pallidum*	1040	24	2.3	+	+	+	−	−
G⁻	*Helicobacter pylori*	1590	29	1.8	+	+	+	−	−
G⁻	*Synechocystis* sp.	1740	11	0.6	+	+	+	−	−
G⁺	*Mycobacterium tuberculosis*	4000	71	1.8	+	+	+	−	−
G⁺	*Bacillus subtilis*	4100	114	2.6	+	+	−	−	−
M	*Mycoplasma genitalium*	400	18	4.5	+	+	−	−	−
M	*Mycoplasma pneumoniae*	680	47	6.9	+	+	−	−	−
A	*Methanococcus jannaschii*	1740	18	1.0	−	−	−	−	−
A	*Archaeoglobus fulgidus*	2440	21	0.9	−	−	−	−	−

[a] To estimate the number of putative lipoproteins, lipoprotein signal peptides were identified with the SignalP algorithm (*16*), or by performing similarity searches with signal peptides of known lipoproteins, using the Blast algorithm (*17*). Similarly, proteins required for lipoprotein modification (Lgt, SPase II, and Lnt; see Section III, B) and sorting (LolA and LolB; see Section III, C) were identified by amino acid sequence similarity searches using the sequences of known proteins (see text for details). "+" indicates the presence of homologous sequences; "−" indicates the absence of homologous sequences in G⁻, gram-negative eubacteria; G⁺, gram-positive eubacteria; M, mycoplasmas; or A, archaea.

lipoprotein-specific signal peptidase (Lsp), also known as SPase II (*1, 18*). In this chapter, we will use the name SPase II in order to make the nomenclature for this class of enzymes consistent with that of the SPases required for the processing of secretory preproteins. The latter enzymes are termed type I SPases (*19*).

II. Type II SPases

A. Membrane Topology and Conserved Domains of Type II SPases

Thus far (April 2000), the nucleotide sequences of 31 *lsp* genes for type II SPases are known (see http://www.tigr.org/tdb/mdb/mdb.html), allowing the detailed comparison of the deduced amino acid sequences of the

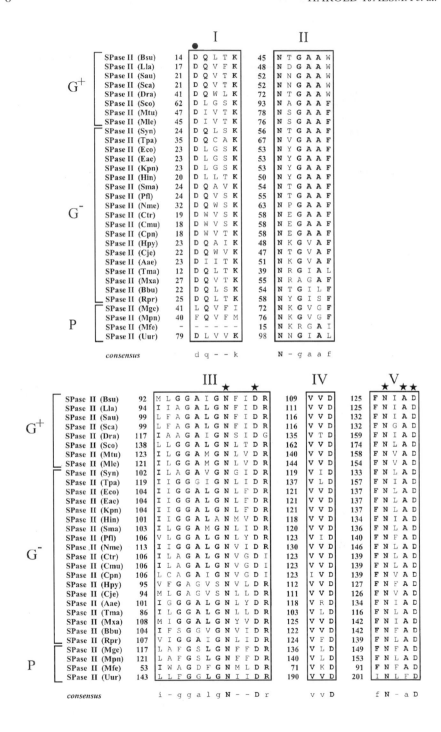

corresponding proteins (Fig. 1). As demonstrated for the SPase II of *E. coli* (*44*), all type II SPases have four predicted transmembrane domains (denoted TM-A to -D; Fig. 2). Two periplasmic (gram-negative eubacteria), or cell wall-exposed (gram-positive eubacteria) regions are localized between the TM-A/TM-B and TM-C/TM-D domains, respectively (Fig. 2). Furthermore, five highly conserved domains (I–V) were detected in these SPases (Figs. 1 and 2): Domain I, containing no strictly conserved residues, is located in TM-A; domain II, containing a strictly conserved Asn residue, is located in the extracytoplasmic region between TM-A and -B; domain III, containing strictly conserved Asn and Asp residues, is located at the junction of TM-C and the extracytoplasmic region between TM-C and -D; domain IV, containing a strictly conserved Asp residue, is located in the extracytoplasmic region between TM-C and -D; and domain V, containing strictly conserved Asn and Asp residues, is located at the junction of TM-D and the extracytoplasmic region between TM-C and -D (Fig. 2). Type II SPases appear to be absent from archaea and eukaryotes (*9*). Nevertheless, several genes for putative lipoproteins can be identified by homology searches in the genomes of archaea, such as *Methanococcus jannaschii* and *Archaeoglobus fulgidus* (see Table I). At present, it is not known whether these putative lipoprotein precursors are indeed lipid-modified and/or processed in archaea.

FIG. 1. Five conserved domains in type II SPases. The deduced amino acid sequences of 31 known type II SPases were compared. These include type II SPases of the gram-positive eubacteria (G+) *B. subtilis* [Bsu; (*20*)], *Lactococcus lactis* (Lla; GenBank Accession U63724), *Staphylococcus aureus* [Sau; (*21*)], *Staphyloccus carnosus* [Sca; (*22*)], *Deinococcus radiodurans* [Dra; (*23*)], *Streptomyces coelicolor* (Sco; GenBank Accession CAB51983), *Mycobacterium tuberculosis* [Mtu; (*24*)], and *Mycobacterium leprae* (Mle; GenBank Accession CAB39579); the gram-negative eubacteria (G−) *Synechocystis* sp. [Syn; (*25*)], *Treponema pallidum* [Tpa; (*26*)], *E. coli* [Eco; (*27, 28*)], *Enterobacter aerogenes* [Eae; (*29*)], *Klebsiella pneumoniae* (Kpn; GenBank Accession AAF19640), *Haemophilus influenzae* [Hin; (*30*)], *Serratia marcescens* (Sma; GenBank Accession AF027768), *Pseudomonas fluorescens* [Pfl; (*31*)], *Neisseria meningitidis* [Nme; (*32*)], *Chlamydia trachomatis* [Ctr; (*33*)], *Chlamydia muridarum* and *Chlamydia pneumoniae* [Cmu and Cpn, respectively; (*34*)], *Helicobacter pylori* [Hpy; (*35*)], *Campylobacter jejuni* [Cje; (*36*)], *Aquifex aeolicus* [Aae; (*37*)], *Thermotoga maritima* [Tma; (*38*)], *Myxococcus xanthus* [Mxa; (*39*)], *Borrelia burgdorferi* [Bbu; (*40*)], *Rickettsia prowazekii* [Rpr; (*41*)]; and the (myco)plasmas (P) *M. genitalium* [Mge; (*8*)], *M. pneumoniae* [Mpn; (*42*)], *Mycoplasma fermentans* (Mfe; GenBank Accession AAF15573; note that the amino-terminal sequences comprising conserved domain I are missing), and *Ureaplasma urealyticum* (Uur; GenBank Accession AAF30723). Five conserved domains (I–V) were identified (*43*). Numbers refer to the position of the first amino acid of each conserved domain in the respective type II SPases. Residues are printed in bold when present in at least 18 of the 31 type II SPases. Consensus sequences of each conserved domain are indicated. Uppercase letters indicate residues that are strictly conserved in all 31 type II SPases. Residues that are present in at least 18 sequences are printed in lowercase letters. Residues important for activity (★) or stability (●) of the *B. subtilis* SPase II (*43*) are indicated.

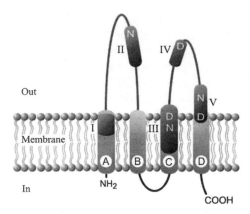

Fig. 2. Membrane topology of SPase II. Model for the membrane topology of type II SPases. The orientation of putative transmembrane regions (A-D) was predicted with the Toppred2 algorithm (*45*). Only the six strictly conserved residues in the conserved domains II–V (see Fig. 1) are indicated. NH_2, amino terminus; COOH, carboxyl terminus.

B.   CRITICAL RESIDUES FOR SPASE II ACTIVITY

The observation that SPase II lacks conserved Ser residues rules out the possibility that type II SPases make use of a catalytic mechanism similar to that of type I SPases (*19*). Furthermore, the lack of conserved Cys and His residues, and the finding that purified SPase II of *E. coli* was active in the absence of metal ions (*46*), demonstrate that type II SPases do not employ the well-defined catalytic mechanisms of thiol- or metalloproteases.

At present, only the SPase II of *B. subtilis* has been used for the systematic mapping of important residues for SPase II-mediated catalysis. These studies showed that Asp-14 in domain I, Asn-99 and Asp-102 in domain III, and Asn-126, Ala-128, and Asp-129 in domain V (see Figs. 1 and 2) are critical for SPase II activity (*43*). Notably, all of these residues are predicted to be localized close to the external surface of the cytoplasmic membrane. Only one residue, Asp-14, is required for the stability of the enzyme, showing that it is an important structural determinant. This conclusion is supported by the fact that the replacement of the equivalent Asp residue in the SPase II of *E. coli* (Asp-23) by Gly merely resulted in temperature sensitivity of the enzyme (*2*). In addition, Asp-14 of the *B. subtilis* SPase II is not conserved in domain I of the type II SPases of *M. genitalium* and *M. pneumoniae* (see Fig. 1), which contain active type II SPases (*47, 48*). In contrast to Asp-14, mutation of the other five residues required for activity of the *B. subtilis* SPase II did not significantly affect the stability of this enzyme, showing that these residues are directly, or indirectly required for catalysis. Interestingly, unlike the SPase II of *E. coli* (*27*), the *B. subtilis* SPase II did not require positively charged residues at the carboxyl terminus for activity (*43*), which implies that these residues are structural, rather than catalytic, determinants for the *E. coli* SPase II.

C. Type II SPases Belong to a Novel Family of Aspartic Proteases

The fact that two strictly conserved Asp residues are essential for SPase II activity (*43*) indicates that this enzyme belongs to the aspartic proteases. This idea is supported by the observation that the SPase II of *E. coli* could be inhibited by pepstatin, a known inhibitor of aspartic proteases (*46*).

Aspartic proteases form a group of proteolytic enzymes of the pepsin family that share the same catalytic mechanism and usually function in acidic environments (*49–51*). The known aspartic proteases of eukaryotes are monomeric enzymes that consist of two subdomains, both containing the conserved sequence Asp-Thr-Gly. All three residues contribute to the active site (*52*). Furthermore, conserved hydrogen bonds between the catalytic Asp residues and conserved Ser or Thr residues that are located at the +3 position relative to the catalytic Asp are present in most pepsin-like aspartic proteases. These hydrogen bonds are, most likely, responsible for the low pK_a values of the active-site Asp residues (*53, 54*). Aspartic proteases encoded by retroviruses and some plant viruses are related to the eukaryotic aspartic proteases, but in these cases the active protease is a homodimer, the active site(s) of which contain the conserved Asp-Thr/Ser-Gly motif (*55, 56*). In contrast to the eukaryotic aspartic proteases, conserved Ser or Thr residues are absent from the +3 position relative to the active site Asp residue of retroviral aspartic proteases. Consequently, the active site Asp residues of retroviral aspartic proteases lack the conserved hydrogen bonding, which explains why these enzymes have a much higher optimum pH than pepsin-like proteases (*57*). Notably, type II SPases lack the conserved Asp-Thr/Ser-Gly motif of previously described (eukaryotic and viral) aspartic proteases (*43*). Moreover, like in the viral aspartic proteases, conserved Ser or Thr residues are absent from the +3 position relative to the putative active site Asp residues. Instead, type II SPases contain strictly conserved Asn residues at the −3 position (see Figs. 1 and 2), which are very important for activity. These observations imply that the type II SPases belong to a novel class of aspartic proteases. As no type II SPases, or otherwise related proteins, have been identified in archaea or eukaryotes, it seems that this novel class of aspartic proteases has evolved exclusively in eubacteria.

D. Model for SPase II-Mediated Catalysis

By analogy to what is known about the catalytic mechanism of aspartic proteases, it seems likely that the two strictly conserved Asp residues in the conserved domains III and V of type II SPases form a catalytic dyad. It is presently unclear whether the pK_a of these residues is reduced by hydrogen bonding, as described for eukaryotic aspartic proteases. If such hydrogen bonds do not exist in type II SPases, this would explain the high optimum pH (7.9) of the SPase II of *E. coli* (*28*). The absence of conserved Ser/Thr residues does not, however, exclude the possibility that the pK_a of the active site Asp residues is modulated by other residues. In fact, this could be one possible role of other residues (i.e., Asn-45, Asn-99, Asp-111, Asn-126, Ala-128 of *B. subtilis* SPase II) required for SPase II activity (*43*). Alternatively, the latter residues could be required for the geometry of the active site of type II SPases, or the specific recognition of the diacylglyceryl-modified Cys residues in the "lipobox" of preproteins.

Based on the catalytic mechanism of the aspartic protease specified by the human immunodeficiency virus type 1 [HIV-1; (*58, 59*)], the following mechanism for type II SPases is proposed. At the start of catalysis, the active site contains a so-called "lytic" water molecule, and only one of the active site Asp residues is protonated (Fig. 3A). Upon binding of a lipid-modified precursor, the carbonyl carbon of the scissile peptide bond becomes hydrated (Fig. 3B), resulting in a tetrahedral intermediate (Fig. 3C). During this event, a proton is transferred (*via* the lytic water molecule) from one active site Asp residue to the other (Figs. 3C and 3D). Next, one hydroxyl group of the tetrahedral intermediate donates a proton to the charged Asp residue and, simultaneously, the nitrogen atom at the scissile peptide bond accepts a proton from the other catalytic Asp residue. The latter event results in peptide bond cleavage (processing), regeneration of the catalytic site of SPase II, and release of the mature lipoprotein and the cleaved signal peptide from the enzyme (Fig. 3D). This model is particularly attractive, because both essential Asp residues are predicted to be located in close proximity to the extracytoplasmic surface of the membrane, similar to the active site Ser residue of type I SPases (*60, 61*). This is the place where C-regions of exported precursors are likely to emerge from the translocation apparatus.

E. TYPE II SPASES AND TYPE IV PREPILIN SPASES USE RELATED
    CATALYTIC MECHANISMS

In various aspects, the processing of type IV pilin and lipoprotein precursors can be regarded as related eubacterial processes (*18*). First, prepilin signal peptides, which consist of an N-, H-, and C-region, show structural similarities to signal peptides of lipoproteins [*1, 18, 62*]. Second, the amino acid at position +1 relative to the SPase cleavage site of pilins is modified, like the +1 Cys residue of mature lipoproteins; although, in the case of type IV pilins, the residue at the +1 position is amino-methylated (*62*). Third, like the type II SPases, type IV prepilin SPases are integral membrane proteins

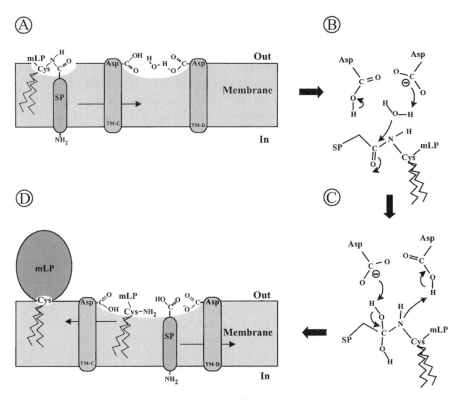

FIG. 3. Model for signal peptide cleavage by SPase II. At the start of the catalytic cycle of SPase II, only one of the two active site Asp residues is protonated, and the active site contains a lytic water molecule (A). Upon binding of the signal peptide (SP) of a lipid-modified preprotein, the carbonyl carbon of the scissile peptide bond is hydrated by the lytic water molecule. This is accompanied by the deprotonation of one active site Asp residue, and the protonation of the other Asp residue (B). Next, one hydroxyl group of the tetrahedral reaction intermediate donates a proton to the charged Asp residue and, simultaneously, the peptidic nitrogen accepts a proton from the other Asp residue (C). The latter event results in cleavage of the scissile peptide bond and regeneration of the initial protonation state of the Asp residues. Finally, the signal peptide (SP) and the mature lipoprotein (mLP) are released (D) and replaced by a new lytic water molecule (not shown). The cytoplasmic (in), and extracytoplasmic (out) sides of the membrane, transmembrane domains C (TM-C) and D (TM-D) of SPase II, and the amino termini (NH_2) of the signal peptide and the mature lipoprotein are indicated.

that have been identified in most gram-positive and gram-negative eubacteria of which the genomes have been sequenced completely (*62, 63*). Fourth and most important, the catalytic mechanisms of prepilin SPases and type II SPases seem to be related, as the potential active sites of both types of SPases contain two catalytic Asp residues (*43, 63*). Furthermore, in both types of SPases, the potential active site residues are predicted to reside in close proximity to the membrane surface.

Despite these similarities, important differences exist between prepilin and lipoprotein processing. First, in contrast to lipoprotein signal peptides, the SPase cleavage site of prepilin signal peptides is located amino-terminally of the hydrophobic H-domain (*1*). Second, consistent with the latter observation, the putative active site of prepilin SPases, such as TcpJ of *Vibrio cholerae,* is localized in the cytoplasm (*63*), whereas that of type II SPases is localized at the extracytoplasmic membrane surface (*43*).

III. Lipoproteins

A. FEATURES OF LIPOPROTEIN SIGNAL PEPTIDES

As the signal peptides of lipoproteins are, in general, shorter than those of secretory proteins, not all lipoproteins can be identified with signal peptide prediction programs, such as SignalP (*16, 18*). To estimate the total number of lipoproteins in a number of organisms with completely sequenced genomes (see Table I), putative lipoprotein signal peptides were also identified through similarity searches using the Blast algorithm (*17*).

Signal peptides from lipoproteins differ in several respects from those of secretory proteins. First, the C-domain of lipoprotein signal peptides contains a lipobox with the consensus sequence L-X-X-C, of which the invariable Cys residue is the target for lipid modification and becomes the first residue of the mature lipoprotein after cleavage by SPase II (Fig. 4). Second, the structural features of lipoprotein signal peptides seem to be more conserved than those of secretory signal peptides (data not shown), as evidenced by the comparison of all predicted lipoprotein signal peptides from the gram-positive eubacterium *B. subtilis,* the gram-negative eubacterium *E. coli,* and the mycoplasmas *M. genitalium* and *M. pneumoniae* (Fig. 5). This suggests that less variation in these signal peptides is allowed by the components involved in lipid modification and the subsequent processing of lipoprotein precursors. Nevertheless, minor species-specific differences in lipoprotein signal peptides can be observed. For example, on average, lipoprotein signal peptides of the mycoplasmas appear to be longer (23 residues) than those of *B. subtilis* (20 residues) and *E. coli* (19 residues). Furthermore, the N-domains of signal peptides of mycoplasmal lipoproteins contain, on average, one positively charged residue more (three residues) than those of *B. subtilis* and *E. coli* (both two residues; Figs. 5 and 6). Also, the hydrophobic H-domains of lipoprotein signal peptides of *E. coli* are on average 2 residues shorter than those of *Mycoplasma* and *B. subtilis* (10 *versus* 12 residues). As

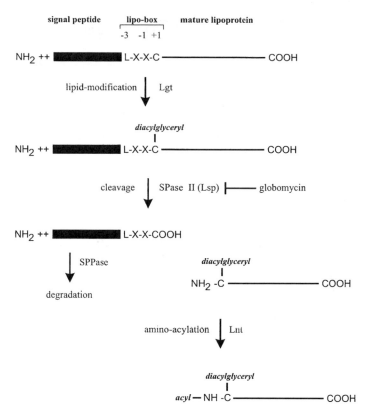

FIG. 4. Subsequent steps in the maturation of lipoproteins. The maturation of lipoproteins comprises three distinct steps. First, the invariable Cys residue of the lipobox (L-X-X-C; positions −3 to +1 relative to the SPase II processing site) is modified by the diacylglyceryl transferase (Lgt), which is a prerequisite for the second step, processing of the lipoprotein precursor by SPase II. Third, the released signal peptide is degraded by signal peptide peptidase (SPPase) while, at least in *E. coli,* mature (apo-) lipoproteins are further modified by fatty acylation of the diacylglyceryl-cysteine amino group by the lipoprotein aminoacyltransferase (Lnt). Notably, the latter modification may not be conserved in certain gram-positive eubacteria and mycoplasmas (see text for details). NH_2, amino terminus; COOH, carboxyl terminus; +, positively charged amino acids.

transmembrane helices seem to require at least 14 hydrophobic residues to span the membrane (*64, 65*), the last findings suggest that not all lipoprotein signal peptides can span the membrane completely. This implies that the active site of SPase II may be embedded in the cytoplasmic membrane, as previously was suggested for SPase I (*66, 67*). Alternatively, the N-domain of the signal peptide may not stay fixed at the cytoplasmic surface of the membrane during translocation. Finally, minor species-specific differences in the

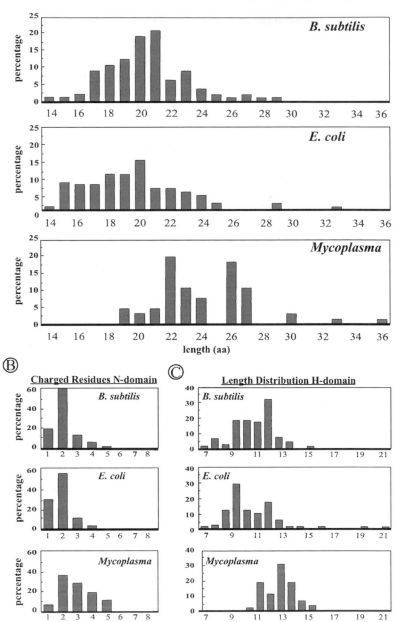

Ⓐ **Length Distribution of Predicted Lipoprotein Signal Peptides**

Ⓑ **Charged Residues N-domain**

Ⓒ **Length Distribution H-domain**

lipobox consensus sequences of *B. subtilis, E. coli,* and mycoplasmal lipoproteins can be observed (Fig. 6; Table II), suggesting that the type II SPases of these organisms have developed slightly different substrate preferences.

B. Lipid Modification

As described earlier, the major difference between signal peptides of lipoproteins and secretory proteins is the presence of the lipobox in the C-domain of the former signal peptides [Fig. 4 (*1, 18, 68*)]. Modification of this Cys residue by the diacylglyceryltransferase (Lgt) is a prerequisite for processing of the lipoprotein precursor by SPase II. This processing step can be inhibited with globomycin, a reversible and noncompetitive peptide inhibitor, structurally resembling a lipid-modified precursor (*2, 69, 70*). In *E. coli,* the processed lipoprotein is further modified by amino acylation of the diacylglyceryl-cysteine amino group [Fig. 4; (*71, 72*)]. It is presently not known whether the latter lipid-modification step is conserved in all eubacteria. For example, *B. subtilis, M. genitalium,* and *M. pneumoniae* lack an *Int* gene for the lipoprotein aminoacyltransferase (*9*). Consistent with the latter finding, a macrophage-stimulating lipoprotein with a nonacylated amino terminus has been isolated from *Mycoplasma fermentans* (*47*). Although the major role of the lipid modification is thought to be anchoring of the lipoprotein to the membrane, it should be noted that lipoprotein acylation also has physiological functions in inflammatory activities of pathogenic bacteria (*12, 14, 73*).

C. Lipoprotein Sorting

Because the lack of an outer membrane in gram-positive eubacteria and mycoplasmas, mature lipoproteins are retained in the cytoplasmic membrane of these organisms (Fig. 7). In contrast, mature lipoproteins of gram-negative eubacteria can either be retained in the cytoplasmic membrane or sorted to the outer membrane. Retention in the cytoplasmic membrane depends on the presence of an additional sorting signal in the form of Asp,

Fig. 5. Features of predicted lipoprotein signal peptides of *B. subtilis, E. coli,* and mycoplasmas (*M. genitalium* and *M. pneumoniae*). To estimate the number of putative lipoproteins, lipoprotein signal peptides were identified with the SignalP algorithm (*16*), or by performing similarity searches with signal peptides of known lipoproteins, using the Blast algorithm (*17*). (A) Length distribution of complete signal peptides (N- H- and C-domains). (B) Distribution of positively charged Lys or Arg residues in the N-domains of predicted signal peptides. (C) Length distribution of the hydrophobic H-domains in predicted signal peptides. Distributions are indicated as percentages of the total number of predicted lipoprotein signal peptides.

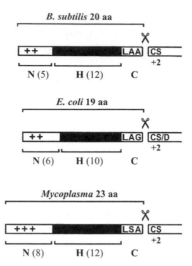

Fig. 6. Average features of predicted signal peptides found in (putative) lipoproteins of *B. subtilis, E. coli,* and mycoplasmas (*M. genitalium* and *M. pneumoniae*). Lipoprotein signal peptides have a tripartite structure: a positively charged N-domain (N), containing Lys and/or Arg residues (indicated with "+"), a hydrophobic H-domain (H, indicated by a black box), and a C-domain (C) containing the lipobox which specifies the cleavage site for SPase II ("scissors"). The average lengths of the complete signal peptide, N-domain and H-domain, and the consensus SPase II recognition sequences are indicated. Finally, the most frequently occuring second amino acid of the mature protein ("+2") is indicated. For *E. coli,* the most abundant amino acids at the +2 position of outer membrane lipoproteins (S) or cytoplasmic membrane lipoproteins (D) are indicated (see text for details).

Gly, Phe, Pro, Trp, or Tyr residues at the +2 position (*3–7*). In *E. coli,* only 13 of the 94 (putative) lipoproteins have such a cytoplasmic membrane retention signal at the +2 position (D, G, or Y; see Table II), suggesting that the other 81 predicted lipoproteins are sorted to the outer membrane. Current models suggest that there are two pathways for lipoprotein transport to and across the outer membrane. First, the release from the cytoplasmic membrane of various lipoproteins of *E. coli* that are targeted to the outer membrane is mediated by the LolCDE complex, which belongs to the family of ATP-binding cassette (ABC) transporters (*74*). These lipoproteins form a complex with the periplasmic LolA protein, which targets them to LolB, the outer membrane receptor for LolA–lipoprotein complexes. LolB, which is a lipoprotein itself, is required for the dissociation of LolA–lipoprotein complexes and incorporation of lipoproteins in the outer membrane [Fig. 7; (*5, 6, 75, 76*)]. Although the LolA–LolB system seems to be conserved in various gram-negative eubacteria [Table I; (*76*)], no homologs could be identified in several gram-negative bacteria, such as *T. pallidum* (Table I) and

TABLE II

Amino Acid Residues Flanking (Putative) SPase II Cleavage Sites[a]

Position −3		Position −2		Position −1		Position +1		Position +2	
aa	%	aa	%	aa	%	aa	%	aa	%
Lipobox *B. subtilis*									
L	65	A	36	A	39	C	100	G	30
V	8	S	24	G	35			S	28
F	7	T	11	L	8			A	9
T	6	I	6	I	4			I	4
I	5	V	6	S	4			T	4
M	2	G	6	V	2			F	4
G	2	M	2	T	2			W	4
S	<1	L	2	F	2			K	4
W	<1	C	2	P	2			L	4
		P	2	M	<1				
		F	2	C	<1				
		L	2	E	<1				
		D	<1						
		N	<1						
Lipobox *E. coli*									
L	74	A	36	G	65	C	100	S	31
V	9	S	27	A	29			A	19
I	9	T	19	S	3			T	12
M	4	V	8	T	1			D[CM]	7
S	2	L	4	I	1			N	6
F	1	Q	3	P	1			Q	5
T	1	G	2					V	5
		M	1					G[CM]	5
								L	3
								K	2
								I	2
								M	1
								H	1
								Y[CM]	1
Lipobox mycoplasmas									
L	79	S	44	A	80	C	100	S	41
F	12	T	25	S	13			G	25
A	7	A	16	G	7			A	24
I	2	V	11					T	8
		M	2					I	2
		I	2						

[a] The frequency of a particular amino acid at the indicated positions is given as the percentage of the total number of predicted lipoprotein signal peptides. Residues at the +2 position of *E. coli* that function as a cytoplasmic membrane retention signal are indicated with superscript CM.

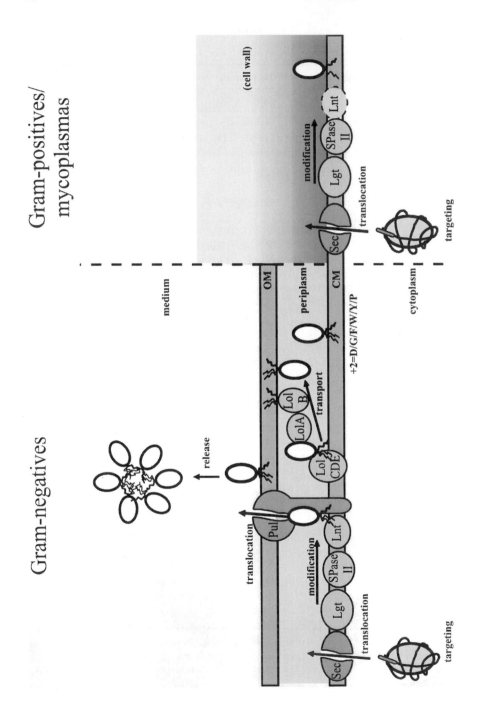

B. burgdorferi (our unpublished observations). As the latter organisms contain surface-exposed lipoproteins (*77*), the absence of LolA and LolB homologs suggests that other systems for sorting of outer membrane lipoproteins are present in these organisms. One well-described LolA/B-independent system is represented by certain Pul proteins of *Klebsiella oxytoca*, which are necessary for the secretion of the lipoprotein pullulanase [PulA; Fig 7; (*78*)]. As PulA contains an Asp residue at the +2 position, export of this lipoprotein is believed to bypass the Lol system completely. The Pul multiprotein complex, necessary for the secretion of PulA, is now known as a type II secretion system or the main terminal branch (MTB) of the general secretion pathway (GSP). Homologous systems are present in many gram-negative eubacteria (*79, 80*).

In accordance with the absence of an outer membrane in gram-positive eubacteria and mycoplasmas, no homologs of LolA and LolB can be found in these organisms (Table I). Although all lipoproteins from *B. subtilis* and *Mycoplasma* have to be retained in the cytoplasmic membrane, no Asp residues are present at the +2 position of these lipoproteins [Table II; (*9*)]. The fact that Asp at the +2 position is absent suggests that this sorting signal has evolved exclusively in gram-negative eubacteria. Nevertheless, Gly, Phe, or Trp residues can be found at the +2 position of various lipoproteins of *B. subtilis,* and Gly can be found at the +2 position of various mycoplasmal lipoproteins (Table II). Like Asp, these residues prevent lipoprotein sorting to the outer membrane of *E. coli,* albeit with reduced efficiency (*7*). Thus, the fact that residues with a potential sorting (retention) function can be found at the +2 position of mature lipoproteins of these organisms might suggest that these are involved in sorting of lipoproteins to specific cytoplasmic membrane locations in organisms lacking an outer membrane. Finally, it

---

FIG. 7. Lipoprotein sorting. Lipoprotein precursors of gram-negative and gram-positive eubacteria and mycoplasmas are targeted into the general (Sec) pathway for protein translocation across the cytoplasmic membrane (CM) by their signal peptides. Next, the precursors are lipid modified by Lgt, cleaved by SPase II, and, in many cases, amino acylated by Lnt, resulting in the retention of mature lipoproteins in the cytoplasmic membrane by their lipid groups. In gram-negative eubacteria, retention of mature lipoproteins in the cytoplasmic membrane depends on the presence of an additional retention signal in the form of an Asp (D), Gly (G), Phe (F), Pro (P), Trp (W), or Tyr (Y) residue at the +2 position of the mature lipoprotein. The absence of these residues from the +2 position results in the LolCDE-dependent release of a mature lipoprotein from the cytoplasmic membrane, LolA-dependent lipoprotein transport across the periplasmic space, and the LolB-dependent lipoprotein incorporation into the inner leaflet of the outer membrane. Alternatively, certain surface-exposed lipoproteins are transported *via* Lol-independent export pathways (e.g., Pul in the case of PulA; see text for details) to the surface of the outer membrane. Upon the formation of micelle-like structures, such lipoproteins can even be released into the medium (*1*).

should be noted that, compared to *B. subtilis* and mycoplasmal lipoproteins, a relatively high degree of variation can be observed at the +2 position of the putative lipoproteins of *E. coli* (Table II). This could indicate that other, as yet undefined, amino acid residues have a role in the sorting of lipoproteins in gram-negative eubacteria.

The apparent lack of *lnt* genes from *B. subtilis* and mycoplasmas (Fig. 7; Table I) indicates that the lipoproteins of these organisms are not amino-acylated. This observation raises the intriguing question whether amino acylation has a particular function in gram-negative eubacteria, for example, in the sorting of lipoproteins to the outer membrane. Alternatively, the incorporation of different fatty acids into one or three of the possible positions, as has been demonstrated for the lipoproteins of *Borrelia recurrentis* and *B. burgdorferi* (*12, 14*), might be important for sorting of lipoproteins.

IV.   The Role of Lipoprotein Processing by SPase II

The importance of lipoprotein processing by SPase II is highlighted by the fact that globomycin can serve as an antibiotic for gram-negative eubacteria (*69*). Consistent with this observation, it was shown that SPase II is essential for cell viability of *E. coli* (*81*). Despite the fact that globomycin can inhibit the processing of certain lipoprotein precursors of gram-positive eubacteria (*9, 82*), this antibiotic displays no cytotoxic activity against gram-positive eubacteria (*69*). One possible reason why SPase II is more important for cell viability of *E. coli* than for cell viability of *B. subtilis* could be that most lipoproteins of *E. coli* are sorted to the outer membrane. Thus, the lethality of *lsp* mutations in *E. coli* may be attributed to the retention of outer membrane lipoproteins in the cytoplasmic membrane. This idea was confirmed by the observation that retention of the outer membrane lipoprotein Lpp in the cytoplasmic membrane is lethal (*83*). Notably, the observation that LolA depletion affects cell viability, even in *E. coli* strains lacking Lpp (*84*), indicates that depletion of at least one other lipoprotein from the outer membrane is lethal, or that its mislocalization has cytotoxic effects.

The importance of lipoproteins for gram-positive eubacteria is underscored by the observation that the most abundant lipoprotein of *B. subtilis*, PrsA, is essential for the folding of various secretory proteins and cell viability (*85–88*). Interestingly, even though the disruption of the *lsp* gene results in cold- and heat-sensitive growth of *B. subtilis*, cells lacking SPase II are viable when grown at 37°C in all media tested (*9*). This finding implies that processing of lipoproteins by SPase II in *B. subtilis* is not strictly required for lipoprotein function (*9, 86*). Cells lacking SPase II accumulate lipid-modified precursor and, surprisingly, mature-like forms of the lipoprotein PrsA. These forms of PrsA appear to be reduced in activity, as the PrsA-dependent secretion of an α-amylase was strongly impaired [Fig. 8; (9)]. It is presently not clear which proteases are responsible for the alternative processing of PrsA in the absence of SPase II. However, the involvement of type I SPases in this process appears to be highly unlikely (9). The cellular level of another lipoprotein of B. subtilis, CtaC, is strongly reduced in the absence of SPase II, indicating that lipoprotein processing in gram-positive eubacteria is important for lipoprotein stability (89). Similar to the disruption of the lsp gene for SPase II, disruption of the B. subtilis lgt gene results in the accumulation of unprocessed (and unmodified) lipoproteins without severe effects on cell viability, at least under laboratory conditions (9, 90). Like SPase II, Lgt is required for the stability of several lipoproteins and efficient protein secretion (89, 90).

V. Outlook

Recent studies in various laboratories have implicated surface-exposed lipoproteins as important mediators in the inflammatory response in human hosts during eubacterial infections. Vaccines based on cell surface-exposed lipoproteins have been successfully tested in clinical trials, but because of the high degree of variability of such lipoproteins, a broad application of these vaccines has not yet been established (15). As the removal of the lipoprotein signal peptide by type II SPases is a critical step in lipoprotein sorting, stability, and function, these SPases might serve as effective drug targets, particularly in gram-negative pathogens. This idea, which would require the development of specific SPase II inhibitors, is especially attractive, because type II SPases seem to be present in all eubacteria. Moreover, these SPases appear to be absent from humans and other higher eukaryotes. In order to develop SPase II-specific inhibitors, more knowledge about the general catalytic mechanism, substrate recognition, and substrate specificity of type II SPases will be needed. Therefore, the elucidation of the tertiary structure of type II SPases is an important challenge for future research.

VI. Summary

Lipoprotein-specific (type II) signal peptidases remove signal peptides from eubacterial lipid-modified preproteins. The comparison of deduced amino acid sequences of 31 known type II signal peptidases of gram-positive and gram-negative eubacteria and mycoplasmas have revealed the presence

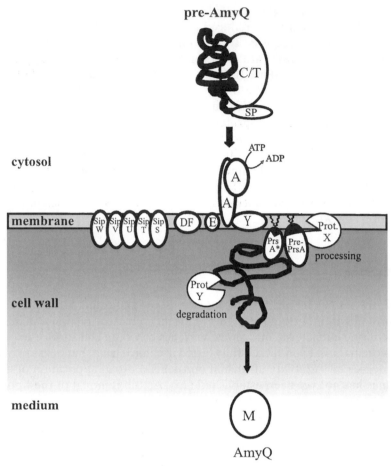

FIG. 8. Model for PrsA-dependent AmyQ secretion in the absence of SPase II in *B. subtilis.*
Pre-AmyQ is synthesized with an amino-terminal signal peptide (SP). Cytoplasmatic chaperones (C) and targeting factors (T) keep the precursor in a translocation-competent conformation and facilitate its targeting to the preprotein translocase in the membrane. Known
components of the *B. subtilis* translocase are SecA (A), SecY (Y), SecE (E), and SecDF [DF;
(*91*)]. SecA acts as a force generator (motor) for protein translocation through cycles of preprotein binding, membrane insertion, preprotein release, and deinsertion from the membrane.
The cycling of SecA is regulated by ATP-binding and hydrolysis (*92*). During or shortly after
the translocation, pre-AmyQ is processed by one of the type I SPases, SipS, SipT, SipU, SipV, or
SipW (*93, 94*). Folding of the mature AmyQ into its protease-resistant conformation depends
on the activity of PrsA which, in the absence of SPase II, is present in the precursor form (pre-
PrsA) and at least two mature-like forms (PrsA*), all of which are lipid-modified and localized
to the outer surface of the membrane. Alternative processing of pre-PrsA and degradation of
AmyQ in the absence of SPase II is catalyzed by unknown proteases (Prot. X and Prot. Y) at
the membrane–cell wall interface. Upon passage through the wall, mature AmyQ is released
into the growth medium.

of five conserved domains, containing two strictly conserved Asp residues. Both Asp residues are critical for signal peptidase II activity. Based on the catalytic mechanisms of known proteases, it is hypothesized that these two residues are directly involved in signal peptidase II–mediated catalysis. This implies that type II signal peptidases belong to a novel family of aspartic proteases. Thus, the type II signal peptidases might be mechanistically related to eubacterial type IV prepilin signal peptidases and eukaryotic aspartic proteases. Computer-assisted analyses indicate that type II signal peptidases are conserved in all eubacteria, in which their prelipoprotein substrates represent about 1–7% of the proteome. In gram-positive eubacteria, such as *Bacillus subtilis*, lipoproteins are retained in the cytoplasmic membrane. In contrast, lipoproteins of gram-negative eubacteria, such as *Escherichia coli*, can be either retained in the cytoplasmic membrane, or sorted to the outer membrane via different mechanisms. The general importance of lipoprotein processing by signal peptidase II is underscored by the fact that the absence of signal peptidase II in gram-positive eubacteria leads to lipoprotein instability and the malfunctioning of uncleaved lipoprotein precursors. Moreover, signal peptidase II is essential for cell viability of gram-negative eubacteria.

ACKNOWLEDGMENTS

We thank Drs. J. D. H. Jongbloed, M. L. van Roosmalen, E. Darmon, A. Pugsley, H. Tokuda, L. Hederstedt, and members of the European *Bacillus* Secretion Group for useful discussions; H.T. was supported by Genencor International (Leiden, the Netherlands) and Gist-brocades B.V. (Delft, the Netherlands); G.Z. was supported by Stichting Technische Wetenschappen (VBI 4837 [98-08]); S.B. and J.M.v.D. were supported by Biotechnology (Bio4-CT95-0278 and Bio4-CT96-0097) and "Quality of Life and Management of Living Resources" Grants (QLK3-CT-1999-00415 and QLK3-CT-1999-00917) from the European Union.

REFERENCES

1. Pugsley, A. P. (1993). *Microbiol. Rev.* **57,** 50–108.
2. Sankaran, K., and Wu, H. C. (1994). In "Signal Peptidases" (G. von Heijne, ed.), pp. 17–29. R. G. Landes Company, Austin, TX.
3. Yamaguchi, K., Yu, F., and Inouye, M. (1988). *Cell* **6,** 423–432.
4. Poquet, I., Kornacker, M. G., and Pugsley, A. P. (1993). *Mol. Microbiol.* **9,** 1061–1069.
5. Matsuyama, S., Tajima, T., and Tokuda, H. (1995). *EMBO J.* **14,** 3365–3372.
6. Matsuyama, S., Yokota, N., and Tokuda, H. (1997). *EMBO J.* **16,** 6947–6955.
7. Seydel, A., Gounon, P., and Pugsley, A. P. (1999). *Mol. Microbiol.* **34,** 810–821.
8. Fraser, C. M., Gocayne, J. D., White, O., Adams, M. D., Clayton, R. A., Fleischmann, R. D., Bult, C. J., Kerlavage, A. R., Sutton, G., Kelley, J. M., *et al.* (1995). *Science* **270,** 397–403.

9. Tjalsma, H., Kontinen, V. P., Pragai, Z., Wu, H., Meima, R., Venema, G., Bron, S., Sarvas, M., and van Dijl, J. M. (1999). *J. Biol. Chem.* **274,** 1698–1707.
10. Sutcliffe, I. C., and Russell, R. R. B. (1995). *J. Bacteriol.* **177,** 1123–1128.
11. Calcutt, M. J., Kim, M. F., Karpas, A. B., Muhlradt, P. F., and Wise, K. S. (1999). *Infect. Immun.* **67,** 760–771.
12. Scragg, I. G., Kwiatkowski, D., Vidal, V., Reason, A., Paxton, T., Panico, M., Dell, A., and Morris, H. (2000). *J. Biol. Chem.* **275,** 937–941.
13. Rawadi, G., Zugaza, J. L., Lemercier, B., Marvaud, J. C., Popoff, M., Bertoglio, J., and Roman-Roman, S. (1999). *J. Biol. Chem.* **274,** 30794–30798.
14. Beermann, C., Lochnit, G., Geyer, R., Groscurth, P., and Filgueira, L. (2000). *Biochem. Biophys. Res. Commun.* **267,** 897–905.
15. Simon, M. M., Bauer, Y., Zhong, W., Hofmann, H., and Wallich, R. (1999). *Int. J. Med. Microbiol. Virol. Parasitol. Infect. Dis.* **289,** 690–695.
16. Nielsen, H., Engelbrecht, J., Brunak, S., and von Heijne, G. (1997). *Protein Eng.* **10,** 1–6.
17. Altschul, S. F., Madden, T. L., Schaffer, A. A., Zhang, J., Zhang, Z., Miller, W., and Lipman, D. J. (1997). *Nucleic. Acids Res.* **25,** 3389–3402.
18. Tjalsma, H., Bolhuis, A., Jongbloed, J. D. H., Bron, S, and van Dijl, J. M., (2000). *Microbiol. Molec. Biol. Rev.* **64,** 515–547.
19. Dalbey, R. E., Lively, M. O, Bron, S., and van Dijl, J. M. (1997). *Prot. Science* **6,** 1129–11385.
20. Pragai, Z., Tjalsma, H., Bolhuis, A., van Dijl, J. M., Venema, G., and Bron, S. (1997). *Microbiology* **143,** 1327–1333.
21. Zhao, X. J., and Wu, H. C. (1992). *FEBS Lett.* **299,** 80–84.
22. Witke, C., and Götz, F. (1995). *FEMS Microbiol. Lett.* **126,** 233–239.
23. White, O., Eisen, J. A., Heidelberg, J. F., Hickey, E. K., Peterson, J. D., Dodson, R. J., Haft, D. H., Gwinn, M. L., Nelson, W. C., Richardson, D. L., *et al.* (1999). *Science* **286,** 1571–1577.
24. Cole, S. T., Brosch, R., Parkhill, J., Garnier, T., Churcher, C., Harris, D., Gordon, S. V., Eiglmeier, K., Gas, S., Barry III, C. E., *et al.*, (1998). *Nature* **393,** 537–544.
25. Kaneko, T., Sato, S., Kotani, H., Tanaka, A., Asamizu, E., Nakamura, Y., Miyajima, N., Hirosawa, M., Sugiura, M., Sasamoto, S., *et al.*, (1996). *DNA Res.* **3,** 109–136.
26. Fraser, C. M., Norris, S. J., Weinstock, G. M., White, O., Sutton, G. G., Dodson, R., Gwinn, M., Hickey, E. K., Clayton, R., Ketchum, K. A., *et al.* (1998). *Science* **281,** 375–388.
27. Innis, M. A., Tokunaga, M., Williams, M. E., Loranger, J. M., Chang, S. Y., Chang, S., and Wu, H. C. (1984). *Proc. Natl. Acad. Sci. USA* **81,** 3708–3712.
28. Tokunaga, M., Loranger, J. M., Chang, S. Y., Regue, M., Chang, S., and Wu, H. C. (1985). *J. Biol. Chem.* **260,** 5610–5615.
29. Isaki, L., Beers, R., and Wu, H. C. (1990). *J. Bacteriol.* **172,** 6512–6517.
30. Fleischmann, R. D., Adams, M. D., White, O., Clayton, R. A., Kirkness, E. F., Kerlavage, A. R., Bult, C. J., Tomb, J. F., Dougherty, B. A., Merrick, J. M., *et al.* (1995). *Science* **269,** 496–512.
31. Isaki, L., Kawakami, M., Beers, R., Hom, R., and Wu, H. C. (1990). *J. Bacteriol.* **172,** 469–472.
32. Tettelin, H., Saunders, N. J., Heidelberg, J., Jeffries, A. C., Nelson, K. E., Eisen, J. A., Ketchum, K. A., Hood, D. W., Peden, J. F., Dodson, R. J., *et al.* (2000). *Science* **287,** 1809–1815.
33. Stephens, R. S., Kalman, S., Lammel, C., Fan, J., Marathe, R., Aravind, L., Mitchell, W., Olinger, L., Tatusov, R. L., Zhao, Q., Koonin, E. V., and Davis, R. W. (1998). *Science* **282,** 754–759.
34. Read, T. D., Brunham, R. C., Shen, C., Gill, S. R., Heidelberg, J. F., White, O., Hickey, E. K., Peterson, J., Utterback, T., Berry, K., *et al.* (2000). *Nucleic. Acids Res.* **28,** 1397–1406.

35. Tomb, J. F., White, O., Kerlavage, A. R., Clayton, R. A., Sutton, G. G., Fleischmann, R. D., Ketchum, K. A., Klenk, H. P., Gill, S., Dougherty, B. A., *et al.* (1997). *Nature* **388,** 539–547.
36. Parkhill, J., Wren, B. W., Mungall, K., Ketley, J. M., Churcher, C., Basham, D., Chillingworth, T., Davies, R. M., Feltwell, T., Holroyd, S., *et al.* (2000). *Nature* **403,** 665–668.
37. Deckert, G., Warren, P. V., Gaasterland, T., Young, W. G., Lenox, A. L., Graham, D. E., Overbeek, R., Snead, M. A., Keller, M., Aujay, M., *et al.* (1998). *Nature* **392,** 353–358.
38. Nelson, K. E., Clayton, R. A., Gill, S. R., Gwinn, M. L., Dodson, R. J., Haft, D. H., Hickey, E. K., Peterson, J. D., Nelson, W. C., Ketchum, K. A., *et al.* (1999). *Nature* **399,** 323–329.
39. Paitan, Y., Orr, E., Ron, E. Z., and Rosenberg, E. (1999). *J. Bacteriol.* **181,** 5644–5651.
40. Fraser, C. M., Casjens, S., Huang, W. M., Sutton, G. G., Clayton, R., Lathigra, R., White, O., Ketchum, K. A., Dodson, R., Hickey, E. K., *et al.* (1997). *Nature* **390,** 580–586.
41. Andersson, S. G. E., Zomorodipour, A., Andersson, J. A., Sicheritz-Pontén, T., Alsmark, U. C. M., Podowski, R. M., Näslund, A. K., Eriksson, A-S., Winkler, H. H., and Kurland, C. G. (1998). *Nature* **396,** 133–143.
42. Himmelreich, R., Hilbert, H., Plagens, H., Pirkl, E., Li, B. C., and Herrmann, R. (1996). *Nucleic. Acids. Res.* **24,** 4420–4449.
43. Tjalsma, H., Zanen, G., Venema, G., Bron, S., and van Dijl, J. M. (1999). *J. Biol. Chem.* **274,** 28191–28197.
44. Muñoa, F. J., Miller, K. W., Beers, R., Graham, M., and Wu, H. C. (1991). *J. Biol. Chem.* **266,** 17667–17672.
45. Sipos, L., and von Heijne, G. (1993). *Eur. J. Biochem.* **213,** 1333–1340.
46. Dev, I. K., and Ray, P. H. (1984). *J. Biol. Chem.* **259,** 11114–11120.
47. Muhlradt, P. F., Kiessm, M., Meyer, H., Sussmuth, R., and Jung, G. (1997). *J. Exp. Med.* **185,** 1951–1958.
48. Pyrowolakis, G., Hofmann, D., and Herrmann, R. (1998). *J. Biol. Chem.* **273,** 24792–24796.
49. Tang, J., and Wong, R. N. (1987). *J. Cell. Biochem.* **33,** 53–63.
50. Rao, J. K., Erickson, J. W., and Wlodawer, A. (1991). *Biochemistry* **14,** 4663–4671.
51. Rawlings, N. D., and Barrett, A. J. (1995). *Methods Enzymol.* **248,** 105–120.
52. Pearl, L. H., and Taylor, W. R. (1987). *Nature* **329,** 351–354.
53. Cooper, J. B., Khan, G., Taylor, G., Tickle, I. J., and Blundell, T. L. (1990). *J. Mol. Biol.* **214,** 199–222.
54. Goldblum, A. (1990). *FEBS Lett.* **261,** 241–244.
55. Miller, M., Jaskolski, M., Rao, J. K., Leis, J., and Wlodawer, A. (1989). *Nature* **337,** 576–579.
56. Weber, I. T., Miller, M., Jaskolski, M., Leis, J., Skalka, A. M., and Wlodawer, A. (1989). *Science* **243,** 928–931.
57. Ido, E., Han, H. P., Kezdy, F. J., and Tang, J. (1991). *J. Biol. Chem.* **266,** 24359–24366.
58. Silva, A. M., Cachau, R. E., Sham, H. L., and Erickson, J. W. (1996). *J. Mol. Biol.* **19,** 321–346.
59. Liu, H., Muller-Plathe, F., and van Gunsteren, W. F. (1996). *J. Mol. Biol.* **261,** 454–469.
60. Tschantz, W. R., Sung, M., Delgado-Partin, V. M., and Dalbey, R. E. (1993). *J. Biol. Chem.* **268,** 27349–27354.
61. van Dijl, J. M., de Jong, A., Venema, G., and Bron, S. (1995). *J. Biol. Chem.* **270,** 3611–3618.
62. Lory, S. (1994). *In* "Signal Peptidases" (G. von Heijne, ed.), pp. 31–48. R. G. Landes Company, Austin, TX.
63. LaPointe, C. F., and Taylor, R. K. (2000). *J. Biol. Chem.* **275,** 1502–1510.
64. Casadio, R., Fariselli, P., Taroni, C., and Compiani, M. (1996). *Eur. Biophys. J.* **24,** 165–178.
65. Bowie, J. U. (1997). *J. Mol. Biol.* **272,** 780–789.
66. van Klompenburg, W., Ridder, A. N. J. A., van Raalte, A. L., Killian, A. J., von Heijne, G., and de Kruijff, B. (1997). *FEBS Lett.* **413,** 109–114.

67. Paetzel, M., Dalbey, R. E., and Strynadka, N. C. (1998). *Nature* **396,** 186–190.
68. von Heijne, G. (1989). *Protein Eng.* **2,** 531–534.
69. Inukai, M., Takeuchi, M., Shimizu, K., and Arai, M. (1978). *J. Antibiot.* **31,** 1203–1205.
70. Giam, C. Z., Chai, T., Hayashi, S., and Wu, H. C. (1984). *Eur. J. Biochem.* **141,** 331–337.
71. Tokunaga, M., Tokunaga, H., and Wu, H. C. (1982). *Proc. Natl. Acad. Sci. USA* **79,** 2253–2259.
72. Sankaran, K., and Wu, H. C. (1994). *J. Biol. Chem.* **269,** 19701–19706.
73. Belisle, J. T., Brandt, M. E., Radolf, J. D., and Norgard, M. V. (1994). *J. Bacteriol.* **176,** 2151–2157.
74. Yakushi, T., Matsuda, K., Narita, S. I., Matsuyama, S. I., and Tokuda, H. (2000). *Nature Cell Biol.* **2,** 212–218.
75. Yakushi, T., Yokota, N., Matsuyama, S., and Tokuda, H. (1998). *J. Biol. Chem.* **273,** 32576–32581.
76. Yokota, N., Kuroda, T., Matsuyama, S., and Tokuda, H. (1999). *J. Biol. Chem.* **274,** 30995–30999.
77. Sellati, T. J., Bouis, D. A., Kitchens, R. L., Darveau, R. P., Pugin, J., Ulevitch, R. J., Gangloff, S. C., Goyert, S. M., Norgard, M. V., and Radolf, J. D. (1998). *J. Immunol.* **160,** 5455–5464.
78. d'Enfert, C., Ryter, A., and Pugsley, A. P. (1987). *EMBO J.* **6,** 3531–3538.
79. Pugsley, A. P., Francetic, O., Possot, O. M., Sauvonnet, N., and Hardie, K. R. (1997). *Gene* **192,** 13–19.
80. Russel, M. (1998). *J. Mol. Biol.* **279,** 485–499.
81. Yamagata, H., Ippolito, C., Inukai, M., and Inouye, M. (1982). *J. Bacteriol.* **152,** 1163–1168.
82. Hayashi, S., and Wu, H. C. (1983). *J. Bacteriol.* **156,** 773–777.
83. Yakushi, T., Tajima, T., Matsuyama, S., and Tokuda, H. (1997). *J. Bacteriol.* **179,** 2857–2862.
84. Tajima, T., Yokota, N., Matsuyama, S., and Tokuda, H. (1998). *FEBS Lett.* **439,** 51–54.
85. Kontinen, V. P., and Sarvas, M. (1988). *J. Gen. Microbiol.* **134,** 2333–2344.
86. Kontinen, V. P., and Sarvas, M. (1993). *Mol. Microbiol.* **8,** 727–737.
87. Kontinen, V. P., Saris, P., and Sarvas, M. (1991). *Mol. Microbiol.* **5,** 1273–1283.
88. Jacobs, M., Andersen, J. B., Kontinen, V. P., and Sarvas, M. (1993). *Mol. Microbiol.* **8,** 957–966.
89. Bengtsson, J., Tjalsma, H., Rivolta, C., and Hederstedt, L. (1999). *J. Bacteriol.* **181,** 685–688.
90. Leskelä, S., Wahlstrom, E., Kontinen, V. P., and Sarvas, M. (1999). *Mol. Microbiol.* **31,** 1075–1085.
91. Bolhuis, A., Broekhuizen, C. P., Sorokin, S. A., van Roosmalen, M. L., Venema, G., Bron, S., Quax, W. J., and van Dijl, J. M. (1998). *J. Biol. Chem.* **273,** 21217–21224.
92. Duong, F., Eichler, J., Price, A., Leonard, M. R., and Wickner, W. (1997). *Cell* **91,** 567–573.
93. Tjalsma, H., Noback, M. A., Bron, S., Venema, G., Yamane, K., and van Dijl, J. M. (1997). *J. Biol. Chem.* **272,** 25983–25992.
94. Tjalsma, H., Bolhuis, A., van Roosmalen, M. L., Wiegert, T., Schumann, W., Broekhuizen, C. P., Quax, W. J., Venema, G., Bron, S., and van Dijl, J. M. (1998). *Genes Dev.* **12,** 2318–2331.

2

Bacterial Type I Signal Peptidases

JOSEPH L. CARLOS* • MARK PAETZEL[†] •
PHILIP A. KLENOTIC[‡] • NATALIE C. J. STRYNADKA[†] •
ROSS E. DALBEY*

*Department of Chemistry
The Ohio State University
Columbus, Ohio 43210

[†]Department of Biochemistry and Molecular Biology
University of British Columbia
Vancouver V6T 1Z3, British Columbia, Canada

[‡]The Cleveland Clinic
Center for Molecular Genetics
Cleveland, Ohio

THE ENZYMES, Vol. XXII

I. Introduction: Bacterial Signal Peptidase, Signal Peptides and Protein Targeting

In the bacterial secretory pathway, proteins that are exported across the plasma membrane are synthesized as higher molecular weight precursors with an N-terminal extension peptide. This extension peptide, called a signal or leader peptide, is proteolytically removed by type I signal peptidase (SPase I or leader peptidase, Lep). Unrelated to type I signal peptidases are the lipoprotein-specific signal peptidases (type II), which recognize lipid-modified eubacterial preproteins, and the (type IV) prepilin signal peptidases (*1, 2*). In the secretory pathway in bacteria, the apparent natural function of signal peptide processing by SPase I is the release of export-targeted and translocated proteins from the cytosolic membrane. Genetic studies in a number of bacteria have shown that SPase I is essential for cell viability (*3–7*). Mechanistic and structural analyses have helped explain how this enzyme binds substrate and how catalysis occurs by a unique mechanism. In this chapter the type I signal peptidase enzymes found in eubacteria with particular emphasis on the most thoroughly studied enzymes in this field, *E. coli* and *B. subtilis* SPase I, will be discussed. The focus will be on the biological and functional enzymology of SPase I. Finally, the three-dimensional X-ray crystal structure of *E. coli* SPase I will also be presented.

A. Bacterial Signal Peptides Are Essential for Preprotein Export

Proteins that are exported across the plasma membrane of bacteria typically use cleavable signal peptides. In bacteria, two preprotein translocation machineries are currently known, the Sec translocase and the Tat translocase systems (see Fig. 1). The Sec machinery in *E. coli* is comprised of the membrane-embedded protein components SecYEGDF and YajC (*8*), a peripheral membrane component SecA (*9–11*), and the cytoplasmic molecular chaperone SecB (*12*). SecB helps target some preproteins to the membrane by interaction the mature regions of the preprotein, as well as with membrane-bound SecA (*13*). The Tat pathway components consist of the integral membrane components TatA, TatB, TatC, and TatE (*14, 15*). TatA, TatB, and TatE have been found to be homologous to the Hcf106 protein (*14, 16, 17*) that is involved in the ΔpH-dependent protein export pathway of plant thylakoids. Resident bacterial inner membrane proteins typically lack cleavable signal peptides (*18*), whereas transient inner membrane proteins encoded by some phage genomes contain signal peptides that are removed during the membrane insertion process. A membrane-embedded protein, YidC, has been found to be essential for the proper insertion of the *M13* procoat protein [Fig. 1, (*19*)]. *M13* procoat was previously thought

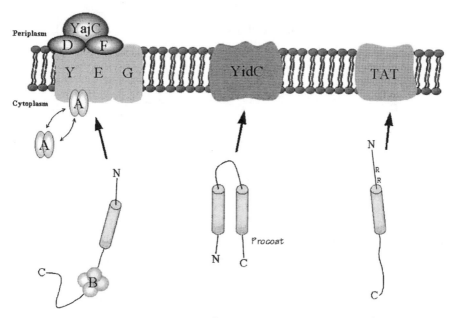

FIG. 1. Schematic representation of the known bacterial protein translocation systems.

to insert into the membrane by a spontaneous and Sec machinery–inde-pendent mechanism.

Although signal peptides do not maintain any overall sequence homol-ogy, they do contain some regions that have conserved features (20, 21) (Fig. 2). At the N-terminus of preproteins involved in the secretory pathway (destined for the outer membrane or cellular export) signal peptides typically contain a basic amino-terminal 1- to 5-residue N-region, a 7- to 15-residue central core hydrophobic H-region, and a polar carboxyl-terminal C-region of 3–7 amino acids (Fig. 2). These regions have been shown to be impor-tant elements for cleavage *in vivo* by signal peptidase (22). In particular, the C-region has been shown to harbor important elements of substrate speci-ficity such as the "Ala-X-Ala" motif that is prevalent at the −1 to −3 position with respect to the cleavage site (20, 21, 23). As shown in Fig. 2, statistical analyses indicate that signal peptides of gram-positive bacteria are on aver-age longer than those found in gram-negative bacteria (24–26). These differ-ences are manifested mainly in longer H- and C-regions for gram-positive bacteria.

Signal peptides can interact directly with the peripheral translocation com-ponent SecA (27) and the SecY/E complex (28). Signal peptides also interact with membrane phospholipids. The positively charged N-region of the signal

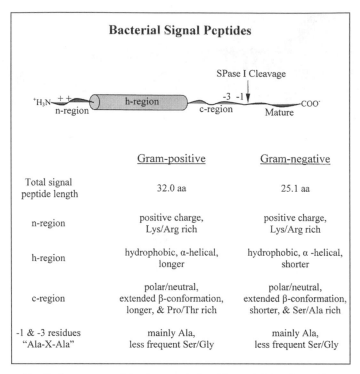

Fig. 2. A summary of the characteristics of bacterial signal peptides.

peptide is believed to function at an early step in the protein export process by interacting with the acidic phospholipid headgroups in the membrane (*29*). On the other hand, the signal peptide H-region likely interacts with the hydrophobic acyl groups of membrane phospholipids. Functional signal peptides have been shown to adopt α-helical conformations in detergent and model membrane systems (*30, 31*).

Early work by the Beckwith and Silhavy laboratories (*32, 33*) demonstrated that signal peptides in bacteria are essential for export of proteins across the inner membrane. Interruption of the H-region with either a charged residue or a deletion led to an export-defective signal peptide. The positively charged N-terminal region also can be important for the efficiency of translocation. Though the positively charged N-region is not as critical for export as the H-region, proteins with signal peptides that contain an acidic N-terminal region are typically exported more slowly. Interestingly, the positive charge(s) in the N-terminal region play a more decisive role when the hydrophobic core is less than optimal (*34*). Defects in the N-region can be compensated for by increased length and/or increased overall hydrophobicity of the corresponding H-region. The C-region contains the sequence elements important for processing, as determined by site-directed mutagenesis experiments (*35, 36*). The sequence elements are small, uncharged amino acids at the −1 and −3 positions relative to the cleavage site.

In the case of signal peptides involved in the TAT translocation pathway (Fig. 1), the twin arginine motif is a critical determinant for export. In addition to a two-arginine consensus (*37*), the export of proteins by this pathway requires a highly hydrophobic residue at the +2 or +3 position (relative to the two arginines) (*38*).

B.   Signal Peptidases in Gram-Negative and Gram-Positive Bacteria

The enzyme now commonly referred to as signal peptidase SPase cleaves signal peptides from preproteins in a wide variety of organisms. The type I SPase from gram-negative *Escherichia coli*, first detected more than two decades ago by Chang *et al.* (*39*), was first purified by Zwizinski and Wickner (*40*). The protease was assayed by its ability to cleave the signal peptide by using *M13* procoat protein as a substrate. The gene encoding *E. coli* type I signal peptidase (lepB) was eventually cloned (*41*) and sequenced (*42*). *Escherichia coli* SPase I has been found to be essential for cell growth (*3*) and its amino acid sequence reveals three apolar stretches: H1 (residues 1–22), H2 (residues 62–77), and H3 (residues 83–91).

The membrane topology of *E. coli* SPase I was determined through several studies. Wolfe *et al.* (*42*) found that treatment of inside-out inner membrane vesicles with trypsin yields a protected *E. coli* SPase I fragment of approximately 32 kDa. The likely cleavage at Lys-57 in the cytoplasmic domain suggests that SPase I spans the membrane *in vivo* from the cytoplasm to the periplasm after this residue. Moreover, Moore and Miura (*43*) found that treatment of right-side-out inner membrane vesicles and spheroplasts with proteinase K or trypsin yields two protected fragments of approximately 80 and 105 amino acids, respectively. Both protected fragments are derived from the amino terminus of the protein and the shorter one is derived from the larger one. Additionally, the same treatment of an amino acid 82–98 deletion mutant (lacking the H3 domain) does not change the size of the smaller protected peptide, but does decrease the size of the larger peptide (*43*). These combined data suggest that the second (H2) apolar domain (residues 62–76) is a membrane-spanning region, whereas the third apolar domain, H3 (residues 82–98) is exposed to the periplasm. Apolar domain 1 also spans the membrane and interacts with the second transmembrane segment. The helix–helix interface was determined by analyzing disulfides formed between pairs of cysteines engineered at the periplasmic ends of the transmembrane regions (*44*). The resulting membrane topology model of *E. coli* SPase I (H3 domain not shown) is shown in Fig. 3.

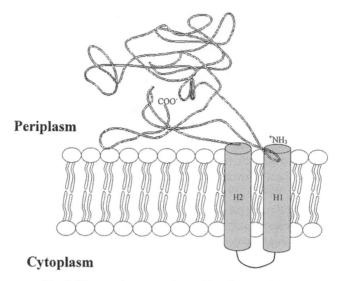

Periplasm

Cytoplasm

FIG. 3. The membrane topology of *E. coli* type I SPase.

After *E. coli,* the next type I signal peptidase to be cloned was from the gram-negative organism *Salmonella typhimurium* (*45*). *Salmonella typhimurium* type I SPase was found to be homologous to the *E. coli* signal peptidase. As with the *E. coli* enzyme, hydropathy and sequence analyses indicated that the enzyme spans the membrane twice with the catalytic domain in the extracellular medium. From whole genome sequencing projects, homologous type I signal peptidase open reading frames have now been and continue to be identified in a wide variety of gram-negative and gram-positive organisms.

Alignment of the known (archeal, eubacterial, and eukaryal) type I SPase sequences revealed that there were five conserved domains in the signal peptidase family. These domains were designated as A–E (*46*). However, our current alignment of sequences retrieved from the results of an *E. coli* SPase I BLAST homology search (http://www.ncbi.nlm.nih.gov/BLAST/) (*47, 48*) reveals that there is a second C-like homology domain. Similar alignments by Tjalsma and co-workers also reveal this sixth domain and they have designated it as C′ (*49, 50*). Additionally, based on differing conservation patterns and the putative general base residue (Lys or His) found in the D domain, the alignments of all currently known type I SPases have resulted in their sub-classification of either P-type (Lys) or ER-type (His) (*49, 50*). The alignment patterns of the sequences found in our BLAST search are shown in Fig. 4. The A domain (not shown), preceding the B domain, corresponds to a region that is within the transmembrane region of the SPase I enzyme.

The P-type B domain contains the putative catalytic Ser in an SM_PTL motif (Fig. 4A). This catalytic Ser (Ser 90 in the *E. coli* signal peptidase) is invariant in the entire family. The C, C′, and D domains contain the residues reported to be important in substrate binding and catalysis (*51–53*).

The putative general base (K145 of domain D in *E. coli*) found in mitochondrial and chloroplast signal peptidases is invariant in all the gram-positive and gram-negative bacteria surveyed. It is replaced by a His residue in the ER-type SPases. Finally, the E domain contains the strictly conserved GDN motif. Domain E has been reported to be important in active site architecture (*54*). The functions of specific residues in the conserved domains will be further discussed below. It is intriguing that SipW, a characterized *B. subtilis* type I SPase (*49*), and Spc21 of *Clostridium perfringens* are gram-positive SPases classified as ER-type signal peptidases (Fig. 4B).

While most bacteria have only one signal peptidase gene, there are exceptions. *B. subtilis,* for example, has five chromosomally encoded type I signal peptidase genes—SipS, SipT, SipU, SipV, and SipW (*55*)—and two plasmid-encoded signal peptidases (*56*). These chromosomal signal peptidases have overlapping substrate specificities. SipS through SipV resemble bacterial, P-type, signal peptidases. But, as we have mentioned earlier, SipW is more like the eukaryotic, ER-type, signal peptidase; it lacks the conserved Lys general base and, instead, has a His at the homologous position. Similarly, multiple signal peptidase genes are found in *Streptomyces lividans* (*57*). These include SipW, SipX, SipY, and SipZ. Three of these genes are in a single operon. Finally, *Staphylococcus aureus* has two signal peptidase genes, one of which is thought to be inactive as it lacks homologous catalytic Ser and Lys residues (*5*).

C.   SIGNAL PEPTIDE PROCESSING IS REQUIRED FOR CELL GROWTH

Signal peptidase is critical for cell viability. In the Date experiments (*3*), integration of an ampicillin-resistant plasmid bearing a promoter and signal peptidase–deficient sequence into the chromosome of a *pol*A mutant (plasmid replication deficient) *E. coli* strain was attempted. However, this technique did not result in the isolation of any signal peptidase–deficient and ampicillin-resistant strains. This suggested that signal peptidase is essential for *E. coli* viability. In another study, construction of an *E. coli* strain with signal peptidase expression under the control of the araB promoter led to arabinose-dependent cell growth (*4*). Signal peptidase I has also been shown to be essential in human pathogens such as *Staphylococcus aureus* (*5*) and *Streptococcus pneumoniae* (*7*). Finally, though the genes encoding SipS or SipT by themselves are not essential for cell viability, deletion of both SipS and SipT genes prevents cell growth in *B. subtilis* (*49*).

A

	B		C		C'		D		E	
Sip_Eco_	IPSGSNMPTLL	96	IGDFLIVEKFA	107	RGDIVVFKYP	136	YIKRAVGLPGDKV	155	VPPGQYFMMGDNRDNSADSR	282
Sip_Ctr_	VPTGSMRPTIL	119	EQDRIIVSKTT	130	RGELVVFTVG	160	YIKRCMGKPGDTV	193	IPEGHVLVLGDNCPMSADSR	562
Sip_Bbu_	IPSGSMENTLQ	78	IGDFLFVDKFS	89	ESDIIIFENP	118	LVKRGAFADGKIV	169	VPDGYILPIGDNRDNSHDGR	299
Sip_Cpn_	VPTGSMRPTIL	114	EQDRIIVSKTT	125	RGGLVVFTVG	155	YIKRCMGRPGDFL	188	VPKGHVLVLGDNYTMSADSR	561
Sip_Hin_	IPSGSMESTLR	121	VGDFLVVNKVA	132	RGDIVVFKAP	161	LIRTGLGATRAAF	177	VPEGQYFVMGDRRDHSDDSR	307
Sip_Hpy_	IPSRSNVGTLY	44	EGDMLFVKKFS	55	RGEVVVFIPP	97	YVKRNFALGGDEV	116	INHDEFFMIGDNRDNSSDSR	227
Sip_Pfl_	IPSESNMKPTLD	96	VGDFLLVNKFS	107	RGDVMVFRYP	136	YIKRVVGLPGDVV	155	VPAGHYFMMGDNRDNSNDSR	235
Sip_Pla_	IPSESMLPTLE	65	VNDRLLVEKIS	76	RGDILVFHPT	93	FIKRVIGLPGETV	119	VPADSFLVLGDNRDNSYDSH	170
Sip_Rca_	IPSGSMKDTLL	46	IGDFLFVNRDA	57	RGDVVVFRHP	97	FIKRLIGLPGDRI	115	VPEGQYFMGDNRDNSEDSR	217
Sip_Sty_	IPSGSNMPTLL	97	IGDFIIVEKFA	108	RGDIVVFKYP	137	YIKRAVGLPGDKI	156	VPFGQYFMMGDNRDNSADSR	283
Sip_Tpa_	IPSESMVPSFW	93	VGDRLLVFKTA	104	RGDIVVFSNP	133	LVKRIVALPGEKV	187	LPEHNYFMMGDNRLNSTDMR	460
Sip_Tma_	VPTGSMIPTIQ	54	IGDRLFVEKTT	55	IGEIVVFWSP	72	YVKRLVGKGGDVL	115	VPEGFYFLMGDNTKESLDCR	276
SipF_Bja_	IPSGSMKATLL	50	VGDYLFVSKYS	61	RGDIVVFRLP	96	YIKRVIGLPGDEV	115	VPAGHFFMMGDNRDNSTDSR	204
Sip_Bja_	VPSGSMEPTLL	55	TRDALLASKFP	66	QGDVVVFRWP	102	WVKRVVGLPGDRI	121	VPAGHLFVLGDNRDNSADSR	205
Sip_Mle_	IPSESMEPTLH	90	VGDRIWVDKTT	106	PGDVVVFKGP	123	LVKRVIAVGGQTV	172	VPQGRLWVMGDNRIHSADSR	235
Sip_Mtu_	IPSGSMEQTIR	102	VGDRIVVDKLS	118	PGDVVVFRGP	135	LVKRVIAVGGQTV	184	VPFGRVVWMGDNRTHSADSR	247
Sip_Sli_	IPSGSMEQTIR	91	IGDRVLVDKLT	102	RGDVVVFRDP	120	LIKRVVGVGGDHV	171	VPEGRLWVMGDHRSNSADSR	226
Sip_Sco_	IPSGSMERGLR	83	IGDRVLVNKLA	94	RGDIVVFDGT	112	YIKRVVGVGGDHV	132	VPDGTLFVLGDHRSDSSDSR	188
Sip_Ssp_	IPSSSMEPTLQ	58	INDRLLIEKIS	69	RGEIVVFNPT	86	FIKRIIGLPGDEV	110	VPDDQYLVLGDNRNNSYDSH	161
Spi_Spn_	VEGHSMDPTLA	44	DGEILFVVKH-	54	RFDIVVAHEE	86	IVKRVIGMPGDTI	86	VPEGEYLLLGDDRLVSSDSR	177
Sip_Sml_	VEGHSMDPTLA	19	DGEILFVVKH-	29	RFDIVVAHEE	43	IVKRVIGMPGDTI	61	VPEGEYLLLGDDRLVSSDSR	152
Sip_Bca_	VEGKSMMPTLE	44	SGNLLIVNKLS	55	RFDIIVFHAN	72	YVKRVIGLPGDRI	89	VPPGCIFVLGDNRLSSWDSR	154
SpsB_Sau_	IKGESMDPTLK	42	DGERVAVNIIG	53	KGNVVVFHAN	70	YVKRVIGVPGDKV	87	IPKGKLIVLGDNREVSKDSR	156
SipB_Sca_	VRGDSMYPTLK	42	DGEKVIVNMIG	53	KGNIVVFHAT	70	YVKRVIGMPGDSI	87	IPKNKLLVLGDNREVSKDSR	155
Sip_Bam_	VEGSSMYPTLH	57	DGERLFVNKSV	68	RGDIVIINGD	85	YVKRLIGKPGETV	103	VPKGKYFVMGDNRLNSMDSR	164
Sip2_Bam_	VDGESMEPTLH	51	DRERLFVNMTV	62	RGQIVVLNGE	79	YVKRIIGLPGDTV	95	VPDDKYFVMGDNRRNSMDSR	156
Sip_Bli_	VEGTSMDPTLH	50	DGERLFVYKTV	61	RGDIVIIDGD	78	YVKRLIGLPGDTV	96	VPEGKYFVMGDNRQRSMDSR	157
Sip_Aae_	IPSASMEPTLH	38	VGDFLLVNKLV	49	RGDMIVFKYP	66	FIKRIIARGGDTV	85	VPEGYYFVMGDNRDNSQDSR	197
Sip_Bsu_	VEGVSMNPTFQ	40	EGNELLVNKFS	51	RFDIVLFKGP	68	LIKRVIGLPGETI	85	VPKGKYFVVGDNRIYSFDSR	150
SipS_Bsu_	VDGDSMYPTLH	49	NRERVFVNMTV	60	RGDIVVLNGD	77	YVKRIIGLPGDTV	93	VPDNKYFVMGDNRRNSMDSR	155

Gram Negative

A

Gram Positive

```
SipT_Bsu    VEGSSMYPTLH 57   DGERLFVNKTV 68   RGDIVIINGE 85   YVKRLIGKPGETV 103   VPKGKYFVMGDNRLNSMDSR 164
SipU_Bsu    IEGSSMAPTLK 52   DSERILVDKAV 63   RGDIIVIHDK 80   FVKRLIGLPGDSI  98   VPSGKYFVMGDNRIANSLDSR 158
SipV_Bsu    VEGVSMNPTFQ 40   EGNELLVNKFS 51   RFDIVLFKGP 68   LIKRVIGLPGETI  85   VPKGKYFVVGDNRIYSFDSR 150
SipP_pTA10  VEGKSMDPTLV 50   DSERLFVNKTV 61   RGDIIILNGK 78   YVKRLIGLPGDTV  96   VPKDKYFVMGDNRQESMDSR 157
SipP_pTA10  VQGESMKPTLF 49   NSERLFVNKFV 60   RGDIVVLNGE 77   YVKRLIGLPGDTI  95   VPKDKYFVMGDNRQNSMDSR 156
```

Consensus   ip SM ptl   gdrl v k   rgdivvf p   yvkr ig pgd v   vp g yfvmGDnr nS Dsr

B

      B   C   C'   D   E

```
SipW_Bsu    TLKSVLSGSMEPFNTGSLILV 60    LAKGDVITFM 78    TAVTHRIV 90    LFKTKGDNNAAADS 113
Spc21_Cpe   RTYSILSGSMEPEINTGDLAIV 35    VKVGDIITFK 52    KVVTHRVL 63    --FITKGDNNNANDT 82
```

Consensus   p vvvlsgSMeP f rgdl fl   vgdivvf   pivHrv   fitkGdnn d

FIG. 4. Amino acid sequence alignment of bacterial SPase I enzymes. This alignment is generated from ClustalX v.1.8 (*103*) amino acid alignments of all P-type (A.) and ER-type (B.) SPases (from a GenBank BLAST search) and are grouped as described by Tjalsma *et al.* (*49*) and Dalbey *et al.* (*46*). The catalytic Ser 90 and Lys 145 residues (*E. coli*) are indicated by a '∗'. In the consensus sequence, uppercase indicates strictly conserved residues and lowercase indicates conserved residues appearing in greater than 50% of the aligned sequences. (A) Alignment of conserved residues in P-type SPase enzymes. I. Eco, *Escherichia coli*; Sty, *Salmonella typhimurium*; Hin, *Haemophilus influenzae*; Pfl, *Pseudomonas fluorescens*; Bja, *Bradyrhizobium japonicum*; Rca, *Rhodobacter capsulatus*; Mle, *Mycobacterium leprae*; Mtu, *Mycobacterium tuberculosis*; Sli, *Streptomyces lividans*; Sco, *Streptomyces coelicolor*; Pla, cyanobacterium *Phormidium laminosum*; Ssf, *Synechocystis* sp. strain PCC6803 II; Aae, *Aquifex aeolicus*; Tpa, *Treponema pallidum*; Hpy, *Helicobacter pylori* J99; Bbu, *Borrelia burgdorferi*; Spn, *Streptococcus pneumoniae*; Smi, *Streptococcus mitis*; Bca, *Bacillus caldolyticus*; Sau, *Staphylococcus aureus*; Sca, *Staphylococcus carnosus*; Bsu, *Bacillus subtilis*; Bam, *Bacillus amyloliquefaciens*; Bli, *Bacillus licheniformis*; Tma, *Thermotoga maritima*; Cpn, *Chlamydopiila pneumoniae*; Ctr, *Chlamydia trachomatis*; Sip, SPase; Spi, *Streptococcus pneumoniae* SPase I; Sps, SPase *Staphylococcus aureus*. (B) Alignment of conserved residues in Eubacterial ER-type SPase I. Bsu, *Bacillus subtilis*; Cpe, *Clostridium perfringens*; Sip, signal peptidase; Spc, signal peptidase complex.

In bacteria, the apparent function of the signal peptidase enzyme is to release translocated preproteins from their membrane attached signal sequences. This allows periplasmic or outer membrane destined proteins to enter the periplasmic space. Using an *E. coli* strain in which synthesis of signal peptidase is under the control of the arabinose operon, it was shown that precursor proteins of *M13* procoat, maltose binding protein (MBP), and outer membrane protein A (OmpA), accumulate in the cell when the synthesis of signal peptidase is repressed (*4*). Protease accessibility assays demonstrated that, with attenuated signal peptidase production, procoat, pre-MBP and pro-OmpA proteins were still found translocated across the plasma membrane. This result is in line with studies showing that noncleavable signal sequence mutants of β-lactamase (*58*) and *M13* procoat (*59*) were also translocated. Interestingly, pre-MBP and pro-OmpA were found remaining on the outer surface of the plasma membrane, suggesting that the proteins were tethered to the membrane by the membrane embedded uncleaved signal sequences (*4*).

It is fascinating that *Mycoplasma genitalium,* a bacteria with the smallest genome known, does not appear to have a gene with any significant homology to type I signal peptidase (*60*). This is the only bacterium known to lack a signal peptidase.

D. TYPE I SIGNAL PEPTIDASE IS AN ANTIBACTERIAL TARGET

Signal peptidase is a potential target for antibacterial compounds and is currently being actively investigated by pharmaceutical companies (*61*). What makes type I signal peptidase an attractive drug target is that it is essential for cell viability for all bacteria. Also, the SPase I enzyme should be readily accessible by small molecules since the protease active site is located in the periplasmic space of gram-negative and in the outside surface of the plasma membrane in gram-positive bacteria. The recently solved 3D crystal structure of the catalytic domain of *E. coli* signal peptidase should provide a useful model for the rational design of potential inhibitors (*52*).

The practicality of using signal peptidase as a drug target has been questioned because SPase I is also an enzyme found in eukaryotic cells. However, there are notable differences between the bacterial versus the eukaryotic paralogs. The prokaryotic signal peptidases are single polypeptide chains, whereas the endoplasmic reticulum signal peptidases are multimeric complexes and contain some nonhomologous polypeptides (*46*). Also, the signal peptidase complex in *Saccharomyces cerevisiae* most likely carries out catalysis using a Ser-His dyad, rather than the Ser-Lys dyad found in prokaryotic signal peptidases (*53*). In mitochondria, signal peptidase is postulated to use a Ser-Lys mechanism, but it is a dimer consisting of the Imp1/Imp2 polypeptides. These differences suggest that it may be possible to design inhibitors that specifically inhibit the bacterial signal peptidases only and not the ER or mitochondrial enzymes.

II. Type I Signal Peptidase Enzymology

A. ENZYME PURIFICATION AND SUBSTRATE ASSAYS

The early isolation work on *E. coli* type I signal peptidase resulted in the characterization of some of its physical properties. Cell extracts overproducing signal peptidase indicate that signal peptidase activity is sensitive to high salt, Mg^{2+} concentration, and pH (*62*). Using purified substrate and purified *E. coli* signal peptidase, a pH optimum of about 9.0 has the greatest level of activity (*63, 64*). Also, a profile of activity versus temperature indicates that the enzyme is stable up to 40°C (*63*).

Overexpression and purification of *E. coli* type I SPase has been accomplished by recombinant techniques. Typically *E. coli* strains bearing plasmids engineered to overexpress the *E. coli lepB* gene are used. One such strain takes advantage of a plasmid (pPS9) bearing the signal peptidase gene under transcriptional control of the *lambda* promoter (*65*). This plasmid also codes for a temperature-sensitive *lambda* repressor causing reduced expression at 30°C and rapid expression at 42°C. Another plasmid, pRD8, expresses the *lepB* gene under the control of the *araB* promoter (*4*). In this system, expression is induced by the addition of arabinose. With both these plasmids the protein is purified by a protocol involving membrane isolation, Triton X-100 detergent extraction, DEAE ion-exchange chromatography (Pharmacia), and final isolation to homogeneity by a polybuffer chromatofocusing technique (Pharmacia) (*65, 66*).

A difficulty with *E. coli* SPase I purification results from an apparent autoproteolysis of the enzyme. Talarico *et al.* (*67*) demonstrated that the purified SPase gets cleaved after an Ala-Gln-Ala sequence (residues 38–40), which is consistent with the "−3, −1" or "Ala-X-Ala" motif of SPase I signal peptide substrate specificity (see Section II,B). With this information, a more efficient scheme for *E. coli* SPase I purification was devised (using pRD8) by inserting a 6-His tag after amino acid residue 35 (cytosolic domain) in the protein sequence. This eliminates yield losses from self-cleavage and enables the use of a nickel-chelate affinity chromatography purification system (*64*). The most productive system, however, utilizes the plasmid pET23Lep (*54*). This method also uses a 6-His/nickel-chelate approach, but also takes advantage of the very high expression levels of the pET vector system (Novagen). With this system, milligram quantities of purified *E. coli* SPase I protein are generated from a few liters of culture in a relatively short amount of time.

The enzymatic activity of the *E. coli* SPase has been assayed with a number of different substrates. These include peptides (*68–71*) and preprotein substrates (*40, 64, 72, 73*). One of the peptide substrates, Phe-Ser-Ala-Ser-Ala-Leu-Ala-Lys-Ile, is based on the cleavage site region of pre-MBP. It is cleaved by SPase to generate the Phe-Ser-Ala-Ser-Ala-Leu-Ala and Lys-Ile fragments. HPLC is used to separate and quantitate the two products. The resulting k_{cat} and K_m values with this substrate are 114 hr^{-1} and 0.8 mM, respectively (*71*). On the other hand, using the preprotein pro-OmpA nuclease A as a substrate for SPase results in much better kinetic constants. In this assay, cleavage of pro-OmpA nuclease A is analyzed by resolving the preprotein from the mature protein with SDS–PAGE. This substrate results in a k_{cat} of 44 s^{-1}, a K_m of 19.2 μM, and k_{cat}/K_m of 2.3×10^6 s^{-1} M^{-1} at pH 8.0. This catalytic efficiency is comparable to that of other Ser proteases such as trypsin and chymotrypsin (*74*). The reason for the dramatic increase in k_{cat}/K_m for the preprotein substrate compared to the peptide substrate is that the preprotein/SPase interactions not available with synthetic peptides may lead to optimal substrate positioning and increased processing efficiency.

Fluorogenic substrates have also been developed as continuous assays of SPase activity (*68, 69*). One example is the internally quenched fluorescent substrate Tyr(NO2)-Phe-Ser-Ala-Ser-Ala-Leu-Ala-Lys-Ile-Lys(Abz) (*68*). This conjugate peptide is also based on the cleavage site region of the preprotein pre-MBP, and *E. coli* SPase cleavage is able to generate the expected products Tyr(NO2)-Phe-Ser-Ala-Ser-Ala-Leu-Ala and Lys-Ile-Lys(Abz). This results in a fluorescence increase that is monitored during the course of the reaction. Unfortunately, like the Dev peptide (*71*), the resulting k_{cat}/K_m for this substrate is also very low (71.1 M^{-1} s^{-1}), indicating it is also a poor substrate. A hydrophobic H-region is a common motif in signal peptides (Fig. 2). The poor catalytic efficiency for this substrate is most likely due to the lack of a hydrophobic core in the primary sequence of the peptide substrate itself (*68*).

With the development of a new fluorogenic substrate, Stein and co-workers (*75*) have addressed some aspects of the function of the H-region in the signal peptide. In this work, the insertion of 10 Leu residues into the N terminus of the peptide used by Zhong and Benkovic (*68*) results in a substrate displaying a dramatic 10^4 increase in k_{cat}/K_m. Stein and co-workers suggest that this increase most likely results not only from the proximity effects gained from anchoring the substrate to micelles (also containing micelle-anchored SPase), but also from specific interactions achieved between the SPase enzyme and the new "H-region-like" domain of the substrate signal peptide itself.

B. SUBSTRATE SPECIFICITY

Statistical analyses of preprotein sequences from gram-negative and gram-positive bacteria have been very useful in the determination of the conservation patterns found in the C-region of the signal peptide. The data indicates patterns that are obligatory for signal peptide processing (*24*) and has led to the so called "−3, −1" or "Ala-X-Ala" rule (*20, 21, 23*), which states that mainly small uncharged residues are found at the −3 and −1 positions relative to the cleavage site. In both gram-negative and gram-positive bacteria, Ala is almost exclusively located at the −1 residue (Gly and Ser are the next most frequent). Ala is also the most common residue found at the −3 position followed by Val and Ser (less frequent). Also, Ala is common at +1 position, while Asp and Glu residues are found in the first few positions of the mature region of prokaryotic secretory proteins.

Using *in vivo* assays, the determinants of substrate specificity have been examined extensively for several preprotein substrates. The results of the studies on *M13* procoat, pre-phoA (pre-phosphatase A), and pre-MBP substrates are summarized in Fig. 5. Site-directed mutagenesis was used to substitute various residues at the +1, −1, −2, −3, −4, −5, and −6 positions of the *M13* procoat protein (*35*). The critical positions in the signal peptide for substrate processing are at the −1, −3, and −6 positions. Processing of procoat only occurs with small residues at the −1 position (Ala, Gly, Ser, and Pro). Some small, uncharged (Ser, Gly, Thr), or aliphatic (Leu and Val) residues at the −3 position result in processing, but others such as Pro, Gln, Lys, or Arg result in no processing. The results also indicate no distinct requirements for *in vivo* processing for residues at the +1, −2, −4, and −5 positions. Almost any residue is tolerated except for a Pro at the +1 position.

As shown in Fig. 5, similar findings were observed in studying the *in vivo* processing of pre-phoA (*36*) and pre-MBP (*76*). As suggested by statistical analyses, the critical positions for processing of these substrates is at the −1 and −3 positions. *In vivo* processing is maintained (*36*) with almost any residue at the +1, −2, −4, −5 positions of pre-phoA.

In addition to the −1 and −3 residue requirements, the presence of a helix breaker or a beta turn residue in the −4 to −6 region has been shown to be important for SPase processing. A Pro and Gly residue is frequently present in this region of bacterial signal peptides (*26*). It is intriguing that almost any residue besides a Pro at the −6 position of the *M13* procoat blocks *in vivo* processing (*35*). Perhaps a helix breaker prevents the C-region from forming a long helix with the hydrophobic core region of the signal peptide and allows the signal peptide C-region to bind to SPase in an extended conformation.

Jain *et al.* (*77*) analyzed the SPase cleavage of a number of phoA signal sequence mutants differing only in the length of the C-region. C-region lengths varied from 3 to 13 residues, and it was found that lengths ranging from 3 to 9 residues are completely and efficiently cleaved whereas those of 11 to 13 residues are not. One interpretation of this data is that since the active site

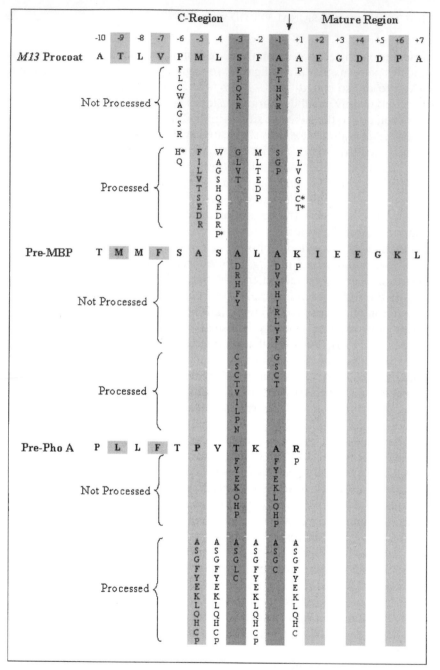

Fig. 5. Signal peptide C-region point mutations affect SPase I processing. The tabulated results of *in vivo* processing by *E. coli* SPase I of C-region signal sequence point mutants of M13 procoat (*35*), pre-MBP (*76*), and pre-Phosphatase A (*36*) substrates are shown. The bold sequences correspond to the wild-type sequences. * indicates processing is <16%.

of *E. coli* SPase is close to the periplasmic surface of the inner membrane, when the C-region of the signal peptide exceeds nine residues, the cleavage site of the preprotein is presented too far away from the active site of the enzyme. This results in the sudden drop-off in activity seen with the insertion of nine or more residues [77].

After von Heijne and co-workers identified the −3, −1 ("Ala-X-Ala") substrate specificity requirement for type I signal peptidases, a computational method was developed to predict whether signal peptides are located within biological sequences using a weight matrix method (*26*). This method has been improved using the neural network and hidden Markov model-based prediction model of Nielsen *et al.* (*24, 25, 78*). Biosequence analysis using this program is now available on the SignalP server (http://www.cbs.dtu.dk/services/SignalP/) to identify signal peptide cleavage sites. The algorithm even discriminates between cleaved signal peptides and uncleaved signal anchors in both prokaryotic and eukaryotic models. Also, the effectiveness of prediction programs such as these have enabled the *de novo* design of artificial signal peptides with demonstrated biological activity. In both the studies of Nilsson and von Heijne (*79*) and Wrede *et al.* (*80*), computer engineered signal peptides, located N-terminal to fusion protein constructs, were shown to be translocated and processed effectively in *E. coli*.

C.   THE ACTIVE SITE AND CATALYTIC MECHANISM: SITE-DIRECTED MUTAGENESIS STUDIES

Type I SPase has the unusual property of being resistant to inhibitors of the classical Ser, Cys, Asp, and metallo-protease classes (*62, 69, 81*). Thus, there is great interest in the protease community to pinpoint its catalytic mechanism. To date, most of the work in this area has been on the *E. coli* enzyme but there has also been some work on the *B. subtilis* type I signal peptidase (SipS). Initial clues to the proteolytic mechanism of SPase were determined using site-directed mutagenesis of the *E. coli* enzyme. SPase maintains activity if each of the Cys and His residues are mutated (*81, 82*), demonstrating that neither of these residues is catalytically important. Substitution of Ser-90 with an Ala completely inactivates the enzyme (*82*) and Lys-145 is also important for activity (*83, 84*). Mutation of Lys-145 to His, Asn, or Ala results in an inactive protease. These data show that *E. coli* SPase has a critical Ser-90 and Lys-145 residue, and support the notion that these are the catalytic residues. Consistent with this is that these residues are conserved in all bacterial type I signal peptidases.

Complementing the loss of function mutagenesis studies, the catalytic roles of Ser-90 and Lys-145 in *E. coli* type I SPase were further substantiated by chemical modification studies. Replacement of Ser-90 with a Cys residue produces an active enzyme that can then be inactivated by the addition of a Cys-specific reagent (*84*). In addition, an inactive Cys-145 *E. coli* SPase mutant can regain activity by modification with bromoethylamine to generate an enzyme with a γ-thia-Lys (*64*). Enzyme activity is also restored, to a lesser extent, by modification of the thio-145 SPase with either bromopropylamine or 2-mercaptoethylamine to generate other Lys analogs. There is no recovery of activity when the Cys-145 mutant derivative is reacted with (2-bromoethyl)trimethylammonium-Br. This finding supports the role of Lys as a general base rather than a positive charge donor in the mechanism.

Guided by amino acid conservation patterns among bacterial, ER, and mitochondrial type I signal peptidases, site-directed mutagenesis studies of the *Bacillus subtilis* type I SPase enzyme (SipS) indicated similar critical roles for some of the homologous residues found to be critical for function in *E. coli* (*85*). *B. subtilis* Ser-43 and Lys-83, homologous to the *E. coli* Ser-90 and Lys-145 residues, are critical for *in vivo* enzymatic activity. In addition, the amino acid homologous to *E. coli* Asp-280, Asp-153, is also essential. Two other residues, Asp-146 and Arg-84 (*E. coli* Asp-273 and Arg-146, respectively), appear to be structural determinants for the *B. subtilis* SPase (*85*).

The crystal structure of the inhibitor-bound, truncated, soluble form of *E. coli* type I SPase (Δ2–75) (*52*) has contributed much to the SPase field. Using this structure as a guide, the work of Klenotic *et al.* (*54*) has shed light onto the roles of most of the conserved residues in the homology domain E-region (see Fig. 4A) of *E. coli* SPase. Most of these residues are in the active site region, as shown in Fig. 6 (see color plate). In contrast to the *B. subtilis* experiments, mutagenesis of *E. coli* Arg-146 does not result in a dramatic loss of function. The crystal structure shows an ionic interaction between Asp-273 and Arg-146, but the mutation of Arg-146 to Ala in *E. coli* results in no reduction in enzymatic activity (*54*). However, like the *B. subtilis* counterpart, there is a marked reduction in activity for Asp-273 mutations. The salt bridge interaction of Asp-280 with Arg-282, also evident in Fig. 6, supports the loss of function resulting from mutagenesis of these residues.

Other conserved *E. coli* Box E residues are Gly-272 and Ser-278 (Fig. 4A). The active site structure (Fig. 6) of the *E. coli* SPase reveals that both Gly-272 and Ser-278 are in close proximity to the catalytic Lys-145. In fact, Ser-278 is within H-bonding distance to Lys-145 (*54*). Mutagenesis studies demonstrate that indeed the Ser-278 is important for activity, as changing this residue to an Ala causes a reduction in activity of approximately 300-fold in processing the substrate pro-OmpA nuclease A (*54*). This suggests that Ser-278 may actually help orient the proposed general base Lys-145. Changing the Gly-272 to Ala reduces the activity of SPase 750-fold relative to the wild-type enzyme (*54*). Consistent with a Gly-272 to Ala mutant with reduced activity, modeling studies suggest that changing the side chain at amino acid 272 from a hydrogen to a methyl, or any other group, causes steric crowding and

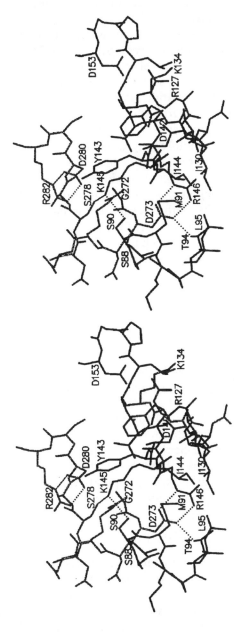

FIG. 6. (See color plate.) The residues conserved among type I SPases (boxes B-E) make up the active site region of *E. coli* SPase I. A ball and stick stereographic representation of the residues that make up box B (88–95, black), box C (127–134, red), box D (142–153, blue), and box E (272–282, green). This figure was made with the program MOLSCRIPT (*104*).

perturbs the positioning of the Lys-145 side chain. In total, the conserved Box E residues may help stabilize the enzyme and are responsible for maintaining the overall architecture of the active site.

In other recent work, the Ser-88 residue has also been shown to be important in the catalytic mechanism of *E. coli* SPase (*86*). The crystal structure of Δ2–75 SPase I reported by Paetzel *et al.* (*52*) indicates that the catalytic Ser-90 amide backbone nitrogen and the Ser-88 hydroxyl may be involved in forming an oxyanion hole, stabilizing a tetrahedral oxyanion transition state intermediate that forms during the course of catalysis. Mutagenesis of Ser-88 to Ala leads to a greater than 2000-fold reduction in the k_{cat} with very little effect on K_m (*86*). Interestingly, sequence alignment studies show that only Ser, Thr, and Gly residues occur at this position in other signal peptidases (Box B of Fig. 4). In signal peptidases surveyed, the Gly residues were present at this homologous position only in gram-positive bacteria (Box B of Fig. 4). It is possible that with this subset of gram-positive bacterial signal peptidases, oxyanion stabilization is mediated by a more conventional backbone amide rather than a side-chain interaction.

From an evolutionary perspective, it is very intriguing that ER- and Archaea-like signal peptidases have been identified in gram-positive bacteria such as *Bacillus subtilis* and *Clostridium perfringens* (Fig. 4B). SipW of *B. subtilis* has been characterized as one of seven signal peptidases found in *B. subtilis* (*49*) while Spc-21 of *C. perfringens* is a putative SPase identified from genome sequencing (accession CAA60213). The overall amino acid conservation patterns and a putative His general base in place of the Lys general base (Lys-145 of *E. coli*) differentiates signal peptidases into the P-type (eubacterial and mitochondria/chloroplast) and ER-type [eukaryal (ER) and archeal] signal peptidases (*49*). Mutagenesis studies show that the conserved Ser and His residues (at the same positions as Ser-90 and Lys-145 in *E. coli*) are critical for the functioning of SipW (*50*). In contrast to similar studies done with the Sec11 homologous subunit in the ER SPase of *S. cerevisiae* where the His cannot be substituted with a lysine residue (*53*), this work on SipW showed that the putative His general base can be substituted by a Lys and still maintain enzyme function.

D.   Inhibitors of Type I Signal Peptidase

It has been very challenging for chemists to synthesize effective inhibitors against bacterial type I signal peptidases because of their unusual mechanism. As mentioned previously, protease inhibitors against the Ser-, Cys-, Asp-, and metalloenzyme classes were ineffective against SPase. Though SPase is a Ser protease, very high concentrations of [³H]diisopropyl fluorophosphate do not inhibit the enzyme (*63*). The first report of an inhibitor was by Kuo and co-workers (*69*), where they showed that certain β-lactams could inhibit the enzyme. β-Lactams had been shown previously to inhibit other Ser proteases and β-lactamases (*87–91*).

The observation that the catalytic mechanism of SPase occurs by a Ser-Lys dyad (*83, 84*) is a significant breakthrough. This mechanism is similar to β-lactamase enzymes, which use Lys as a general base in the acylation step (*92*). With this information, researchers at Smithkline Beecham Pharmaceuticals focused on β-lactam type compounds. Several types of effective compounds were identified with an IC$_{50}$ in the 0.260 to 50 μM region (*61*). As shown in Fig. 7, the best inhibitors found include clavams, thioclavams,

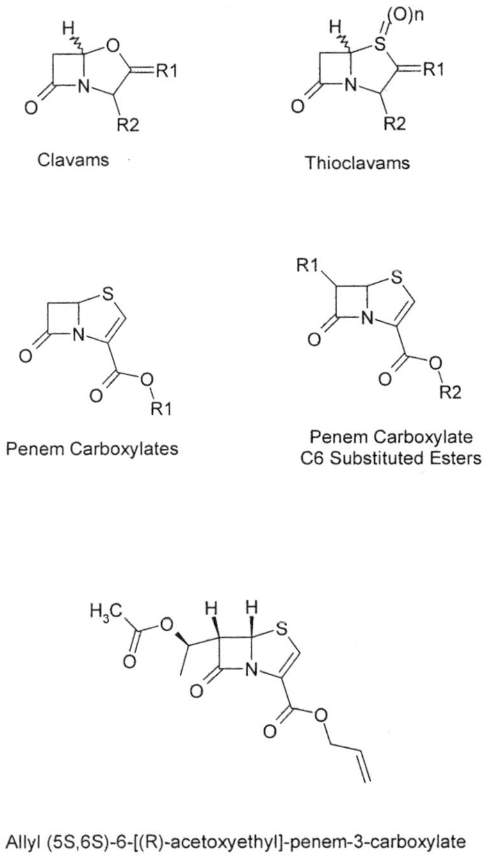

Clavams

Thioclavams

Penem Carboxylates

Penem Carboxylate
C6 Substituted Esters

Allyl (5S,6S)-6-[(R)-acetoxyethyl]-penem-3-carboxylate

FIG. 7. Inhibitors of *E. coli* SPase I. A listing of some of the types of inhibitors that have been designed to inhibit *E. coli* SPase I (*61*). The most effective inhibitor, synthesized by researchers at SmithKline Beecham (*61*), is the C6 substituted penem carboxylate ester, allyl (5*S*, 6*S*)-6-[(*R*)-acetoxyethyl]penem-3-carboxylate as shown.

and penem carboxylates. The 5S-penem derivatives are the most potent (*61*). The compound allyl(5S,6S)-6-[(R)-acetoxyethyl]penem-3-carboxylate (Fig. 7) is the most potent inhibitor developed to date and has been shown to inhibit *E. coli* and *S. aureus* as well as chloroplast signal peptidases. From a mechanistic point of view, it is interesting that the penem inhibitors of signal peptidases are of the 5S stereochemistry. This is the opposite to that of the 5R β-lactams that are recognized by β-lactamases and penicillin binding proteins (*61*).

Besides small molecule inhibitors, *E. coli* type I SPase is also inhibited by the signal peptide of the *M13* procoat protein (*72*). Also, the substitution of a Pro residue into the +1 position of pre-MBP prevents its processing by *E. coli* SPase (*93*). Expression of this pre-MBP mutant *in vivo* leads to the accumulation of preproteins normally processed by type I SPase but not proteins processed by lipoprotein (type II) SPase (*93*). This suggests that the pre-MBP +1 Pro mutant acts as a competitive inhibitor of type I SPase.

III. Three-Dimensional Structure

A. A NOVEL PROTEIN FOLD

The SPase crystal structure of the soluble, catalytic domain fragment of *E. coli* SPase I, Δ2–75 (*52*), reveals a unique protein fold (see Fig. 8). It consists mainly of two large antiparallel β-sheet domains (I and II), two small 3_{10} helices (residues 246–250 and 315–319), and one small α-helical region (residues 280–285). There is one disulfide bond between Cys-170 and Cys-176 located immediately before a beta turn between the outer strands of β-sheet domain II. An extended β ribbon protrudes from domain I. In conjunction with the N-terminal strand, this ribbon gives the overall molecule a conical shape with dimensions of 60 Å × 40 Å × 70 Å.

Another protease that has been proposed to use a Ser-Lys dyad is UmuD, a member of the LexA family of proteases (*94*). UmuD maintains 23.4% sequence identity (residues 40–139) with *E. coli* type I SPase (75–202). The crystal structure of the fragment of UmuD, UmuD′, reveals a fold similar to SPase and is mostly β sheet (*94, 95*). In UmuD′, however, there are no structural counterparts to the β-sheet domain II and the extended hairpin (between residues 108 and 124) found in SPase. In fact, sequence homology and modeling studies of other signal peptidases in gram-positive bacteria, mitochondria, and endoplasmic reticulum indicate that the extended hairpin is also missing and that most of β-sheet domain II is missing. Whether these differences manifest themselves through variations in substrate binding or specificity is yet to be determined.

FIG. 8. General fold of *E. coli* Δ2–75 type I SPase. A MOLSCRIPT ribbon diagram of *E. coli* Δ2–75 type I SPase. The domain that appears [from sequence alignments (*105*)] to be conserved across all type I SPases is shown in black.

B. VIEW OF A UNIQUE ACTIVE SITE AND CATALYTIC MECHANISM

A GRASP molecular surface representation of the Δ2–75 crystal structure is shown in Fig. 9. The substrate binding pockets S1 and S3 are labeled. The dark gray areas in Fig. 9 represent the exposed hydrophobic surfaces. The large exposed surface is formed by antiparallel β strands consisting of residues 81–85, 99–105, 292–307, and 321–314, and includes the hydrophobic residues Tyr-81, Phe-100, Leu-102, Trp-300, Met-301, Trp-310, and Leu-314 within the β strand, and the nearby residues Phe-79, Ile-80, Leu-316, and Ile-319. Studies have shown that Δ2–75 can bind to the inner membrane vesicles of *E. coli* and insert into membrane monolayers (*96*). The insertion of the catalytic domain into the lipid phase suggests that the active site may be partially buried in the membrane. Thus the extended hydrophobic patch of Δ2–75 seen in the crystal structure may constitute the membrane association surface (*52*).

The crystal structure also revealed a covalent bond from the active site Ser-90 Oγ to the carbonyl carbon, C7 of the 5*S*,6*S*-penem inhibitor (Fig. 10). This is the first direct evidence of the nucleophilic nature of the catalytic Ser-90. In addition, the Ser-90 Oγ oxygen is within 2.9 Å of the Nε of Lys-145.

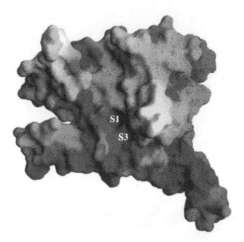

Fig. 9. A representation of the molecular surface of *E. coli* Δ2–75 type I SPase made with the program GRASP. The dark gray areas indicate hydrophobic surfaces. The location of the S1 and S3 substrate binding sites are indicated.

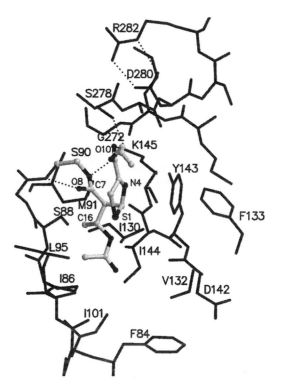

Fig. 10. A MOLSCRIPT (*104*) ball-and-stick representation of the active site of *E. coli* Δ2–75 type I SPase that is bound to inhibitor.

Lys-145 is the only titratable amino acid residue within H-bonding distance of Ser-90 (Fig. 10). This is further evidence that it may act as a general base during catalysis. Also, the main chain nitrogen of Ser-90 forms a hydrogen bond (2.9 Å) to the carbonyl oxygen of the cleaved β-lactam ring within the penem inhibitor. This suggests that the Ser-90 N stabilizes the tetrahedral transition state oxyanion intermediate. The Ser-88 side-chain hydroxyl is also a likely candidate for oxyanion stabilization provided it is able to rotate about its χ_1 angle. Because of steric conflicts, the covalently bound inhibitor prevents the Ser-88 hydroxyl from being in position to contribute to the oxyanion stabilization. As described previously, the mutational studies of Carlos et al. (86) suggest that, as with the classical serine proteases, an oxyanion hole is operational in E. coli SPase.

The crystal structure shows that E. coli SPase I Lys-145 is buried and makes van der Waals contacts with the side chains of Phe-133, Tyr-143, and Met-270, as well as the main-chain atoms of Met-270, Met-271, Ala-279, and Gly-272. This hydrophobic environment may altogether be responsible for lowering of the Lys-145 pK_a. The results of studies on temperature-sensitive mutants of B. subtilis SPase I (SipS) are consistent with these observations of the E. coli SPase I structure. The mutations of B. subtilis SPase I (SipS) Tyr-81 (E. coli Tyr-143) to an Ala or Leu-74 to Ala (E. coli Phe-133) result in reduced SipS SPase I activity at 37°C and almost no activity at 42°C (51). These results are consistent with the hypothesis that the side chains of these residues reduce the pK_a of the general base Lys that is important for catalysis.

Also with B. subtilis SipS, the homologs of E. coli Arg-146 and Asp-273 (B. subtilis Arg-84 and Asp-146) are important for activity (51). SipS R84A and SipS D146A mutants display little activity at 37°C but sufficient to replace the chromosomally encoded enzyme (49). However, these SPase I mutants are temperature-sensitive for growth with evidence that they are prone to proteolytic degradation in vivo. In the crystal structure of E. coli SPase I, Arg-146 and Asp-273 form an ionic salt bridge interaction and the results from B. subtilis SipS indicate an analogous interaction is essential for optimal activity. In contrast, mutation of the Arg-146 in the E. coli full-length SPase I does not impair activity at all, suggestive of a much less catalytically important interaction in the E. coli enzyme (54).

The active site geometry from the crystal structure also reveals that Lys-145 is hydrogen bonded to Ser-278 which in turn is also hydrogen bonded to Asp-280 (see Fig. 6). Ser-278 is also held firmly in place by an interaction with the main chain amide of Gly-272. Ser-278 may help orient the Lys-145 residue similar to the manner in which the Asp residue functions to orient the His residue in the classical Ser/His/Asp catalytic triads of serine proteases. Asp-280 may also help orient Ser-278 in addition to playing a structural role by forming a salt bridge with Arg-282. Asp-273 maintains a bifurcated interaction by forming a salt bridge with Arg-146 and a hydrogen bond with Thr-94 (Fig. 6). In total, these residues may help stabilize the enzyme and are perhaps responsible for maintaining the overall architecture of the active site (*54*).

Finally, it is noteworthy that Ser-278 is invariant in not just bacterial but all P-type SPase I enzymes [bacterial (see Fig. 4), mitochondrial (Imp1/Imp2), and chloroplast type I signal peptidases (thylakoid processing peptidases)] that are proposed to utilize a Ser-Lys dyad catalytic mechanism. The ER-type (endoplasmic reticulum and archaeal) SPase I enzymes, proposed to use a Ser-His dyad, instead do not have a homologous *E. coli* Ser-278 residue at this position. This implies that the Ser-278–Lys-145 interaction mentioned earlier is critical for the functioning of the Ser-90–Lys-145 dyad mechanism.

C. A Binding Site Consistent with General Substrate Specificity

The structure of the Δ2–75 SPase-penem inhibitor complex suggests an S1 pocket [Schechter and Berger notation (*97*)] that binds the P1 residue of a preprotein substrate (*52*). This hypothesis is based on the observation that the methyl group (C16) on the 6-[acetoxethyl] side chain of the penem inhibitor is critical for the effectiveness of the inhibitor and presumably mimics the P1 (−1 relative to the cleavage site) Ala side chain of a preprotein substrate (*61*). The residues making direct van der Waals contact with the P1 methyl group of the inhibitor in the crystal structure are Met-91, Ile-144, Leu-95, and Ile-86, which are all conserved residues (Fig. 10).

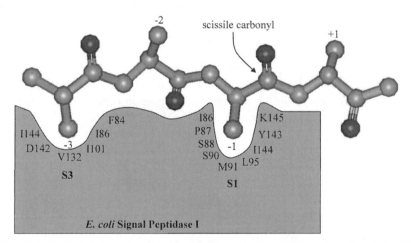

Fig. 11. *E. coli* SPase I substrate binding. A schematic representation of the S1 and S3 subsite interactions of *E. coli* SPase I with the −1 (P1) and −3 (P3) residues of a modeled tetraalanine peptide substrate. The substrate is shown in an extended (β) conformation.

An S3 subsite is also extrapolated by modeling an extended tetra-Ala peptide into the active site of SPase. This model uses the methyl group of the inhibitor and the carbonyl group of the cleaved inhibitor as a template. Hydrogen-bonding interactions between the peptide and the SPase I β strand containing the Lys145 are obtained with this model. The residues that form the S3 site are Phe-84, Ile-144, Val-132, and Ile-86. A schematic diagram of the S1 and S3 interactions with the modeled tetra-Ala peptide is shown in Fig. 11. As was revealed from surface analysis (Fig. 9), S3 is shallower and broader than the S1 hydrophobic depression. This result is consistent with computational analyses of naturally occurring bacterial signal peptide substrates indicating that, although Ala is most frequently found at both −1 (P1) and −3 (P3) positions, larger aliphatic residues are sometimes observed at the −3 position (*24, 25*).

IV. Other Ser-Lys Dyad Proteases and Amidases

LexA, a protein involved in the SOS response in DNA repair (*98*), was the first protease suggested to carry out catalysis using a Ser Lys dyad (*99*). UmuD is another member of the LexA family and also utilizes a Ser-Lys dyad catalytic mechanism. In the crystal structure of the UmuD′ fragment, the homologous critical Ser and Lys SPase I residues are within hydrogen bonding distance of each other (*94*). Interestingly, members of the LexA family of proteases undergo proteolysis at sites that follow the "−3, −1" rule (*85*).

Sauer and colleagues (*100*), using site-directed mutagenesis techniques, showed that Tsp protease also has Ser and Lys residues critical for catalysis. Similarly, mutagenesis studies suggest that a noncanonical (viral) Lon protease as well as the bacterial and organeller Lon proteases most likely utilize a Ser-Lys dyad catalytic mechanism (*101*).

Although they are not proteases, some amidases are also reported as utilizing a Ser-Lys dyad mechanism. An example is *E. coli* RTEM-1 β-lactamase. The crystal structure of a complex with penicillin G shows a Lys residue found within H-bonding distance of a Ser Oγ group of the cleaved penicillin complex (*92*). Mutagenesis studies of another amidase, mammalian fatty acid amide hydrolase (FAAH), strongly suggest a catalytic Ser residue and a Lys acid/base catalyst are present (*102*). FAAH is a hydrolase that is widespread in many different organisms, including mammals.

V. Conclusions and Perspective

Type I SPase belongs to a family of membrane-bound proteases that remove signal sequences from exported proteins after they are translocated across the membrane. These proteases are found in all domains of life: eubacteria, archaea, and eukarya. Because they appear to be essential for bacterial viability, there is a great deal of interest in studying SPase as a drug target. To date, efforts to obtain effective inhibitors have been mixed, but researchers at SmithKline Beecham have isolated an inhibitor with an IC50 of 260 nM. SPase is a very challenging target for medicinal chemists focusing on drug design, especially given the fact that SPase I homologs are found in mitochondria and in the endoplasmic reticulum.

It is striking that the proteolytic mechanism used by type I bacterial signal peptidases is not the Ser/His/Asp triad of prototypical serine proteases, but rather, a Ser-Lys dyad. In this paradigm, the general base Lys, in a reduced pK_a environment, deprotonates the catalytic Ser hydroxyl group creating a strong nucleophile for subsequent catalysis. Evidence for this Ser–Lys mechanism has also been reported in other proteases such as LexA, Tsp, Lon, as well as amidases such as β-lactamase and fatty acid amide hydrolase.

Though this field has advanced considerably in recent years, other questions regarding the mechanism and substrate specificity of type I signal peptidases remain unanswered. For example, the pK_a of the putative general base Lys residue has never been directly measured and more precise studies to pinpoint microenvironmental factors responsible for its pK_a shift are necessary. How can SPase I accurately cleave its substrate when other potential sites are available nearby within the signal peptide and mature region of the substrate? Additional interactions likely occur during catalysis such as interactions between the H- or C-region of the substrate signal peptide and SPase enzyme. Also, studies have shown that there is a detergent or phospholipid requirement for the optimal catalytic activity of SPase I, but it is not known precisely which regions of the catalytic domain of the *E. coli* SPase physically interact with detergents or the actual membrane bilayer. Continuing efforts to address these and other issues will undoubtedly enlighten our understanding of this novel class of enzyme.

ACKNOWLEDGMENTS

Ross E. Dalbey thanks his students who have contributed to the signal peptidase work over the years. Work from RED has been supported by a grant from the National Institutes of Health (GM 08512) and NSF (MCB-9808843). Joe Carlos was funded by the National Institutes of Health Predoctoral Training Grant GM 08512; Mark Paetzel was supported by a Medical Research Council of Canada postdoctoral fellowship. Natalie Strynadka was supported by the Medical Research Council of Canada, the Canadian Bacterial Diseases Network of Excellence, and the Burroughs Wellcome Foundation grant.

REFERENCES

1. Pugsley, A. P. (1993). *Microbiol. Rev.* **57,** 50.
2. LaPointe, C. F., and Taylor, R. K. (2000). *J. Biol. Chem.* **275,** 1502.
3. Date, T. (1983). *J. Bacteriol.* **154,** 76.
4. Dalbey, R. E., and Wickner, W. (1985). *J. Biol. Chem.* **260,** 15925.
5. Cregg, K. M., Wilding, I., and Black, M. T. (1996). *J. Bacteriol.* **178,** 5712.
6. Klug, G., Jager, A., Heck, C., and Rauhut, R. (1997). *Mol. Gen. Genet.* **253,** 666.
7. Zhang, Y. B., Greenberg, B., and Lacks, S. A. (1997). *Gene* **194,** 249.
8. Duong, F., and Wickner, W. (1997). *EMBO J.* **16,** 2756.
9. Cabelli, R. J., Chen, L., Tai, P. C., and Oliver, D. B. (1988). *Cell* **55,** 683.
10. Cabelli, R. J., Dolan, K. M., Qian, L. P., and Oliver, D. B. (1991). *J. Biol. Chem.* **266,** 24420.
11. Oliver, D. B., and Beckwith, J. (1982). *J. Bacteriol.* **150,** 686.
12. Kumamoto, C. A., and Beckwith, J. (1983). *J. Bacteriol.* **154,** 253.
13. Hartl, F. U., Lecker, S., Schiebel, E., Hendrick, J. P., and Wickner, W. (1990). *Cell* **63,** 269.
14. Sargent, F., Bogsch, E. G., Stanley, N. R., Wexler, M., Robinson, C., Berks, B. C., and Palmer, T. (1998). *EMBO J.* **17,** 3640.
15. Weiner, J. H., Bilous, P. T., Shaw, G. M., Lubitz, S. P., Frost, L., Thomas, G. H., Cole, J. A., and Turner, R. J. (1998). *Cell* **93,** 93.
16. Chanal, A., Santini, C., and Wu, L. (1998). *Mol. Microbiol.* **30,** 674.
17. Santini, C. L., Ize, B., Chanal, A., Muller, M., Giordano, G., and Wu, L. F. (1998). *EMBO J.* **17,** 101.
18. Boyd, D., Schierle, C., and Beckwith, J. (1998). *Protein Sci.* **7,** 201.
19. Samuelson, J. C., Chen, M., Jiang, F., Moller, I., Wiedmann, M., Kuhn, A., Phillips, G. J., and Dalbey, R. E. (2000). *Nature* **406,** 637.
20. von Heijne, G. (1983). *Eur. J. Biochem.* **133,** 17.
21. Perlman, D., and Halvorson, H. O. (1983). *J. Mol. Biol.* **167,** 391.
22. von Heijne, G. (1990). *J. Membr. Biol.* **115,** 195.
23. von Heijne, G. (1985). *J. Mol. Biol.* **184,** 99.
24. Nielsen, H., Engelbrecht, J., Brunak, S., and von Heijne, G. (1997). *Protein Eng.* **10,** 1.
25. Nielsen, H., Engelbrecht, J., Brunak, S., and von Heijne, G. (1997). *Int. J. Neural Syst.* **8,** 581.
26. von Heijne, G. (1986). *Nucleic Acids Res.* **14,** 4683.
27. Akita, M., Sasaki, S., Matsuyama, S., and Mizushima, S. (1990). *J. Biol. Chem.* **265,** 8164.
28. van der Wolk, J. P., Fekkes, P., Boorsma, A., Huie, J. L., Silhavy, T. J., and Driessen, A. J. (1998). *EMBO J.* **17,** 3631.
29. Nesmeyanova, M. A., Karamyshev, A. L., Karamysheva, Z. N., Kalinin, A. E., Ksenzenko, V. N., and Kajava, A. V. (1997). *FEBS Lett.* **403,** 203.
30. Hoyt, D. W., and Gierasch, L. M. (1991). *J. Biol. Chem.* **266,** 14406.
31. Rizo, J., Blanco, F. J., Kobe, B., Bruch, M. D., and Gierasch, L. M. (1993). *Biochemistry* **32,** 4881.
32. Michaelis, S., Hunt, J. F., and Beckwith, J. (1986). *J. Bacteriol.* **167,** 160.
33. Stader, J., Benson, S. A., and Silhavy, T. J. (1986). *J. Biol. Chem.* **261,** 15075.
34. Puziss, J. W., Fikes, J. D., and Bassford, P. J., Jr. (1989). *J. Bacteriol.* **171,** 2303.
35. Shen, L. M., Lee, J. I., Cheng, S. Y., Jutte, H., Kuhn, A., and Dalbey, R. E. (1991). *Biochemistry* **30,** 11775.
36. Karamyshev, A. L., Karamysheva, Z. N., Kajava, A. V., Ksenzenko, V. N., and Nesmeyanova, M. A. (1998). *J. Mol. Biol.* **277,** 859.
37. Chaddock, A. M., Mant, A., Karnauchov, I., Brink, S., Herrmann, R. G., Klosgen, R. B., and Robinson, C. (1995). *EMBO J.* **14,** 2715.

38. Brink, S., Bogsch, E. G., Edwards, W. R., Hynds, P. J., and Robinson, C. (1998). *FEBS Lett.* **434,** 425.
39. Chang, C. N., Blobel, G., and Model, P. (1978). *Proc. Natl. Acad. Sci. USA* **75,** 361.
40. Zwizinski, C., and Wickner, W. (1980). *J. Biol. Chem.* **255,** 7973.
41. Date, T., and Wickner, W. (1981). *Proc. Natl. Acad. Sci. USA* **78,** 6106.
42. Wolfe, P. B., Wickner, W., and Goodman, J. M. (1983). *J. Biol. Chem.* **258,** 12073.
43. Moore, K. E., and Miura, S. (1987). *J. Biol. Chem.* **262,** 8806.
44. Whitley, P., Nilsson, L., and von Heijne, G. (1993). *Biochemistry* **32,** 8534.
45. van Dijl, J. M., van den Bergh, R., Reversma, T., Smith, H., Bron, S., and Venema, G. (1990). *Mol. Gen. Genet.* **223,** 233.
46. Dalbey, R. E., Lively, M. O., Bron, S., and van Dijl, J. M. (1997). *Protein Sci.* **6,** 1129.
47. Altschul, S. F., Madden, T. L., Schaffer, A. A., Zhang, J., Zhang, Z., Miller, W., and Lipman, D. J. (1997). *Nucleic Acids Res.* **25,** 3389.
48. Jeanmougin, F., Thompson, J. D., Gouy, M., Higgins, D. G., and Gibson, T. J. (1998). *Trends Biochem. Sci.* **23,** 403.
49. Tjalsma, H., Bolhuis, A., van Roosmalen, M. L., Wiegert, T., Schumann, W., Broekhuizen, C. P., Quax, W. J., Venema, G., Bron, S., and van Dijl, J. M. (1998). *Genes Dev.* **12,** 2318.
50. Tjalsma, H., Stover, A. G., Driks, A., Venema, G., Bron, S., and van Dijl, J. M. (2000). *J. Biol. Chem.* **275,** 25102.
51. Bolhuis, A., Tjalsma, H., Stephenson, K., Harwood, C. R., Venema, G., Bron, S., and van Dijl, J. M. (1999). *J. Biol. Chem.* **274,** 15865.
52. Paetzel, M., Dalbey, R. E., and Strynadka, N. C. (1998). *Nature* **396,** 186.
53. Van Valkenburgh, C., Chen, X., Mullins, C., Fang, H., and Green, N. (1999). *J. Biol. Chem.* **274,** 11519.
54. Klenotic, P. A., Carlos, J. L., Samuelson, J. C., Schuenemann, T. A., Tschantz, W. R., Paetzel, M., Strynadka, N. C., and Dalbey, R. E. (2000). *J. Biol. Chem.* **275,** 6490.
55. Tjalsma, H., Noback, M. A., Bron, S., Venema, G., Yamane, K., and van Dijl, J. M. (1997). *J. Biol. Chem.* **272,** 25983.
56. Meijer, W. J., de Jong, A., Bea, G., Wisman, A., Tjalsma, H., Venema, G., Bron, S., and van Dijl, J. M. (1995). *Mol. Microbiol.* **17,** 621.
57. Parro, V., Schacht, S., Anne, J., and Mellado, R. P. (1999). *Microbiology* **145,** 2255.
58. Koshland, D., Sauer, R. T., and Botstein, D. (1982). *Cell* **30,** 903.
59. Kuhn, A., and Wickner, W. (1985). *J. Biol. Chem.* **260,** 15914.
60. Fraser, C. M., Gocayne, J. D., White, O., Adams, M. D., Clayton, R. A., Fleischmann, R. D., Bult, C. J., Kerlavage, A. R., Sutton, G., Kelley, J. M., Fritchman, J. L., Weidman, J. F., Small, K. V., Sandusky, M., Fuhrman, J., Nguyen, D., Utterback, T. R., Saudek, D. M., Phillips, C. A., Merrick, J. M., Tomb, J.-F., Dougherty, B. A., Bott, K. F., Hu, P.-C., and Lucier, T. S. (1995). *Science* **270,** 397.
61. Black, M. T., and Bruton, G. (1998). *Curr. Pharm. Des.* **4,** 133.
62. Zwizinski, C., Date, T., and Wickner, W. (1981). *J. Biol. Chem.* **256,** 3593.
63. Kim, Y. T., Muramatsu, T., and Takahashi, K. (1995). *J. Biochem. (Tokyo)* **117,** 535.
64. Paetzel, M., Strynadka, N. C., Tschantz, W. R., Casareno, R., Bullinger, P. R., and Dalbey, R. E. (1997). *J. Biol. Chem.* **272,** 9994.
65. Wolfe, P. B., Silver, P., and Wickner, W. (1982). *J. Biol. Chem.* **257,** 7898.
66. Tschantz, W. R., and Dalbey, R. E. (1994). *Methods Enzymol* **244,** 285.
67. Talarico, T. L., Dev, I. K., Bassford, P. J., Jr., and Ray, P. H. (1991). *Biochem. Biophys. Res. Commun.* **181,** 650.
68. Zhong, W., and Benkovic, S. J. (1998). *Anal. Biochem.* **255,** 66.
69. Kuo, D., Weidner, J., Griffin, P., Shah, S. K., and Knight, W. B. (1994). *Biochemistry* **33,** 8347.

70. Dierstein, R., and Wickner, W. (1986). *EMBO J.* **5,** 427.
71. Dev, I. K., Ray, P. H., and Novak, P. (1990). *J. Biol. Chem.* **265,** 20069.
72. Wickner, W., Moore, K., Dibb, N., Geissert, D., and Rice, M. (1987). *J. Bacteriol.* **169,** 3821.
73. Chatterjee, S., Suciu, D., Dalbey, R. E., Kahn, P. C., and Inouye, M. (1995). *J. Mol. Biol.* **245,** 311.
74. Suciu, D., Chatterjee, S., and Inouye, M. (1997). *Protein Eng.* **10,** 1057.
75. Stein, R. L., Barbosa, M. D. F. S., and Bruckner, R. (2000). *Biochemistry* **39,** 7973.
76. Fikes, J. D., Barkocy-Gallagher, G. A., Klapper, D. G., and Bassford, P. J., Jr. (1990). *J. Biol. Chem.* **265,** 3417.
77. Jain, R. G., Rusch, S. L., and Kendall, D. A. (1994). *J. Biol. Chem.* **269,** 16305.
78. Nielsen, H., Brunak, S., and von Heijne, G. (1999). *Protein Eng.* **12,** 3.
79. Nilsson, I., and von Heijne, G. (1991). *J. Biol. Chem.* **266,** 3408.
80. Wrede, P., Landt, O., Klages, S., Fatemi, A., Hahn, U., and Schneider, G. (1998). *Biochemistry* **37,** 3588.
81. Black, M. T., Munn, J. G., and Allsop, A. E. (1992). *Biochem J.* **282,** 539.
82. Sung, M., and Dalbey, R. E. (1992). *J. Biol. Chem.* **267,** 13154.
83. Black, M. T. (1993). *J. Bacteriol.* **175,** 4957.
84. Tschantz, W. R., Sung, M., Delgado-Partin, V. M., and Dalbey, R. E. (1993). *J. Biol. Chem.* **268,** 27349.
85. van Dijl, J. M., de Jong, A., Venema, G., and Bron, S. (1995). *J. Biol. Chem.* **270,** 3611.
86. Carlos, J. L., Klenotic, P. A., Paetzel, M., Strynadka, N. C., and Dalbey, R. E. (2000). *Biochemistry* **39,** 7276.
87. Chabin, R., Green, B. G., Gale, P., Maycock, A. L., Weston, H., Dorn, C. P., Finke, P. E., Hagmann, W. K., Hale, J. J., MacCoss, M., Shah, S. K., Underwood, D., Doherty, J. B., and Knight, W. B. (1993). *Biochemistry* **32,** 8970.
88. Faraci, W. S., and Pratt, R. F. (1986). *Biochemistry* **25,** 2934.
89. Knight, W. B., Maycock, A. L., Green, B. G., Ashe, B. M., Gale, P., Weston, H., Finke, P. E., Hagmann, W. K., Shah, S. K., and Doherty, J. B. (1992). *Biochemistry* **31,** 4980.
90. Pratt, R. F., Surh, Y. S., and Shaskus, J. J. (1983). *J. Am. Chem. Soc.* **105,** 1006.
91. Wilmouth, R. C., Kassamally, S., Westwood, N. J., Sheppard, R. J., Claridge, T. D., Aplin, R. T., Wright, P. A., Pritchard, G. J., and Schofield, C. J. (1999). *Biochemistry* **38,** 7989.
92. Strynadka, N. C., Adachi, H., Jensen, S. E., Johns, K., Sielecki, A., Betzel, C., Sutoh, K., and James, M. N. (1992). *Nature* **359,** 700.
93. Barkocy-Gallagher, G. A., and Bassford, P. J., Jr. (1992). *J. Biol. Chem.* **267,** 1231.
94. Peat, T. S., Frank, E. G., McDonald, J. P., Levine, A. S., Woodgate, R., and Hendrickson, W. A. (1996). *Nature* **380,** 727.
95. Paetzel, M., and Strynadka, N. C. (1999). *Protein Sci.* **8,** 2533.
96. van Klompenburg, W., Paetzel, M., de Jong, J. M., Dalbey, R. E., Demel, R. A., von Heijne, G., and de Kruijff, B. (1998). *FEBS Lett.* **431,** 75.
97. Schechter, I., and Berger, A. (1967). *Biochem. Biophys. Res. Commun.* **27,** 157.
98. Shinagawa, H. (1996). *Exs* **77,** 221.
99. Little, J. W. (1993). *J. Bacteriol.* **175,** 4943.
100. Keiler, K. C., and Sauer, R. T. (1995). *J. Biol. Chem.* **270,** 28864.
101. Birghan, C., Mundt, E., and Gorbalenya, A. E. (2000). *EMBO J.* **19,** 114.
102. Patricelli, M. P., and Cravatt, B. F. (2000). *J. Biol. Chem.* **275,** 19177.
103. Thompson, J. D., Gibson, T. J., Plewniak, F., Jeanmougin, F., and Higgins, D. G. (1997). *Nucleic Acids Res.* **25,** 4876.
104. Kraulis, P. K. (1991). *J. Appl. Cryst.* **24,** 946.
105. Paetzel, M., Dalbey, R. E., and Strynadka, N. C. (2000). *Pharmacol. Ther.* **87,** 27.

3

Structure and Function of the Endoplasmic Reticulum Signal Peptidase Complex

NEIL GREEN* • HONG FANG* • STEPHEN MILES[†] •
MARK O. LIVELY[†]

Department of Microbiology and Immunology
School of Medicine, Vanderbilt University
Nashville, TN

[†]*Department of Biochemistry*
Wake Forest University School of Medicine
Winston-Salem, NC

THE ENZYMES, Vol. XXII
Copyright © 2001 by Academic Press

I. Introduction

Animal cells and other cells from eukaryotic organisms contain two multisubunit enzymes that belong to the type I signal peptidase (SP) family, one located within the endoplasmic reticulum (ER) membrane and the other in the mitochondrial inner membrane. The chloroplasts of plants also have a type I signal peptidase. Analysis of the SP derived from the ER has revealed a wealth of structural and functional information. This chapter summarizes studies of this SP that have been performed during the past approximately 15 years. Descriptions of the mitochondrial and chloroplast SPs can be found in other chapters in this volume.

A.   EUKARYOTIC TYPE I SIGNAL PEPTIDASES: A MEMBRANE PROTEIN COMPLEX

The SP located within the ER was the first enzyme of its type whose activity was identified experimentally (*1*). Using an *in vitro* protein translocation system derived from canine microsomal membranes, Blobel and Dobberstein observed the proteolytic processing of a portion of a newly synthesized secretory protein immediately upon its insertion into the translocation across the ER membrane (*2*). This proteolytic event occurred very early during translocation, proceeding even before the precursor was released from the ribosome. According to the signal hypothesis proposed by Blobel (*3*) and based on numerous experimental observations that followed, we know that the processing event observed was the removal of a signal peptide from the amino terminus of the polypeptide chain as it was translocated across the lipid bilayer.

The signal peptide targets the protein to which it is attached to the ER membrane and is then removed in most cases (there are examples of uncleaved signal peptides) by SP, an endoprotease bound to the ER membrane. In eukaryotic cells, signal peptides are synthesized as part of many soluble proteins including those that are secreted from the cell, those that remain resident in one of the compartments of the secretory pathway, or those that reside within lysosomes. Signal peptides are also initially present on many integral membrane proteins, particularly those that contain a single transmembrane segment and exhibit a topology in which only their amino-terminal regions are translocated, such as that observed for the epidermal growth factor receptor (*4*).

Cleavage of signal peptides of precursor proteins synthesized in the *in vitro* translocation systems often proceeds in a cotranslational manner. That is, movement of the protein into the microsomal compartment and cleavage by the peptidase occurs as the protein is being synthesized (Fig. 1). Because

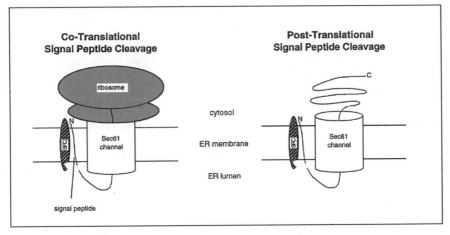

FIG. 1. Different modes of signal peptide cleavage. This diagram depicts both cotranslational signal peptide cleavage and posttranslational signal peptide cleavage. In the cotranslational mode, a signal peptide is cleaved by the SPC while a portion of the polypeptide chain is in the process of traversing the Sec61 channel. Still another portion remains attached to the ribosome. In the posttranslational mode, the polypeptide containing the signal peptide is released from the ribosome before being transported through the Sec61 channel. Cleavage of the signal peptide by the SPC is shown in this drawing to occur cotranslocationally; however, this has not been demonstrated directly in the posttranslational mode, as cleavage may occur immediately after the polypeptide chain has passed fully through the Sec61 channel.

translation and signal peptide cleavage occur on opposite sides of the ER membrane, a signal peptide cleaved in the cotranslational mode must also be in the process of traversing the protein translocation apparatus (Sec61p channel) within the ER membrane (5, 6). Because of this topography, signal peptides must also be cleaved cotranslocationally.

In more recent studies, precursor proteins have been found that can be cleaved by SP after their synthesis has been terminated, as demonstrated *in vitro* using microsomal membranes derived from yeast cells (7–9). In this posttranslational form of signal peptide cleavage, translation of the precursor is terminated prior to entry of the polypeptide chain into the Sec61 channel. The signal peptide is then cleaved during (cotranslocationally) or immediately after translocation. Figure 1 shows a signal peptide being cleaved cotranslocationally in the posttranslational mode of cleavage, but this has not been demonstrated experimentally.

Whether cleavage proceeds co- or post-translationally appears more to be a function of the mechanism by which a given polypeptide is translocated across the membrane, rather than resulting from a link between translation and signal peptide cleavage event. Evidence suggests, instead, that SP activity may be coupled to the process of translocation (*10, 11*). Indeed, studies *in vivo* and *in vitro* demonstrate that cleavage of the signal peptide is one of the first events in the maturation of a newly synthesized secretory protein as it enters the secretory pathway. Despite its being coupled to the translocation process, the SPC is not required for translocation.

The enzyme mediating the removal of signal peptides from proteins inserted into the ER is often called the signal peptidase complex (SPC) because, unlike all known eubacterial type I signal peptidases, it contains multiple nonidentical subunits. The SPC is found in eukaryotic cells from organisms as diverse as yeast and mammals. As many as five different proteins are known to be associated with the SPC, but the stoichiometry and relationships of each subunit in the complex are not yet clearly known. Each of the protein subunits is anchored to the lipid bilayer with the active site for proteolysis positioned in the lumen of the ER. Figure 2 represents the relative membrane positions of the SPC protein subunits, but is not intended to describe the stoichiometry or relative positions of the subunit proteins.

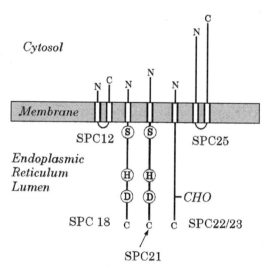

Fɪɢ. 2. Topology of SPC subunit proteins. This image represents the five different subunits of the mammalian SPC. The subunits are named "SPC," for signal peptidase complex, followed by a number that represents the apparent molecular mass of the subunit as observed in preparations of purified canine pancreas SPC (*20*). The open rectangles in the membrane domain represent the membrane anchors for each protein. The circles with the one-letter amino acid abbreviations show the relative positions of the active site serine (S), the active site histidine (H), and a conserved aspartic acid residue (D).

B. ROLE OF SIGNAL PEPTIDASE IN PROTEIN MATURATION

The SPC plays a crucial role in the maturation of newly synthesized polypeptide chains during or immediately following translocation. Proteins normally secreted from the cell have been observed to accumulate within the ER when their signal peptides are not cleaved (*12, 13*). The propensity for the ER to exercise quality control over polypeptides entering the secretory pathway is a commonly observed phenomenon, and it is seen when mutant, unfolded, and other abnormal proteins enter the lumenal space (*14*). Further evidence for the role of the SPC in the maturation process comes from the fact that inhibiting the SPC, usually by a mutation affecting one of its subunits, causes an activation of the unfolded protein response pathway (*15–17*). This intracellular signal transduction pathway is responsible for an increase in the transcription of genes encoding certain ER resident proteins in response to an accumulation of abnormal polypeptides in the ER lumen.

The role of SP in protein maturation is underscored by the fact that the SPC functions in a coordinated manner with oligosaccharyltransferase, the enzyme that adds the preassembled oligosaccharide $Glc_3Man_9GlcNAc_2$ to Asn-X-Ser/Thr tripeptides in proteins entering the lumenal space (*18*). The SPC and oligosaccharyltransferase interact in a sequential and dependent manner, as a signal peptide must be cleaved before an oligosaccharide can be added to an acceptor tripeptide juxtaposed to the signal peptide cleavage site (*19*). By virtue of its ability to cleave signal peptides, the SPC thus plays a vital role in the maturation and subsequent transport through the secretory pathway of a large number of newly synthesized lumenal and membrane-bound proteins.

II. Structure of the Signal Peptidase Complex

A. IDENTIFICATION OF SP AS A COMPLEX OF MEMBRANE PROTEINS

The three best characterized microsomal SPCs are those purified from canine pancreas, chicken oviduct, and yeast. Canine pancreas SPC, first purified by Evans and co-workers (*20*), consists of five proteins. Cloned cDNAs have been characterized for each subunit: SPC12 (*21*); SPC18 (*22*); SPC21 (*23*); SPC22/23 (*24*); and SPC25 (*25*). The numbers refer to the apparent molecular weights of each subunit observed following SDS–PAGE. SPC22/23 is a glycoprotein that migrates as a doublet, even in the absence of added carbohydrate (*26*). The reason for this electrophoretic behavior is unknown. SPC18 and SPC21 are homologous isoforms that are 80% identical (*22*). The SPC18 and SPC21 subunits contain the active site for peptide bond hydrolysis (see below).

Microsomal SPC has also been purified from the magnum region of hen oviduct (*27, 28*). Enzymatically active preparations of chicken SPC appear by SDS–PAGE analysis to have only two proteins: a glycoprotein, gp23, and p19 (*29*). However, subsequent peptide mapping and amino acid sequencing of the two protein bands confirmed the presence of homologs of the canine SPC18, SPC21, SPC22/23, and SPC25 subunits (Miles, S. M., and Lively, M. O., unpublished). Unlike the canine SPC18 and SPC21 subunits that differ in size by 12 amino acids, the avian subunits differ by only three amino acids and thus migrate as a single band during electrophoresis. Similarly, peptides corresponding to the avian homolog of SPC25 have been identified in peptide maps of a protein that comigrates with the SPC22/23 glycoprotein upon electrophoresis. The SPC12 homolog has not yet been detected in preparations of the hen oviduct SPC, but it is likely to be present in oviduct ER. Thus, four of the five subunits present in the canine SPC have also been found associated with chicken SPC.

Yeast SPC was the first identified genetically in *Saccharomyces cerevisiae* (*30*). When the active yeast enzyme was first isolated, it was found to be part of a complex of four proteins (*31, 32*). DNA clones encoding the four yeast SP subunits have been characterized: Spc1p and Spc2p (*21, 33*); Sec11p (*30*), and Spc3p (*34*). Each of these proteins is homologous to the SPC subunits identified in higher eukaryotes: Spc1p to SPC12; Spc2p to SPC25; Sec11p to both SPC18 and SPC21; and Spc3p to SPC22/23.

Genetic studies using this system are beginning to define the cellular roles of the different subunits (see later discussion). Experiments of this type have demonstrated that only two of the yeast SPC genes are essential for survival of yeast. As originally demonstrated by Böhni (*30*), the *SEC11* gene is essential for cleavage of signal peptides and for survival of the yeast. Similarly, the glycoprotein subunit Spc3p is essential for SP activity (*34, 35*).

B. MEMBRANE TOPOLOGY

Nondenaturing detergents are required for solubilization of the SPC from microsomal membranes. The SPC is not dissociated from the membrane by high concentrations of salt or by treatment under alkaline conditions that solubilize peripheral membrane proteins (*36*). Attempts to dissociate the detergent-solubilized SPC into a smaller complex of subunits or individual proteins through chromatographic or harsher denaturing methods consistently produces an inactive enzyme. The canine SPC22/23 (chicken gp23 and yeast Spc3p) subunit is the major glycosylated SP subunit. In addition, a small amount of glycosylated Sec11p has been seen in yeast cells (*37, 38*). The presence of glycosylated SPC22/23 (and Spc3p) has facilitated further purification of intact enzyme by concanavalin A (ConA) glycoprotein affinity chromatography. When isolated in this manner, the nonglycosylated subunits are isolated in addition to the glycoprotein, giving further evidence that the SPC subunits are tightly associated (27, 39). Most of the SPC22/23 homologs have a single N-linked glycosylation site near the C-terminal end of the protein sequence. Yeast Spc3p, on the other hand, has two N-linked glycosylation sites (35). The carbohydrate attached to SPC22/23 is believed to have a high percentage of mannose residues because of affinity of the enzyme for ConA (specific for binding proteins with a high percentage of mannose) and because of the size of the added carbohydrate (~2.5 kDa).

Three of the SP subunits, SPC18, SPC21, and SPC22/23, are membrane proteins with a single membrane-spanning domain (Fig. 2). None of the SPC proteins is synthesized with a cleavable signal peptide. A single hydrophobic segment near the N terminus of each protein is predicted to provide a membrane anchor, and this prediction is supported by experimental evidence demonstrating the relative orientations of each protein chain (26). The relatively large C-terminal domains of SPC18, SPC21, and SPC22/23 are positioned within the lumen. With this orientation, the SPC is able to efficiently recognize precursors as they emerge from the Sec61 channel (Fig. 1).

Both SPC12 and SPC25 have two hydrophobic membrane-spanning domains that position the soluble domains primarily in the cytoplasm of the cell (Fig. 2) (40). Organization of the complex in the lipid bilayer is most likely determined by specific interactions of the transmembrane helical domains, but this has not yet been probed experimentally. The SPC physically interacts with the translocation channel consisting of the heterotrimer Sec61p complex as the β subunit of the Sec61 complex can be cross-linked to SPC25 in the canine system (10), and Spc2p coimmunoprecipitates with yeast Sec61β (11).

C. HOMOLOGOUS ISOFORMS OF THE CATALYTIC SUBUNIT

The canine SPC18 and SPC21 subunits, chicken p19 (two isoforms), and yeast Sec11p are members of a common family that is homologous to all known type I SPs in eubacteria as well as archeabacteria. Curiously, available genomic evidence suggests that all higher eukaryotes have two homologs of the Sec11p subunit, whereas the yeast S. cerevisiae contains only a single SEC11-like gene. This family of proteins, sometimes called the "Sec11 family," includes the catalytic site for peptide bond hydrolysis by the SPC. This conclusion is based on extensive active site studies on the E. coli type I SP (see Chapter 2 by Carlos et al., this volume), mutagenesis studies on yeast Sec11p (38), and analysis of the alignments of extensive amino acid sequences now available. Combined, these studies suggest that the SEC11 family of genes shares a common ancestor with all bacterial type I signal peptidases (see later discussion).

One of the most intriguing aspects of the SP structure is the existence of homologous isoforms of the catalytic subunit, the Sec11family of proteins. DNA sequence data in GenBank and the EST databases (http://www. ncbi.nlm.nih.gov/) suggest that all higher eukaryotes have at least two forms of the active site subunit. In contrast to the presence of these homologous forms, yeast cells contain only a single *SEC11* gene in their entire genome. Additional sequence data will be required to determine at what point during evolution the second form of this subunit appeared.

The biological reason for the two forms of the active site subunit is not known. Interestingly, the related mitochondrial SP known as the inner membrane protease has two closely related subunits, Imp1p and Imp2p, that possess catalytic activity (*41*). A study of this enzyme concluded that Imp1p and Imp2p have separate, nonoverlapping substrate specificities. It is therefore plausible that the two homologous isoforms of the microsomal SPC subunits play a similar role in the ER, but there is currently no evidence to support that hypothesis. It is not known whether the complex of SP proteins contains a single Sec11-like subunit or whether each enzyme complex contains both Sec11 homologs. The stoichiometry of the SPC subunits has not been established.

D.   EVOLUTIONARY RELATIONSHIPS AND ENZYME MECHANISM

Extensive analyses of the eubacterial SPs have shown that their mechanism of action involves a Ser and Lys dyad (*42– 44*). An amino acid sequence alignment that includes some of these sequences is presented in the chapter on the *E. coli* leader peptidase (Lep) (Carlos *et al.*, Chapter 2 in this volume). As observed in the alignment of the eubacterial, eukaryotic, and archaebacterial sequences, there are a few amino acid residues that are strictly conserved among all of the sequences aligned. These residues cluster into six different highly conserved regions or "boxes." Figure 3A shows a view of the active site of Lep (*45*) created from the protein databank crystal structure file designated 1b12 (http://www.rcsb.org/pdb/). The first, Box A, includes the membrane anchor for each sequence and is not shown in the Figure. Box B includes the invariant serine residue (Ser-90 in Lep) that is the active site serine of the eubacterial sequences (*46, 47*). The side chain of the Ser-90 is shown in the top center of Fig. 3A in the ball-and-stick representation. The conserved amino acid sequence in this region is Ser-Met-Xaa-Pro, where the Ser is Ser-90 in *E. coli* Lep.

The next conserved box is divided into two parts called C and C', each containing a highly conserved Gly-Asp sequence. In the view represented in Fig. 3A, Boxes C and C' are positioned in a region surrounding the active-site Ser-90. These conserved residues are thought to play roles in substrate

A

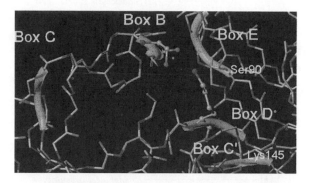

B

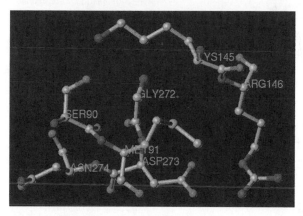

Fig. 3. (A) Leader peptidase active site. View of the active site of *E. coli* Lep created by the X-ray crystal structure coordinates (1b12) from the protein databank using Sybyl molecular modeling software (Tripos, Inc.). The polypeptide backbone is represented by capped sticks. The five conserved regions are designated Boxes B, C, C', D, and E are shown. The side chains of the active-site Ser-90 and Lys-145 residues are represented by ball-and-stick structures. (B) Close-up view of the most highly conserved amino acid residues of *E. coli* Lep.

binding (*38, 45, 48*). Box E contains a highly conserved sequence: Gly-Asp-Asn-Asn. Its position near the active site also suggests that it plays a role in substrate binding.

Box D includes the active site Lys-145 that is the general base in the cleavage mechanism (*43, 47*). The lysine side chain (shown in a ball-and-stick representation in the right center of Fig. 3A) extends toward the active site Ser where it is positioned to accept a proton during catalysis. In addition to Lys-145, Arg-146 is also strictly conserved in alignments of all SP sequences,

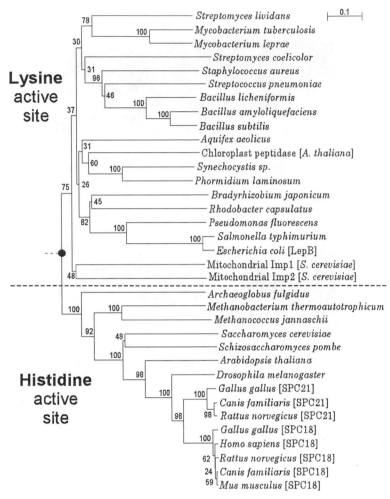

FIG. 4. Phylogenetic tree of 35 selected signal peptidase amino acid sequences. All amino acid sequence data for both prokaryotic and eukaryotic SPs were obtained from the SwissProt, PIR, EMBL, and GenBank databases with the exception of *G. gallus* (M. O. Lively, unpublished). The multiple sequence alignment was created by the Clustal W program (*49*) using the Gonnet weight matrix. The phylogenetic tree was constructed using the neighbor-joining algorithm (*76*) in Clustal W. The deepest root of the tree (black circle) was determined by the use of the sequence of the human signal recognition particle 19 kDa subunit (accession no. P09132) as an outgroup. Reliablity of the tree was determined by a bootstrap analysis of 1000 replicates. The estimated bootstrap probabilities (in percentages) are shown for each of the nodes. The nodes are considered to be reliable if they have a bootstrap value of 70% or greater (*77*).

SP accession numbers: *H. sapiens* [SPC18] (O75957); chloroplast peptidase [*A. thaliana*]. (O04348); *S. pneumoniae* (O07344); *S. aureus* (P72365); *P. laminosum* (Q51876); *Synechocystis* including archaebacteria and eukaryotes. Figure 3B shows a related view of the active site of Lep depicting only the strictly conserved amino acid side chains from these conserved regions. In this representation of the structure, Arg-146 (from Box D) can be seen to be positioned close enough to the side chain of Asp-273 (from Box E) to form hydrogen bonds and a salt bridge with Arg-146. From this view it appears that this salt bridge must play an important role in defining the geometry of the active site. One can also see the conserved Gly-272 on the back side of the active-site pocket. This residue is likely to be conserved because any other amino acid side chains placed at this position would block substrate binding at the active site.

Comparative amino acid sequence analysis of the known SP genes reveals a puzzling fact. The active site Lys-145 found in the conserved Box D is not found in the Sec11 family of proteins or in known archaebacterial SP sequences. These sequences contain an absolutely conserved His residue in place of the lysine corresponding to Lys-145. Figure 4 shows a phylogenetic tree generated by alignment of 35 type I SPs from the three main branches of life. Amino acid sequences were predicted from DNA sequences available in the databases and were aligned using Clustal W (49). The phylogenetic analysis was performed using the neighbor-joining software provided by the Biology Workbench Web site (http://workbench.sdsc.edu/). This analysis reveals an interesting distribution. Those sequences with lysine aligned at the position corresponding to *E. coli* Lys-145 are found in eubacteria and include mitochondrial and chloroplast peptidases. Only the ER SP sequences and archaebacteria sequences, with two known exceptions, are found with histidine aligned at the active site position.

There are two eubacterial exceptions that do not fit this pattern. SipW, one of the seven *Bacillus subtilis* SPs (50, 51), has a histidine at this alignment position. Similarly, Spc21 from *Clostridium perfringens* also has His instead of Lys. Experimental evidence demonstrates that this histidine residue is required for SP activity in the Sec11p subunit of the yeast SPC (38) and in SipW (52).

---

sp. (P73157); *M. jannaschii* (Q57708); *R. norvegicus* [SPC18] (P42667); *R. norvegicus* [SPC21] (Q9WTR7); *M. musculus* [SPC18] (Q9R0P6); *D. melanogaster* (AE001574); *A. fulgidus* (O28616); *E. coli* [LepB] (P00803); *A. thaliana* (Q9SSR2); *S. pombe* (O74323); *A. aeolicus* (O67088); *M. thermoautotrophicum* (O27497); *S. cerevisiae* (P15367); *C. familiaris* [SPC18] (A35309); *C. familiaris* [SPC21] (P13679); mitochondrial Imp1 [*S. cerevisiae*] (P28627); mitochondrial Imp2 [*S. cerevisiae*] (P46972); *M. leprae* (O33021); *B. japonicum* (S70840); *B. subtilis* (P28628); *B. licheniformis* (P42668); *B. amyloliquefaciens* (P41026); *S. coelicolor* (T34784); *R. capsulatus* (S66597); *P. fluorescens* (P26844); *M. tuberculosis* (Q10789); *S. lividans* (O54237); *S. typhimurium* (P23697); *G. gallus* [SPC18] (M. O. Lively, unpublished); *G. gallus* [SPC21] (M. O. Lively, unpublished).

Multiple analyses of the available sequences suggest strongly that there are two mechanistic SP families. The vast majority of eubacterial SPs use a mechanism based on an active site serine–lysine dyad (*43, 47*). The phylogenetic analysis shows that a very early gene duplication event must have resulted in the replacement of the active site lysine with an active site histidine (Fig. 4). The switch from lysine to histidine corresponds to a time prior to the origin of archaea and eukaryotes. Thus, the SP is an extremely old protein with origins that date back to the beginnings of life itself.

III. Function of SPC

A. Signal Peptide Cleavage Reaction

Inhibiting the yeast SPC by a mutation affecting one of its two essential subunits leads to a defect in protein maturation, a deficit in the trafficking of secretory proteins to the plasma membrane, and cell death (*30, 34, 35*). The signal peptide cleavage reaction can occur both cotranslationally and posttranslationally (Fig. 1). In the cotranslational mode and perhaps in the posttranslational mode, signal peptide cleavage proceeds before the incoming polypeptide chain has fully traversed the Sec61 complex, indicating the SPC gains access to the signal peptide in a cotranslocational manner. Cleavage of signal peptides is thus an early process, which can occur before the polypeptide attached to the signal peptide is fully synthesized. Furthermore, the cleavage event is highly conserved through evolution, as signal peptides of eubacterial proteins can often be cleaved correctly by the SPC when expressed in the presence of ER membranes (*53*).

The strong functional conservation of various SPs raises the issue of why the SPC and the eubacterial SPs exhibit major structural and catalytic differences (discussed above). One idea is that the SPC encounters a more diverse substrate profile than its eubacterial counterparts because of the greater number of proteins synthesized in eukaryotic cells and the diversity exhibited by different signal peptides at the primary structural level (*54–56*). Additional subunits may therefore be needed to improve efficiency of the SPC (see below). It is also plausible that the SPC has additional subunits to accommodate special environmental or structural requirements necessary for function within the ER lumen.

Microsomal signal peptidase was initially found to be resistant to inhibition by the usual reagents known to inhibit proteolytic enzymes from the different mechanistic classes (*57, 58*). Consequently, little can be deduced about the mechanism of the enzyme based upon an inhibition profile. Based on the amino acid sequences of the presumed catalytic domains, it is clear that the microsomal enzyme uses a mechanism very closely related to the catalytic dyad found in the eubacterial enzymes. Although microsomal SP is clearly not a classical serine protease, the enzyme is inhibited by selected peptidyl chloromethyl ketones (*59, 60*). The best inhibitor is methoxysuccinyl-Ala-Ala-Pro-ValCH$_2$Cl, a chloromethylketone developed as an inhibitor of leukocyte elastase (*61*). Although inhibition of microsomal SP by this compound appears to be irreversible, as it is with elastase, efforts to identify a modified amino acid in the SP active site by peptide mapping and mass spectrometry have not yet recovered a covalently modified peptide (Miles, S. M., and Lively, M. O., unpublished). This compound inhibits peptidase activity in intact microsomal vesicles and has proven to be useful in experiments probing the role of SP in translocation of secretory proteins (*59, 62, 63*).

B. MEMBRANE PROTEIN FRAGMENTATION

In addition to its role in cleaving signal peptides, the SPC can fragment some abnormal membrane proteins at internal sites within the polypeptide chains (*37, 64–67*). The internal sequences that are fragmented by the SPC share at least some structural similarities with signal peptide cleavage sites. As shown in Fig. 5, signal peptides possess a general motif consisting of a positively charged N-terminal region (N-region), a hydrophobic core (H-region), and a polar amino acid stretch containing the cleavage site (C-region). The first amino acid and, to a lesser extent, the third amino acid N-terminal to the cleavage site (−1 and −3 residues) usually have small, uncharged side chains, and the amino acids in these positions are important for proper cleavage by the SPC (*56, 68*). For example, the signal peptide of pre-chorionic gonadotropin α subunit (*69*) has a −1 serine and a −3 leucine (Fig. 5).

The similarities between signal peptide cleavage sites and sites within membrane proteins that are fragmented by the SPC are probably best illustrated by the H2 subunit of the human asialoglycoprotein receptor. When H2 is synthesized in the absence of its partner subunit H1, a distinct proteolytic fragment is generated (*64*). This proteolytic fragment results from fragmentation immediately after serine (the −1 residue), and substitution of this serine with bulky or charged amino acids inhibits fragmentation at this site. The −3 amino acid from the fragmentation site of H2 is also serine (Fig. 5). Upstream of the fragmentation site is the transmembrane segment of H2, which may be analogous to the H-region of cleavable signal peptides. H2 is a single spanning membrane protein, the transmembrane segment exhibiting a type II orientation (N_{cyt}–C_{lum}). Similarly, cleavable signal peptides enter the membrane in a N_{cyt}–C_{lum} orientation (*70, 71*). Direct evidence for

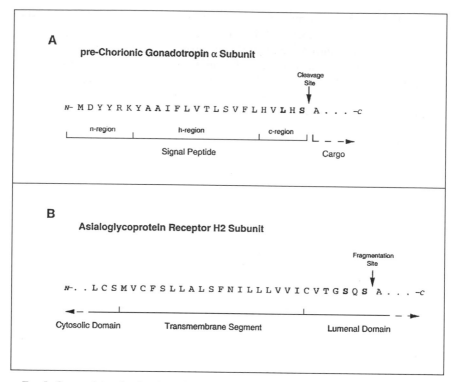

Fɪɢ. 5. Comparison of a signal peptide cleavage site and a membrane protein fragmentation site. (A) The sequence of the signal peptide of pre-chorionic gonadotropin α subunit is shown along with a small portion of the cargo sequence that is attached to the signal peptide. The signal peptide cleavage site marks the boundary between the signal peptide and the cargo. Within the signal peptide, three sequence motifs found in most cleavable signal peptides are indicated. These motifs consist of a charged N-region, a hydrophobic H-region, and a polar C-region that contains the cleavage site. The –1 and –3 residues upstream from the cleavage site are indicated using bold letters. (B) The relevant sequence near the fragmentation site of the human asialoglycoprotein receptor H2 subunit is shown. This subunit is a type II membrane protein, having a cytosolic N-terminal domain separated from a lumenal C-terminal domain by a single transmembrane segment. The fragmentation site, which is processed by the SPC, is indicated. The first and third residues upstream from the fragmentation site are highlighted using bold letters.

the involvement of the SPC in membrane protein fragmentation has come from studies in the yeast system. Mutations affecting either of the two essential SPC subunits, Sec11p or Spc3p, inhibit fragmentation and subsequent degradation of an abnormal type II membrane protein (*34, 37*).

Interestingly, the kinetics of fragmentation often differ from that of signal peptide cleavage. Whereas signal peptides are cleaved during or immediately after precursor translocation, membrane proteins fragmented by the SPC *in vivo* generally experience a postintegration delay in processing that can last as long as 30 mins (*37, 64*). The molecular basis for the fragmentation delay is not known; however, this pause may be important in regulating the SPC, thus preventing it from immediately fragmenting most, if not all, incoming polypeptide chains, which enter the ER in an unfolded state.

What keeps normal membrane proteins from being fragmented by the SPC? For the H2 subunit of the asialoglycoprotein receptor, assembly with H1 is clearly a key factor in the protection of H2 (*64*), suggesting that the fragmentation site may be buried within the tertiary structure of the heterodimeric complex or hidden within a region binding to H1. Thus, the correctly folded protein structure probably provides a general means by which membrane proteins escape fragmentation.

IV. Roles of the SPC Subunits

A. ESSENTIAL SUBUNITS

Most of the functional studies of the SPC subunits have focused on the yeast system. The yeast SPC contains four nonidentical subunits, named Spc1p, Spc2p, Spc3p and Sec11p. Sec11p and Spc3p are the only subunits in this complex that are essential for cell viability (*30, 34, 35*). Sec11p was given its name because mutations in the *SEC11* gene were originally found in a selection for temperature-sensitive secretion mutants (*72*). As previously mentioned, mutations that cause a general signal peptide cleavage defect lead to an accumulation of precursor proteins within the ER lumen. The gene encoding Spc3p was isolated based on the fact that overexpression of Spc3p partially corrects a temperature-sensitive defect in the *SEC11* gene (*34*). Temperature-sensitive *sec11* and *spc3* mutations inhibit both of the known SPC activities, namely cleavage of signal peptides and fragmentation of abnormal membrane proteins (*30, 34, 37*).

Sec11p is the catalytic subunit of the SPC. Site-directed mutagenesis studies have narrowed the probable catalytic residues to four amino acids: Ser-44, His-83, Asp-103, and Asp-109 (*38*). It follows, therefore, that Sec11p may contain either the canonical catalytic triad (Ser/His/Asp) (*73*), with Asp-103 or Asp-109 serving a catalytic role, or a Ser/His catalytic dyad, similar to that seen in an esterase from *Streptomyces scabies* (*74*). As mentioned earlier, SipW of *Bacillus subtilis* probably also contains a catalytic histidine. However, SipW differs from Sec11p in that replacement of the histidine of Sec11p with lysine results in an inactive enzyme (*38*). SipW retains its function when the catalytic histidine is replaced by lysine (*52*).

Although essential for SP activity, Spc3p probably does not contain a catalytic amino acid (*38*). In light of the fact that the eubacterial type I SPs, such as Lep from *E. coli*, are monomeric, why does the SPC have two essential subunits? A model that has been proposed (*34, 38*) is that Spc3p and Sec11p may form a core enzyme analogous to Lep. The crystal structure of Lep reveals that it consists of two distinct domains, only one of which contains the catalytic site (*45*). Thus, Sec11p and Spc3p may be functionally equivalent to the catalytic and noncatalytic domains, respectively, of Lep. It should be emphasized, however, that the precise role of Spc3p has not been determined.

B. Nonessential Subunits

The yeast SPC has two nonessential subunits, Spc1p and Spc2p (*21, 33*). Overexpression of Spc1p partially corrects a temperature-sensitive defect in Sec11p, which served as the basis for cloning the *SPC1* gene (*21*). The *SPC2* gene was identified in the *Saccharomyces* genome Database (http://genome-www.stanford.edu/Saccharomyces/) because of its homology to the gene encoding canine SPC25 (*33*). Cells lacking Spc1p or Spc2p individually or lacking both of these subunits grow well under normal laboratory conditions. Yeast lacking either or both Spc1p and Spc2p exhibit almost normal SP and membrane protein fragmentation activities *in vivo* (*21, 33*), although reduced signal peptide cleavage activity is seen *in vitro* when Spc2p is missing from the SPC (*11*). Spc1p and Spc2p improve the efficiency of the SPC under certain "extreme" conditions *in vivo*. When Spc1p is missing, uncleaved precursors accumulate in cells expressing high levels of an abnormal membrane protein fragmented by the SPC (*21*). Spc2p greatly improves SP efficiency at relatively high temperatures (42°C) (*33*). Spc2p also helps to maintain cell viability at high temperatures. Physical studies have revealed that Spc2p is required for the association of Spc1p with the SPC and for normal levels of Sec11p and Spc3p to be expressed in yeast cells (*11*).

Analyses in both the mammalian and yeast systems argue that SPC25 (Spc2p) functions to link the SPC to the Sec61 channel. SPC25 can be cross-linked to the β subunit of the Sec61 complex (*10*). The amount of SPC25 that cross-links to Sec61β is greatly reduced in microsomal membranes that have been stripped of ribosomes, suggesting a recruitment of SPC25 to the Sec61 complex during cotranslational translocation. A physical interaction has also been detected between Spc2p and yeast Sec61β (*11*). This latter interaction was detected in the absence of added cross-linker. Although these data point to a physical association existing between the SPC and the Sec61 complex, the importance of this interaction is unclear, as Spc2p is not needed for SP activity under normal laboratory conditions (*33*).

V. Future Directions

Although numerous investigations of the SPC subunits have been conducted, almost nothing is known about the roles of the Spc3p (SPC22/23) and Spc1p (SPC12) subunits. The fact that higher eukaryotic cells contain two Sec11p homologs is also a major puzzle. Are these homologs functionally equivalent or dissimilar? Does one homolog perform a role other than the cleavage of signal peptides? These important questions require a combination of genetic and biochemical approaches involving appropriate model systems that may very well yield unexpected answers.

How the SPC-mediated fragmentation process interfaces with the ER-associated degradation pathway (75) is a major question. The SPC may generate fragments of certain abnormal membrane proteins that are then processed by the cytosolic proteasome or are transported to the lysosome for degradation. The mechanism by which most normal membrane proteins avoid fragmentation by the SPC is not clear. Although some membrane proteins probably lack appropriate fragmentation sites, it is highly likely that others have such sites and yet evade the SPC even though proteins enter the ER in an unfolded state. These issues raise the intriguing possibility that access to the SPC is regulated.

REFERENCES

1. Milstein, C., Brownlee, G. G., Harrison, T. M., and Mathews, M. B. (1972). *Nature New Biol.* **239,** 117–120.
2. Blobel, G., and Dobberstein, B. (1975). *J. Cell Biol.* **67,** 835.
3. Blobel, G., Walter, P., Chang, C. N., Goldman, B. M., Erickson, A. H., and Lingappa, V. R. (1979). *Symp. Soc. Exp. Bio.* **33,** 9.
4. Clements, J. M., Catlin, G. H., Price, M. J., and Edwards, R. M. (1991). *Gene* **106,** 267.
5. Hanein, D., Matlack, K. E., Jungnickel, B., Plath, K., Kalies, K. U., Miller, K. R., Rapoport, T. A., and Akey, C. W. (1996). *Cell* **87,** 721.
6. Mothes, W., Jungnickel, B., Brunner, J., and Rapoport, T. A. (1998). *J. Cell Biol.* **142,** 355.
7. Hansen, W., Garcia, P. D., and Walter, P. (1986). *Cell* **45,** 397.
8. Rothblatt, J. A., and Meyer, D. I. (1986). *Cell* **44,** 619.
9. Waters, M. G., and Blobel, G. (1986). *J. Cell Biol.* **102,** 1543.
10. Kalies, K. U., Rapoport, T. A., and Hartmann, E. (1998). *J. Cell Biol.* **141,** 887.
11. Antonin, W., Meyer, H. A., and Hartmann, E. (2000). *J. Biol. Chem.* **275,** 34068.
12. Racchi, M., Watzke, H. H., High, K. A., and Lively, M. O. (1993). *J. Biol. Chem.* **268,** 5735.
13. Ito, M., Oiso, Y., Murase, T., Kondo, K., Saito, H., Chinzei, T., Racchi, M., and Lively, M. O. (1993). *J. Clin. Invest.* **91,** 2565.
14. Ng, D. T., Spear, E. D., and Walter, P. (2000). *J. Cell Biol.* **150,** 77.
15. Sidrauski, C., Chapman, R., and Walter, P. (1998). *Trends Cell Biol.* **8,** 245.
16. Stroud, R. M., and Walter, P. (1999). *Curr. Opin. Struct. Biol.* **9,** 754.

17. Travers, K. J., Patil, C. K., Wodicka, L., Lockhart, D. J., Weissman, J. S., and Walter, P. (2000). *Cell* **101,** 249.
18. Yan, Q., and Lennarz, W. J. (1999). *Biochem. Biophys. Res. Commun.* **266,** 684.
19. Chen, X., Van Valkenburgh, C., Liang, H., Fang, H., and Green, N. (2001). *J. Biol. Chem.* **276,** 2411.
20. Evans, E. A., Gilmore, R., and Blobel, G. (1986). *Proc. Nat. Acad. Sci. USA* **83,** 581.
21. Fang, H., Panzner, S., Mullins, C., Hartmann, E., and Green, N. (1996). *J. Biol. Chem.* **271,** 16460.
22. Shelness, G. S., and Blobel, G. (1990). *J. Biol. Chem.* **265,** 9512.
23. Greenburg, G., Shelness, G. S., and Blobel, G. (1989). *J. Biol. Chem.* **264,** 15762.
24. Shelness, G. S., Kanwar, Y. S., and Blobel, G. (1988). *J. Biol. Chem.* **263,** 17063.
25. Greenburg, G., and Blobel, G. (1994). *J. Biol. Chem.* **269,** 25354.
26. Shelness, G. S., Lin, L., and Nicchitta, C. V. (1993). *J. Biol. Chem.* **268,** 5201.
27. Baker, R. K., and Lively, M. O. (1987). *Biochemistry* **26,** 8561.
28. Lively, M. O., Newsome, A. L., and Nusier, M. (1994). *Methods Enzymol.* **244,** 301.
29. Newsome, A. L., McLean, J. W., and Lively, M. O. (1992). *Biochem. J.* **282,** 447.
30. Bohni, P. C., Deshaies, R. J., and Schekman, R. W. (1988). *J. Cell Biol.* **106,** 1035.
31. YaDeau, J. T., and Blobel, G. (1989). *J. Biol. Chem.* **264,** 2928.
32. YaDeau, J. T., Klein, C., and Blobel, G. (1991). *Proc. Natl. Acad. Sci. USA* **88,** 517.
33. Mullins, C., Meyer, H. A., Hartmann, E., Green, N., and Fang, H. (1996). *J. Biol. Chem.* **271,** 29094.
34. Fang, H., Mullins, C., and Green, N. (1997). *J. Biol. Chem.* **272,** 13152.
35. Meyer, H. A., and Hartmann, E. (1997). *J. Biol. Chem.* **272,** 13159.
36. Lively, M. O., and Walsh, K. A. (1983). *J. Biol. Chem.* **258,** 9488.
37. Mullins, C., Lu, Y., Campbell, A., Fang, H., and Green, N. (1995). *J. Biol. Chem.* **270,** 17139.
38. Van Valkenburgh, C., Chen, X., Mullins, C., Fang, H., and Green, N. (1999). *J. Biol. Chem.* **274,** 11519.
39. Gorlich, D., and Rapoport, T. A. (1993). *Cell* **75,** 615.
40. Kalies, K. U., and Hartmann, E. (1996). *J. Biol. Chem.* **271,** 3925.
41. Nunnari, J., Fox, T. D., and Walter, P. (1993). *Science* **262,** 1997.
42. Paetzel, M., and Strynadka, N. C. (1999). *Protein Sci.* **8,** 2533.
43. Black, M. T. (1993). *J. Bacteriol.* **175,** 4957.
44. Paetzel, M., and Dalbey, R. E. (1997). *Trends Biochem. Sci.* **22,** 28.
45. Paetzel, M., Dalbey, R. E., and Strynadka, N. C. (1998). *Nature* **396,** 186.
46. Sung, M., and Dalbey, R. E. (1992). *J. Biol. Chem.* **267,** 13154.
47. Tschantz, W. R., Sung, M., Delgado-Partin, V. M., and Dalbey, R. E. (1993). *J. Biol. Chem.* **268,** 27349.
48. Bolhuis, A., Tjalsma, H., Stephenson, K., Harwood, C. R., Venema, G., Bron, S., and van Dijl, J. M. (1999). *J. Biol. Chem.* **274,** 15865.
49. Thompson, J. D., Higgins, D. G., and Gibson, T. J. (1994). *Nucleic Acids Res.* **22,** 4673.
50. Tjalsma, H., Noback, M. A., Bron, S., Venema, G., Yamane, K., and van Dijl, J. M. (1997). *J. Biol. Chem.* **272,** 25983.
51. Meijer, W. J., de Jong, A., Bea, G., Wisman, A., Tjalsma, H., Venema, G., Bron, S., and van Dijl, J. M. (1995). *Mol. Microbiol.* **17,** 621.
52. Tjalsma, H., Stover, A. G., Driks, A., Venema, G., Bron, S., and van Dijl, J. M. (2000). *J. Biol. Chem.* **275,** 25102.
53. Lingappa, V. R., Chaidez, J., Yost, C. S., and Hedgpeth, J. (1984). *Proc. Natl. Acad. Sci. USA* **81,** 456.
54. von Heijne, G., and Abrahmsen, L. (1989). *FEBS Lett.* **244,** 439.
55. von Heijne, G. (1985). *J. Mol. Biol.* **184,** 99.

56. von Heijne, G. (1983). *Eur. J. Biochem.* **133,** 17.
57. Jackson, R. C., and Blobel, G. (1980). *Ann. N.Y. Acad. Sci.* **343,** 391.
58. Evans, E., Shelness, G., and Blobel, G. (1986). *J. Cell Biol.* **103,** 291a.
59. Miles, S., and Lively, M. (1998). *FASEB J.* **12,** A1423.
60. Nusier, M., Kam, C.-M., Powers, J., and Lively, M. (1996). *FASEB J.* **10,** A1406.
61. Nakajima, K., Powers, J. C., Ashe, B. M., and Zimmerman, M. (1979). *J. Biol. Chem.* **254,** 4027.
62. van Geest, M., Nilsson, I., von Heijne, G., and Lolkema, J. S. (1999). *J. Biol. Chem.* **274,** 2816.
63. Nilsson, I., and von Heijne, G. (2000). *J. Biol. Chem.* **275,** 17338.
64. Yuk, M. H., and Lodish, H. F. (1993). *J. Cell Biol.* **123,** 1735.
65. Liljestrom, P., and Garoff, H. (1991). *J. Virol.* **65,** 147.
66. Schmid, S. R., and Spiess, M. (1988). *J. Biol. Chem.* **263,** 16886.
67. Lipp, J., and Dobberstein, B. (1986). *Cell* **46,** 1103.
68. Folz, R. J., Nothwehr, S. F., and Gordon, J. I. (1988). *J. Biol. Chem.* **263,** 2070.
69. Birken, S., Fetherston, J., Canfield, R., and Boime, I. (1981). *J. Biol. Chem.* **256,** 1816.
70. Eusebio, A., Friedberg, T., and Spiess, M. (1998). *Exp. Cell Res.* **241,** 181.
71. Shaw, A. S., Rottier, P. J., and Rose, J. K. (1988). *Proc. Natl. Acad. Sci. USA* **85,** 7592.
72. Novick, P., Field, C., and Schekman, R. (1980). *Cell* **21,** 205.
73. Carter, P., and Wells, J. A. (1988). *Nature* **332,** 564.
74. Wei, Y., Schottel, J. L., Derewenda, U., Swenson, L., Patkar, S., and Derewenda, Z. S. (1995). *Nature Struct. Biol.* **2,** 218.
75. Plemper, R. K., and Wolf, D. H. (1999). *Trends Biochem. Sci.* **24,** 266.
76. Saitou, N., and Nei, M. (1987). *Mol. Biol. Evol.* **4,** 406.
77. Hillis, D., and Bull, J. (1993). *Syst. Biol.* **42,** 182.

4

Mitochondrial Processing Peptidase/Mitochondrial Intermediate Peptidase

JIRI ADAMEC* • FRANTISEK KALOUSEK† •
GRAZIA ISAYA*

*Department of Pediatric and Adolescent Medicine and Biochemistry and Molecular Biology
Mayo Clinic and Foundation
Rochester, Minnesota 55905

†Institute of Microbiology
Academy of Sciences of the Czech Republic
Prague, Czech Republic

THE ENZYMES, Vol. XXII
Copyright © 2001 by Academic Press

I. Introduction

Mitochondria originate from growth and division of preexisting mitochondria. Central to this process is a steady supply of proteins, only a handful of which are encoded by mitochondrial DNA and synthesized within the organelle. The bulk of mitochondrial proteins (probably more than 500 in *S. cerevisiae* and close to 1000 in mammals) are encoded by nuclear genes, translated on free cytosolic ribosomes, and posttranslationally targeted to mitochondria. Most proteins of the mitochondrial matrix and many proteins of the intermembrane space and the inner membrane are translated as larger precursors with N-terminal targeting peptides that are removed within mitochondria by specific processing peptidases. Most outer membrane proteins and carrier proteins of the inner membrane are synthesized without cleavable presequences and possess N-terminal, C-terminal, and/or internal targeting signals.

Mitochondrial processing peptidase (MPP; EC 3.4.24.64) and mitochondrial intermediate peptidase (MIP; EC 3.4.24.59) account for most precursor processing activities within the mitochondrial matrix. MPP cleaves most if not all of the precursors that are fully or partially translocated to this compartment, including precursors in transit to the inner membrane or the intermembrane space. Precursors destined for the intermembrane space, for instance, have bipartite presequences consisting of a matrix-targeting signal followed by an intermembrane space-sorting signal. The matrix-targeting signal is cleaved by MPP, whereas the intermembrane space-sorting signal is removed by peptidases in the inner membrane. A single cleavage by MPP is otherwise sufficient for the maturation of most matrix and inner membrane protein precursors, with the exception of the octapeptide-containing precursors, which require two sequential cleavages by MPP and MIP.

II. Overview of Mitochondrial Protein Import

Depending on their final destination within the mitochondrion, precursor proteins may follow at least three different import pathways (*1–3*). Figure 1 illustrates the main steps and components involved in mitochondrial import and processing of precursor proteins with positively charged matrix-targeting presequences. Proteins of the outer membrane and intermembrane space and carrier proteins of the inner membrane follow variations of this

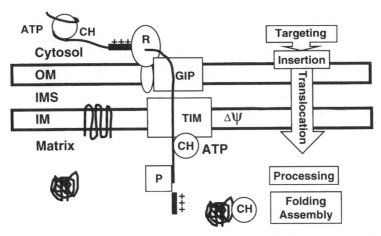

Fig. 1. Steps and molecules involved in mitochondrial import of precursor proteins with matrix-targeting presequences. OM, mitochondrial outer membrane; IMS, intermembrane space; IM, inner membrane. The presequence is shown as a positively charged region at the precursor N terminus. CH, cytoplasmic or mitochondrial chaperones; R, surface receptors; GIP, general insertion pore; TIM, translocase of the inner membrane; P, processing peptidases.

pathway and will not be considered here. Matrix-targeting presequences generally consist of 20–40 amino acids N-terminal to the "mature" protein, that is, the form found in mitochondria under native conditions at steady state (4, 5). Presequences have been shown to be necessary and sufficient to drive mitochondrial import of the mature protein as well as a variety of passenger molecules, including cytoplasmic proteins and even nucleic acids (6, 7). In the cytoplasm, the presequence mediates recognition of the nascent precursor protein by molecular chaperones of the Hsp70 family that protect the precursor from aggregation and degradation, and present it to the mitochondria in an "import competent" form (8–11). Import is initiated by interaction of the presequence with specialized outer membrane proteins, globally known as the translocase of the outer membrane or TOM complex. Matrix-targeting presequences bind to the cytoplasmic domains of the Tom20 and Tom22 receptor proteins (12–14) and are subsequently inserted into the general import pore (GIP), consisting of five integral outer membrane proteins (Tom5, Tom6, Tom7, Tom22, and Tom40) (15–18). Translocation across the outer membrane is driven by electrostatic interactions between the positively charged presequence and acidic domains of the Tom22 and Tom40 proteins in the intermembrane space (16). Translocation across the inner mitochondrial membrane requires an intact electrochemical potential ($\Delta\Psi$) (19, 20), a set of integral inner membrane proteins known as the translocase of the inner membrane or TIM complex, and the pulling force of mitochondrial

Hsp70 on the matrix side of the inner membrane (*21–23*). The presequence is inserted into the TIM translocation channel, and as it begins to emerge into the matrix, Tim44 passes it to mitochondrial Hsp70, the motor that pulls the bulk of the precursor molecule across both membranes at the expense of ATP hydrolysis (*24*). At this point, the presequence is no longer necessary and actually represents a potential impediment to proper protein folding. Hence, it is removed by specific processing peptidases, MPP and MIP, that will be the subject of the rest of this chapter.

III. Mitochondrial Processing Peptidase

A. NOMENCLATURE

MPP was originally isolated from different sources by independent groups, and a wide variety of names were assigned to its two subunits. According to a new nomenclature established in 1993, the larger subunit is now designated α-MPP and the smaller subunit β-MPP, regardless of the source of the enzyme (*25*). One exception is MPP from higher plants, which consists of one α-MPP subunit and two larger β-MPP subunits, β^1-MPP and β^2-MPP (*26*). The names MPP, MPP-1, MAS2, MIF2, P-55, and cytochrome *c* reductase subunit III have been previously used for α-MPP, and the names PEP, MAS1, MIF1, P-52, and cytochrome *c* reductase subunit I and II for β-MPP (*25*).

B. CLEAVAGE SPECIFICITY

Initial surveys of mitochondrial precursor proteins revealed three main features that characterize most presequences cleaved by MPP across species: an overall positive charge and a predicted tendency to form amphiphilic α-helices (*4, 5, 27*), and frequently, the presence of an arginine residue at the −2 position (the "R-2 rule") from the cleavage site (*27, 28*). Gavel and von Heijne (*29*) later identified four cleavage site motifs: xRx↓x(S/x) (R-2 motif); xRx(Y/x)↓(S/A/x)x (R-3 motif); xRx↓(F/L/I)xx(S/T/G)xxxx↓ (R-10 motif, which is cleaved sequentially by MPP and MIP); xx↓x(S/x) (R-none motif) (Fig. 2). Branda and Isaya (*30*) analyzed a total of 71 mitochondrial matrix-targeting presequences from *S. cerevisiae,* the precise length of which was known from N-terminal sequencing of the corresponding native proteins. An R-2, R-3, or R-10 motif was found in 65% of the presequences analyzed and no conserved residues in the remaining 35%. This relatively high degeneracy limits the degree of confidence with which MPP cleavage sites can be predicted using the canonical motifs. Algorithms based on comparisons with already known presequences and secondary structure predictions allow for

Two-step cleavage: R-10

			↓MPP	↓MIP
HOTC	/ K02100	MLFNLRILLNNAAFRNGHNFMV**R**N	**F**RC**G**QPLQ	NKVQ..
MMDH	/ M16229	MLSALARPVGAALR**R**S	**F**STSAQNN	AKVA..
NCPB	/ P10255	MFGPRHFSVLKTTGSLVSSTFSSSLKPTATFSCAR**A**	**F**S**Q**TSSIM	SKVF..
YAT14	/ Q12349	MFPIASRRILLNASVLPLRLCN**R**N	**F**TTTRISY	NVIQ..
YCOXIV	/ P04037	MLSLRQSIRFFKPAT**R**T	**L**CS**S**RYLL	QQKP..
YDLDH	/ P09624	MLRIRSLLNNK**R**A	**F**SS**T**VRTL	TINK..
YFe/S	/ P08067	MLGIRSSVKTCFKPMSLTSK**R**L	**IS**Q**S**LLAS	KSTY..
YRM07	/ P36519	MQRFSLVTH**R**S	**FSHS**CVKP	KSAC..

One-step cleavage: R-2 or R-3

HMUT	/ A59145	MLRAKNQLFLLSPHYLRQVKESSGSRLIQ**Q**RL	LHQQ..
HPDHE3	/ P09622	MQSWSRVYCSLAKRGHFNRISHGLQGLSAVPL**R**TY	ADQP..
MAAT	/ J02622	MALLHSGRVLSGVASAFHPGLAAAASA**R**A	SSWW..
RSCSA	/ J03621	MVSGSSGLAAARLLSRTFLLQQNGI**R**H	GSYT..
YCOX8	/ P04039	MLCQQMIRTTAKRSSNIMTRPIIMK**R**S	VHFK..
YDHA7	/ P46367	MFSRSTLCLKTSASSIGRLQL**R**YF	SHLP..
YIDH1	/ P28834	MLNRTIAK**R**TL	ATAA..
YRM02	/ P12687	MWNPILLDTSSFSFQKHVSGVFLQV**R**N	ATKR..

One-step cleavage: No R-2, R-3 or R-10

BMPC	/ X05340	..NAPHLQLVHDGLAGPRSDPAGPPGPPRRSRNLAAA	AVEE..
HF0A	/ M16453	..GALRRLTPSAALPPAQLLLRAVRRRSHPVRDYAAQ	TSPS..
PCS	/ P00889	MALLTAAARLFGAKNASCLVLAARHAS	ASST..
YATP6	/ P00854	MFNLLNTYIT	SPLD..
YADH3	/ P07246	MLRTSTLFTQRVQPSLFSRNILRLQST	AAIP..
YDHA1	/ P22281	MLATRNLVPIIRASIKWRIKL	SALH..
YNDI1	/ P32340	MLSKNLYSNKRLLTSTNTLVRFASTR	STGV..
YRM04	/ P36517	MWKRSFHSQGGPLR	ARTK..

FIG. 2. Comparison of leader peptide cleavage sites of mitochondrial protein precursors. The abbreviated names and GenBank accession numbers are indicated. Conserved amino acids in the R-2, R-3, and octapeptide motives are in bold. BMPC, bovine mitochondrial phosphate carrier; HF0A, human F_0-ATPase proteolipid subunit; HMUT, human methylmalonyl-CoA mutase; HOTC, human ornithine transcarbamylase; HPDHE3, human pyruvate dehydrogenase E3 subunit; MAAT, mouse aspartate aminotranferase; MMDH, mouse malate dehydrogenase; NCPB, *Neurospora* cyclosporin-binding protein; PCS, pig citrate synthase; RSCSA, rat succinyl-CoA synthetase α subunit; YADH3, *S. cerevisiae* alcohol dehydrogenase III; YAT14, *S. cerevisiae* ATP synthase H chain; YATP6, *S. cerevisiae* ATP synthase A chain; YCOX4, *S. cerevisiae* cytochrome *c* oxidase subunit IV; YCOX8, *S. cerevisiae* cytochrome *c* oxidase subunit VIII; YDHA1, *S. cerevisiae* aldehyde dehydrogenase; YDHA7, *S. cerevisiae* potassium-activated aldehyde dehydrogenase; YDLDH, *S. cerevisiae* dihydrolipoamide dehydrogenase; YFe/S, *S. cerevisiae* ubiquinol–cytochrome *c* reductase Rieske iron–sulfur protein; YIDH1, *S. cerevisiae* isocitrate dehydrogenase subunit 1; YNDI1, *S. cerevisiae* rotenone-insensitive NADH-ubiquinone oxidoreductase; YRM02, *S. cerevisiae* 60S ribosomal protein MRP2; YRM04, *S. cerevisiae* 60S ribosomal protein L4; YRM07, *S. cerevisiae* 60S ribosomal protein L7.

a higher level of confidence (*29*). Whenever accuracy is of importance, however, an unknown MPP cleavage site should always be determined experimentally, either by purification and N-terminal sequencing of the native protein or by processing of ^{35}S-labeled precursor *in vitro* and N-terminal radiosequencing of the processed form (*31*).

The R-2 requirement was first addressed using intact rat liver mitochondria or mitochondrial fractions as the source of MPP, and *in vitro* translated mitochondrial protein precursors as the substrate. Substitution of R-2 with glycine abolished processing of the human ornithine transcarbamylase precursor (pHOTC), while replacement of R-2 with other amino acids was associated with partial (R-2 replaced by asparagine or alanine) or even normal (R-2 replaced by lysine) processing, suggesting that R-2 participates in a α-helical structure (*32, 33*) (note that pHOTC possesses a R-10 motif, where R-10 is actually R-2 relative to the MPP cleavage site). Similar results were obtained when R-2 was replaced by leucine in the mitochondrial aspartate aminotransferase precursor (*34*), or by glycine, lysine, or alanine in a chimeric precursor consisting of residues 1–167 of the cytochrome b_2 precursor fused to cytosolic dehydrofolate reductase (*35*). Likewise, substitution of glycine for either of two arginine residues at position –2 and –3 from the MPP cleavage site of the yeast iron–sulfur protein of the cytochrome bc_1 complex completely abolished cleavage of this precursor by MPP (*36*). Interestingly, in the mouse OTC precursor, replacement of R-2 by either glycine or aspartate resulted in a novel cleavage site between residues 16 and 17, that is, eight residues upstream of the usual site (*37*).

The role of other amino acid residues in the presequence is much less clear. In pHOTC, all arginine residues could be replaced by glycine without significantly affecting processing, unless R-2 was changed at the same time (*33, 38*). Substitution of neutral amino acids for multiple arginine and lysine residues in the aspartate aminotransferase presequence had minor effects (*34*), whereas similar replacements in the middle of the adrenodoxin presequence reduced (two changes) or eliminated (four and six changes) cleavage by MPP (*39*). A number of replacements of nonbasic amino acids had no significant effect on the processing of pHOTC (*38*), the alcohol dehydrogenase III precursor (*40*), or synthetic leader peptides (*41*). Collectively, these findings have led to the conclusion that the R-2 and R-3 motifs, or any other motifs found at MPP cleavage sites, play a role in the context of the overall secondary structure of the presequence, which probably represents the most important determinant for substrate recognition by MPP. NMR and circular dichroism have revealed that synthetic presequences may form specific helix–linker–helix structures in solution (*42, 43*). Whether this structure is specifically recognized by MPP remains to be established.

Additional observations suggest that the enzyme may also recognize some higher order structure on the C-terminal side of the cleavage site. Processing by MPP is affected by deletions or mutations in the intermediate octapeptide, which is found downstream of the MPP cleavage site in precursors cleaved by MPP and MIP (as will be discussed in detail in the part of this chapter related to MIP). When the pHOTC octapeptide was deleted, MPP cleavage was abolished even though the R-2 motif had not been changed (*31*), and similar results were obtained with point mutations in the octapeptide of the malate dehydrogenase precursor (*44*). The role played by the region C-terminal to the cleavage site was also confirmed in studies using synthetic peptides (*45*). In addition, point mutations in the mature portion of the human frataxin precursor have been shown to inhibit processing by MPP even if they are far away from the cleavage site (*46, 47*). These findings underscore the importance of structural elements in the substrate specificity of MPP.

C. Purification and Characterization

That one or more leader peptidases might exist within mitochondria became obvious when different laboratories used total polysomal RNA and rabbit reticulocyte lysate to translate mitochondrial proteins *in vitro*, and the protein products obtained under these conditions turned out consistently larger than the corresponding native forms (*48–52*). Shortly thereafter, MPP was partially purified from *S. cerevisiae* (*53, 54*), *Neurospora*, and rat liver (*55–57*) mitochondria, representing the first component of the mitochondrial import apparatus to be identified. MPP was first purified to homogeneity from *Neurospora* by Hawlitschek *et al.* in 1988 (*58*). Two subunits of different size were separately isolated, and because the larger subunit (57 kDa) exhibited some processing activity that was strongly enhanced by addition of the smaller subunit (52 kDa), the former was designated matrix processing peptidase (MPP, currently α-MPP) and the latter processing enhancing protein (PEP, currently β-MPP). Subsequent studies, however, showed that both subunits are actually inactive when separate from each other (*35, 59, 60*), and that the activity associated with the larger subunit in the original purification probably originated from contamination by the smaller subunit. Meanwhile, MPP was also purified from *S. cerevisiae* (*61*), rat liver (*62, 63*) and potato (*64*), and more recently, bovine liver (*41*), wheat (*65*); and spinach (*66*). The original protocol for purification of *Neurospora* MPP included DEAE-cellulose, zinc chelate, polyethyleneimine cellulose, polyethylene glycol precipitation, and ecteola–cellulose chromatography. A whole-cell extract (∼20 g total protein) was used as the starting material with a final yield of 0.11 mg MPP, corresponding to ∼6% of the activity in the total extract (*58*). Modified protocols and mitochondrial matrix as the starting material were used to isolate MPP from *S. cerevisiae* (*61*) and rat liver (*62, 63*). A method for the isolation of cytochrome *c* reductase was used for potato MPP, as the peptidase is an integral component of the cytochrome bc_1 complex in higher plants (*67*).

Partially purified MPP from different sources exhibited optimal activity between pH 7 and 8 and was inactivated at pH 6.5 or lower. Furthermore, the enzyme was inhibited by EDTA and o-phenanthroline, and stimulated by divalent metals, especially Co^{2+} and Mn^{2+} (53, 54, 56, 68, 69). Even though these properties characterized MPP as a neutral metallopeptidase, early studies showed no metal ions associated with purified native enzyme (70). This was probably due to metal loss during purification, however, as Zn : MPP and Zn : β-MPP molar ratios of 1.05 and 0.86 were later determined for yeast MPP under milder purification conditions (71). Addition of low concentrations of metals to metal-free apoenzyme showed that the activity is restored by $Zn^{2+} > Mn^{2+} > Co^{2+}$ (71). As described for other zinc-dependent metallopeptidases (72), excess Zn^{2+} inhibited yeast MPP (50% inhibition at a Zn : MPP molar ratio of 20 : 1), whereas Mn^{2+} and Co^{2+} did not at ratios of up to 1000 metal ions per mole of enzyme (71). Similar results were obtained with crude preparations of yeast, Neurospora, and rat MPP (60) (Fig. 3).

The kinetics of the MPP reaction were measured using purified yeast (70, 73) and Neurospora (35) MPP, and natural or artificial precursors as substrates. The k_{cat} ranged between 0.3 and 6.75 min^{-1} and the K_M between 0.1 and 1.0 μM. A higher k_{cat} and a similar K_M were determined for bovine ($k_{cat} = 30\ min^{-1}$; $K_M = 0.7–1.2\ \mu M$) (41), and rat liver MPP ($k_{cat} = 84\ min^{-1}$; $K_M = 0.074\ \mu M$) (74), using synthetic peptides as substrates.

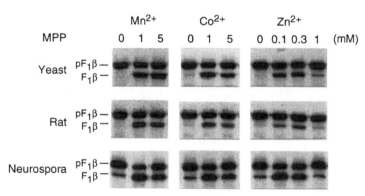

Fig. 3. Effect of various metals on the activity of MPP from different species. Bacterial lysates containing recombinant α-MPP and β-MPP from yeast, rat, or Neurospora were treated with 5 mM EDTA, mixed together in the presence of Mn^{2+} or Co^{2+} at a final concentration of 1 mM and 5 mM, respectively, or in the presence of Zn^{2+} at a final concentration of 0.1, 0.3, or 1 mM, and incubated on ice for 30 min. [^{35}S]Methionine-labeled precursor for the yeast F_1ATPase subunit β (p$F_1\beta$) was added and the incubation continued for 30 min 27°C. Processing products were analyzed directly by SDS–PAGE and fluorography. Reprinted from (60).

D. RECOMBINANT ENZYME

The main limitation of purifying MPP from natural sources is a relatively low yield (*58, 61–63*). Larger amounts of enzyme have been obtained by use of different overexpression systems. Overexpression of yeast MPP was first achieved by use of yeast strains in which the genes encoding the two MPP subunits were placed under the control of a strong galactose-inducible promoter (*75*). This system coupled with a purification method slightly different from that described by Yang (*61*) yielded almost 10 times more active enzyme than wild-type yeast (*75*). For expression in *E. coli*, cDNAs specifying only the mature portion of α-MPP or β-MPP were cloned downstream of different inducibile promoters in various prokaryote expression vectors. Both subunits were expressed as fusion proteins either with the maltose binding protein (*59*) or a polyhistidine tag (*74, 76, 77*). These tags allowed for rapid purification by affinity chromatography, and whereas the maltose binding protein tag was proteolytically removed upon purification, the histidine tags were not removed and did not seem to interfere with enzyme function. A mild, two-step purification procedure based on chelating chromatography and gel filtration was developed for histidine-tagged yeast MPP subunits by Luciano and co-workers (*71, 76*).

Although overexpression in *E. coli* can result in very high protein yields (20–30% of the total *E. coli* proteins), a common problem is that recombinant MPP subunits are largely recovered in an insoluble form inside "inclusion bodies." Kalousek and co-workers used coexpression of the molecular chaperones GroEL and GroES (*78*) and a very low postinduction incubation temperature (12°C for 24 hr) (*79*) and were able to improve the solubility of both yeast and rat β-MPP (unpublished results). The solubility of α-MPP was not improved, however, and the denaturation–renaturation procedure described by Geli *et al.* (*75*) was necessary to obtain soluble protein. Once soluble α-MPP and β-MPP were obtained, they were mixed together in a 1 : 1 molar ratio in the presence of 1 mM MnCl$_2$. Subsequent purification through a Mono Q column yielded highly-purified active MPP heterodimers (J.A. and F.K., unpublished results). *E. coli* has also been used to coexpress yeast (*76, 77*) or rat (*74, 78*) MPP subunits, with yields of active MPP heterodimers comparable to those of the expression systems described above.

E. CLONING OF MPP SUBUNITS

The two genes encoding the α and β subunit of yeast MPP were isolated by complementation of temperature-sensitive mutants defective in mitochondrial precursor processing (*mas1/mif1* and *mas2/mif2*). Sequence analysis of the wild-type alleles revealed that *MAS1/MIF1* and *MAS2/MIF2* encode two related proteins, Mas1p/Mif1p corresponding to β-MPP, and Mas2p/Mif2p corresponding to α-MPP (*61, 80–83*). Screening of an expression library with polyclonal antibodies against *Neurospora* MPP identified two cDNAs encoding a Mas2p/Mif2p- and a Mas1p/Mif1p-related protein (*58, 84*). The cDNAs for the rat (*63, 85*) and potato (*64, 86*) MPP subunits were cloned by conventional PCR screens of cDNA libraries using degenerate primers derived from the amino acid sequence of the purified subunits. Two cDNAs for the human α (*87*) and β (*88*) MPP subunits were identified by analysis of expressed sequence tags. MPP subunits have also been cloned from *Lentinula edodes* (*89, 90*) and *Blastocladielle emersonii* (*91, 92*). Interestingly, these molecular studies revealed that MPP subunits are synthesized as larger precursor proteins with N-terminal matrix targeting presequences, meaning that newly synthesized subunits require the presence of active MPP in order to be processed to the mature form upon import. This is part of the reason why mitochondria can only originate from preexisting organelles.

F. STRUCTURE–FUNCTION ANALYSES

Both α- and β-MPP subunits exhibit structural similarity to the core I and core II proteins of the cytochrome bc_1 complex. Moreover, in *Neurospora*, β-MPP and the core I protein are encoded by the same gene and are identical (*93*). *Neurospora* mitochondria contain approximately 15 times more β-MPP than α-MPP, and more than 80% of β-MPP is isolated from the inner membrane fraction (*35*). These findings suggest that *Neurospora* β-MPP is a bifunctional protein. In complex with α-MPP, β-MPP fulfills a processing peptidase function, whereas in complex with the other subunits of the cytochrome bc_1 complex, it participates in electron transfer (*35, 93*). A different configuration is found in yeast and rat liver mitochondria, where MPP is a soluble matrix-localized heterodimer with processing function only (*61–63*). In potato, wheat, and probably all higher plants, however, both α- and β-MPP are integral components of the cytochrome bc_1 complex and carry out the precursor processing function of MPP in the complex (*26, 65, 86*). Thus, the configuration of *Neurospora* MPP represents an intermediate stage between soluble matrix forms (yeast and rat MPP) and membrane integrated forms (MPPs of higher plants) (*94*).

Alignments of α- and β-MPP subunits from different species show an overall identity of 30–43% among α-MPPs (*63, 64, 84*) and of 42–52% among β-MPPs (*85, 86, 95, 96*). The homology is greater among the N-terminal 200 amino acids of the mature sequences than among their C-terminal regions. All known MPP subunits share a negatively charged region and a putative metal ion binding motif that are important for enzyme function (Fig. 4).

A

α-MPPs

Rat / P20069	104	L	S	G	I	A	**H**	**F**	**L**	**E**	**K**	L	A	F	S	S	T	119
Neurospora / P23955	90	V	R	G	A	S	**H**	I	M	**D**	**R**	L	A	F	K	S	T	105
Yeast / P11914	57	L	K	G	C	T	**H**	I	L	**D**	**R**	L	A	F	K	S	T	72
Potato / P29677	113	S	Y	G	A	T	**H**	L	L	**E**	**R**	M	A	F	K	S	T	128

β-MPPs

Rat / Q03346	96	N	N	G	T	A	**H**	**F**	**L**	**E**	**H**	M	A	F	K	G	T	111
Neurospora / P11913	79	T	N	G	T	A	**H**	**F**	**L**	**E**	**H**	L	A	F	K	G	T	94
Yeast / P10507	65	N	N	G	T	A	**H**	**F**	**L**	**E**	**H**	L	A	F	K	G	T	80
Potato (β¹) / X80237	138	N	N	G	V	A	**H**	**F**	**L**	**E**	**H**	M	I	F	K	G	T	153
Potato (β²) / X80235	135	T	N	G	T	A	**H**	**F**	**L**	**E**	**H**	M	I	F	K	G	T	150

B

α-MPPs

Rat	177	**D**	**E**	**E**	I	**E**	M	T	R	M	A	V	Q	F	**E**	L	**E**	**D**	L	N	195
Neurospora	162	**D**	**E**	**E**	L	**E**	G	Q	⊥	M	T	A	Q	Y	**E**	V	N	**E**	I	W	180
Yeast	129	**E**	Q	**E**	L	Q	**E**	Q	K	L	S	A	**E**	Y	**E**	I	**D**	**E**	V	M	148
Potato	185	**D**	W	**E**	V	K	**E**	Q	L	**E**	K	V	K	**E**	I	S	**E**	Y	S	K	204

β-MPPs

Rat	168	**E**	A	**E**	I	**E**	R	**E**	R	G	V	I	L	R	**E**	M	Q	**E**	V	**E**	186
Neurospora	151	**E**	S	A	I	**E**	R	**E**	R	**D**	V	I	L	R	**E**	S	**E**	**E**	V	**E**	169
Yeast	137	N	S	A	I	**E**	R	**E**	R	**D**	V	I	I	R	**E**	S	**E**	**E**	V	**D**	155
Potato (β¹)	210	**E**	**D**	K	I	I	R	**E**	R	S	V	I	L	R	**E**	M	**E**	**E**	V	**E**	228
Potato (β²)	207	**E**	R	K	I	**E**	R	**E**	R	**D**	V	I	L	R	**E**	M	**E**	**E**	V	**E**	225

C

α-MPPs

Rat	342	M	M	**G**	**G**	**G**	**G**	**S**	**F**	**S**	**A**	**G**	**G**	**P**	**G**	**K**	**G**	**M**	F	**S**	360
Neurospora	382	L	L	**G**	**G**	**G**	**G**	**S**	**F**	**S**	**A**	**G**	**G**	**P**	**G**	**K**	**G**	**M**	Y	**S**	400
Yeast	282	L	L	**G**	**G**	**G**	**G**	**S**	**F**	**S**	**A**	**G**	**G**	**P**	**G**	**K**	**G**	**M**	Y	**S**	300
Potato	332	L	M	**G**	**G**	**G**	**G**	**S**	**F**	**S**	**A**	**G**	**G**	**P**	**G**	**K**	**G**	**M**	Y	**S**	350

Fig. 4. Conservation of MPP functional domains. (A) Putative metal binding site. (B) Negatively charged region. (C) Glycine-rich region. Conserved residues are in bold; residues directly involved in metal binding are underlined. The numbers indicate the position of the first and last amino acid of each stretch of sequence relative to the first amino acid of the precursor protein. GenBank accession numbers are also indicated.

The putative metal ion binding site consists of an inverted zinc-binding motif, H-F-L-E-H, which is present in all known β-MPP subunits (*85, 97*). This motif is much less conserved among α-MPP subunits, varying from H-F-L-E-K in rat to H-I-L-D-R in yeast, with a *histidine–hydrophobic–hydrophobic–acidic–basic* consensus. The H-X-X-E-H motif is an inversion of the H-E-X-X-H motif, commonly found in zinc-dependent metallopep-tidases (*98*). Together with the human and *Drosophila* insulin-degrading enzyme (IDE) (*99, 100*) and the *E. coli* protease III (pitrilysin) (*101*), MPP belongs to the pitrilysin family of zinc-dependent proteases (*102, 103*). The region around the metal-binding motif is the only highly conserved region among these proteins, however, with an overall homology of only 20–24% (*102*). In both MPP subunits, the negatively charged region is located down-stream of the putative metal-binding motif, and in the case of β-MPP, it includes a highly conserved glutamic acid 76 residues from the H-X-X-E-H motif. The catalytic zinc atom of IDE and pitrilysin is coordinated by the two histidine residues in the H-X-X-E-H motif, and a glutamic acid 76 residues downstream of this motif. Replacement of any of these three residues with other amino acids completely abolishes zinc binding and eliminates the pep-tidase activity of IDE and pitrilysin (*103–106*). Similarly, the two histidines and the glutamic acid in the H-F-L-E-H motif of β-MPP, and the conserved glutamic acid 76 residues downstream of the motif are required for activity (*73, 78, 107*). In fact, when these residues were mutated in rat (*74*) or yeast (*71*) β-MPP, the mutant enzymes contained less than 0.6 atoms of nickel or zinc per molecule, indicating that these residues are involved in metal ion coordination. In addition, mutational analysis of rat β-MPP identified two additional glutamic acid residues that are also important for catalysis (*74*). The first (E124 in Fig. 5; see color plate.) is between the the H-F-L-E-H motif and the negatively charged region, whereas the second (E174) is within this region. Replacement of histidine 109 with lysine, arginine, or phenylalanine in the H-F-L-E-K motif of rat α-MPP nearly abolishes enzyme activity, and the same is true when glutamate 179 in the negatively charged region of rat α-MPP is replaced with glutamine (*78*). Other single substitutions within these two regions have only mild effects on activity, however. Collectively, these observations have led to the conclusion that β-MPP, not α-MPP, is re-sponsible for catalysis (*76, 78, 108*). On the other hand, α-MPP appears to be involved in substrate recognition, probably through its negatively charged region. This region is predicted to form an amphiphilic α helix and could interact with positively charged α helices formed by the leader peptides (*63, 78*). In addition, deletion and site-directed mutagenesis of rat α-MPP have shown that the C-terminal region is also important for substrate bind-ing, especially two residues, a glutamic acid and an aspartic acid, at positions 446 and 447 (*109*).

The exact role played by each subunit in the ability of MPP to specifically recognize and cleave many different precursor proteins is still unclear. In yeast MPP, surface plasmon resonance and cross-linking experiments showed that the holoenzyme or α-MPP alone bind a peptide substrate with the same efficiency, suggesting that β-MPP does not participate in substrate binding (76). On the other hand, studies with fluorescence-labeled substrates and point-mutated β-MPP suggested a cooperative formation of the substrate-binding pocket by both subunits (77). Crystallographic analysis of MPP and MPP–precursor complexes will be required to better understand enzyme–substrate interactions. Although many attempts to crystallize MPP have been made by several different groups, no three-dimensional structure is as yet available. The structures of the chicken and bovine cytochrome bc_1 complexes have been determined (110–112), and the structure of their core proteins used to generate an energy-minimized three-dimensional model of MPP (108) (Fig. 5). Based on this model, the metal-binding site is predicted to be arranged around a cavity, the surface of which is delineated by mostly hydrophilic amino acid residues. A highly conserved glycine-rich segment in α-MPP (Figs. 4 and 5) is predicted to form a loop that may function to present the leader peptide to the active center or to release the cleaved peptide from the enzyme (108).

IV. Mitochondrial Intermediate Peptidase

A. NOMENCLATURE

MIP (EC 3.4.24.59) was originally characterized in rat liver mitochondria and called P_2 (113) or MPP-2 (63). It was later designated mitochondrial intermediate peptidase, as the enzyme cleaves *intermediate-size* proteins that originate from cleavage of certain mitochondrial precursors by MPP (31). The symbol MIP was later changed to RMIP in order to distinguish the rat enzyme from the enzymes of *S. cerevisiae* (YMIP) (114), *Schizophillum commune* (*SMIP*) (115), and man (*HMIP*) (116). The gene encoding YMIP was initially designated *MIP1*, but because this name had also been assigned to a gene encoding the catalytic subunit of mitochondrial DNA polymerase, a new name, *OCT1,* was adopted in 1999 (117). The name *MIPEP* was assigned to the gene encoding HMIP (116).

B. TWO-STEP PROCESSING

Whereas most precursors are processed to their mature form in a single step by MPP, many precursors targeted to the mitochondrial

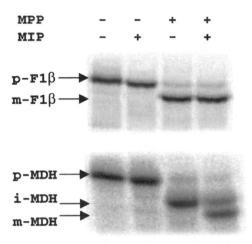

FIG. 6. One- vs. two-step processing. Radiolabeled $pF_1\beta$ or malate dehydrogenase precursor (pMDH) was incubated with *E. coli* lysates containing MPP and/or MIP for 30 min at 27°C. Reaction products were analyzed directly by SDS–PAGE and fluorography.

matrix and the inner membrane require an additional cleavage by MIP (Fig. 6). These precursors are characterized by the motif R-X↓(F/L/I)-X-X-(T/S/G)-X-X-X-X↓ at the C terminus of the leader peptide (R-10 motif) (*28, 29*). MPP first cleaves the motif two peptide bonds from the arginine residue, yielding a processing intermediate with a typical N-terminal octapeptide that is subsequently cleaved by MIP to produce mature-size protein (Fig. 2). As discussed previously, MPP appears to recognize a higher order structure rather than specific amino acid residues. Thus, we hypothesized that two-step processing *via* formation of an octapeptide-containing intermediate reflects some unique structural characteristics of a subset of mitochondrial proteins, and that the octapeptide supplies an MPP-compatible structure in these precursors (*31*). This hypothesis was tested by asking whether the octapeptide is necessary for formation of mature-size protein or whether an arginine placed at −2 from the MIP cleavage site might be able to direct cleavage of a twice-cleaved precursor by MPP alone. Processing was indeed abolished when the octapeptide was deleted from twice-cleaved precursors. The same result was obtained when presequences of once-cleaved precursors were fused to the mature N termini of twice-cleaved precursors. Because the general requirements for cleavage by MPP (i.e., an R-2 motif) had been maintained in all these constructs, it was concluded that a higher order structure at the N termini of twice-cleaved precursors is incompatible with cleavage by MPP (*31*). Indeed, MPP was able to cleave constructs in which presequences of twice-cleaved precursors (without the octapeptide) had been directly joined to the mature N termini of once-cleaved precursors. Although a construct carrying an artificial octapeptide consisting of four leucine–glutamine repeats was not cleaved by either MPP or MIP, octapeptide swaps among twice-cleaved precursors yielded constructs that were efficiently processed in two steps (*31*). These findings landed experimental support to the hypothesis that structural determinants in the mature N termini of once-cleaved precursors or the octapeptides of twice-cleaved precursors are required for cleavage by MPP (*31*). Gavel and von Heijne (*29*) found that the mature N termini of twice-cleaved precursors contain more positively and fewer negatively charged amino acids than the mature N termini of once-cleaved precursors, and suggested that the octapeptide provides a spacer to separate the MPP cleavage site from a positively charged structure that would otherwise interfere with processing. However, more recent studies have shown that this may not be a general requirement. When the MPP and MIP cleavage sites of the yeast iron–sulfur protein of the cytochrome bc_1 complex were eliminated, and the latter replaced with a new MPP cleavage site, the resulting construct was correctly processed in a single step by MPP *in vitro*, and able to generate a functional iron-sulfur protein when expressed in yeast (*36, 118, 119*).

C. CLEAVAGE SPECIFICITY

All known octapeptides contain a large hydrophobic residue (phenylalanine, leucine, or isoleucine) at position –8, and a small hydroxylated residue (serine or threonine) or glycine at position –5, relative to the N terminus of the mature protein (*28, 29*) (Fig. 2). In addition, a survey of mitochondrial protein precursors of *S. cerevisiae* showed that yeast octapeptides often have a small hydroxylated residue at positions –7 and –6,   R-X↓ (F/L/I)-(S/X)-(S/T/X)-(T/S/G)-X-X-X-X↓ (*30*). Octapeptide-containing precursors cannot be processed to the mature form by mitochondrial fractions containing MIP activity but devoid of MPP activity (*31, 120*) (see also Fig. 6). However, a construct translated from a methionine replacing the phenylalanine at the N terminus of the pHOTC octapeptide (M-iOTC) was correctly processed to mature form by MIP alone (*31*). This result indicates that the octapeptide must be at the protein N terminus in order to be accessible to MIP, and indeed, cleavage of M-iOTC was abolished when the initiating methionine was added in front of the phenylalanine, thereby placing nine instead of eight amino acids upstream of the MIP cleavage site (*120*). Such findings led to the conclusion that MIP acts as an octapeptidyl-amino peptidase. Accordingly, synthetic octapeptides were shown to inhibit MIP activity at micromolar concentrations *in vitro*, whereas higher concentrations of random octapeptides or decapeptides had no significant effect (*121*). Deletion of the N-terminal phenylalanine abolished inhibition of MIP by the octapeptide, whereas replacement of phenylalanine with tyrosine decreased it 10-fold. Likewise, deletion of the C-terminal amino acid also reduced inhibition 10-fold (*121*). Similar results were obtained by mutational analysis of the octapeptide of the yeast iron–sulfur protein of the cytochrome bc_1 complex (*118*). Substitution of glycine for the N-terminal isoleucine resulted in partial inhibition of cleavage by MIP without affecting the initial cleavage by MPP. When this mutation was coupled with substitution of phenylalanine for the canonical serine at–5, however, the ability of MIP to cleave the intermediate was completely abolished (*118*). These findings suggest that some hydrophobic interaction may occur between the octapeptide N terminus and the MIP substrate binding site, and that residues 2–8 in the octapeptide may be required to stabilize this interaction.

Extensive surveys of twice-cleaved precursors have not revealed any obvious consensus sequence on the C-terminal half of MIP cleavage sites (*28–30*). As discussed earlier for MPP, however, it appears that structural determinants in the mature portion of twice-cleaved precursors play a role in the substrate specificity of MIP. Octapeptides were cleaved by MIP only when fused to the mature N terminus of a twice-cleaved precursor, but not a once-cleaved precursor (*120*). When N-terminal fragments of the yeast iron–sulfur protein of the cytochrome bc_1 complex were fused to the cytosolic enzyme dehydrofolate reductase, the presequence plus the octapeptide was sufficient for cleavage of this construct by MPP (*122*). However, at least 42 N-terminal residues of the mature iron–sulfur protein had to be placed between the MPP cleavage site and dehydrofolate reductase to restore cleavage of the octapeptide by MIP (*122*). Moreover, cleavage by MIP was inhibited by mutations in the mature iron–sulfur protein as far as 70 residues from the cleavage site (*123*).

D. PURIFICATION AND CHARACTERIZATION

MIP was first purified to homogeneity from rat liver mitochondrial matrix and shown to exist as a soluble monomer of 75 kDa. The purification procedure involved five steps including DEAE Bio Gel, heparin–agarose, hydroxyapatite, ω-aminooctyl-agarose, and Mono-Q chromatography, and was consistently characterized by a very low final yield of only 2% (*124, 125*). This depended on the enzyme relatively low abundance ($\sim$0.04% of total mitochondrial proteins) and its instability, which resulted in significant losses especially during the first two steps of purification (*124*). In addition, MPP and MIP activities could not be separated during the first four purification steps, such that only fractions with the highest MIP activity could be used. Over-expression of YMIP was later achieved by use of an *S. cerevisiae* strain in which the *OCT1* gene was cloned into a 2-μm vector under the control of the constitutive yeast glyceraldehyde-3-phosphate dehydrogenase promoter (*126*). This system yielded mitochondrial levels of active enzyme about 100 times higher than physiological levels. The fusion of a *c*-myc tag to the YMIP C terminus did not have any obvious deleterious effects on enzyme function, but instead enabled a rapid purification from total mitochondrial matrix by immunoaffinity chromatography, thereby reducing potential losses caused by the enzyme's instability. In a typical purification, 100 μl of mitochondrial matrix (20 mg/ml, total protein) is treated with 50 μl of an agarose-bead conjugate of a monoclonal anti-*c*-myc antibody. After elution with glycine-HCl buffer at pH 3.5 and reequilibration to pH 7.4, the final purified fraction (200 μl) has a YMIP-myc protein concentration of about 10 ng/μl, with a total yield of about 1 μg of YMIP-myc per mg of total matrix protein ($\sim$10 μg of YMIP-myc are obtained from 5 g of yeast). The entire procedure can be carried out in less than 4 hr and is reproducible regarding purity, recovery, and enzyme activity (*126*).

The requirements for MIP activity were tested using Met-iOTC as the substrate. Purified RMIP showed a pH optimum between 6.6 and 8.9, was strongly inhibited by 0.01 mM EDTA, and was fully reactivated by addition of 0.1–1 mM Mn^{2+}. Other metal ions either stimulated (Mg^{2+} or Ca^{2+}) or fully inhibited (Co^{2+}, Fe^{2+}, or Zn^{2+}), the enzyme at a metal concentration of 1 mM. This inhibition could be reversed by higher concentrations of Mn^{2+}. Interestingly, RMIP was inactivated by *N*-ethylmaleimide (NEM) and other thiol-group reagents (*124*). Similar to native RMIP, purified recombinant YMIP-myc was also stimulated by Mn^{2+} but not Zn^{2+}, reversibly inhibited by EDTA, and inactivated by *N*-ethylmaleimide and other thiol-blocking agents (*126*). As it turned out, MIP sequences are related to the thimet oligopeptidase (TOP) family of metalloproteases, whose members are often characterized by thiol-dependence (*127*). Amino acid potentially responsible for metal- or thiol-dependence were identified by sequence comparisons among evolutionarily distant MIP proteins (*115*). All known MIPs contain a highly conserved domain, F-H-E-X-G-H-(X)$_2$-H-(X)$_{12}$-G-(X)$_5$-D-(X)$_2$-E-X-P-S-(X)$_3$-E, also shared by TOPs (*102, 115*). This domain includes a zinc-binding motif, H-E-X-X-H, as well as a histidine (H565) and two glutamic acids (E587 and E594), predicted to participate in the zinc coordination (Fig. 7). As for the thiol dependence, RMIP, HMIP, YMIP, and SMIP contain 18, 16, 17, and 8 cysteine residues, respectively; two of these residues (C131 and C581) are conserved in all four proteins, and one of them (C581) is located within the metal-binding domain (*126*). Single amino

```
HMIP / U80034    491 ENLFHEMGHAMHSMLGRTRYQHVTGTRCPTDFAEVPSILMEYFAND 536
RMIP / M96633    488 ENLFHEMGHAMHSMLGRTRYQHVTGTRCPTDFAEVPSILMEYFSND 533
SMIP / L43072    554 MTLFHEMGHAMHSMIGRTEYQNVSGTRCPTDFVELPSILMEHFLNS 599
YMIP / U10243    554 ETLFHEMGHAMHSMLGRTHMQNISGTRCATDFVELPSILMEHFAKD 599

EDCP / P24171    465 ITLFHEFGHTLHGLFARQRYATLSGTNTPRDFVEFPSQINEHWATH 510
PMOP / Q02038    493 RTYFHEFGHVMHQICAQTDFARFSGTNVETDFVEVPSQMLENWVWD 538
RTOP / P24155    468 ETYFHEFGHVMHQLCSQAEFAMFSGTHVERDFVEAPSQMLENWVWE 513
```

```
    Consensus          xxxFHExGHxxHxxxxxxxxxxxxxGxxxxxDxxExPSxxxExxxxx
```

FIG. 7. Sequence conservation at the putative active sites of MIPs and TOPs. HMIP, RMIP, SMIP and YMIP, mitochondrial intermediate peptidase from human, rat, *Schizophyllum commune,* and *S. cerevisiae*; EDCP, *E. coli* peptidyl-dipeptidase Dcp; PMOP, pig mitochondrial oligopeptidase; RTOP, rat thimet oligopeptidase. Identical amino acids are in bold. Amino acids predicted to be involved in metal binding are underlined. GenBank accession numbers are also indicated.

acid substitutions within the H-E-X-X-H motif or replacement of E587 by an aspartic acid residue caused loss of YMIP function *in vivo*. In contrast, replacements of the two conserved cysteine residues did not affect the function of YMIP *in vivo* or decrease its sensitivity to thiol-blocking agents *in vitro* (*126*). The essential role of highly conserved amino acids in the MIP metal binding domain is therefore consistent with the enzyme's metal dependence. On the other hand, the thiol dependence of MIP is not associated with its only two conserved cysteines, suggesting that thiol groups play an important but probably noncatalytic role. It is possible that the thiol dependence of RMIP and YMIP simply reflects the high number of cysteine residues (18 and 17) in their sequences. However, purified recombinant SMIP, which contains only eight cysteines, exhibited a pattern of inhibition by thiol-blocking agents very similar to that of RMIP and YMIP (*126*).

E.   CLONING AND GENOMIC STRUCTURE ANALYSIS OF MIP HOMOLOGS

RMIP was cloned by RT-PCR using degenerate primers based on tryptic peptides of purified native RMIP, coupled with a cDNA library screen. A 2337-bp cDNA was isolated and found to encode a protein of 710 amino acids including a presequence of 33 amino acid (*121*). A typical R-2 motif was identified at the mature N terminus, and *in vitro* processing showed that the MIP precursor is cleaved to the mature form by MPP. Low stringency screening of an *S. cerevisiae* genomic library using the RMIP cDNA as a probe yielded a new gene, *MIP1* (later redesignated *OCT1*) that specifies a mitochondrial precursor of 772 amino acids with a predicted presequence of 37 amino acids and a typical MPP cleavage site (YMIP) (*114*). The enzyme was later cloned for the basidiomycete fungus *Schizophyllum commune* (SMIP) (*115*) and human liver (*116*) (HMIP). The MIP genomic structure was also characterized in these two organisms. The SMIP gene spans approximately 2 kbp of genomic DNA and consists of four exons of 523, 486, 660, and 629 bp, separated by three small introns of 79, 50, and 53 bp, respectively (*115*). The HMIP gene (designated *MIPEP*) spans 57 kbp and consists of 19 exons encoding a mitochondrial precursor protein of 713 amino acids (*116*). A 35 amino acid long presequence is encoded by exon 1, and the entire metal-binding domain by exon 13. The exon–intron structure of *MIPEP* is surprisingly different from those of mitochondrial oligopeptidase (EC 3.4.24.16) and thimet oligopeptidase (EC 3.4.24.15) (*128*), in spite of the sequence homology among these proteins and their almost identical catalytic domains (*115, 127*). Primer extension analysis identified a major transcript of *MIPEP* expressed differentially and predominantly in tissues with high oxygen consumption [heart, skeletal muscle, and several different regions of the brain (*116, 129*)]. Using a polymorphic $(CA)_n$ repeat in intron 4, *MIPEP* has been genetically mapped within a 7 cM interval between markers D13S283 and D13S217 on chromosome 13q12 (*129*).

F. BIOLOGICAL ROLE

In *S. cerevisiae*, octapeptide-containing proteins include (i) subunits of pyridine- and flavin-linked dehydrogenases, (ii) iron–sulfur proteins and other nuclear encoded subunits of respiratory chain complexes, (iii) proteins required for replication and expression of mitochondrial DNA (mtDNA), and (iv) ferrochelatase, the enzyme that catalyzes iron attachment in the last step of heme biosynthesis (*30*). The composition of this group of proteins suggests that YMIP is important for oxidative metabolism, and accordingly, YMIP-deficient yeast show multiple respiratory enzyme defects, loss of mtDNA, and mitochondrial iron depletion (*30, 117*). RMIP, SMIP, and HMIP can rescue this phenotype, indicating that the role of MIP is conserved in eukaryotes (*115, 129*).

G. STRUCTURE–FUNCTION ANALYSES

The protein sequences of yeast and mammalian MIPs show approximately 35% overall identity and 55% similarity. Homology is particularly high around the putative active site toward the protein C terminus. MIP sequences show 20–24% identity and 40–47% similarity to the thimet oligopeptidases (TOPs), including thimet oligopeptidase (*130, 131*), mitochondrial oligopeptidase (*127*), and the bacterial dipeptidyl-carboxypeptidase (*132*) and oligopeptidase A (*133, 134*). MIPs and TOPs share the unusual property of being thiol-dependent metallopeptidases [although thiol dependence is not associated with certain members of the TOP family (*127, 135*)] and together constitute the M3 family of metallopeptidases (*102*). TOPs cleave short peptide substrates, and whereas mitochondrial oligopeptidase resides in the mitochondrial intermembrane space, most TOPs are localized to the cytoplasm (*127*).

As already mentioned, MIPs and TOPs share the motif F-H-E-X-G-H-$(X)_2$-H-$(X)_{12}$-G-$(X)_5$-D-$(X)_2$-E-X-P-S-$(X)_3$-E-X. In a mutational analysis of YMIP, four single amino acid substitutions, H558R, E559D, H562R, and E587D, led to loss of enzyme function *in vivo*, suggesting that these residues are part of a thermolysin-like active site, H-E-X-X-H-$(X)_n$-E. Substitutions of the other amino acids in the metal-binding domain affected YMIP stability or substrate specificity. All known MIP sequences contain two highly conserved cysteine residues, corresponding to C131 and C581 in the YMIP sequence (*126*). C581 is located within the putative metal-binding domain, 19 residues from the H-E-X-X-H motif. However, single or double replacements of C131 and C581 with serine or valine residues (C130S/V, C581S/V, and C130S/V + C581S/V) had no effect on YMIP activity *in vivo*, indicating that C131 and C581 are not required for global peptidase function. To test the effects of the C131S, C581S, and C131S + C581S mutations *in vitro*, the corresponding proteins were overexpressed in yeast and affinity-purified from mitochondrial matrix. Before elution from the affinity column, the agarose-bound forms of the C131S, C581S, and C131S + C581S variants exhibited the same specific activity as wild-type YMIP (*126*). Upon elution at acidic pH, however, wild-type YMIP retained 70% activity while the C131S, C581S, and C131S + C581S variants retained only 37%, 40%, and 15% activity, respectively, suggesting that C131 and C581 are important for protein stability. All four enzymes showed 100% and 50% inhibition by 2 and 0.2 mM NEM, respectively, and similar levels of inhibition were observed with 2 and 0.2 mM *p*-hydroxymercuribenzoate. In-addition, the three mutant enzymes were almost fully inhibited by 2 mM iodoacetamide or iodoacetic acid, which otherwise had essentially no effect on wild-type YMIP (*126*). These findings suggest that cysteine residues may be important for MIP function in at least two ways: by forming one or more S–S bonds, which may be important to MIP stability, or by forming one or more Mn^{2+} binding sites, which may play a role in MIP activity. The latter possibility is suggested by the fact that recombinant YMIP is activated by Mn^{2+}, and that this effect is enhanced by DTT, an SH-reducing agent (G.I., unpublished results).

ACKNOWLEDGMENT

We thank Dr. Akio Ito for providing the simulation model of rat MPP.

REFERENCES

1. Voos, W., Martin, H., Krimmer, T., and Pfanner, N. (1999). *Biochim. Biophys. Acta* **1422**, 235.
2. Herrmann, J. M., and Neupert, W. (2000). *Curr. Opin. Microbiol.* **3**, 210.
3. Koehler, C. M., Merchant, S., and Schatz, G. (1999). *Trends. Biochem. Sci.* **24**, 428.
4. Allison, D. S., and Schatz, G. (1986). *Proc. Natl. Acad. Sci. USA* **83**, 9011.
5. Roise, D., and Schatz, G. (1988). *J. Biol. Chem.* **263**, 4509.
6. Hurt, E. C., Allison, D. S., Muller, U., and Schatz, G. (1987). *J. Biol. Chem.* **262**, 1420.
7. Vestweber, D., and Schatz, G. (1989). *Nature* **338**, 170.
8. Endo, T., Mitsui, S., Nakai, M., and Roise, D. (1996). *J. Biol. Chem.* **271**, 4161.
9. Deshaies, R. J., Koch, B. D., Werner-Washburne, M., Craig, E. A., and Schekman, R. (1988). *Nature* **332**, 800.
10. Sheffield, W. P., Shore, G. C., and Randall, S. K. (1990). *J. Biol. Chem.* **265**, 11069.
11. Roise, D. (1997). *J. Bioenerg. Biomembr.* **29**, 19.
12. Moczko, M., Gartner, F., and Pfanner, N. (1993). *FEBS Lett.* **326**, 251.
13. Moczko, M., Ehmann, B., Gartner, F., Honlinger, A., Schafer, E., and Pfanner, N. (1994). *J. Biol. Chem.* **269**, 9045.
14. Ramage, L., Junne, T., Hahne, K., Lithgow, T., and Schatz, G. (1993). *EMBO J.* **12**, 4115.
15. Vestweber, D., Brunner, J., Baker, A., and Schatz, G. (1989). *Nature* **341**, 205.
16. Dietmeier, K., Honlinger, A., Bomer, U., Dekker, P. J., Eckerskorn, C., Lottspeich, F., Kubrich, M., and Pfanner, N. (1997). *Nature* **388**, 195.
17. Alconada, A., Kubrich, M., Moczko, M., Honlinger, A., and Pfanner, N. (1995). *Mol. Cell Biol.* **15**, 6196.
18. Honlinger, A., Bomer, U., Alconada, A., Eckerskorn, C., Lottspeich, F., Dietmeier, K., and Pfanner, N. (1996). *EMBO J.* **15**, 2125.
19. Schleyer, M., Schmidt, B., and Neupert, W. (1982). *Eur. J. Biochem.* **125**, 109.
20. Martin, J., Mahlke, K., and Pfanner, N. (1991). *J. Biol. Chem.* **266**, 18051.
21. Dekker, P. J., Martin, F., Maarse, A. C., Bomer, U., Muller, H., Guiard, B., Meijer, M., Rassow, J., and Pfanner, N. (1997). *EMBO J.* **16**, 5408.
22. Blom, J., Kubrich, M., Rassow, J., Voos, W., Dekker, P. J., Maarse, A. C., Meijer, M., and Pfanner, N. (1993). *Mol. Cell Biol.* **13**, 7364.
23. Kubrich, M., Keil, P., Rassow, J., Dekker, P. J., Blom, J., Meijer, M., and Pfanner, N. (1994). *FEBS Lett.* **349**, 222.
24. Horst, M., Azem, A., Schatz, G., and Glick, B. S. (1997). *Biochim. Biophys. Acta.* **1318** 71.
25. Kalousek, F., Neupert, W., Omura, T., Schatz, G., and Schmitz, U. K. (1993). *Trends Biochem. Sci.* **18**, 249.
26. Braun, H. P., and Schmitz, U. K. (1995). *J. Bioenerg. Biomembr.* **27**, 423.
27. von Heijne, G., Steppuhn, J., and Herrmann, R. G. (1989). *Eur. J. Biochem.* **180**, 535.
28. Hendrick, J. P., Hodges, P. E., and Rosenberg, L. E. (1989). *Proc. Natl. Acad. Sci. USA* **86**, 4056.
29. Gavel, Y., and von Heijne, G. (1990). *Protein Eng.* **4**, 33.
30. Branda, S. S., and Isaya, G. (1995). *J. Biol. Chem.* **270**, 27366.
31. Isaya, G., Kalousek, F., Fenton, W. A., and Rosenberg, L. E. (1991). *J. Cell Biol.* **113**, 65.
32. Horwich, A. L., Kalousek, F., and Rosenberg, L. E. (1985). *Proc. Natl. Acad. Sci. USA* **82**, 4930.
33. Horwich, A. L., Kalousek, F., Fenton, W. A., Pollock, R. A., and Rosenberg, L. E. (1986). *Cell* **44**, 451.
34. Nishi, T., Nagashima, F., Tanase, S., Fukumoto, Y., Joh, T., Shimada, K., Matsukado, Y., Ushio, Y., and Morino, Y. (1989). *J. Biol. Chem.* **264**, 6044.

35. Arretz, M., Schneider, H., Guiard, B., Brunner, M., and Neupert, W. (1994). *J. Biol. Chem.* **269,** 4959.
36. Nett, J. H., Denke, E., and Trumpower, B. L. (1997). *J. Biol. Chem.* **272,** 2212.
37. Sztul, E. S., Hendrick, J. P., Kraus, J. P., Wall, D., Kalousek, F., and Rosenberg, L. E. (1987). *J. Cell Biol.* **105,** 2631.
38. Horwich, A. L., Kalousek, F., Fenton, W. A., Furtak, K., Pollock, R. A., and Rosenberg, L. E. (1987). *J. Cell Biol.* **105,** 669.
39. Ou, W. J., Kumamoto, T., Mihara, K., Kitada, S., Niidome, T., Ito, A., and Omura, T. (1994). *J. Biol. Chem.* **269,** 24673.
40. Mooney, D. T., Pilgrim, D. B., and Young, E. T. (1990). *Mol. Cell Biol.* **10,** 2801.
41. Niidome, T., Kitada, S., Shimokata, K., Ogishima, T., and Ito, A. (1994). *J. Biol. Chem.* **269,** 24719.
42. Jarvis, J. A., Ryan, M. T., Hoogenraad, N. J., Craik, D. J., and Hoj, P. B. (1995). *J. Biol. Chem.* **270,** 1323.
43. Hammen, P. K., Gorenstein, D. G., and Weiner, H. (1994). *Biochemistry* **33,** 8610.
44. Shimokata, K., Nishio, T., Song, M. C., Kitada, S., Ogishima, T., and Ito, A. (1997). *J. Biochem. (Tokyo)* **122,** 1019.
45. Song, M. C., Ogishima, T., and Ito, A. (1998). *J. Biochem. (Tokyo)* **124,** 1045.
46. Koutnikova, H., Campuzano, V., and Koenig, M. (1998). *Hum. Mol. Genet.* **7,** 1485.
47. Cavadini, P., Adamec, J., Gakh, O., Taroni, F., and Isaya, G. (2000). *J. Biol. Chem.* **275,** 41469.
48. Conboy, J. G., Kalousek, F., and Rosenberg, L. E. (1979). *Proc. Natl. Acad. Sci. USA* **76,** 5724.
49. Cote, C., Solioz, M., and Schatz, G. (1979). *J. Biol. Chem.* **254,** 1437.
50. Maccecchini, M. L., Rudin, Y., Blobel, G., and Schatz, G. (1979). *Proc. Natl. Acad. Sci. USA* **76,** 343.
51. Michel, R., Wachter, E., and Sebald, W. (1979). *FEBS Lett.* **101,** 373.
52. Harmey, M. A., and Neupert, W. (1979). *FEBS Lett.* **108,** 385.
53. Bohni, P., Gasser, S., Leaver, C., and Schatz, G. (1980). *In* "The Organization and Expression of the Mitochondrial Genome" (A. M. Kroon and C. Saccone, eds.). Elsevier/North-Holland, Amsterdam.
54. McAda, P. C., and Douglas, M. G. (1982). *J. Biol. Chem.* **257,** 3177.
55. Conboy, J. G., and Rosenberg, L. E. (1981). *Proc. Natl. Acad. Sci. USA* **78,** 3073.
56. Miura, S., Mori, M., Amaya, Y., and Tatibana, M. (1982). *Eur. J. Biochem.* **122,** 641.
57. Schmidt, B., and Neupert, W. (1984). *Biochem. Soc. Trans.* **12,** 920.
58. Hawlitschek, G., Schneider, H., Schmidt, B., Tropschug, M., Hartl, F. U., and Neupert, W. (1988). *Cell* **53,** 795.
59. Saavedra-Alanis, V. M., Rysavy, P., Rosenberg, L. E., and Kalousek, F. (1994). *J. Biol. Chem.* **269,** 9284.
60. Adamec, J., Gakh, O., Spizek, J., and Kalousek, F. (1999). *Arch. Biochem. Biophys.* **370,** 77.
61. Yang, M., Jensen, R. E., Yaffe, M. P., Oppliger, W., and Schatz, G. (1988). *EMBO J.* **7,** 3857.
62. Ou, W. J., Ito, A., Okazaki, H., and Omura, T. (1989). *EMBO J.* **8,** 2605.
63. Kleiber, J., Kalousek, F., Swaroop, M., and Rosenberg, L. E. (1990). *Proc. Natl. Acad. Sci. USA* **87,** 7978.
64. Braun, H. P., Emmermann, M., Kruft, V., and Schmitz, U. K. (1992). *EMBO J.* **11,** 3219.
65. Braun, H. P., Emmermann, M., Kruft, V., Bodicker, M., and Schmitz, U. K. (1995). *Planta* **195,** 396.
66. Eriksson, A. C., Sjoling, S., and Glaser, E. (1996). *J. Bioenerg. Biomembr.* **28,** 285.

67. Braun, H. P., Kruft, V., and Schmitz, U. K. (1994). *Planta* **193,** 99.
68. Conboy, J. G., Fenton, W. A., and Rosenberg, L. E. (1982). *Biochem. Biophys. Res. Commun.* **105,** 1.
69. Schmidt, B., Wachter, E., Sebald, W., and Neupert, W. (1984). *Eur. J. Biochem.* **144,** 581.
70. Yang, M. J., Geli, V., Oppliger, W., Suda, K., James, P., and Schatz, G. (1991). *J. Biol. Chem.* **266,** 6416.
71. Luciano, P., Tokatlidis, K., Chambre, I., Germanique, J. C., and Geli, V. (1998). *J. Mol. Biol.* **280,** 193.
72. Auld, D. S. (1995). *Methods Enzymol.* **248,** 228.
73. Luciano, P., and Geli, V. (1996). *Experientia* **52,** 1077.
74. Kitada, S., Kojima, K., Shimokata, K., Ogishima, T., and Ito, A. (1998). *J. Biol. Chem.* **273,** 32547.
75. Geli, V., Yang, M. J., Suda, K., Lustig, A., and Schatz, G. (1990). *J. Biol. Chem.* **265,** 19216.
76. Luciano, P., Geoffroy, S., Brandt, A., Hernandez, J. F., and Geli, V. (1997). *J. Mol. Biol.* **272,** 213.
77. Kojima, K., Kitada, S., Shimokata, K., Ogishima, T., and Ito, A. (1998). *J. Biol. Chem.* **273,** 32542.
78. Striebel, H. M., Rysavy, P., Adamec, J., Spizek, J., and Kalousek, F. (1996). *Arch. Biochem. Biophys.* **335,** 211.
79. Janata, J., Kogekar, N., and Fenton, W. A. (1997). *Hum. Mol. Genet.* **6,** 1457.
80. Yaffe, M. P., and Schatz, G. (1984). *Proc. Natl. Acad. Sci. USA* **81,** 4819.
81. Yaffe, M. P., Ohta, S., and Schatz, G. (1985). *EMBO J.* **4,** 2069.
82. Witte, C., Jensen, R. E., Yaffe, M. P., and Schatz, G. (1988). *EMBO J.* **7,** 1439.
83. Pollock, R. A., Hartl, F. U., Cheng, M. Y., Ostermann, J., Horwich, A., and Neupert, W. (1988). *EMBO J.* **7,** 3493.
84. Schneider, H., Arretz, M., Wachter, E., and Neupert, W. (1990). *J. Biol. Chem.* **265,** 9881.
85. Paces, V., Rosenberg, L. E., Fenton, W. A., and Kalousek, F. (1993). *Proc. Natl. Acad. Sci. USA* **90,** 5355.
86. Emmermann, M., Braun, H. P., Arretz, M., and Schmitz, U. K. (1993). *J. Biol. Chem.* **268,** 18936.
87. Nagase, T., Seki, N., Tanaka, A., Ishikawa, K., and Nomura, N. (1995). *DNA Res.* **2,** 167.
88. Mao, M., Fu, G., Wu, J. S., Zhang, Q. H., Zhou, J., Kan, L. X., Huang, Q. H., He, K. L., Gu, B. W., Han, Z. G., Shen, Y., Gu, J., Yu, Y. P., Xu, S. H., Wang, Y. X., Chen, S. J., and Chen, Z. (1998). *Proc. Natl. Acad. Sci. USA* **95,** 8175.
89. Zhang, M., Xie, W., Leung, G. S., Deane, E. E., and Kwan, H. S. (1998). *Gene* **206,** 23.
90. Leung, G. S., Zhang, M., Xie, W. J., and Kwan, H. S. (2000). *Mol. Gen. Genet.* **262,** 977.
91. Costa Rocha, C. R., and Lopes Gomes, S. (1998). *J. Bacteriol.* **180,** 3967.
92. Rocha, C. R., and Gomes, S. L. (1999). *J. Bacteriol.* **181,** 4257.
93. Schulte, U., Arretz, M., Schneider, H., Tropschug, M., Wachter, E., Neupert, W., and Weiss, H. (1989). *Nature* **339,** 147.
94. Braun, H. P., and Schmitz, U. K. (1997). *Int. J. Biochem. Cell Biol.* **29,** 1043.
95. Jansch, L., Kruft, V., Schmitz, U. K., and Braun, H. P. (1995). *Eur. J. Biochem.* **228,** 878.
96. Emmermann, M., Braun, H. P., and Schmitz, U. K. (1994). *Mol. Gen. Genet.* **245,** 237.
97. Rawlings, N. D., and Barrett, A. J. (1991). *Biochem. J.* **275,** 389.
98. Vallee, B. L., and Auld, D. S. (1990). *Biochemistry* **29,** 5647.
99. Affholter, J. A., Fried, V. A., and Roth, R. A. (1988). *Science* **242,** 1415.
100. Kuo, W. L., Gehm, B. D., and Rosner, M. R. (1990). *Mol. Endocrinol.* **4,** 1580.
101. Finch, P. W., Wilson, R. E., Brown, K., Hickson, I. D., and Emmerson, P. T. (1986). *Nucleic Acids Res.* **14,** 7695.
102. Rawlings, N. D., and Barrett, A. J. (1995). *Methods Enzymol.* **248,** 183.

103. Becker, A. B., and Roth, R. A. (1992). *Proc. Natl. Acad. Sci. USA* **89,** 3835.
104. Ebrahim, A., Hamel, F. G., Bennett, R. G., and Duckworth, W. C. (1991). *Biochem. Biophys. Res. Commun.* **181,** 1398.
105. Becker, A. B., and Roth, R. A. (1993). *Biochem. J.* **292,** 137.
106. Ding, L., Becker, A. B., Suzuki, A., and Roth, R. A. (1992). *J. Biol. Chem.* **267,** 2414.
107. Kitada, S., Shimokata, K., Niidome, T., Ogishima, T., and Ito, A. (1995). *J. Biochem. (Tokyo)* **117,** 1148.
108. Ito, A. (1999). *Biochem. Biophys. Res. Commun.* **265,** 611.
109. Shimokata, K., Kitada, S., Ogishima, T., and Ito, A. (1998). *J. Biol. Chem.* **273,** 25158.
110. Xia, D., Yu, C. A., Kim, H., Xia, J. Z., Kachurin, A. M., Zhang, L., Yu, L., and Deisenhofer, J. (1997). *Science* **277,** 60.
111. Zhang, Z., Huang, L., Shulmeister, V. M., Chi, Y. I., Kim, K. K., Hung, L. W., Crofts, A. R., Berry, E. A., and Kim, S. H. (1998). *Nature* **392,** 677.
112. Iwata, S., Lee, J. W., Okada, K., Lee, J. K., Iwata, M., Rasmussen, B., Link, T. A., Ramaswamy, S., and Jap, B. K. (1998). *Science* **281,** 64.
113. Kalousek, F., Hendrick, J. P., and Rosenberg, L. E. (1988). *Proc. Natl. Acad. Sci. USA* **85,** 7536.
114. Isaya, G., Miklos, D., and Rollins, R. A. (1994). *Mol. Cell Biol.* **14,** 5603.
115. Isaya, G., Sakati, W. R., Rollins, R. A., Shen, G. P., Hanson, L. C., Ullrich, R. C., and Novotny, C. P. (1995). *Genomics* **28,** 450.
116. Chew, A., Buck, E. A., Peretz, S., Sirugo, G., Rinaldo, P., and Isaya, G. (1997). *Genomics* **40,** 493.
117. Branda, S. S., Yang, Z. Y., Chew, A., and Isaya, G. (1999). *Hum. Mol. Genet.* **8,** 1099.
118. Nett, J. H., Schagger, H., and Trumpower, B. L. (1998). *J. Biol. Chem.* **273,** 8652.
119. Nett, J. H., and Trumpower, B. L. (1999). *J. Biol. Chem.* **274,** 9253.
120. Isaya, G., Kalousek, F., and Rosenberg, L. E. (1992). *J. Biol. Chem.* **267,** 7904.
121. Isaya, G., Kalousek, F., and Rosenberg, L. E. (1992). *Proc. Natl. Acad. Sci. USA* **89,** 8317.
122. Japa, S., and Beattie, D. S. (1992). *FASEB J.* **6,** Abstr. No 1318.
123. Graham, L. A., Brandt, U., Sargent, J. S., and Trumpower, B. L. (1993). *J. Bioenerg. Biomembr.* **25,** 245.
124. Kalousek, F., Isaya, G., and Rosenberg, L. E. (1992). *EMBO J.* **11,** 2803.
125. Isaya, G., and Kalousek, F. (1995). *Methods Enzymol.* **248,** 556.
126. Chew, A., Rollins, R. A., Sakati, W. R., and Isaya, G. (1996). *Biochem. Biophys. Res. Commun.* **226,** 822.
127. Barrett, A. J., Brown, M. A., Dando, P. M., Knight, C. G., McKie, N., Rawlings, N. D., and Serizawa, A. (1995). *Methods Enzymol.* **248,** 529.
128. Kato, A., Sugiura, N., Saruta, Y., Hosoiri, T., Yasue, H., and Hirose, S. (1997). *J. Biol. Chem.* **272,** 15313.
129. Chew, A., Sirugo, G., Alsobrook, J. P., 2nd., and Isaya, G. (2000). *Genomics* **65,** 104.
130. Pierotti, A., Dong, K. W., Glucksman, M. J., Orlowski, M., and Roberts, J. L. (1990). *Biochemistry* **29,** 10323.
131. McKie, N., Dando, P. M., Rawlings, N. D., and Barrett, A. J. (1993). *Biochem. J.* **295,** 57.
132. Hamilton, S., and Miller, C. G. (1992). *J. Bacteriol.* **174,** 1626.
133. Conlin, C. A., Vimr, E. R., and Miller, C. G. (1992). *J. Bacteriol.* **174,** 5869.
134. Conlin, C. A., and Miller, C. G. (1992). *J. Bacteriol.* **174,** 1631.
135. Buchler, M., Tisljar, U., and Wolf, D. H. (1994). *Eur. J. Biochem.* **219,** 627.

5

Chloroplast and Mitochondrial Type I Signal Peptidases

CHRISTOPHER J. HOWE • KEVIN A. FLOYD

Department of Biochemistry and Cambridge Centre for Molecular Recognition
University of Cambridge, UK

I. Chloroplast Type I Signal Peptidase: The Thylakoidal Processing Peptidase

A. INTRODUCTION

Chloroplasts are organelles bounded by a pair of membranes called the envelope. Within the organelle are a soluble phase, the stroma, and a series of connected, flattened membrane sacs called the thylakoids. The thylakoids are the site of the light reactions of oxygenic photosynthesis. These membranes

THE ENZYMES, Vol. XXII
Copyright © 2001 by Academic Press

are often stacked closely together as structures visible under the light micro-scope and termed grana. The membranes connecting the grana are termed intergranal lamellae (1). Chloroplasts contain a type I signal peptidase activ-ity that removes cleavable signal sequences from proteins targeted through or to the thylakoid membrane. The peptidase is therefore usually referred to as the thylakoidal processing peptidase. Early studies on the effects of pea chloroplast extracts on translation products from poly A^+-RNA of wheat and barley suggested that cytoplasmically synthesized proteins destined for the chloroplast (including plastocyanin, which is found in the thylakoid lumen) are synthesized as precursors that are cleaved to their mature form in a single step by a soluble, metal-ion dependent protease located in the stroma and perhaps also in the thylakoid lumen (2). Smeekens et al. (3) cloned and sequenced a cDNA for plastocyanin from Silene pratensis (white campion) and identified an N-terminal region of the pri-mary translation product that was not present in the mature protein. They identified a hydrophobic stretch within this N-terminal extension that was not found in stromal proteins and suggested this might be in-volved in thylakoid targeting. Subsequent experiments on the import of radiolabeled precursors into isolated chloroplasts (4) indicated that plas-tocyanin was processed to the mature form through a stromal interme-diate rather than in a single step. Experiments on the processing in vitro of the Silene plastocyanin precursor confirmed that incubation of the pre-cursor with a processing peptidase partially purified from the stroma gave a product of the same size as the stromal intermediate. This intermedi-ate could be further processed to a form with the mature size by a pro-tease that was released from thylakoids by treatment with Triton X-100 (5). The conflict between the initial experiments suggesting a single pro-cessing event to the mature size by a stromal enzyme (2) and the later experiments demonstrating a stromal intermediate processed by a sec-ond, thylakoidal, protease was accounted for by the anomalous mobility in SDS–PAGE of wheat plastocyanin in the former experiments. Hage-man et al., noting that the thylakoid transfer domain resembled a prokary-otic signal peptide, suggested that the thylakoidal protease was function-ing analogously to the prokaryotic signal peptidase (5). Thus the model would be for nuclear encoded polypeptides of the thylakoid lumen to be synthesized with a bipartite presequence. The first stage would direct im-port into the stroma, where it would be removed by a stromal peptidase (referred to as the stromal processing peptidase). The second stage, anal-ogous to a bacterial leader sequence, would direct passage across the thy-lakoid membrane and be removed by a peptidase in the thylakoid. The latter activity is thus routinely referred to as the thylakoidal processing peptidase or TPP.

B. BIOCHEMICAL FEATURES

1. *Assay and Purification*

TPP activity is routinely assayed by the ability to cleave precursor proteins *in vitro*. These may be full precursors, carrying the bipartite sequence, or stromal intermediates. The cleavage is assessed by the increase in mobility under SDS–PAGE. Radiolabeled substrates are routinely used, generated by transcription *in vitro* of sequences encoding precursors of thylakoid lumen proteins and translation of the RNA *in vitro*, usually in a wheat germ extract in the presence of ^{35}S-methionine or other appropriately labeled amino acid. The most commonly used precursors are those of plastocyanin and the 23 kDa extrinsic polypeptide of Photosystem II. The precursor to the extrinsic 33 kDa polypeptide of Photosystem II is also efficiently cleaved *in vitro*, but the precursor for the extrinsic 16 kDa is not (6), perhaps through adopting an unfavorable conformation. Cleavage of the full-length precursors (bearing the chloroplast targeting sequence as well as the thylakoid signal sequence) may be slower than cleavage of the stromal intermediate forms (7). For some substrates it is necessary for the peptidase to be present while they are being synthesized for cleavage to occur. It is possible that these substrates adopt a conformation that is unfavorable for cleavage. The plastid-encoded cytochrome *f* is synthesized with a leader sequence that can be cleaved by the *E. coli* leader peptidase (8), so the enzyme may also cleave precursors that are synthesized inside the chloroplast, as well as those that are imported. Whether or not this is the case is an important outstanding question.

A partial purification of the enzyme was described by Kirwin *et al.* (7). Membranes from lysed chloroplasts were extracted in the dark with Triton X-100 added to a ratio of 2.5 mg Triton/mg of chlorophyll. A 30,000 *g* supernatant was then applied to a hydroxylapatite column and eluted with buffer containing 0.15% v/v Triton X-100. The activity was further purified on DEAE-Sephacel. Overall a 36-fold purification was achieved with a 7% yield. The majority of other thylakoid proteins were removed, although the resulting preparation contained 10–15 polypeptides as judged by silver staining.

2. *Characteristics*

The purified enzyme was found to have a pH optimum of 6.5 to 7, but to retain activity outside this range. This pH optimum is rather lower than the optimum of 8.5 to 9 reported for the stromal processing peptidase (7), presumably reflecting a lowering of the pH of the thylakoid lumen during photosynthetic electron transfer. The enzyme was shown to be resistant to a wide range of protease inhibitors, including the serine protease inhibitor phenylmethylsulfonyl fluoride as well as leupeptin, iodoacetic acid, iodoacetamide, pepstatin, bestatin, N-ethylmaleimide, benzamidine hydrochloride, and the chelating agents 1,10-phenanthroline, EDTA, and EGTA. Indeed, the last three stimulated enzyme activity (7, 9). The enzyme was later shown to be inhibited by a 19-residue synthetic consensus leader peptide that also inhibited the E. coli leader peptidase (10). The enzyme was also shown to be inhibited by the penem compound SB214357 (11). This family of compounds contains a beta-lactam ring and many of its members are inhibitors of the Escherichia coli leader peptidase (12).

The activity shows high specificity for thylakoid lumen proteins. It was shown to be inactive against a range of other proteins and the purified activity did not degrade the Silene pratensis plastocyanin beyond the cleavage site for the mature protein (7). The peptidase activity from one plant species will in general process polypeptides from other species. The thylakoidal processing peptidase from the cyanobacterium (oxygenic photosynthetic bacterium) Phormidium laminosum was also shown to process lumen protein precursors from higher plants, although it was not clear if the plant enzyme accurately processed cyanobacterial precursors (13). Nevertheless, the plant enzyme has also been shown to cleave precursors of proteins exported either through the endoplasmic reticulum (yeast prepro-alpha factor) or the bacterial plasma membrane (10). Halpin et al. (10) examined the proposal of Hageman et al. (5) that the TPP activity was analogous to a bacterial signal peptidase. This was done by synthesizing the precursor of the 23 kDa extrinsic polypeptide of Photosystem II from wheat in the presence of ^{3}H-lysine, cleaving samples of the radiolabeled polypeptide with the thylakoidal processing peptidase or with E. coli leader peptidase and subjecting the cleavage products to sequential Edman degradation. Identification of the cycles in which radiolabel was released and comparison with the known sequence allowed the cleavage sites to be identified and showed that both activities cleaved the protein at the same site. Subsequently, Howe and Wallace (14) suggested that there may be minor differences in the cleavage specificities of the bacterial and thylakoidal enzymes, on the grounds that an algorithm for predicting bacterial leader peptidase cleavage sites (15) performed much less well with chloroplast thylakoid lumen precursors than with bacterial sequences. The algorithm correctly predicted 89% of bacterial cleavage sites, but only 58% of thylakoid peptidase cleavage sites. Howe and Wallace (14) produced a modified algorithm for prediction of thylakoid lumen precursor cleavage sites. (Interestingly, even this algorithm did not perform well in predicting cleavage sites for chloroplast-encoded intrinsic polypeptides such as cytochrome f, raising the possibility that they might be processed by a different enzyme.) Shackleton and Robinson (16) carried out a more systematic survey of the cleavage site requirements of the thylakoidal processing peptidase by studying the processing of a series of site-directed mutants at the -3 and -1 positions (both occupied by Ala in the wild-type protein) of the precursor of the 33 kDa extrinsic Photosystem II polypeptide from wheat. Radiolabeled precursors were incubated with isolated chloroplasts, which were subsequently fractionated. Protease protection was used to determine whether proteins had been targeted to the thylakoid lumen. Replacement of -3Ala by Val had no effect on targeting of the precursor to the lumen and processing to the mature size. Replacement of -3Ala by Lys, Leu, or Glu still allowed the precursor to be targeted to the lumen but gave rise to varying amounts of the mature 33 kDa polypeptide and a 36 kDa species, which was interpreted as the result of processing at a site upstream from the wild-type one, with a reasonable fit to a consensus cleavage site. Fractionation data suggested that the 36 kDa species was more tightly bound to the thylakoid membrane than the wild-type protein, consistent with a part of the signal sequence remaining as a membrane anchor. For the Glu substitution roughly similar amounts of the 33 kDa and 36 kDa forms were found at the end of the assay. For the Lys and Leu substitutions the 36 kDa form predominated, suggesting that their effect on the thylakoidal processing peptidase was more significant. All the -1 substitutions tested affected processing. Substitution to Gly or Ser generated significant amounts of the 33 kDa and 36 kDa forms (the Gly substitution having a less pronounced effect than the Ser). Substitution to Leu, Lys, Glu, or Thr abolished production of the 33 kDa form. More detailed analysis of the kinetics of processing over shorter periods confirmed that the rate of processing was greatly slowed in all cases other than the substitution to valine at -3. These results therefore indicated a much more stringent sequence requirement for the thylakoidal processing peptidase than is observed for the *E. coli* leader peptidase or for the eukaryotic signal peptidase, with even relatively conservative substitutions at the -1 position (G,S) significantly slowing processing by the thylakoid enzyme, and others (E,K,L,T) stopping it almost completely. [On the basis of a proteomic analysis of the cyanobacterium *Synechocystis* sp. PCC6803 it has been suggested that Ala is much more common at the -1 position in polypeptides cleaved by the cyanobacterial signal peptidase than is the case with *E. coli* signal peptides (17). This suggests that the tight stringency of the cleavage site compared to the *E. coli* enzyme predates the origin of chloroplasts.] Although the predicted sequence of the *Arabidopsis thaliana* enzyme is now available from the DNA sequence (18), it is not clear what the basis of the increased stringency is in the absence of detailed structural information for the plant enzyme. The two residues in the *E. coli* enzyme that seem to be most directly associated with the -1 residue of the cleavage site, Ser-88 and Leu-95 (19), are conserved in the *A. thaliana* enzyme. However, the replacement of Gly-89 and Met-92 of the *E. coli* enzyme by Thr and Tyr in the otherwise conserved active site region may be significant. Understanding how the increased stringency is determined is an important outstanding question, which will presumably require detailed 3D structural information for the thylakoidal enzyme.

3. *Organization*

Two approaches have been taken to studying the organization of the TPP in the membrane. The first has been to study the enzymatic activity, and the second has been to use polyclonal antibodies raised to protein expressed using the cDNA clone isolated by Chaal *et al.* (*18*). Early experiments indicated that the TPP was an intrinsic membrane protein with the active site directed toward the thylakoid lumen, since isolated intact thylakoids did not show TPP activity and detergents such as Triton X-100 are required to solubilize the enzyme (*5*). A more detailed study of the organization of the TPP in pea was described by Kirwin *et al.* (*20*). Prolonged sonication of thylakoid membranes caused the alpha and beta subunits of ATP synthase, the 33, 23, and 16 kDa polypeptides of Photosystem II, and plastocyanin to be dislodged from the membranes, while the TPP activity remained associated with the membrane fraction, consistent with an intrinsic location. The supernatant after Triton X-100 extraction and centrifugation at 30,000 g was analyzed by sucrose density gradient centrifugation. The TPP activity did not cofractionate with any of the major membrane protein complexes and remained close to the top of the sucrose density gradient, indicating that (under these circumstances at least) it was not associated with a large complex. However, it is possible that the Triton extraction disrupts other interactions of the enzyme. Interestingly, a comparison of the behavior in sucrose gradients of TPP activity from pea and wheat thylakoids showed that with the latter the TPP moved further down the gradient (*9*). It is also worth noting that pelleting of the TPP activity by centrifugation at 200,000 g led to some loss of activity, which may have been due to loss of some associated protein (*20*). Given the organization of the mitochondrial type I signal peptidase into a heterooligomeric complex, and the observation that there are multiple TPP genes in *A. thaliana* (see later discussion), establishing the organization of the enzyme in the thylakoid membrane remains an important goal.

Prolonged sonication of thylakoids in the presence of the precursor of *S. pratensis* plastocyanin resulted in the processing of the precursor to the mature size (*20*), consistent with an exposure of the active site of the enzyme in the thylakoid lumen (with the precursor gaining access to the lumen through transient shearing of the membrane occurring during sonication). This was consistent with experiments using trypsin treatment, which inactivated the TPP activity of thylakoid membranes only if they were treated with Triton X-100 to permeabilize them at the same time. Fractionation of thylakoids into granal lamellae and stromal lamellae by digitonin treatment followed by assay of TPP activity indicated that it was restricted to the stromal lamellae (20).

The availability of sequence data from a cDNA clone of the TPP from *A. thaliana* (as described later) and antibodies raised against protein expressed from the cDNA has provided further insights into the organization of the protein in the membrane (18, 21). The polypeptide is predicted to be synthesized as a precursor of ca. 38 kDa, with an N-terminal region likely to function as a chloroplast import domain. When Western blots of chloroplast fractions were probed with antibodies raised against expressed proteins, reaction with a polypeptide of ca. 30 kDa was observed. This polypeptide was absent from a stromal preparation and could be partially, but not completely, extracted from thylakoids by the standard treatment with Triton X-100 used for extracting TPP activity. Inspection of the predicted amino acid sequence indicated that the protein possessed a single transmembrane span close to the N terminus of the mature protein (assuming cleavage of a chloroplast import signal). It would therefore be expected that the N terminus would be exposed in the stroma for the C terminus to be in the thylakoid lumen. Treatment of thylakoids with increasing concentrations of thermolysin reduced the apparent size of the cross-reacting protein to ca. 28 kDa and then ca. 23 kDa, consistent with thermolysin cleavage close to the N terminus of the transmembrane span, and indicating that some 50–60 residues of the N-terminal region of the mature protein might be exposed in the stroma. Interestingly this region contains a stretch of 19 residues containing a large fraction of charged amino acids (8 aspartate, 1 glutamate, 1 arginine, and 2 lysine residues). It is possible that this may be involved in interaction with other polypeptides. Some 160 resides of the C terminus of the protein would be exposed in the thylakoid lumen, including the presumed catalytic site. The catalytic site is located close to the transmembrane domain, as discussed later, and therefore close to the membrane surface, and is presumably consequently well placed to cleave polypeptide chains as they emerge through the membrane.

4. *Mechanism and Structure*

The amino acid sequence predicted from the *A. thaliana* cDNA sequence shows clearly that the TPP is homologous to other type I signal peptidases and belongs to the serine–lysine catalytic dyad group of peptidases. This is consistent with its inhibition by the penem SB214357. A number of sequence motifs found in other signal peptidases (22) are present in the *A. thaliana* enzyme (Fig. 1) (18). These have been denoted domains A–E. A is the transmembrane spanning domain. B contains a completely conserved dipeptide -SM-. The equivalent serine residue has been shown in other organisms, such as *F. coli* (Ser-90), to be essential for catalytic activity (23), and it seems likely

```
E.coli
Chr III                                                 MVNKITNNYIFPRKPLFQLMMVMISLHF----    28
Chr II     MAIRITFTYSTHVARNLV---GTRVGPGGY--CFESLVRPRFFSHKRD---FDRSP----    48
Chr I      MAIRVTFTYSSYVARSIASSAGTRVGTGDVRSCFETWVRPRFCGHNQIPDIVDKSPGSNT    60


E.coli                                                         MANMFALILVI    11
Chr III    -------STPPLAFLKSDSNSRFLKNPNPNFIQFTPKSQLLFPQRLNFNTGTNLNRRTLS    81
Chr II     -------RNRPAS-MYGSIARELIGEGSQSPLVMGLISILKSTTGHESSTMNVLGVSSFK   100
Chr I      WGPSSGPRARPASSMYSTIAREILEEGCKSPLVLGMISLMNLTGAPQFSGMTGLGISPFK   120
                                                              .  :       .


E.coli     ATLVTGILWCVDKFFFA----------PKRRERQAA------AQAARDSLDKATLKKVAP    55
Chr III    CYGIKDSSETTKSAPSL-------DSGDGGGGDGGD-------DDKGEVEEKNR---LFP   124
Chr II     ASSIIPFLQGSKWIKN--PPVID------DVDKGGTVCD-----DDDDKESRNGGSGWVN   147
Chr I      TSSVIPFLRGSKWMPCSIPATLST--DIAEVDRGGKVCDPKVKLELSDKVS-NGGNGWVN   177
               :        .                .                  .


E.coli     K--PGWLETGASVFPVLAIVLIVRSFIYEPFQIPSGSMMPTLLIGDFILVEKFAYGIKDP   113
Chr III    EWLDFTSDDAQTVFVAIAVSLAFRYFIAEPRYIPSLSMYPTFDVGDRLVAEKVSY-----   179
Chr II     KLLSVCSEDAKAAFTAVTVSILFRSALAEPKSIPSTSMYPTLDKGDRVMAEKVSY-----   202
Chr I      KLLNICSEDAKAAFTAVTVSLLFRSALAEPKSIPSTSMLPTLDVGDRVIAEKVSY-----   232
               :       :  .  :.*  .:::  :  .*     :  **    *** ** **:   ** ::.**.:*
                                          IPS-SM-PTldvG
                                          DOMAIN B


E.coli     IYQKTLIENGHPKRGDIVVFKYPEDP----------KLDYIKRAVGLPGDKVTYDPVSKE   163
Chr III    -------YFRKPCANDIVIFKSPPVLQEV---GYTDADVFIKRIVAKEGDLVEVH--NGK   227
Chr II     -------FFRKPEVSDIVIFKAPPILLEYPEYGYSSNDVFIKRIVASEGDWVEVR--DGK   253
Chr I      -------FFRKPEVSDIVIFKAPPILVEH---GYSCADVFIKRIVASEGDWVEVC--DGK   280
                   :*  .***:** *               ::*** *.  ** *      .
                        -DIViFK-P                      vfIKRiVa-eGD
                        DOMAIN C                          DOMAIN D


E.coli     LTIQPGCSSGQACENALPATYSNVEPSDFVQTFSRRNGGEATSGFFEVPKNETKENGIRL   223
Chr III    LMVN-GVARNEKFILEPPGYEMT-------------------------------------   249
Chr II     LFVN-DIVQEEDFVLEPMSYEMEP------------------------------------   276
Chr I      LLVN-DTVQAEDFVLEPIDYEMEP------------------------------------   303
               * ::. .     :


E.coli     SERKETLGDVTHRILTVPIAQDQVGMYYQQPGQQLATWIVPPGQYFMMGDNRDNSADSRY   283
Chr III    ----------------------------------------PVPENSVFVMGDNRNNSYDSHV   271
Chr II     ------------------------M-----------FVPKGYVFVLGDNRNKSFDSHN   299
Chr I      ------------------------M-----------FVPEGYVFVLGDNRNKSFDSHN   326
                                                    **   .  *::****::* *:
                                                            GDNRn-S-DSh
                                                            DOMAIN E


E.coli     WGFVPEANLVG--RATAIWMSFDK--QEGEWPTGLRLSRIGGIH                 323
Chr III    WGPLPLKNIIG--RSVFRYWPPNRVSGTVLEGGCAVDKQ                      308
Chr II     WGPLPIENIVG--RSVFRYWPPSKVSDTIYHDQAITRGPVAVS                  340
Chr I      WGPLPIKNIIG--RSVFRYWPPSKVSDIIHHEQVSQKRAVDVS                  367
           ** :*  *::*  *:.   :  . :
```

Fig. 1. Aligned predicted amino acid sequences of TPPs from *A. thaliana*. The figure shows the amino acid sequences inferred from coding sequences on chromosomes I, II, and III aligned together with the predicted amino acid sequence of the *E. coli* enzyme. The figure also shows the conserved features of domains B–D (*22*).

to provide a nucleophile for proteolytic attack. Domain C is less well conserved and typically contains a -GD- dipeptide. In the published *A. thaliana* sequence (*18*) the Gly is replaced by Ser. Domain D contains a highly conserved -KR- motif, which is also present in *A. thaliana*. The Lys may function as a base for abstraction of the proton from the catalytic serine. Domain E has a highly conserved -GDN- tripeptide followed shortly by -S-DS-. The role of residues in this region in *E. coli* has been probed by site-directed mutagenesis (*24*). Gly-272 (in the *E. coli* numbering) is proposed to play a structural role, being adjacent in the three-dimensional structure to the Ser-Lys dyad. Substitution of Gly-272 by Ala greatly reduced activity, perhaps through distortion of the region by increasing the size of the sidechain. The equivalent position of Gly-272 is conserved in the *A. thaliana* enzyme. The first of the Asp residues of domain E (Arg-273) is also essential for catalytic activity in the *E. coli* enzyme, possibly to form a salt bridge with Arg-146 in domain D (*23, 24*). When Arg-146 is mutated, Asp-273 becomes less important. The equivalent positions in the *A. thaliana* enzyme are conserved, so the salt bridge is likely to be present also. An Asn-274-Ala substitution in domain E had little effect on the *E. coli* enzyme. A Ser-278-Ala substitution led to a major loss of activity, possibly because of the loss of the hydrogen bond to the catalytic lysine. Again, these residues are conserved in *A. thaliana*. Asp-280 *E. coli* mutants also had reduced activity, indicating an important role for the Asp-280–Arg-282 interaction, though the *A. thaliana* enzyme has His in the corresponding position to Arg-282.

The expressed catalytic domain of *A. thaliana* TPP exhibited rather low activity, even when five residues of the transmembrane span were included (*21*) in the protein and refolding techniques were employed. This contrasts with the situation in *E. coli,* where an active catalytic domain could be obtained (*25*). The activity of the *E. coli* catalytic domain was enhanced by the presence of phospholipids (*25*) and phosphatidylethanolamine has been shown to mediate insertion of the catalytic domain into membranes (*26*). However, a soybean phospholipid preparation did not enhance the activity of the chloroplast domain (*21*). The reason for the lack of activity of the expressed TPP catalytic domain is not clear. Other than simple explanations such as incorrect folding (and a failure to refold), it is possible that the association of the TPP with the thylakoid membrane is critical, perhaps to allow the substrate molecule to be presented in the correct orientation. It is worth noting that the proposed catalytic serine is only a few residues from the end of the transmembrane domain. It is also possible that the chloroplast enzyme functions in concert with other proteins. As discussed earlier, the behavior of the Triton-extracted protein in sucrose density gradient centrifugation would be consistent with the protein being part of a small complex, and the loss of activity when the TPP is pelleted by centrifugation at 200,000 *g* would also be consistent with the peptidase requiring some other factor for activity. The mitochondrial homolog, the inner membrane peptidase, comprises two subunits (both with peptidase activity) (27) as described later. Additional genes for TPP homologs have been identified in *A. thaliana,* which would be consistent with such a model. Even if the TPP does not form a stable complex with other polypeptides, it is of course still possible that a transient interaction would be needed for function.

C. MOLECULAR GENETICS

1. *A. thaliana cDNA Cloning*

Attempts to isolate a cDNA for the TPP were unsuccessful for a long time, owing to the difficulty of purifying the enzyme. However, Chaal *et al.* (*18*) identified a clone from an EST library of *A. thaliana* that showed clear sequence similarity to a cyanobacterial type I signal peptidase (*28*) and subsequently obtained a cDNA covering the complete open reading frame. A polypeptide synthesized *in vitro* from the cDNA was imported into isolated pea chloroplasts and processed to a lower molecular weight form, antibodies raised against overexpressed protein cross-reacted with a thylakoid membrane protein, and the overexpressed protein cleaved the stromal intermediate form of the 23 kDa extrinsic protein of wheat Photosystem II to its mature size. These observations, along with a phylogenetic analysis, indicated that the polypeptide encoded by the cDNA was indeed the thylakoidal processing peptidase. Possible TPP cDNAs have subsequently been reported from a number of plant species.

Exactly which initiation codon in the TPP message is used was not clear. The first ATG in the cDNA corresponds to an open reading frame of six codons, yet has a favorable context for initiation (*29*). The TPP open reading frame begins at the second ATG, although it is possible that it initiates further still from the 5′ end of the mRNA. Chaal (*21*) suggested that, because of the favorable context of the first ATG, it was unlikely that leaky scanning was responsible for initiation at the second ATG, and that reinitiation after the short open reading frame was more likely. Short open reading frames upstream of major ones have been reported before in plants (*30*), and Chaal suggested that the short open reading frame in the TPP message might be involved in modulation of translation.

2. *A. thaliana TPP Genes*

The gene corresponding to the cDNA described by Chaal (*21*) maps to chromosome II. More recently, two further putative TPP sequences have been reported from *A. thaliana.* One has been identified both from cDNA and genomic DNA (KAF, unpublished) and maps to chromosome III. The third is from genomic DNA alone and maps to chromosome I. The predicted aligned amino acid sequences are shown in Fig. 1. All the conserved residues referred to in the discussion of mechanism and structure are found in the additional sequences. The positions of introns in the genes can be determined directly by comparison with cDNA or by alignment of predicted amino acid sequences and they are shown in Fig. 2. They occur in exactly the same positions in relation to the encoded amino acid sequence in all three genes, although the second intron is absent from the chromosome III gene. The chromosome II and chromosome I sequences appear to have diverged from each other more recently than either has from the chromosome III sequence (see Fig. 3). Determining whether they have different roles within the chloroplast is an important outstanding question. Given the observation that the mitochondrial signal peptidase functions as a heterodimer (see later discussion), it is tempting to suggest that the different genes may allow the same for the chloroplast enzyme. Other possibilities are that the different forms function in different plastid types or in different developmental stages. It is also possible that one of the forms may be located in the inner chloroplast envelope (the topologically equivalent location to the thylakoid membrane), although there is as yet no clear evidence of a TPP-like activity being present there.

3. *TPP Sequences from Other Plants*

A number of TPP sequences are now available, generally as ESTs, from other plants. These include cDNA sequences from *Medicago trunculata* (developing stem), *Solanum tuberosum* (stolon), *Lotus japonicus* (whole plant), *Lycopersicon esculentum* (callus), and *Glycine max* (whole plant). Phylogenetic analysis of these sequences suggest that they all group with the chromosome III gene from *A. thaliana* (Fig. 3), but this may be simply a reflection of the limited sample size. It is interesting that cDNA sources include nonphotosynthetic material, indicating that TPP or a close homolog is present in other plastid types.

II. Mitochondrial Type I Signal Peptidase: The Mitochondrial Inner Membrane Protease

A. INTRODUCTION

Like the chloroplast, the mitochondrion is bounded by two membranes. The inner one is generally invaginated to form cristae and encloses a soluble phase called the matrix, although there also exist contact sites between

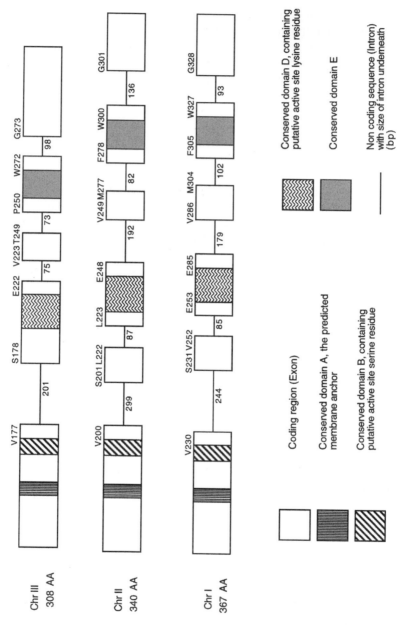

Fig. 2. Positions of introns in the TPP genes from *A. thaliana*. The figure shows the position of introns relative to the regions encoding conserved domains, and the positions in the coding sequence where the introns are located. Not to scale.

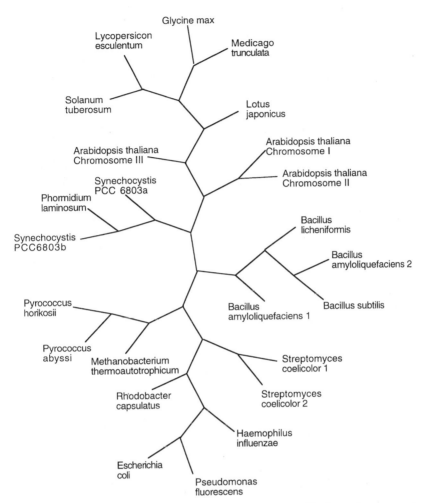

FIG. 3. Phylogenetic analysis of TPP sequences from *A. thaliana*. The figure shows an analysis of TPP amino acid sequences from plants, and homologs from cyanobacteria (*Phormidium laminosum, Synechocystis* sp. PCC6803—both genes) and other bacteria. It was constructed using a distance matrix method under PAUP4.0 (*31*) on aligned sequences with gaps excluded. A parsimony analysis gave the same topology. Branch lengths are arbitrary (K.A.F., unpublished).

the inner and outer membranes. The compartments can be fractionated by procedures such as osmotic shock or digitonin treatment (*32*). The organelle contains a type I signal peptidase activity that removes signal-like targeting sequences from precursors of proteins of the intermembrane space. Gasser *et al.* (*33*) reported experiments on the biogenesis of cytochrome b_2 in

Saccharomyces cerevisiae. This cytochrome is a soluble enzyme containing heme and flavin, which functions as a L-lactate cytochrome c oxidoreductase. It is encoded in the nucleus and synthesized outside the mitochondrion, but ultimately located in the intermembrane space, and Gasser *et al.* (*33*) demonstrated that it was processed to the mature form in two stages. Processing to an intermediate was carried out by a protease located in the matrix that could be inhibited by chelating agents such as *o*-phenanthroline. Processing to the mature form was due to a different protease. Cytochrome c_1, a component of the cytochrome bc_1 complex of the inner membrane, which is also nuclear encoded, was likewise shown to be processed to an intermediate by the matrix protease before being converted to the mature form. That the intermediates of both cytochromes were formed *in vivo* was demonstrated by pulse-chase experiments. Similar observations were made using isolated mitochondria (*34*). A model was proposed in which proteins of the intermembrane space were synthesized with a bipartite presequence. The first part of the presequence directed import into the matrix of the mitochondrion and the second part directed the protein to the intermembrane space. Targeting was proposed to involve an initial "detour" (*33*) to expose at least a part of the protein in the matrix, where the first cleavage took place, followed by a second cleavage by a protease, probably located in the inner membrane, to generate the mature protein. Gasser *et al.* (*33*) also noted a similarity to the export of bacterial proteins. [Interestingly, for cytochrome c_1, the second cleavage is dependent on heme (*35*). The reason for this dependence is not clear. It may be that the incorporation of heme is required for the polypeptide to achieve the correct structure for cleavage. An alternative explanation is that the polypeptide is sequestered by the heme lyase before it becomes available to the peptidase.]

A similar model was proposed for another nuclear-encoded component of the intermembrane space, cytochrome c peroxidase, a heme-containing protein transferring electrons to peroxides from cytochrome c (*36*). In this case, however, the intermediate was not detected, although it was suggested this might be a consequence of the kinetics of processing and import. The nucleotide sequences of genes for these three proteins were consistent with the bipartite presequence model in that the mature proteins were preceded by hydrophobic stretches preceded by one or more basic residues (*37*). Experiments on the targeting of fusion proteins have also supported the model (e.g., *38*).

The exact pathway of sorting to the intermembrane space remains controversial. It has been discussed elsewhere (*39*) and so will not be discussed in detail here. There are two models, which differ in whether the protein in question fully enters the mitochondrial matrix. In the "stop-transfer" model the region corresponding to the mature part of the protein passes only across the outer membrane, and passage across the inner membrane is stopped by the inner membrane targeting signal peptide (40). Cleavage of the latter releases the polypeptide into the intermembrane space. In the second model the protein is imported fully into the matrix and then reexported across the inner membrane by the inner membrane signal. This is called the "conservative sorting" model because of the resemblance to the export of bacterial proteins. It is possible that a single model is inadequate to describe the translocation of all proteins.

Studies on subunit II of *S. cerevisiae* cytochrome oxidase, which is encoded in the mitochondrion, show that it is certainly possible for proteins in the matrix to be exported across the inner membrane. This polypeptide is synthesized as a higher molecular weight precursor and processed as it reaches the intermembrane space. Cleavage to the mature form was found to be inhibited in the nuclear mutant strain *pet ts*2858, which was able to grow on lactate at the permissive temperature, but not at the nonpermissive temperature (41). However, the cytochrome oxidase subunit II presequence was less similar to a conventional signal sequence than the presequences for the nuclear-encoded polypeptides referred to earlier, as it contains several polar residues. It was proposed that the *pet ts*2858 mutation lay in a gene either for the processing enzyme or a polypeptide regulating its activity. Surprisingly, the cytochrome oxidase II precursor also accumulated when the cells were grown at the permissive temperature, so the presence of the extension presumably did not block protein function. The presence of an N-terminal presequence in cytochrome oxidase subunit II of *S. cerevisiae* is somewhat unexpected, as the mammalian homolog lacks a leader peptide. However, deletion of the region encoding the leader peptide led to a loss of the subunit II protein in yeast, so the leader peptide seems to be indispensable in this system (42).

It was later shown (43) that the *pet ts*2858 mutation also blocked the processing of cytochrome b_2 from the intermediate to the mature form. At the nonpermissive temperature the intermediate form accumulated, whereas at the permissive temperature a mixture of intermediate and processed forms was seen. Although this mutation was therefore shown to affect processing of both a nuclear-encoded and a mitochondrially encoded polypeptide, it did not affect processing of all intermembrane space proteins, since cytochrome *c* peroxidase was unaffected at the nonpermissive temperature. Pratje and Guiard (43) noted the similarity of the cleavage sites for the cytochrome oxidase subunit II (Asn-'Asp) and the cytochrome b_2 (Asn-'Glu) and suggested that, in accordance with the effects of the *pet ts*2858 mutation, the same enzyme was responsible for processing both precursors. They also noted that the cytochrome *c* peroxidase had a rather different cleavage site (Ser-'Thr).

There are several other polypeptides that may be substrates for the inner membrane peptidase. One is the (mitochondrial) NADH-cytochrome b_5 reductase (Mcr1p) characterized from yeast mitochondria. This is the product of a single gene, but the 47 amino acid cleavable targeting sequence directs the protein both to the outer membrane and the intermembrane space and contains a hydrophobic stretch reminiscent of signal peptides (*44*). Another candidate polypeptide for cleavage by the inner membrane peptidase is a mitochondrial form of creatine kinase, which is imported into mitochondria and directed to the intermembrane space. It is synthesized as a precursor of higher molecular weight and processed to the mature form through an intermediate (*45*). The mitochondrial nucleoside diphosphate kinase of *Dictyostelium discoideum* and other organisms is also believed to be localized in the intermembrane space and is synthesized with a cleaved presequence that may include a signal peptide domain (*46*). However, not all intermembrane space polypeptides have a cleavable presequence. The bovine adenylate kinase isoform AK2, which is present in the intermembrane space and probably also the cytosol, does not appear to be synthesized with a cleavable presequence (*47*).

B.   BIOCHEMICAL FEATURES

1.   *Assay, Purification, and Characteristics*

To emphasize that the enzyme that cleaves the intermembrane space precursors (and is believed to be located in the inner membrane) is different from those functioning in the matrix (termed the mitochondrial processing peptidase MPP and the mitochondrial intermediate peptidase MIP), it has been termed the inner membrane protease (or peptidase), Imp (*48*). Characterization of an Imp activity, denoted Imp1, was reported by Schneider *et al.* (*49*). Treatment of either the precursor or matrix intermediate forms of cytochrome b_2, synthesized *in vitro*, with extracts of yeast mitochondria made with nonionic detergents did not result in processing to the mature forms. It was suggested that this was because the substrates synthesized *in vitro* were not in a suitable conformation for processing. Instead the assay used extracts from the *pet ts*2858 mutant strain, which contained the intermediate form of cytochrome b_2, as a substrate. Processing to the mature form, indicating Imp activity, was followed by immunodetection of the two forms of the cytochrome. However, this necessitated the strain used for peptidase preparation being deficient in cytochrome b_2 (otherwise the mature form would already be present), so a mutant was used that did not produce the cytochrome. Because of the complexity of the assay, many other studies on Imp have followed processing *in vivo*. The Imp activity could be solubilized by Triton

X-100, CHAPS, deoxycholate, octylglucoside, or octylpolyoxyethylene. A minimum concentration of 0.4% v/v octylpolyoxyethylene was required for solubilization, with concentrations in excess of 1% causing irreversible loss of activity. Divalent cations were required for activity, since it was inhibited by EDTA and stimulated by Mg^{2+} and Ca^{2+}. This is in contrast to TPP, although there is no evidence that the cations are directly involved in the catalytic mechanism. Acidic phospholipids such as phosphatidylserine, phosphatidylinositol, and cardiolipin were also required, especially when preparations were diluted. Phosphatidyl choline and phosphatidylethanolamine had no effect. The enzyme was inhibited by adriamycin. This inhibition can be explained by the propensity of adriamycin to bind phospholipids such as cardiolipin (50) and the inhibition of Imp could be relieved by addition of either phosphatidylserine or cardiolipin. When different mitochondrial fractions were assayed, the activity was found to be located in the inner membrane, and protease protection experiments indicated that the enzyme was exposed to the intermembrane space. It was active over a pH range of at least 5.5 to 8.5.

The *PET2858* gene presumed to encode the enzyme responsible for cleavage of cytochrome b_2 to the mature form was cloned by complementation and found to encode a 21.4 kDa polypeptide with sequence similarity to the *E. coli* leader peptidase. The protein was designated Imp1 (51), and the *ts pet*2858 mutation was found to be a G–A transition resulting in the substitution of Gly at position 88 by aspartate (52).

Subsequent purification work (53) allowed the purification of Imp as a heterodimer, consistent with other studies suggesting it functioned as a heterooligomer (see later discussion). This was achieved by a combination of octylpolyoxyethylene extraction, chromatography on Mono-Q–Sepharose, glycerol gradient centrifugation, and immunopurification (using antibodies raised against the expressed Imp1 polypeptide). The heterodimeric preparation had an apparent molecular weight of 35,000 and contained two polypeptides, of 21.4 kDa and 19 kDa, Imp1 and Imp2, respectively. Curiously, the Imp1 band was less intense than Imp2 under silver staining, but this could have been simply due to different staining properties of the proteins.

There are as yet no reports of Imp activity in plant mitochondria (54). However, targeting of intermembrane space proteins such as cytochrome c_1 still seems to be by a bipartite presequence, so the presence of Imp activity is to be expected (55).

2. Organization

A notable feature of the inner membrane peptidase is that it is organized as a heterooligomer, comprising two catalytic subunits with separate, nonoverlapping cleavage functions. This became apparent from a number of lines of evidence. One was the observation that the $ts2858$ mutation, which blocked processing of cytochrome oxidase subunit II and cytochrome b_2, did not interfere with processing of cytochrome c peroxidase, suggesting that a second peptidase was involved (43). A second followed from the cloning by complementation of the $PET2858$ gene, encoding the Imp1 peptidase (51). When this was overexpressed in yeast, no increase in Imp activity in the cytochrome b_2 processing assay was detected, indicating that a second protein as well as Imp1 was required (49). Perhaps the major line of evidence was from the identification of a mutation, designated $imp2$-1, which interfered with processing of cytochrome b_2 and with processing of cytochrome c_1 (27). Although $pet2858$ also interfered with processing of cytochrome b_2, it did not interfere with cytochrome c_1 processing and complemented the $imp2$-1 mutation, indicating that the two mutations were in different genes. In the $imp2$-1 mutant, both cytochrome b_2 and cytochrome c_1 were exposed to the intermembrane space, as judged by protease sensitivity, and cytochrome c_1 had also incorporated its heme moiety. The $IMP2$ gene was cloned by complementation, and the predicted translation product was found to have 25% sequence identity to the Imp1 protein and thus to bacterial leader peptidases (27). The purification of the Imp complex by Schneider $et\ al.$ (53), with amino acid sequencing of the constituent subunits, confirmed its heterodimeric nature. Nunnari $et\ al.$ (27) proposed that the two subunits had nonoverlapping cleavage specificities, with Imp1 processing cytochrome oxidase subunit II and cytochrome b_2 and Imp2 processing cytochrome c_1. The observation that cytochrome b_2 was not processed in the $imp2$ mutant was explained by a requirement for Imp2 in the stable expression of Imp1, suggested by the observation that the amount of Imp1 protein was much reduced in a $\Delta imp2$ strain (although the $imp1$ mRNA level was unaffected). This model was further supported by experiments using a modified $IMP2$ gene where the active site serine had been mutated to alanine to produce a stable but catalytically inactive protein. Cells with this modified gene in a $\Delta imp2$ background were able to process cytochrome oxidase subunit II and cytochrome b_2 to the mature form, consistent with a functional Imp1 activity. Noting the difference in sequence around the cleavage sites of Imp1 and Imp2 (as discussed later), Nunnari $et\ al.$ (27) suggested that the two peptidases had arisen by a gene duplication that allowed the cell to broaden their substrate specificity. How the Imp polypeptides are anchored in the membrane remains unclear, with proposed models including a single N-terminal transmembrane span for Imp1 and either a single N-terminal span or both N-terminal and C-terminal spans for Imp2 (22, 27, 56).

A number of other mutations have been identified that lead to the accumulation of incompletely processed forms of intermembrane space proteins. $S.\ cerevisiae$ cells with the $pet\ ts1402$ mutation, which affects the protein

Oxa1p, accumulate the precursor of cytochrome oxidase subunit II at nonpermissive temperatures (*57*). However, Imp1 activity is not affected, as indicated by processing of cytochrome b_2 to its mature form, and although Oxa1p is probably involved in cytochrome oxidase subunit II export, there is no evidence as yet for an association between Oxa1p and Imp1 or Imp2 (*58*). Similarly, the Cox20p protein, which is required for cytochrome oxidase biogenesis, including the processing of subunit II, is probably a chaperone rather than being intimately involved in the processing itself (*59*). The Som1 protein, by contrast, is likely to be more closely associated with the Imp1/Imp2 complex (*60*). The *SOM1* gene was originally identified in *S. cerevisiae* as a multicopy suppressor of the *pet ts*2858 mutation, and a Δ*som1* strain showed defects in processing of cytochrome oxidase subunit II, cytochrome b_2, and Mcr1, but not cytochrome c_1 or cytochrome c peroxidase (*52, 60*). The gene encodes a primary translation product of 74 residues exposed to the intermembrane space (*61*). The Som1 protein depends on Imp1 for its stability (but not vice versa) and the two proteins can be cross-linked and coimmunoprecipitated, indicating a physical association (*60*). The function of Som1 is unclear, although the presence of cysteine residues conserved in homologs in a number of species has been taken to suggest a metal ion binding function.

3. *Mechanism and Structure*

Partial or complete sequence data for Imp1 and Imp2 are available from a range of mammals, fungi, and plants. Behrens *et al.* (*51*) suggested that the N terminus of the predicted Imp1 protein had features of a matrix targeting sequence, and they showed that substitution of the N-terminal 97 residues for the targeting sequence of cytochrome oxidase subunit IV could direct the latter into the mitochondrial matrix. There was no evidence of cleavage of the targeting signal, however. The purified Imp1 of *S. cerevisiae* had an N-terminal amino acid sequence beginning at residue 2 of the predicted primary translation product (*53*), also indicating that the protein is imported into the mitochondrion without cleavage. Nunnari *et al.* (*27*) identified a putative matrix targeting signal in *S. cerevisiae* Imp2, although the apparent molecular weight of the polypeptide seen in purified preparations is consistent with the primary translation product, again suggesting that the targeting signal is not cleaved. The sequence of the Imp1 and Imp2 polypeptides shows clear similarity to other type I signal peptidases (Fig. 4). As usual for the type I signal peptidases, the polypeptides have a membrane anchor, with the catalytic domain exposed to the soluble phase, but with the active site close to the membrane anchor. A number of important residues are conserved in both Imp1 and Imp2 (*22*). These include the -SM- motif of the B domain. The C domain is less well conserved, with a -GD- motif present in *S. cerevisiae* Imp1, but -RD- being found in Imp2. The -KR- of the D domain

```
IMP1 YEAST                                                          MTVGTLPIWS      10
IMP2 YEAST                                                          MFRAGSSKRFL     11
E.coli      MANMFALILVIATLVTGILWCVDKFFFAPKRRERQAAAQAARDSLDKATLKKVAPKPGWL           60
                                                                        ...


IMP1 YEAST  KTFSYAIRSLCFLHIIHMYAYEFTETRGESMLPTLS-----ATNDYVHVLKN-FQNGRGI           64
IMP2 YEAST  RNTLIAISWVPVLLTINNNVVHIAQVKGTSMQPTLNPQTETLATDWVLLWKFGVKNPSNL           71
E.coli      ETGASVFPVLAIVLIVRSFIYEPFQIPSGSMMPTLLIGDFILVEKFAYGIKDPIYQKTLI          120
             ..   .:  : .:  :.        ** ***        . .:.    *   . :  :
                                      sgSM-ptl
                                      DOMAIN B


IMP1 YEAST  K-----MGDCIVALK-PTDPNHRICKRVTGMPGDLVLVDPS-------------------           99
IMP2 YEAST  S------RDDIILFKAPTNPRKVYCKRVKGLPFDT--ID-------------------          102
E.coli      ENG-HPKRGDIVVFKYPEDPKLDYIKRAVGLPGDKVTYDPVSKELTIQPGCSSGQACENA          179
            .       . *:  :* *  :*.   **. *;:* *        *
                     rgdi                 KR    gpgd
                     DOMAIN C             DOMAIN D


IMP1 YEAST  --------------------------------------------------------TI          101
IMP2 YEAST  --------------------------------------------------------TK          104
E.coli      LPATYSNVEPSDFVQTFSRRNGGEATSGFFEVPKNETKENGIRLSERKETLGDVTHRILT          239
                                                                     :


IMP1 YEAST  VNYVGDVL-VDEERFGTYIK---VPEGHVWVTGDNLSHSLDSRTYNALPMGLIMGKIVAA          157
IMP2 YEAST  FPYP----------KPQVN---LPRGHIWVEGDNYFHSIDSNTFGPISSGLVIGKAITI          150
E.coli      VPIAQDQVGMYYQQPGQQLATWIVPPGQYFMMGDNRDNSADSRYWGFVPEANLVGRATAI          299
            :       *:  ::    :* *: :: * **. :.  :. .  ::*::  ..
                                       GDn  ns Dsr
                                       DOMAIN E


IMP1 YEAST  NN-FDKPF--WDGSIRNIWGFKWINN-TFLDVQAKSN                              190
IMP2 YEAST  V--WPPSR--WGTDLKLSTGRDCISKRAILE                                    177
E.coli      WMSFDKQEGEWPTGLRLSR----IGG---IH                                    323
            :         *    .::     *.     :.
```

FIG. 4. Aligned Imp1 and Imp2 sequences from *S. cerevisiae*. The inferred amino acid sequences are aligned with that from *E. coli,* together with the conserved domains B–D (*22*).

is conserved in both polypeptides, and the Gly–Asp substitution responsible for the *pet ts*2858 mutant phenotype is just C-terminal to the -KR- motif (*52*). The -GDNXXXSXDS- of the E domain is also conserved. That the serine residue in the B domain is required for catalytic activity was shown by Nunnari *et al.* (*27*). Chen *et al.* (*62*) showed that mutations affecting the five key residues of the B, D, and E domains of Imp1 or Imp2 abolished the production of mature cytochrome b_2 and cytochrome c_1, respectively, in pulse-chase experiments (Table I). Chen *et al.* also studied the unusual specificity of the Imp1 peptidase, which cleaves precursors with an asparagine at the −1 position, in contrast to the more usual alanine (which is found at the −1 position in the cytochrome c_1 substrate of Imp2). Substitution of all 19 other amino acids for the Asn at −1 of cytochrome b_2 showed that Ser, Cys, Met, Ala, and Leu were consistent with cleavage. Gln and Thr allowed some cleavage, and other substitutions were not tolerated. Substitution of the wild-type Ile of cytochrome b_2 at position −3 with Ala was also consistent with cleavage by Imp1, but the Asn −1Ala and Ile −3Ala double mutant could not be processed. Surprisingly Imp2

TABLE I

EFFECTS OF MUTATIONS AT EQUIVALENT POSITIONS ON IMP1 AND IMP2 PROCESSING
OF CYTOCHROME b_2 AND CYTOCHROME c_1, RESPECTIVELY $(62)^a$

Imp1										
substitution:	S40A	S40T	K84R	K84H	R85A	D131Y	D131N	D131E	D138N	D138E
Effect:	—	—	—	—	(−)	—	(−)	(−)	—	—
Imp2										
substitution:	S41A	S41T	K91R	K91H	R92A	D124Y	D124N	D124E	D131N	D131E
Effect:	—	—	—	—	(−)	—	—	(−)	—	—

^{a}A dash indicates that processing was abolished, and (−) indicates that it was merely reduced. Numbering is as used in (22), differing from (27).

could not cleave any of the −1 substituted cytochrome b_2 polypeptides (including −1 Ala, as in its usual substrate cytochrome c_1) or even the Asn −1Ala and Ile −3Ala double mutant with the same −1, −3 residues as cytochrome c_1. Thus, a simple model in which Asn at −1 of cytochrome b_2 is needed to ensure its cleavage by Imp1 rather than Imp2 is inaccurate. However, what the precise requirements are for cleavage by either peptidase, and why there should be two separate peptidases in any case, remain important outstanding questions. It will be particularly interesting to understand how the insertion of heme into cytochrome c_1 influences its ability to undergo cleavage.

III.   Origins of the Chloroplast and Mitochondrial Type I Signal Peptidases

The idea that organelles such as chloroplasts and mitochondria can be traced back ultimately to endosymbiotic bacteria had its origins as far back as 1883 (63). Although there is still discussion over details of the hypothesis, such as how many different endosymbioses were responsible (64), the principles are generally accepted. Mitochondria almost certainly arose initially from a purple photosynthetic bacterium, probably from the alpha subgroup, and chloroplasts from some kind of oxygenic photosynthetic bacterium (64). Phylogenetic analysis of the mitochondrial Imp sequences is consistent with a eubacterial origin for the genes, although it does not place them especially close to the type I signal peptidase from the purple photosynthetic bacterium *Rhodobacter capsulatus* (Fig. 5). The chloroplast thylakoidal processing peptidases are clearly placed as a sister group to the cyanobacterial enzymes (Figs. 3, 5). It is interesting to note that there are also multiple type I signal peptidase genes in at least one cyanobacterium (*Synechocystis* sp. PCC6803). However, the duplications giving rise to the

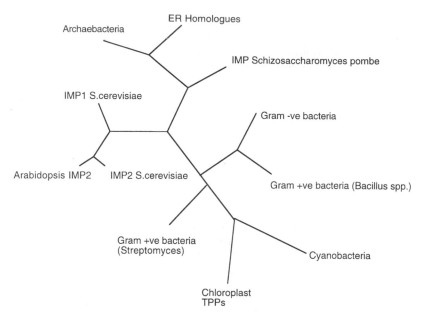

FIG. 5. Phylogenetic analysis of Imp sequences. The figure shows a simplified analysis of Imp sequences and homologs. It was constructed using a distance matrix method under PAUP4.0 (*31*) on aligned sequences with gaps excluded. A parsimony analysis gave the same topology. Branch lengths are arbitrary (K.A.F., unpublished).

different TPP genes in *A. thaliana* appear to be independent of the duplication giving rise to the multiple genes in *Synechocystis* (Figs. 3, 5), and the presence of multiple genes in prokaryotes is not uncommon (e.g., *65*). Thus we can be reasonably confident in saying that the chloroplast and mitochondrial type I signal peptidases are another remnant of the cellular machinery existing in the free-living ancestors of these organelles (*66*). How the bacterial enzymes originated is unknown. There are other bacterial enzymes that function with serine–lysine catalytic dyad mechanisms, such as LexA of *E. coli* (*67*), but none with compelling sequence homology to the type I signal peptidases. It is particularly interesting that the beta-lactamases and the peptidases that function in bacterial cell wall synthesis also probably use a serine–lysine catalytic mechanism, although there is again no obvious sequence homology to the type I signal peptidases (*68*). The observation that the bacterium *Mycoplasma genitalium,* which lacks a cell wall, apparently has no type I signal peptidase gene, although it has a number of proteins with possible type I signal peptidase cleavage sites and it also has a lipoprotein signal peptidase gene, may also prove illuminating (*69*).

Acknowledgment

We thank the Biotechnology and Biological Sciences Research Council for financial support and Dr. P. E. Reynolds and Dr. D. S. Bendall for helpful discussions and comments.

References

1. Heldt, H.-W., (1997). "Plant Biochemistry and Molecular Biology." Oxford University Press, Oxford.
2. Robinson, C., and Ellis, R. J. (1984). *Eur. J. Biochem.* **142,** 337.
3. Smeekens, S., de Groot, M., van Binsbergen, J., and Weisbeek, P. (1985). *Nature* **317,** 456.
4. Smeekens, S., Bauerle, C., Hageman, J., Keegstra, K., and Weisbeek, P. (1986). *Cell* **46,** 365.
5. Hageman, J., Robinson, C., Smeekens, S., and Weisbeek, P. (1986). *Nature* **324,** 567.
6. James, H. E., Bartling, D., Musgrove, J. E., Kirwin, P. M., Herrmann, R. G., and Robinson, C. (1989). *J. Biol. Chem* **264,** 19573.
7. Kirwin, P. M., Elderfield, P. D., and Robinson, C. (1987). *J. Biol. Chem.* **262,** 16386.
8. Anderson, C. M., and Gray, J. C. (1991). *FEBS Lett.* **280,** 383.
9. Musgrove, J. E., Elderfield, P. D., and Robinson, C. (1989). *Plant Physiol.* **90,** 1616.
10. Halpin, C., Elderfield, P. D., James, H. E., Zimmerman, R., Dunbar, B., and Robinson, C. (1989). *EMBO J.* **8,** 3917.
11. Barbrook, A. C., Packer, J. C. L., and Howe, C. J. (1996). *FEBS Lett.* **398,** 198.
12. Allsop, A. E., Brooks, G., Bruton, G., Coulton, S., Edwards, P. D., Hatton, I. K., Kaura, A. C., McLean, S. D., Pearson, N. D., Smalet, T. C., and Southgate, R. (1995). *Bioorg. Med. Chem. Lett.* **5,** 443.
13. Wallace, T. P., Robinson, C., and Howe, C. J. (1990). *FEBS Lett.* **272,** 141.
14. Howe, C. J., and Wallace, T. P. (1990). *Nucleic Acids Res.* **18,** 3417.
15. von Heijne, G. (1986). *Nucleic Acids Res.* **14,** 4683.
16. Shackleton, J. B., and Robinson, C. (1991). *J. Biol. Chem.* **266,** 12152.
17. Sazuka, T., Yamaguchi, M., and Ohara, O. (1999). *Electrophoresis* **20,** 2160.
18. Chaal, B. K., Mould, R. M., Barbrook, A. C., Gray, J. C., and Howe, C. J. (1998). *J. Biol. Chem.* **273,** 689.
19. Paetzel, M., Dalbey, R. E., and Strynadka, N. C. J. (1998). *Nature* **396,** 186.
20. Kirwin, P. M., Elderfield, P. D., Williams, R. S., and Robinson, C. (1988). *J. Biol. Chem.* **263,** 18128.
21. Chaal, B. K. (1998). Ph.D. thesis, University of Cambridge, U.K.
22. Dalbey, R. E., Lively, M. O., Bron, S., and van Dijl, J. M. (1997). *Protein Sci.* **6,** 1129.
23. Sung M., and Dalbey, R. E. (1992). *J. Biol. Chem.* **267,** 13154.
24. Klenotic, P. A., Carlos, J. L., Samuelson, J. C., Schuenemann, T. A., Tschantz, W. R., Paetzel, M., Strynadka, N. C. J., and Dalbey, R. E. (2000). *J. Biol. Chem.* **272,** 6490.
25. Tschantz, W. R., Paetzel, M., Cao, G., Suciu, D, Inouye, M., and Dalbey, R. A. (1995). *Biochemistry* **34,** 3935.
26. van Klompenburg, W., Paetzel, M., de Jong, J. M., Dalbey, R. E., Demel, R. A., von Heijne, G., and de Kruiff, B. (1998). *FEBS Lett.* **431,** 75.
27. Nunnari, J., Fox, T. D., and Walter, P. (1993). *Science* **262,** 1997.
28. Packer, J. C. L., André, D., and Howe, C. J. (1995). *Plant Mol. Biol.* **27,** 199.
29. Cavener, D., and Ray, S. C. (1991). *Nucleic Acids Res.* **19,** 3185.
30. Fütterer, J., and Hohn, T. (1996). *Plant Mol. Biol.* **32,** 159.

31. Swofford, D. (1998). PAUP*. Sinauer Associates, Sunderland, MA.
32. Daum, G., Böhni, P. C., and Schatz, G. (1982). *J. Biol. Chem.* **257,** 13028.
33. Gasser, S. M., Ohashi, O., Daum, G., Böhni, P. C., Gibson, J., Reid, G. A., Yonetani, Y., and Schatz, G. (1982). *Proc. Natl. Acad. Sci. USA* **79,** 267.
34. Daum, G., Gasser, S. M., and Schatz (1982). *J. Biol. Chem.* **257,** 13075.
35. Nicholson, D.W., Stuart, R. A., and Neupert, W. (1989). *J. Biol. Chem.* **264,** 10156.
36. Reid, G. A., Yonetani, T., and Schatz, G. (1982). *J. Biol. Chem.* **257,** 13068.
37. Kaput, J., Goltz, S., and Blobel, G. (1982). *J. Biol. Chem.* **257,** 15054; Sadler, I., Suda, K., Schatz, G., Kaudewitz, F., and Haid, A. (1984). *EMBO J.* **3,** 2137; Guiard, B. (1985). *EMBO J.* **4,** 3265.
38. van Loon, A. P. G. M., Brändli, A. W., Pesold-Hurt, B., Blank, D., and Schatz, G. (1987). *EMBO J.* **6,** 2433.
39. Neupert, W. (1997). *Annu. Rev. Biochem.* **66,** 863.
40. van Loon, A. P. G. M., and Schatz, G. (1987). *EMBO J.* **6,** 2441.
41. Pratje, E., Mannhaupt, G., Michaelis, G., and Beyreuther, K. (1983). *EMBO J.* **2,** 1049.
42. Torello, A. T., Overholtzer, M. H., Cameron, V. L., Bonnefoy, N., and Fox, T. D. (1997). *Genetics* **145,** 903.
43. Pratje, E., and Guiard, G. (1986). *EMBO J.* **5,** 1313.
44. Haucke, V., Ocana, C. S., Hönlinger, A., Tokatlidis, K., Pfanner, N., and Schatz, G. (1997). *Mol. Cell. Biol.* **17,** 4024.
45. Haas, R. C., Korenfeld, C., Zhang, Z., Perryman, B., and Roman, D. (1989). *J. Biol. Chem.* **264,** 2890.
46. Troll, H., Winckler, T., Lascu, I., Müller, N., Saurin, W., Véron, M., and Mutzel, R. (1993). *J. Biol. Chem.* **268,** 25469; Gahris, M. L. E., Hakansson, G., Alexciev, K., and Knorpp, C. (1999). *Biochimie* **81,** 1089.
47. Kishi, F., Tanizawa, Y., and Nakazawa, A. (1987). *J. Biol. Chem.* **262,** 11785; Nobumoto, M., Yamada, M., Song, S. C., Inouye, S., and Nakazawa, A. (1998). *J. Biochem.* **123,** 128.
48. Kalousek, F., Neupert, W., Omura, T., Schatz, G., and Schmitz, U. K. (1993). *Trends Biochem. Sci.* **18,** 249.
49. Schneider, A., Behrens, M., Scherer, P., Pratje, E., Michaelis, G., and Schatz, G. (1991). *EMBO J.* **10,** 247.
50. Goormaghtigh, E., and Ruysschaert, J.-M. (1984). *Biochim. Biophys. Acta* **779,** 271.
51. Behrens, M., Michaelis, G., and Pratje, E. (1991). *Mol. Gen. Genet.* **228,** 167.
52. Esser, K., Pratje, E., and Michaelis, G. (1996). *Mol. Gen. Genet.* **252,** 437.
53. Schneider, A., Oppliger, W., and Jenö P. (1994). *J. Biol. Chem.* **269,** 8635.
54. Braun, H.-P., and Schmitz, U. K. (1999). *Planta* **209,** 267.
55. Braun, H.-P., Emmermann, M., Kruft, V., and Schmitz, U. K. (1992). *Mol. Gen. Genet.* **231,** 217.
56. Klug, G., Jäger, A., Heck, C., and Rauhut, R. (1997). *Mol. Gen. Genet.* **253,** 666.
57. Bauer, M., Behrens, M., Esser, K., Michaelis, G., and Pratje, E. (1994). *Mol. Gen. Genet.* **245,** 272.
58. Hell, K., Herrmann, J., Pratje, E., Neupert, W., and Stuart, R. A. (1997). *FEBS Lett.* **418,** 367; Hell, K., Merrmann, J. M., Pratje, E., Neupert, W., and Stuart, R. A. (1998). *Proc. Natl. Acad. Sci. USA* **95,** 2250.
59. Hell, K., Tzagoloff, A., Neupert, W., and Stuart, R. A. (2000). *J. Biol. Chem.* **275,** 4571.
60. Jan, P.-S., Esser, K., Pratje, E., and Michaelis, G. (2000). *Mol. Gen. Genet.* **263,** 483.
61. Bauerfeind, M., Esser, K., and Michaelis, G. (1998). *Mol. Gen. Genet.* **257,** 635.
62. Chen, X., Van Valkenburgh, C., and Green, N. (1999). *J. Biol. Chem.* **274,** 37750.
63. Schimper, A. F. W. (1883). *Bot. Zeit.* **41,** 105.
64. Howe, C. J., Beanland, T. J., Larkum, A. W. D., and Lockhart, P. J. (1992). *Trends Ecol. Evol.* **7,** 378; Lang, B. F., Gray, M. W., and Burger, G. (1999). *Annu. Rev. Genet.* **33,** 351.

65. Tjalsma, H., Noback, M. A., Bron, S., Venema, G., Yamane, K., and van Dijl, J. M. (1997). *J. Biol. Chem.* **272,** 25983.
66. Howe, C. J., Barbrook, A. C., and Packer, J. C. L. (1996). *Biochem. Soc. Trans.* **24,** 750.
67. Paetzel, M., and Dalbey, R. E. (1997). *Trends Biochem. Sci.* **22,** 28.
68. Strynadka, N. C. J., Adachi, H., Jensen, S. E., Johns, K., Sielecki, A., Betzel, C., Sutoh, K., and James, M. N. G. (1992). *Nature* **359,** 700; Goffin, C., and Ghuysen, J.-M. (1998). *Microbiol. Mol. Biol. Rev.* **62,** 1079.
69. Fraser, C. M., Gocayne, J. D., White, O., Adams, M. D., Clayton, R. A., Fleischmann, R. D., Bult, C. J., Kerlavage, A. R., Sutton, G., Kelley, J. M., Fritchman, J. L., Weidman, J. F., Small, K. V., Sandusky, M., Fuhrmann, J., Nguyen, D., Utterback, T. R., Saudek, D. M., Phillips, C. A., Merrick, J. M., Tomb, J.-F., Dougherty, B. A., Bott, K. F., Hu, P.-C., Lucier, T. S., Peterson, S. N., Smith, H. O., Hutchison III, C. A., and Venter, J. C. (1995). *Science* **270,** 397.

6

Type IV Prepilin
Leader Peptidases

MARK S. STROM* • STEPHEN LORY†,1
*Northwest Fisheries Science Center
National Marine Fisheries Service, National Oceanic and Atmospheric Administration
Seattle, Washington 98112
†Department of Microbiology
University of Washington
Seattle, Washington 98195

I. Introduction

Filamentous surface organelles called pili or fimbriae have been described on a variety of bacterial cells. Based on common features of these

[1]Present address: Dept. of Microbiology and Molecular Genetics, Harvard Medical School,
200 Longwood Ave, Boston, MA 02115.

THE ENZYMES, Vol. XXII
Copyright © 2001 by Academic Press

structures, they have been implicated in numerous cellular processes, including the recognition of receptors on eukaryotic cell membranes and interactions with other bacteria or with a variety of inanimate surfaces. There are distinct families of pili, based on functional relatedness and several conserved features, including extensive sequence similarities of the subunit and accessory proteins. The latter also suggests a conserved pathway of organelle assembly within these families. One class of specialized pili, termed type IV pili, are filamentous surface structures that have been shown to be expressed only by gram-negative bacteria with a strong bias toward pathogens. Their prominent location on the bacterial surfaces is consistent with their function as adhesive organelles that act by anchoring the bacteria to host-cell surfaces or during formation of microcolonies and biofilms (1). Type IV pili also serve in conjugal DNA transfer of plasmids (2). Less clear is their role in gliding movement on surfaces, termed twitching motility (in *Pseudomonas aeruginosa* and other gram-negative bacteria) or social motility (in *Myxococcus xanthus*) (3, 4). The basis of the requirement for pili in natural transformation in several gram-negative bacteria, such as *Neisseria* and *Pseudomonas* species is not understood, nor is the function of the type IV pilin homologs during natural transformation in *Bacillus subtilis*. Finally, the striking similarities between type IV pilin sequences, the sequences of the protein constituents of the pilus biogenesis machinery, and the components of the type II protein secretion apparatus in a variety of gram-negative bacteria (5, 6) suggest that extracellular protein secretion may take place though cellular complexes resembling pili. Such organelles composed of the pilin homologs in the type II secretion pathway, termed "pseudopili," have been identified (7).

The subunits of the type IV pili are subject to two sequential posttranslational modification reactions. Each subunit is synthesized as a precursor with a short, amino-terminal extension, which is usually rich in basic amino acids. Based on the accumulation of sequences from individual laboratories working on different microorganisms and from the various microbial genome sequencing projects, it is apparent that there is a more extensive variation in the size and the amino acid composition of the N-terminal leader peptide. This leader peptide does not appear in the final organelle, since it is proteolytically removed during or shortly after the completion of synthesis of the subunits. The second feature of the mature pilin subunit is the N-methylation of the amino terminus created by leader peptide cleavage. Both of these reactions are carried out by the same bifunctional enzyme, first described in *P. aeruginosa* as a product of the *pilD* gene (8, 9). The enzyme is localized in the cytoplasmic membrane, has a broad specificity of substrates, and can process related homologs required for type II secretion and DNA uptake (9–17). In this chapter we will refer to this bifunctional enzyme as type IV prepilin peptidase (abbreviated TFPP) with the understanding that its bifunctionality is preserved in bacteria that express it, and will discuss in depth the structure–function of the enzyme and its importance in expression of a number of factors important for normal bacterial processes and often virulence.

II. The Substrates of Type IV Prepilin Peptidases

Numerous studies on the expression of virulence factors secreted through the type II secretion pathway and discoveries of type IV pili in a variety of bacteria have resulted in a rapid increase in the number of proteins identified that require a TFPP for processing. However, this updated list of such proteins has not significantly altered the basic conserved features as they were described for the first few members of the type IV pilin subunit family, based on the homology of the primary amino acid sequence in the N-terminal portions of pilins from *P. aeruginosa, Neisseria gonorrhoeae,* and *Moraxella nonliquefaciens* (*18–20*). The majority of the pilin subunits contain a gradient of sequence similarity, consisting of a very high level (ca. 90% identity) sequence conservation over the first 20–30 amino acids, with a progressive sequence divergence toward the carboxy terminus. Cloning of the structural genes for pili from various microorganisms revealed the presence of the uncommon leader peptide in the precursors of the pilin subunits. Although the precise role of the conserved region of the pilins is unknown, it is very likely that it is involved in two steps of pilus biogenesis. The first 30 amino acids, a relatively hydrophobic segment, probably function as an inner membrane targeting sequence (*21*), a role that may be analogous to the signal peptides of most secreted proteins, including those found on most *E.coli* pilin precursors. In such proteins the signal peptides serve as the recognition sequences for the Sec machinery and display a three-domain arrangement, consisting of a small cluster of basic N-terminal amino acids. This is followed by a 25–35 amino acid segment containing hydrophobic amino acids almost exclusively, and a consensus signal-peptidase cleavage site. Although the sequences at the extreme N termini of type IV prepilins are similar to typical signal peptides, they apparently bypass the Sec pathway, because they do not undergo cleavage by signal peptidase I in spite of the presence of a reasonable consensus recognition sequence for this enzyme. Instead, a short peptide is cleaved from the pilin precursor by the TFPP. Second, the conserved N terminus of the mature pilins has been suggested to represent the domains where individual pilins interact in the mature organelle, presumably via hydrophobic interactions with neighboring subunits.

The leader sequences present on most of the TFPP substrates are short, typically 5–7 amino acids in length, with a net basic charge due to the presence

A. Type IV pilins

Group A

```
                                                              ▶
P. aeruginosa PilA        MetLys    AlaGlnLysGly            PheThrLeuIleGluLeuMetIleValVal-
P. syringae PilA          Met       AsnAlaGlnLysGly         PheThrLeuIleGluLeuMetIleValVal-
M. bovis TfpQ             Met       AsnAlaGlnLysGly         PheThrLeuIleGluLeuMetIleValIle-
M. nonliquefaciens TfpA   Met       AsnAlaGlnLysGly         PheThrLeuIleGluLeuMetIleValIle-
D. nodosus FimA           MetLysSerLeuGlnLysGly             PheThrLeuIleGluLeuMetIleValIle-
N. gonorrhoeae PilE       MetAsnThrLeuGlnLysGly             PheThrLeuIleGluLeuMetIleValIle-
N. meningitidis PilE      MetAsnThrLeuGlnLysGly             PheThrLeuIleGluLeuMetIleValIle-
E. corrodens EcpA         MetLysGlnValGlnLysGly             PheThrLeuIleGluLeuMetIleValIle-
X. campestris PilA        MetLys    LysGlnAsnGly            PheThrLeuIleGluLeuMetIleValVal-
A. hydrophila TapA        MetLys    LysGlnSerGly            PheThrLeuIleGluLeuMetIleValVal-
A. salmonicida TapA       MetLys    LysGlnSerGly            PheThrLeuIleGluLeuMetIleValVal-
A. veronii bv sobria TapA MetLys    LysGlnSerGly            PheThrLeuIleGluLeuMetIleValVal-
P. putida PilA            MetLys    GlyGlnArgGly            IleThrLeuIleGluLeuMetIleValVal-
V. cholerae PilA   MetLys AlaTyrLysAsnLysGlnGlnLysGly       PheThrLeuIleGluLeuMetIleValVal-
V. vulnificus VvpA        MetLysLysLeuAspLysThrLysLysGlnGlnGly PheThrLeuIleGluLeuMetIleValVal-
```

Group B

```
V. cholerae TcpA   MetGln..16..LysLysThrGlyGlnGluGly  MetThrLeuLeuGluValIleIleValLeu-
E. coli BfpA       MetVal.. 9..AsnLysLysTyrGluLysGly  LeuSerLeuIleGluSerAlaMetValLeu-
```

B. Pilin-like proteins required for type IV pilus biogenesis

```
P. aeruginosa PilE            MetArgThrArgGlyLysGly        PheThrLeuLeuGluMetValValValVal-
P. aeruginosa PilV   MetLeu..5..ArgSerLeuHisGlnSerGly     PheSerMetIleGluValValLeuValAlaLeu-
P. aeruginosa FimU       MetSerTyrArgSerAsnSerThrGly       PheThrLeuIleGluLeuLeuIleIleVal-
P. aeruginosa FimT           MetValGluArgSerGlnArgAla      LeuThrLeuThrGluLeuLeuPheAlaLeu-
N. meningitidis FimT          MetCysThrArgLysGlnGlnGly     PheThrLeuThrGluLeuLeuIleValMet-
N. meningitidis PilV   MetAsn..7..ArgLeuLysSerGlnSerGly   MetAlaLeuIleGluValLeuValAlaMet-
```

of one or more lysine residues. This family, called group A (22), is easily distinguishable from a more heterogeneous family of type IV prepilins
(group B), which is typified by the subunits of the bundle-forming pili of
enteropathogenic *Escherichia coli* and the toxin coregulated pili (TCP) of
Vibrio cholerae. The pilins that make up these pili have leader peptides that
are longer than those of group A, but still lack the hydrophobic character of
signal sequences as a whole. Cleavage of the leader peptides from type IV
pilin precursors occurs between an invariant glycine residue at the −1 position
and a hydrophobic amino acid (usually phenylalanine) at the +1 position.
A glutamate residue at the +5 position relative to the cleavage site is also
invariant. Mutagenesis of Gly^{-1} in *P. aeruginosa* PilA showed that it is essential for proteolytic cleavage by the PilD TFPP, whereas the +1 position
can tolerate a number of changes (23). The overall features of the type IV
pilin leader peptides are shown in Fig. 1A, along with the first 10 amino acid
residues of the mature proteins.

   The precise function of the leader peptide and the basis for the requirement for prepilin processing by the TFFPs is unknown. It is conceivable that
removal of the leader peptide initiates the entry of the mature pilins into a
specific biogenesis pathway. An interesting difference in the two groups of
type IV prepilins is that prepilins of group A are often readily interchangeable between the various bacteria expressing such pili, since subunits from
one species can be correctly processed and assembled into a pilus structure in a heterologous host (24–28). However, assembly of group A pili in
a group B host has not been successfully demonstrated, in spite of efficient
processing of the precursors by the heterologous TFFP (13). In striking contrast to *P. aeruginosa*, which can assemble most heterologous group A pili,
attempts to assemble the *V. cholerae* TcpA pilin into functional pili in enteropathogenic *E. coli* using the bundle-forming pilus biogenesis machinery

---

FIG. 1. Amino acid sequences of the amino-terminal domains of some type IV pilins and type
IV pilin-like proteins involved in pilus biogenesis. The ▼ denotes the cleavage site where TFPPs
proteolytically remove the leader peptides. (A) The group A and group B type IV prepilins, with
Genbank protein accession number in parentheses: *P. aeruginosa* PilA (P02973), *P. syringae*
PilA (AAA25974), *Moraxella bovis* TcpQ (A55851), *M. nonliquefaciens* TfpA (AAA25310),
Dichelobacter nodosus FimA (P04953), *Neisseria gonorrhoeae* PilE (AAC38436), *N. menin-
gitidis* PilE (S55496), *Eikenella corrodens* EcpA (CAA78250), *Xanthomonas campestris*
PilA (S52692), *Aeromonas hydrophila* TapA (P45791), *A. salmonicida* TapA (AAC23566),
A. veronii biovar *sobria* TapA (AAD09352), *Pseudomonas putida* PilA (D36961), *Vibrio
cholerae* PilA (AAD21029), *V. vulnificus* VvpA (M. Strom, unpublished), *V. cholerae* TcpA
(P23024), *Escherichia coli* BfpA (P33553). (B) Pilin-like proteins involved in type IV pilus bio-
genesis, with Genbank protein accession number in parentheses: *P. aeruginosa* PilE (S54700),
P. aeruginosa PilV (S77594), *P. aeruginosa* FimU (AAB39271), *P. aeruginosa* FimT
(AAB39270), *N. meningitidis* FimT (AAF62321), *N. meningitidis* PilV (AAF41297).

failed (29). These studies suggest that the primary sequence of the mature portion of the group B pilins function in a highly specific way during the interaction of the subunits with the assembly machinery. Alternatively, the pilin precursors may interact with the components of the biogenesis apparatus prior to the removal of the signal peptide, and cleavage of the peptide may be necessary for the next series of interactions in the pilus assembly pathway.

The type IV pilins are also characterized by the presence of a posttranslationally modified amino acid residue at the N terminus of the mature protein, first recognized in the pilin subunit of *N. gonorrhoeae* (19). This amino acid is always modified by methylation of the free amino group (termed N-methylation), and thus all of the group A pilins (with the exception of *P. putida*) have *N*-methylphenylalanine at the amino terminus. The *N*-methylmethionine of the TCP subunit is similarly N-methylated. Although reversible protein methylation of internal glutamate residues of bacterial chemotaxis receptors is a common signaling mechanism, the modification of amino-terminal residues is a rather rare modification in bacteria. It has only been described for CheZ, a component of the chemotactic sensory apparatus of *E. coli* and *Salmonella typhimurium*, and for protein components of bacterial ribosomes (L11, L16, and S11) and translational initiation factor IF3 (30). N-Terminal amino acids of primary translational products usually contain formylmethionine, which necessitates the removal of the blocking group before a modification such as N-methylation can take place. This can occur by deformylation or by internal cleavage of a peptide bond, generating a free amino group. The processing carried out by the TFPPs on type IV prepilins and related proteins provides this function and suggests that the two reactions are likely sequential and may be coupled.

Studies on type IV pilus biogenesis in *P. aeruginosa* have identified at least four additional proteins with TFPP consensus cleavage sites that form a second set of TFPP substrates. These proteins do not appear to be part of the assembled pilus structure, but are required for assembly of the processed pilin pools into pili. Because of the high homology seen in the type IV pilus assembly genes identified to date, it is likely that all bacteria that synthesize these pili have homologs of the *P. aeruginosa* pilin-like proteins involved in pilus assembly. The first one of these identified is encoded by the *pilE* gene and is located on different region of the *P. aeruginosa* chromosome from the *pilA–D* gene cluster that contains the pilin and TFPP genes (31). The deduced amino acid sequence for PilE shows it to be a 15.3 kDa protein with a seven-residue leader peptide including the invariant glycine residue at the −1 position and phenylalanine at +1 (Fig. 1B). The other three homologs, *pilV, fimU,* and *fimT,* are also clustered together in a different region of the chromosome from *pilE* and *pilA–D* (32, 33). The deduced PilV polypeptide has a molecular weight of 20 kDa, and although its encoded leader peptide is

14 amino acid residues, it maintains the net positive charge of the structural type IV pilins with glycine and phenylalanine residues at the -1 and $+1$ positions of the cleavage site. Site-directed mutagenesis of the glycine to aspartic acid resulted in loss of proteolytic cleavage of the leader peptide *in vivo,* suggesting specific processing by PilD. FimU and FimT are 18.5 kDa proteins with slightly different leader peptides. Both are nine amino acids in length, and whereas FimU has glycine–phenylalanine at the consensus cleavage site, FimT has alanine–leucine. FimU is necessary for assembly of pilin into pili, whereas FimT does not appear to be required. However, the *fimU* mutant phenotype can be restored to wild-type by overexpression of FimT. This suggestes a subtler role for FimT in pilus biogenesis.

One of the functions performed by pili on gram-negative bacteria is in conjugation, where they serve in the recognition of recipient cells during the transfer of conjugative plasmids. To date, only the R64 and IncI1 conjugative plasmid utilizes type IV pili for its transfer, and all of the various genes encoding the pilus structural and assembly components are encoded on this plasmid. R64 specifies two proteins that contain the consensus TFPP peptidase cleavage site, as well as a TFPP encoded by the *pilU* gene. Interestingly, the major subunit of the R64 pilin undergoes an N-terminal modification that has yet to be characterized but that is not methylation (*34*).

The final process associated with functional type IV pili is a form of surface translocation or gliding motility (also referred to as twitching motility) [reviewed in (*4*)]. Fully processed and assembled pili are necessary for this motility in *P. aeruginosa* and in *Myxococus xanthus,* where type IV pili mediate social motility of this organism [see (*3, 4*) for reviews]. Although the assembly of pili requires the full complement of the accessory proteins involved in the biogenesis of the pilus structure, including action of TFPPs on the precursor subunits, the products of additional genes mediate twitching motility, which is independent of other pili-associated functions such as adhesion or the ability to act as receptors for pilus-specific phages.

The third family of substrates processed by the TFPPs is a group of proteins that are part of an apparatus in the gram-negative cell envelope responsible for extracellular targeting of a class of proteins that are first secreted into the periplasm. This secretion machinery is referred to as the terminal branch of the general secretion pathway, or alternatively, the type II secretion machinery (*5*). The N-terminal regions of these proteins are highly homologous to the same region on the type IV pilins (*35*). These proteins were identified by the linkage of their structural genes to those of other members of the secretion machinery (*5*), by demonstration of a pleiotrophic secretion defect in bacteria carrying mutated genes in their chromosome, or by low stringency hybridization with heterologous pilin gene probes. The genes encoding the four proteins of the type II secretion pathway in *P. aeruginosa* were first identified by using mixed oligonucleotide probes spanning the recognized consensus cleavage site recognized by the TFPP. Within the 14 genes that have now been identified to encode the type II secretion machinery, four are homologs of the type IV pilins, with the most extensive homology in the N-terminal region where there is extensive similarity to the group A prepilins. A fifth member of this family (XcpX/GspK) has also been identified to be a substrate of TFPP. XcpX shares a rather modest level of sequence similarity with the other prepilins; however, like the other members of this family of TFFP substrates, it also requires the presence of glycine at the cleavage site (36). There are a large number of bacterial species with such pilin-like groups of proteins, including P. aeruginosa, Klebsiella oxytoca, Erwinia carotovora, E. chrysanthemi, A. hydrophila, A. salmonicida, V. cholerae, and Xanthomonas campestris. A number of these also express type IV pili, including P. aeruginosa, V. cholerae, V. vulnificus, and A. hydrophila. In these organisms, a mutation in the TFPP gene results in a pleiotrophic mutation abolishing expression of type IV pili and extracellular secretion of proteins by the type II pathway (15, 16, 37–39).

In addition to the requirement for TFPP there appears to be a functional overlap between the pilus biogenesis and type II secretion pathways. In P. aeruginosa, and perhaps in other gram-negative bacteria that express type IV pili and carry out extracellular protein secretion via the type II mechanism, the major subunit of the type IV pili in this organism, PilA, forms stable complexes with the homologous proteins of the type II secretion apparatus (40). Expression of PilA is required for optimal secretion of proteins via this pathway. Interestingly, efficient interaction of PilA with the components of the secretion apparatus does not require the removal of the leader peptide from prelipin, suggesting that the leader peptide cleavage may not take place until an intermediate of the type II secretion apparatus has been assembled.

The fourth group of proteins that show sequence similarity to the conserved regions of type IV pilins are required for the process of natural competence and DNA uptake in gram-positive bacteria. In B. subtilis, these proteins are involved in binding of the transforming DNA to the surface of the recipient bacteria. The pilin homologs are encoded by the comG operon, and these genes specify four related products, ComGC, ComCD, ComGE, and ComGG (41–43). A corresponding TFPP has also been identified as the product of the comC gene, where mutation of comC results in a processing defect for ComGC–GE. It is unclear whether any of the potential substrates of ComC form a pilus-like structure required for DNA uptake, since such organelles have not been detected on the surface of B. subtilis. Interestingly, a DNA uptake or transformation mechanism in the gram-negative bacteria N. gonorrhoeae and Legionella pneumophila is dependent on expression of type IV pili (*44–46*), whereas in *Acinetobacter*, the analogous transformation system is distinct from pili expressed by this bacterium (*47*).

III. Genetic Characterization of Type IV Prepilin Peptidases

Major progress in the development of bacterial genetic methods for microorganisms other than *E. coli* K12 resulted in the identification of various novel protein-targeting processes in a variety of bacteria. One of the outcomes of this work was the identification of the type II secretion machinery in *K. oxytoca* [reviewed in (*5*)] and the discovery that the machinery contains homologues of type IV pili (*48*), as well as the finding of the similar proteins in *B. subtilis* (*41*). It was also obvious from the comparison of the primary sequence of type IV pilin precursors that these related proteins are posttransitionally modified by a previously uncharacterized proteolytic enzymatic activity.

One of the initial analyses of the phenotypes of nonpiliated mutants in *P. aeruginosa* provided clues about the existence and function of TFPPs. During a screen of a collection of *P. aeruginosa* transposon insertions that were located within the vicinity of the structural gene for the pilin subunit, Nunn *et al.* (*49*) observed that several transposon mutants lost the ability to assemble functional pili, as judged by their resistance to a pilus-specific bacteriophage PO4. These insertions defined a pilus biogenesis cluster consisting of three genes, *pilB*, *pilC*, and *pilD*, located 5′ to the *pilA* gene and oriented in the opposite reading frame. Analysis of membrane fractions of mutants in these open reading frames revealed that they all synthesized wild-type levels of pilin. However, the pilin antigen in strains with transposon insertions in *pilD* showed altered electrophoretic mobility on SDS–polyacrylamide gels when compared to wild-type pilin or pilin found in strains with transposon insertions in the *pilB* and *pilC* open reading frames. The slower migration of pilin in the *pilD* mutant suggested that it represents the unprocessed form of pilin, which was subsequently confirmed by N-terminal amino acid sequence analysis of the prepilin isolated from the *pilD* mutant. This result suggested either that the *pilD* gene product is a proteolytic enzyme responsible for the cleavage of the leader peptide from prepilin, or alternatively, that PilD is indirectly required for leader peptide cleavage by another *P. aeruginosa* protease. Using a protocol consisting of detergent solubilization of *P. aeruginosa* membranes and immunoaffinity purification of the *pilD* gene product and development of an *in vitro* cleavage assay provided the final proof that PilD is indeed a leader peptidase (*8*). This is discussed in detail in Section V of this chapter. At the same time a new gene, designated *xcpA*, was identified and shown to be essential for extracellular protein secretion in *P. aeruginosa;* its sequence was identical to *pilD* (*39*).

In studies characterizing the genes required for pullulanase secretion by *Klebsiella oxytoca,* one gene, *pulO,* was shown to encode a product that was essential for this secretion and that shared considerable sequence homology to PilD (*12*). PulO was shown to process a prepilin homolog PulG (*5, 50*). Very likely, the remaining pilin-like proteins encoded by the linked genes *pulH–J* are also substrates of PulO (*51*). The gene *comC* from *B. subtilis* was also found to share considerable homology with *pilD.* This gene was identified by a competence-defective phenotype during characterization of transposon insertions (*52*), and mutants in *comC* failed to process ComGC–ComGE (*43*).

Characterization and sequencing of a gene cluster (the *out* genes) in *E. chrysanthemi* required for extracellular protein secretion showed that these genes were homologous to the pullulanase secretion gene cluster of *K. oxytoca* (*53, 54*). This cluster contained a homolog of *P. aeruginosa pilD* and was named *outO.* The genes specifying the putative substrates of OutO were located within the same gene cluster and were designated *outG–J.* Similarly, secretion-defective mutants of *E. carotovora* were analyzed by cosmid complementation of gene segments from wild-type bacteria, and the molecular analysis of such cloned resulted in the identification of a homologous secretion gene cluster, organized similarly to the one in *E. chrysanthemi* (*55*). Based on the sequence similarity with the respective homologs in other bacteria, these genes were also designated *outO,* which encodes the TFPP, and *outG–H,* specifying the pilin-like protein substrates of the leader peptidase.

Over the past several years, a number of other approaches have been used to isolate the genes encoding components of type IV pilus biogenesis and TFPPs. These include Southern hybridization, polymerase chain reaction (PCR), and complementation of heterologous mutants with DNA libraries. In some of these studies, the results also led to the discovery of type II secretion systems. In others, characterization of genes required for type II secretion led to the discovery of type IV pili. A few of these studies are described next.

The TFPP gene of *N. gonorrhoeae* was identified by Lauer *et al.* (*56*) after Southern hybridization with the *pilB* gene probe from *P. aeruginosa* identified a similar pilus biogenesis gene cluster in the *N. gonorrhoeae* chromosome. In this species it was found that the *pilB* homolog was adjacent to the gonoccocal *pilD* homolog. The *N. gonorrhoeae* PilD was able to cleave prepilin when expressed in *E. coli,* confirming its expected enzymatic activity. Dupuy and Pugsley (*11*) used PCR to isolate the *pilD* homolog from *N. gonorrhoeae,* and to also show its presence in a broad range of different *Neisseria* species. This finding was especially interesting as many of these

Neisseria species do not express type IV pili, even though the TFPP gene is transcribed. This suggests that there are substrates other than pilin in these species. They further showed that coexpression of the gonococcal prepilin and peptidase in *E. coli* resulted in processing of the former, and that the *N. gonorrhoeae* TFPP was able to complement a *pulO* mutation in *K. oxytoca.*

De Groot *et al.* (*57*) cloned a region from *P. putida* that contained homology to a probe from *P. aeruginosa pilD* (*xcpA*). This homologous region contained a gene cluster similar but not identical to that specifying *pilA–D* and includes a homolog of *pilC, pilD,* and an open reading frame with high homology to type IV prepilin. When expressed in *P. aeruginosa,* the *P. putida* PilD is capable of processing the heterologous substrate pre-XcpT, a component of the *P. aeruginosa* type II secretion system, again demonstrating that the enzyme can process a variety of substrates. Although pili are not observable on the surface of *P. putida,* a number of proteins are secreted extracellularly, and other studies revealed the presence of a number of homologous genes related to the *xcp* (type II secretion) genes of *P. aeruginosa* (*58*), some of which are likely substrates for *P. putida* PilD.

The group B type IV pili differ from the group A on the basis of longer signal or leader peptides and an amino acid residue other than phenylalanine at the cleavage site. However, the bundle-forming pili (BFP) of enteropathogenic *E. coli* and the toxin coregulated pili of *V. cholerae* (TCP) do share considerable homology in the domain flanking the leader peptide cleavage site with pilins belonging to group A. The candidate TFPP in *V. cholerae* was identified genetically by analysis of *TnphoA*-generated mutations in genes co-regulated with the TCP gene. This analysis demonstrated that a gene (*tcpJ*) homologous to other TFPP genes was located adjacent to the pilin structural gene *tcpA* (*59*). Mutations that interrupt expression of TcpJ result in failure to process and assemble TcpA into TCP pili but do not affect extracellular secretion of proteins. TcpJ appears to be a TFPP that is limited in substrate range. *In vitro* experiments demonstrated that TcpJ is unable to process pre-PilA from *P. aeruginosa,* and conversely *P. aeruginosa* PilD is unable to process pre-TcpA (M. Strom and S. Lory, unpublished results). Recently, another four gene cluster was identified in *V. cholerae* that includes a second type IV pilin (*vcpA*) that is a typical member of group A, as well as a second TFPP (*vcpD*) (*38*). In a subsequent report from Fullner and Mekalonos (*60*), this gene cluster (also designated *pilA–D*) was found to be widespread in both classical and El Tor strains of *V. cholerae.* Mutations in *vcpD/pilD* resulted in a pleiotrophic protein secretion defect including the inability to secrete cholera toxin. The finding that *V. cholerae* TcpJ only processes TcpA, whereas VcpD/PilD presumably process both prepilin and type II secretion-specific prepilin-like proteins (*61*), is perhaps not surprising in light of recent evidence demonstrating that TCP pili serve as receptors for the bacteriophage CTXϕ that contains the cholera toxin gene, and that in turn the gene for TCP itself may be part of a filamentous bacteriophage (62). This suggests a somewhat more specialized function for TcpA, and subsequently TcpJ.

Conversely, the TFPP gene from enteropathogenic *E. coli* responsible for processing the group B BfpA prepilin was cloned by complementation of a *P. aeruginosa pilD* mutant (13). Subsequent sequence analysis of the gene encoding this enzyme, *bfpP*, demonstrated its high homology to other TFPPs. However, a role for BfpP in type II protein secretion has not been demonstrated.

Two additional studies demonstrated that a TFPP gene could be isolated by direct complementation of the *P. aeruginosa pilD* mutant strain. Cosmid libraries on compatible vectors from *A. hydrophila* (16) and *V. vulnificus* (15) were conjugated into a *pilD* mutant of *P. aeruginosa*, and the resulting colonies were screened for restoration of twitching motility and by zones of hemolysis on blood agar. Twitching motility is a function of type IV pili and can easily be detected in *P. aeruginosa* by the formation of colonies with a rough-looking surface and slightly irregular and spread-out edges. Hemolysis on blood agar is the result of secretion of *P. aeruginosa* phospholipase C via the type II secretion pathway (37). In both cases complementing clones were isolated. Sequencing demonstrated the presence of a four-gene cluster in each that encode homologs of *pilA*, *pilB*, *pilC*, and *pilD*. These genes (designated *tapABCD* in *A. hydrophila* and *vvpABCD* in *V. vulnificus*) are highly homologous to their *P. aeruginosa* counterparts. The only difference in genetic organization is that where the direction of transcription of the *pilA* gene in *P. aeruginosa* is oriented in the opposite direction from *pilB–D*, all four genes in both *A. hydrophila* and *V. vulnificus* are oriented in the same direction. TapD was shown to cleave and N-methylate *P. aeruginosa* prepilin *in vitro*, again demonstrating that it has the expected enzymatic activity. A *tapD* mutant is unable to secrete several exoenzymes, notably aerolysin, confirming that it plays a central role in type II secretion. TapD is also responsible for processing the TapA prepilin, although *A. hydrophila* produces few TapA-specific pili (63). VvpD from *V. vulnificus* can also cleave *P. aeruginosa* prepilin *in vitro*, and a *vvpD* mutant exhibits a pleiotrophic secretion defect and does not express surface pili. This mutant is less adherent to cultured epithelial cells and is significantly less virulent in animal models. Interestingly, a *V. vulnificus* strain carrying a mutation in the type IV pilin gene *vvpA* still has pili on the cell surface (R. N. Paranjpye and M. S. Strom, unpublished results). This suggests that this species expresses more than one type IV pilin (and perhaps pili), both of which must be processed by the VvpD TFPP, along with the (presumably) four pilin-like proteins that are part of the type II secretion apparatus.

These studies demonstrate that there is considerable interchangeability of the peptidases between bacteria that express somewhat less related type IV prepilins. A comparison of 18 TFPPs by protein parsimony and protein distance matrix analyses using the PHYLIP package (*64, 65*) in part supports the results demonstrated biochemically or genetically. In an examination of a consensus phylogenetic tree generated using the protein parsimony method, the majority of TFPPs that have been shown to process Group A type IV prepilins, and in most cases components of the type II secretion machinery as well, cluster together (Fig. 2). This group includes the TFPPs from *P. aeruginosa, P. putida, N. gonorrhoease, A. hydrophila, A. salmonicida, L. pneumophila, D. nodosus, V. vulnificus,* and *V. cholerae* (VcpD/PilD). TFPPs known to be involved in type II secretion in species that do not express type IV pili cluster on a separate branch (*E. carotovora, E. chrysanthemi, K. oxytoca,* and *B. pseudomallei*). The TFPPs responsible for processing the Group B type IV prepilins (TcpJ and BfpP from *V. cholerae* and enteropathogenic *E. coli,* respectively) appear closely related and are placed on a branch along with *M. xanthus* PilD (gliding motility) and the grampositive *B. subtilis* (ComC) (see Fig. 2).

IV. Cellular Localization and Topology of Type IV Prepilin Peptidases

The majority of TFPPs characterized to date are similar in size (249–290 amino acids), and the sequence conservation is reflected in similar predicted secondary structure based on computer analysis of the polypeptide sequences. Hydropathy analysis shows that the TFPPs are extremely hydrophobic, with 5–8 stretches of hydrophobic amino acids that are typical of transmembrane-spanning segments (*66*). This analysis predicts that TFPPs are integral membrane proteins, and this hypothesis was confirmed by the demonstration that PilD of *P. aeruginosa* is localized to the cytoplasmic membrane by Western immunoblotting of fractionated cells (*8*). Using an *in vitro* prepilin cleavage assay, Nunn and Lory demonstrated that the peptidase activity as well as the prepilin substrate are present in the cytoplasmic membrane fraction (*8*). A topological map for one TFPP, OutO of *E. carotovora,* was determined by the use of membrane topology probes. An analysis of a series of truncated OutO proteins that were fused at their C termini to β-lactamase was used as the indicator of external surface localization of the particular OutO domains by the ability of the hybrid protein to protect the bacteria against antibiotic killing (*67*). This study confirmed that hydrophobic segments made up eight transmembrane spanning segments. The most conserved region of the TFPPs is a segment which is very likely cytoplasmic (*9, 67, 68*). The most notable feature of this region is the

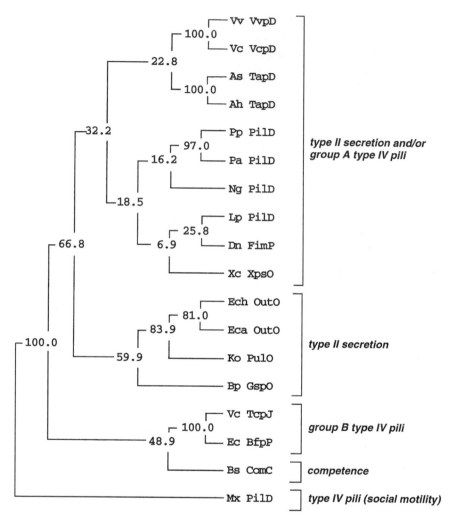

Fig. 2. Consensus phylogenetic tree showing relationships of 18 TFPPs as constructed by the protein sequence parsimony method included in the Phylogeny Inference Package (PHYLIP) of Felsentein (64). The number at each node is the percentage of bootstrapped trees in which the TFPPs to the right of the node were clustered. Species abbreviations and Genbank protein accession number of each TFPP is as follows (in order from top to bottom of tree): *Vibrio vulnificus* VvpD (AAA91206), *V. cholerae* VcpD (AAC63504), *Aeromonas salmonicida* TapD (AAC23569), *A. hydrophila* TapD (P45794), *Pseudomonas putida* PilD (P36642), *P. aeruginosa* PilD (P22610), *Neiserria gonorrhoeae* PilD (P33566), *Legionella pneumophila* PilD (AAC12718), *Dichelobacter nodosus* FimP (AAB65807), *Xanthomonas campestris* XpsO (AAC43571), *Erwinia chrysanthemi* OutO (P31711), *E. carotovora* OutO (P31712), *Klebsiella oxytoca* P410 (P15754), *Burkholderia pseudomallei* GspO (AAD05189), *V. cholerae* TcpJ (AAA69695), *Escherichia coli* BfpP (S47153), *Bacillus subtilis* ComC (P15378), and *Myxococcus xanthus* PilD (AAC36157).

TABLE I

Effect of Various Enzyme Inhibitors on TFPP Activity

Class	Inhibitor	Type	TFPP	Result	Ref.
Protease[a,b]	NEM	Cysteine protease	PilD	68% at 0.5 mM	68
	PMSF	Cysteine protease	PilD	little, at 500 μg/ml	8
	PCMB	Cysteine protease	PilD	74% at 0.5 mM	68
	PCMPS	Cysteine protease	PilD	90% at 0.5 mM	68
	Iodacetamide	Sulfydryl reagent	PilD	72% at 2 mM	68
	E-64	Cysteine protease	TcpJ	4% inhibition at 28 μM	77
	Calpain inhibitor I	Cysteine protease	TcpJ	13% inhibition at 45 μM	77
	NEM	Cysteine protease	TcpJ	85% inhibition at 1 mM	77
	Aprotinin	Serine protease	TcpJ	0% inhibition at 0.3 μM	77
	3,4-DCI	Serine protease	TcpJ	35% inhibition at 100 μM	77
	Pefabloc SC	Serine protease	TcpJ	50%$ inhibition at 4 mM	77
	Leupeptin	Ser and Cys protease	TcpJ	15% at 10 μM	77
	PMSF	Ser and Cys protease	TcpJ	11% at 1 mM	77
	Phosphoramidon	Metalloprotease	TcpJ	11% at 0.6 mM	77
	EDTA-Na	Metalloprotease	TcpJ	0% at 1.3 mM	77
	Bestatin	Amino metalloprotease	TcpJ	19% at 130 μM	77
	Pepstatin	Aspartic protease	TcpJ	5% at 1 μM; 26% at 100 μM	77
	EDAC/ Glycinamide	Acid protease	TcpJ	98%	77
Methyltransferase[c]	Sinefungin	AdoMet analogue	PilD	>90% inhibition at 1 μM	9
	S-Adenosyl-L-homocysteine	AdoMet analogue	PilD	>90% inhibition at 1 mM	9
	S-Adenosyl-L-ethionine	AdoMet analogue	PilD	>90% inhibition at 1 mM	9
	NEM	Sulfhydryl reagent	PilD	95% at 1 mM	68
	PCMB	Sulfhydryl reagent	PilD	91% at 2 mM	68
	PCMB + DTT	Sulfhydryl reagent	PilD	45% at 2 mM	68
	PCMPS	Sulfhydryl reagent	PilD	94% at 2 mM	68
	PCMPS + DTT	Sulfhydryl reagent	PilD	55% at 2 mM	68
	Iodacetamide	Sulfhydryl reagent	PilD	64% at 1 mM	68
	Iodacetamide + DTT	Sulfhydryl reagent	PilD	72% at 1 mM	68

[a] PilD protease inhibition was determined by mixing the given concentration of inhibitor with an amount of membrane material known to contain ~75 ng PilD and incubating for 30 min at 25°C. Substrate (purified PilA isolated from a *pilD P. aeruginosa* mutant strain) was then added and the reaction incubated at 37°C for 15 min. Cleavage was measured by electrophoresis on SDS (Tricine)-polyacrylamide gels (8, 68, 76).

[b] TcpJ protease inhibition was determined by mixing the given concentration of inhibitor with an amount of membrane material known to contain 1 unit of TcpJ activity (1 unit cleaves 50% of TcpA prepilin in 1 hour at 37°C) and then added to 1 μg TcpA prepilin. Cleavage of the prepilin was analyzed by Western immunoblot analysis (77).

[c] Methyltransferase inhibition of PilD was carried out by mixing the inhibitor at the concentration shown with 20 pmol purified PilD, 60 nmol prepilin, and 50 pmol [^{3}H]Ado Met; the percent inhibition was determined by the decrease in addition of the labeled methyl group onto processed pilin measured on autoradiographs after electrophoresis SDS–polyacrylamide gels (9).

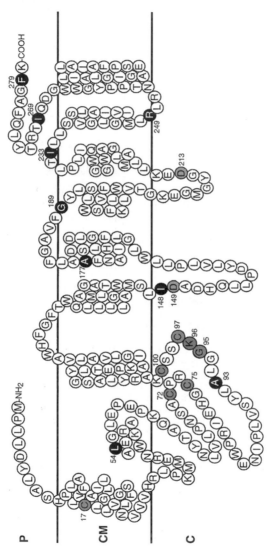

Fig. 3. A predicted topological model of *P. aeruginosa* PilD based on its hydrophobicity profile (66), alkaline phosphatase and β-galactosidase fusion data (68, 76), and the β-lactamase fusion data of *Erwinia carotovora* (67). Amino acid residues shaded black denote those positions where alkaline phosphatase and β-galactosidase were fused (68, 76). Residues shaded dark gray are those that were mutagenized and shown to be required for peptidase or methyltransferase activity (73, 76). The corresponding aspartic acid residues shown to be responsible for the peptidase activity of the TFPP TcpJ from *V. cholerae* (77) are shown shaded light gray.

presence of two pairs of cysteine residues found in the majority of TFPPs and implicated in enzymatic activities. However, other domains on the cytoplasmic face have also been shown to be involved in catalysis of TFPP substrates, and this topic will be more completely discussed later in this chapter (Section VII). Since hydrophobicity profiles of the TFPPs are similar, membrane topology is probably also highly conserved. A topological map using PilD as a model is shown in Fig. 3.

There are several reasons why an enzyme capable of processing a secreted protein should be localized in the membrane. The prepilin substrate inserts into the cytoplasmic membrane very rapidly after synthesis, and the localization of the enzyme in the same compartment may facility the cleavage reaction. Moreover, the correct orientation of the peptidase active site and the cleaved peptide bond may be most readily accomplished in the lipid environment of the membrane (see later discussion). Finally, the membrane location of the substrates should allow them to fold into a conformation where only one out of several possible cleavage sites is accessible to the enzymes, and other potential sites are protected. Analyses of mutant prepilin substrates with a variety of amino acid substitutions in the regions surrounding the cleavage site have demonstrated that many different substitutions can be tolerated in the regions flanking the cleavage site, especially at the +1 position (23). However, cleavage of the peptide bond only occurs between the signal peptide and the mature protein. Therefore, although the overall context of the peptide bond within the conserved N-terminal region of prepilin may be important, TFPPs may recognize a cleavage site that is presented only after prepilin (or other TFPP substrate) attains a specific conformation in the membrane lipid bilayer. *In vitro* cleavage of substrate by TFPPs solubilized with nonionic detergents requires the addition of cardiolipin, suggesting that a lipophilic environment is necessary for the cleavage reaction. Absence of cardiolipin does not lead to incorrect cleavage, suggesting that the role of the lipid is just as important for the enzymatic activity as it is for the ability of prepilin to serve as a substrate. Whether prepilin is inserted into membrane during or following its synthesis, the ability of the prepilin to become associated with the membrane does not appear to be linked to removal of the leader peptide (49).

V. Characterization of Type IV Prepilin Peptidase Activity

Proving that the family of proteins now known as type IV prepilin peptidases are indeed enzymes with endopeptidase activities, and not simply cofactors of other proteases, was made possible with the biochemical purification of PilD from *P. aeruginosa* and subsequent demonstration of its activity using purified substrate (8). Because PilD and other TFPPs are integral cytoplasmic membrane proteins, attempts to purify the enzyme using conventional schemes were unsuccessful. Purification of the enzyme was eventually accomplished by detergent solubilization of membranes from a *P. aeruginosa* strain that overexpresses PilD from an inducible plasmid vector, followed by immunoaffinity chromatography (8, 17). The successful procedure employed mild solubilization of total membrane material with Triton X-100 (4% v/v) in a solution of 25 m*M* triethanolamine hydrochloride, pH 7.5, and 10% (v/v) glycerol. The solubilized membranes were clarified by centrifugation and protein in the supernatant captured by passage through a DEAE–Sephacel column. The column was washed several times with the same buffer except with Triton X-100 reduced to 1%, and eluted with the addition of 200 m*M* NaCl in 25 m*M* triethanolamine hydrochloride, pH 7.5, 10% glycerol, and 1% Triton X-100.

For subsequent immunoaffinity purification, specific antiserum was prepared by immunizing rabbits with a synthetic PilD peptide (based on the deduced amino acid sequence and corresponding to a 21 amino-acid region in the large hydrophilic domain deduced to lie on the cytoplasmic side of the membrane) coupled to keyhole limpet hemocyanin. Immunoglobulin G (IgG) was purified from whole serum by ammonium sulfate precipitation and DEAE chromatography. Specific PilD-reactive IgG was further purified by passing the IgG through a Sulfo-link column containing conjugated PilD peptide. This antibody was then cross-linked to protein A–Sepharose CL-4B.

The collected protein from the membranes was then mixed with the anti-PilD IgG–protein A Sepharose. Bound PilD protein was eluted from the Sepharose with 100 m*M* glycine hydrochloride, pH 2.5, 1% Triton X-100 and immediately neutralized with one-fifth volume of 1 *M* triethanolamine hydrochloride, pH 7.5. Purity of the enzyme preparation was analyzed using SDS–PAGE and Western blot (the purified enzyme was enriched 109-fold from the starting material and migrated as a single band on SDS–PAGE). The eluted enzyme could be stored at −20°C in 50% (w/v) glycerol retaining activity for several months.

Demonstration of enzyme activity by purified PilD was accomplished by analyzing its ability to process purified *P. aeruginosa* precursor PilA. Overexpression of PilA in a *P. aeruginosa pilD* mutant results in the accumulation of unprocessed prepilin in the membrane fraction, where it accounts for up to half of the total membrane protein. Purification of this substrate is easily accomplished by gel filtration of detergent-solubilized membrane extracts. Initial experiments showed that mixing the substrate with crude membrane preparations from wild-type but not a *pilD* mutant of *P. aeruginosa* resulted in processed PilA, as measured by SDS–PAGE. For optimal cleavage, especially with the purified enzyme, it was found that the reaction had to be supplemented with detergents (Triton X-100 or n-octylglucosides) and acidic phospholipids, with cardiolipin showing the highest stimulatory effect. Specific cleavage of the leader peptide between the glycine and phenylalanine residues in the reactions involving purified PilD and pre PilA was verified by microsequencing the aminoterminus of the processed pilin (8). These experiments verified that PilD is a specific type IV prepilin peptidase.

A number of known protease inhibitors and sulfhydryl-reactive reagents were tested for inhibition of the protease activity of PilD (8). The protease inhibitors phenylmethylsulfonyl fluoride (PMSF) N-tosyl-L-phenylalanine chloromethyl ketone (TPCK; serine protease inhibitors), pepstatin (acid protease inhibitor), and ethylenediaminetetraacetic acid (EDTA; metalloprotease inhibitor) had little or no effect on PilD activity. However, the sulfhydryl reagents N-ethylmaleimide (NEM) or p-chloromercuribenzoate (PCMB) significantly reduced PilD enzymatic activity. In addition, PCMB inhibition could be reversed by the subsequent addition of dithiothreitol (DTT), suggesting a role for the cysteine residues in the activity or PilD. This will be discussed further in a subsequent section.

Analysis by amino-terminal sequencing of cleaved *P. aeruginosa* PilA expressed by various pilus biogenesis gene mutants showed that PilA was not completely N-methylated when extracted from mutants containing lesions in other pilus biogenesis genes, specifically *pilB* and *pilC*, the two assembly related genes immediately upstream of *pilD*. Overall expression of PilD was decreased in these mutant strains and led to the hypothesis that PilD may also function as the N-methyltransferase of type IV pilins following endopeptidic cleavage. Addition of radiolabeled S-adenosyl-L-methionine (AdoMet) to an *in vitro* prepilin cleavage reaction using purfied PilD and prepilin resulted in transfer of the radiolabeled methyl group from AdoMet to the processed pilin (9). Methyltransferase activity could be inhibited by AdoMet inhibitors such as sinefungin, S-adenosyl-L-homecysteine, and S-adenosyl-L-methionine. Amino-terminal sequencing of purified pilin processed *in vitro* in the absence of methyl donor and comparison to pilin processed in the presence of AdoMet demonstrated specific conversion of the amino-terminal phenylalanine to N-monomethylphenylalanine. And lastly, purified PilD is unable to further methylate processed pilin isolated from wild-type *P. aeruginosa,* suggesting that under the experimental conditions used, either strictly mono N-methylation occurs or that other potential methylation sites are masked during pilus biogenesis. Purified PilD has also been shown to methylate a range of substrates other than pilin, including one of the other TFPP substrates in *P. aeruginosa,* pre-XcpT, as well as prepilin of *N. gonorrhoeae* (pre-PilE) (9, 69). Thus, it appears that all PilD substrates are probably N-methylated after proteolytic cleavage. With the demonstration that

A. hydrophila TapD is also capable of N-methylation of a heterologous substrate (*P. aeruginosa* PilA) (*16*), it is likely that all TFPPs are bifunctional enzymes with peptidase and N-methyltransferase activities.

VI. Role of Leader Peptide Cleavage and N-Methylation in Pilus Biogenesis and Type II Secretion

Early work on the location of specific secretion signals within the prepilin of *P. aeruginosa* demonstrated that the short, relatively hydrophilic signal peptide of type IV prepilins does not serve the same function as the typical, hydrophobic secretion signals found in proteins destined for secretion into the periplasm or outer membrane, or for export out of the bacterium (*21*). Instead, it is very likely that the highly conserved and hydrophobic N-terminal region of processed pilins and related proteins serve this role in promoting the translocation of proteins across the cytoplasmic membrane (*21*). That is not to say there is no role for the prepilin basic leader peptide, since it is possible that it performs a function similar to the cluster of positively charged amino acid residues at the N terminus of signal peptides of secretory proteins. However, although to date the exact role of the leader peptides of type IV pilins or related proteins has not been precisely defined, there are several possibilities for their function. They may have a protective function against proteolytic degradation while the precursor subunits are in a susceptible, monomeric form. They may be required to maintain subunits in a specific conformation required for specific interactions with other components of the biogenesis machinery, or may contain information that specifically targets the pilin to one or more steps in the assembly pathway. Alternatively, the leader peptide could serve to prevent subunit polymerization until translocation across the membrane can occur.

Little is known about the mechanisms involved in initiation and assembly of the type IV pilin monomers into pili. Cleavage of the leader peptide from prepilin probably occurs on the cytoplasmic face of the inner membrane. This possibly results in release of the pilin monomers into the periplasm (unlikely since no periplasmic pilin pools have been demonstrated) or to the periplasmic surface of the inner membrane. Initiation of assembly could then take place either by self-assembly, through interaction of the processed monomers with other biogenesis components, or with other structural components of the pilus filament. In this sense, the TFPP acts much the same as signal peptidase I, which is responsible for release of secreted proteins from the membrane by cleaving their membrane-bound anchors. This leads one to speculate why cleavage of type IV prepilins is not simply processed by signal peptidase I, since all of the type IV prepilins and pilin-like proteins that are substrates of TFPPs contain reasonable signal peptidase recognition and cleavage sites (5, 21). Obviously, endopeptidic cleavage of the leader peptides at any site within the hydrophobic core other than the correct site could lead to the loss of a domain or conformation essential for assembly or function of the mature proteins. Proper folding of the TFPP substrates as influenced by the positively charged leader peptide and hydrophobic domain may ensure that the cleavage site of the TFPP substrates is the correct one.

The role of N-methylation of type IV pilins and pilin-like proteins is less clear. Although the necessity of proteolytic processing of the leader peptide for subsequent type IV pilus biogenesis or type II protein secretion is unequivocal (8, 13, 15, 37, 38), there has been no experimental evidence demonstrating that methylation of the TFPP substrates is absolutely required for an essential biological function. Studies to determine the role of N-methylation are difficult to design, since proteolytic cleavage of the leader peptide is required to generate methyltransferase substrate for the same bifunctional enzyme. In studies where the substrate has been mutagenized (i.e., *P. aeruginosa* PilA or *K. oxytoca* PulG), it has been clearly demonstrated that change of a single amino acid (the glutamic acid at position +5 of the cleavage site) results in no pili being assembled in *P. aeruginosa* or loss of pullulanase secretion in *K. oxytoca*, respectively, with a complete or partial undermethylation of the substrate monomers (23, 70–72). However, the leader peptides are completely cleaved in these mutated substrates. It is difficult therefore to assess whether the methylation defect leads to the assembly defect, or whether lack of methylation is secondary to loss of protein function. In mutants of *P. aeruginosa* prepilin where the phenylalanine at the +1 position was substituted with a serine, the new mature N terminus was not methylated even though the substrate was fully processed and assembled into pili (23). What is needed to resolve the role of N-methylation in type IV pilus biogenesis and/or type II protein secretion are mutants in the methyltransferase domain of TFPPs that are not altered in their ability to proteolytically cleave TFPP substrate leader peptides. As more clearly described in Section VII concerning location of the methyltransferase active site, such mutants have been constructed and demonstrate that pili are assembled and the type II protein secretion machinery is functional when little (if any) of the respective substrates are N-methylated (73). The identification of specific methyltransferase inhibitors that would function *in vivo* to block N-methylation in growing bacteria would also help determine the answer to this question.

There are several speculative roles for N-methylation of the TFPP substrates. It is thought that type IV pili have the ability to retract via the assembly and disassembly of subunits (74). A form of motility ascribed to the type IV pili, namely twitching motility, may be a function of the constant formation and dissociation of pilin oligomers. It is also possible that the type IV pilin-like components of the type II secretion machinery and competence for DNA uptake may also involve similar assembly/disassembly actions on multimeric protein complexes. Reversible methylation and demethylation of pilin and/or pilin-like proteins may play an integral role in the formation of the oligomeric protein complexes and their subsequent dissolution. However, experimental evidence demonstrating disassembly of subunits is mostly indirect, and no experimental data exists to suggest that methylation of TFPP substrates is anything but irreversible.

Other roles for N-methylation of TFPP substrates include a possibility that it may be a mechanism to prevent proteolytic degradation prior to assembly of subunits into oligomeric complexes. Alternatively, N-methylated amino acids may aid the recognition of these proteins by assembly components or by specific chaperonins during normal biogenesis of type IV pili, extracellular protein secretion, or with the apparatus involved in DNA uptake. However, no studies have been performed to determine whether N-methylated proteins are inherently more stable then their nonmethylated counterparts. Moreover, there is no genetic or biochemical data that suggests specific chaperonins are required for type IV pilus biogenesis or assembly of the type II secretion apparatus.

VII. Structure–Function of the Type IV Prepilin Peptidases

Understanding the mechanisms underlying the catalysis of leader peptide cleavage and N-methylation, and whether methylation of processed substrates is required for maturation of type IV pili or the type II secretion machinery, is a necessary prerequisite for designing inhibitors of the enzyme. Such inhibitors have the potential for attenuating virulence of a number of pathogens. Comparison of TFPPs to known proteases shows them to be in a class by themselves, unrelated to any known families of proteases. The current MEROPS protease database [(75) and http://www.bi.bbsrc.ac.uk/Merops/merops.htm] places the TFPPs in a single family of proteases with unknown catalytic mechanisms. The TFPPs contain at least three potential functional domains: an anchoring domain to localize the enzyme in the cytoplasmic membrane; a domain responsible for leader peptide cleavage; and a domain for N-methylation of the newly exposed terminal amino acid residue. To date a true three-dimensional structure analysis of TFPPs has not been performed, so it is difficult to determine the exact location or extent of overlap of these domains. However, a variety of biochemical and site-directed mutagenesis studies have been carried out and are giving some insights into the structure–function of these enzymes.

As described previously, a TFPP has several hydrophobic transmembrane-spanning segments that are important for anchoring the enzyme within the membrane (Fig. 3). These hydrophobic residues undoubtedly are required for positioning the catalytic sites of the enzyme in the optimal orientation relative to the substrates for proteolytic cleavage and methylation. Theoretical considerations suggest that the active sites for these processes must be on the cytoplasmic face of the cytoplasmic membrane. First, it is known that substrates of the enzyme will associate with the inner membrane in its absence. It is likely that these substrates insert into the membrane in a conformation that allows only the correct peptide bond to be cleaved while masking all other cleavage sites. Because the basic leader peptide likely anchors the protein within the membrane, its cleavage can only be accomplished from the cytoplasmic side. In addition, the subsequent methylation step also has a topological requirement for the cytoplasmic side of the membrane. This is because the amino-terminal amino acid to be methylated must remain facing the cytoplasm, as the methyl-group donor, S-adenosyl-L-methionine (AdoMet), is synthesized and remains within the cytoplasm. Therefore, both reactions carried out by a TFPP must occur facing the cytoplasm. As described in the previous section, topological studies using protein reporter fusions bear out this model (Fig. 3).

In studies aimed at characterizing the two active sites of this bifunctional enzyme using biochemical tools and by mutant analysis, researchers have tried to determine whether the two sites are adjacent, whether the two reactions are coupled, or conversely, whether each reaction is carried out at distal domains independent of each other. Studies using the purified *P. aeruginosa* TFPP demonstrated that the rate of methylation of prepilin is essentially the same as cleavage, with methylation taking place as soon as there are free α-amino groups. This rate of methylation was the same as that observed with purified and already cleaved pilin, demonstrating that the two functions were not coupled (*9*). Effective methylation of substrate does not require that prepilin must first be bound to the enzyme, which implies that the same enzyme molecule that cleaves off the leader peptide is not necessarily the same one that methylates the substrate. Therefore, the methylation site can be distal to the cleavage site.

Another method used to determine the location of the AdoMet binding site was to examine the effect of interference with AdoMet binding on cleavage (*9*). This experiment took advantage of the methyltransferase inhibitor sinefungin, a natural structural analog of AdoMet that blocks the methyltransferase activity of TFPPs, presumably by binding to the AdoMet binding site, thereby excluding access of the methyl donor. Adding excess sinefungin with purified *P. aeruginosa* PilD in the presence of AdoMet completely blocked methylation of pilin without affecting leader peptide cleavage, even when the inhibitor was used at relatively high concentrations. Therefore, the AdoMet binding site is distinct from the catalytic site for leader peptide cleavage.

Analysis of the activity of TFPPs containing either engineered amino acid mutations or after reaction with specific protease inhibitors was carried out to determine the specific location of the catalytic sites. The first such study targeted the largely hydrophilic domain near the N terminus of the protein, approximately amino acids spanning 40–110 in PilD of *P. aeruginosa*. This region contains most or all of the cysteine residues of the TFPPs, four of which are arranged in a pairwise fashion, each pair two amino acids apart, with 21 amino acids separating the cysteine pairs (C-X-X-C ... X_{21} ... C-X-X-C, with the cysteine residues at postions 72, 75 and 97, 100 of PilD). This region was implicated in the reactions catalyzed by the enzyme based on the observations that both peptidase and *N*-methyltransferase activities were inhibited *in vitro* by thiol-specific reagents, including the cysteine-protease inhibitor *N*-ethylmaleimide (NEM) (*8, 21, 68, 76*). The role of the cysteines in enzyme function of PilD was confirmed by specific site-directed mutagenesis of individual cysteine codons and subsequent assessment of enzyme activity (*76*). Modification of the four cysteine residues making up the pairwise motif resulted in substantial reduction of both peptidase and *N*-methyltransferase activities *in vitro,* with the latter showing a greater decrease than the former. Mutagenesis of a fifth cysteine residue, not part of the two-pair motif (Cys[17]), had no effect on enzyme function. This finding seemed to argue for proximity of the two catalytic sites. Alternatively, cysteine modification by chemical reagents or by mutagenesis could lead to protein conformation changes that could adversely affect both activities, although the mutagenesis strategy involved conservative substitutions of serine for cysteine residues. An additional experiment demonstrated the close proximity of the peptidase and *N*-methyltransferase sites. Blocking the methyltransferase activity with sinefungin prior to alkylation with NEM resulted in no loss of proteolytic activity (*76*). This is further evidence that the catalytic sites are very likely adjacent, if not on the linear sequence, then in the final folded form of the active enzyme.

Examination of PilD and other TFPPs has failed to identify any amino acid sequence motifs associated with typical proteases, such as those associated with serine, histidine, acid, or metalloproteases (*75*). In the results described earlier where mutagenesis of the Cys residues in the pairwise Cys motif of PilD decreased both activities, *N*-methyltransferase activity was actually more reduced. This suggested that the Cys residues may form part of the methyltransferase active site. Indeed, a motif similar to the *E. coli* *Eco*RI (INGKCP) and *Eco*RII (INGKCS) methyltransferases was noted in PilD that included Cys-97 (i.e., LGGKC[97]S) (*76*). Interestingly, by this time the TFPP of *X. campestris* was cloned and characterized and found to not contain the two-pair cysteine motif, while still maintaining the ability to complement a *P. aeruginosa pilD* mutant strain (*10*). This suggested that the Cys residues may not be directly involved in enzyme catalysis but instead may be important for enzyme conformation. Therefore, a followup study to examine the role of the potential "methyltransferase box" in TFPP activity was undertaken (*73*). When 14 different TFPPs were aligned in this region, it was found that the Gly residue (Gly-95 in PilD) was invariant, while the residues aligning with Lys-96 were present in half of the sequences, with all but one of the others having a basic amino acid substitution at this position. Using site-directed mutagenesis, Gly-95 and Lys-96 were altered with a number of different substitutions creating both single and double mutations at these positions. *In vitro* cleavage assays showed no reduction in peptidase activity in all of the mutants, indicating that these two amino acids are not essential for prepilin cleavage activity. However, *in vitro* N-methyltransferase assays showed that versions of PilD with mutations at Gly-95 had no detectable activity, while PilD with mutations at Lys-96 had low to wild-type levels of methyltransferase activity. Introduction of PilD with the Gly-95 mutations into a *pilD P. aeruginosa* mutant strain resulted in the expression of apparently normal pili. Extracellular secretion via the type II pathway was similarly unaffected. Pilin monomers isolated from *P. aeruginosa* expressing PilD with the Gly-95 mutation were not methylated as expected from the *in vitro* data, although it is difficult to determine whether there is a complete methyltransferase block. In any event, these data strongly suggest that complete methylation of TFPP substrates is not essential for pilus biogenesis or extracellular protein secretion, and it remains to be determined why N-methylation of the TFPP substrates is a property that has been so highly conserved.

LaPointe and Taylor (*77*) have carried out a systematic study to identify the TFPP protease active site residues and at the same time have attempted to classify the protease. In this study they focused on specific amino acids that have been identified as important active site residues in other proteases, specifically Cys, Ser, Asp, Lys, His, and Glu amino acids. When they aligned 27 TFPP homologs using the DNASTAR MegAlign/Clustal alignment method, they found that 16 such potential residues (7 Ser, 4 Cys, 3 Asp, 1 Lys, and 1 Glu) were present in the majority of TFPPs, were predicted to be in domains that were located on the cytoplasmic face, and were aligned with the same residues of TcpJ, the TFPP of *V. cholerae* that processes the class B type IV prepilin TcpA. Only two residues, Asp-125 and Asp-189 (numbering relative to TcpJ) were found to be conserved in all TFPPs. The possible active site residues were mutated by site directed mutagenesis, and the ability of the mutated TcpJ to proteolytically cleave the leader peptide from TcpA was assessed *in vivo*. This was accomplished by cloning the mutated *tcpJ* gene into a plasmid construct already carrying *tcpA* and analyzing expression and prepilin processing in *E. coli* after SDS–polyacrylamide gel electrophoresis of whole cell extracts followed by Western blotting with anti-TcpA antiserum. Of all the residues altered in this manner, only mutations in Asp-125 and Asp-189 resulted in a complete block in peptidase activity, with only Asp-125 to Glu substitutions resulting in any residual peptidase activity. A mutation in Cys-48 (analogous to Cys-72 of PilD) resulted in partial peptidase activity, whereas other Cys mutations did not measurably decrease peptidase activity *in vivo*, and neither did any mutations in other potential protease-active site residues.

Further analysis of the importance of the Asp-125 and Asp-189 residues showed that they were required for peptidase cleavage of TcpA and pilus biogenesis in an *in vitro* cleavage assay and *in vivo* in a *V. cholerae* strain carrying mutations in *tcpJ* and *vcpD*, the gene encoding the second TFPP required for processing PilA/VcpA (a group A type IV pilin) and components of the *V. cholerae* type II secretion machinery (77). Furthermore, examination of TcpA processing with TcpJ containing a deletion of the large cytoplamic domain 1 eliminating the two cysteine-pair motif showed that this region was not required for peptidase activity, as such a construct exhibited wild-type peptidase activity. This construct was also able to functionally complement the *tcpJ, vcpD* double mutant *V. cholerae* strain, restoring expression of complete and apparently functional TCP pili. The requirement for the aspartic acid residues was also shown to be necessary for the peptidase activity of VcpD (77). Classification of the TFPPs as an aspartic acid protease was further strengthened by the demonstration that *in vitro* enzyme activity was substantially decreased (98%) by the specific aspartic acid protease inhibitor combination, 1-ethyl-3-(3-dimethylaminopropyl) carbodiimide hydrochloride (EDAC) plus glycinamide (77). Other specific cysteine, serine, and metalloprotease inhibitors had much less effect on peptidase activity. TcpJ activity was also resistant to pepstatin, an aspartic protease inhibitor. Interestingly, the cysteine protease inhibitor NEM, which substantially reduced the *in vitro* peptidase and methyltransferase activities of *P. aeruginosa* PilD (76), also reduced the peptidase activity of TcpJ by 85%. In contrast, the serine–cysteine inhibitor PMSF only reduced the peptidase activity of TcpJ by 11%. While these results suggest that the TFPPs are aspartic acid proteases, it should be pointed out that they differ from known acid proteases in that they have a requirement for a neutral pH and do not carry the amino acid motif (D(T/S)G) found in at least half of the aspartic acid proteases described to date. However, resistance to pepstatin is common to many aspartic acid proteases, including pepsin and the immunodeficiency virus type 1 protease. Indeed, the latter two proteases are bilobed molecules in which the active site cleft is located between two domains or lobes with each lobe containing one of the active site aspartic acid residues. An examination of the predicted topology of TFPPs using hydrophobicity profiles and protein fusions with β-lactamase, alkaline phosphatase, and β-galactosidase (*66, 67, 77*) shows that, in addition to the main cytoplasmic domain containing the cysteine residues, there are at least two additional peptide loops that extend into the cytoplasm (Fig. 3). Asp-125 of TcpJ is predicted to be in the middle loop/lobe and Asp-189 is on the third. The peptidase active site for TFPPs can then be envisioned as a cleft between the second and third lobes with one active site aspartic acid residue in each lobe.

One can conclude from the results presented by LaPointe and Taylor (*77*) that the TFPP family consists of bilobed aspartic acid proteases, and because of their insensitivity to pepstatin, they should be regarded as a novel family of non-pepsin-like acid proteases. Yet to be determined is the location of the active site for the TFPP N-methyltransferase activity, although the results presented to date appear to indicate that it lies in domain 1 near the cysteine motif (*73, 76*). Analysis of domain 1 deletion mutants for methyltransferase activity will help answer this question, and whether methyltransferase activity is an evolutionary holdover with no current necessary function, or whether there are substrates within the type IV pilus biogenesis or type II protein secretion pathways that require methylation for function. Since attenuation of virulence of several gram-negative pathogens carrying mutations in TFPP genes has been demonstrated (*15, 38*), understanding the location and type of enzyme active sites will aid the rational development of specific therapeutic inhibitors that may have practical value in the treatment of infections caused by these bacteria.

VIII. Substrate Specificity of Type IV Peptidases and Enzyme Kinetics

The peptide bond of the prepilin substrates that is cleaved by TFPPs contains an invariable glycine at the −1 position and a hydrophobic amino acid at the +1 position. The +1 position contains a phenylalanine in over 70% of the potential TFPP substrates, and this residue is in all type IV pilins placed in group A (*22*). The other amino acids at this position include methionine (found in 10%) followed by individual proteins with serine, tyrosine, and leucine. All prepilins that contain these amino acids at the +1 position are efficiently cleaved during all growth conditions. No accumulation of unprocessed pilin has ever been demonstrated that contain these residues, or in strains that are naturally hyperpiliated.

In a systematic study to determine the requirement for specific amino acid residues flanking the cleavage site for correct processing by a TFPP, random and site-specific mutagenesis of *P. aeruginosa* pilin was performed (*23*). Substitution of the glycine at position −1 to the cleavage site with a variety of amino acid residues showed that this residue was necessary and essentially invariant. Only a substitution with alanine resulted in any proteolytic processing by the *P. aeruginosa* TFPP PilD. All other substitutions resulted in no processed pilin *in vivo* and no expression of pili. However, it is interesting to note that natural variants of glycine at the −1 position in pilin expressed by *N. gonorrhoeae* have a different phenotype (*78, 79*). These natural variants have a serine in place of the glycine and are nonpiliated, but the pilin is processed at an alternate site. This aberrantly processed pilin is then released from the bacterium in a soluble form. The alternate processing site has been mapped to be between a leucine at +38 and an alanine at position +39, still within the hydrophobic N terminus of gonococcal pilin (*80*). The catalytic enzyme for this cleavage has not been identified and it seems doubtful that the gonococcal TFPP (PilD) is responsible because of the absence of the −1 glycine. However, this type of alternative processing at a distal location because of a mutation in the prepilin cleavage site was not observed in the *P. aeruginosa* study (*23*) or for any other TFPP substrate.

The number of different amino acids at the +1 position is highly restricted in the TFPP substrates, which implies that this position might be equally important in enzyme recognition of substrate or in hydrolysis of the peptide bond at the substrate cleavage site. However, this has proven not to be the case, as the mutagenesis study with *P. aeruginosa* prepilin demonstrated that the +1 position can tolerate a relatively high number of substitutions (*23*). The amino acid changes at the +1 position replacing phenylalanine that had no effect on leader peptide catalysis included several that should have drastically changed the local peptide environment at the cleavage site because of charge and polarity modifications. Substituting serine, cysteine, threonine, asparagine, histidine, and tyrosine resulted in normal processing and assembly of pilin into typical appearing pili. Placing adjacent glycines at the −1 and +1 position had no effect on processing of prepilin, but having glycine as the N-terminal amino acid prevented assembly of the cleaved monomers into pili. Replacement of the phenylalanine with aspartic acid resulted in a prepilin substrate that was only partially processed by PilD, appeared to be either expressed in lower amounts or was unstable, and was not assembled *in vivo* into pili.

A similar study on PulG, a pilin-like protein that is part of the type II secretion machinery in *K. oxytoca,* confirmed the requirement for glycine at the −1 position for TFPP processing. Substitution with alanine, valine, and glutamate led to a leader peptide processing defect and subsequently, an extracellular secretion defect (*72*). The +1 position in PulG also tolerated several different amino acids, with changes from phenylalanine to isoleucine,

leucine, or valine having no effect on processing or function. However, it is possible that both in *P. aeruginosa* PilA and in *K. ocytoca* PulG precursors, the substitutions may slow the rate of leader peptide catalysis without causing any observable phenotypic changes.

Other mutations generated in the leader peptide of *P. aeruginosa* prepilin also did not effect cleavage or pilus assembly. Wild-type prepilin has a leader peptide with a net charge of $+2$ due to the presence of two lysine residues. Altering the net charge of the leader peptide to -1, neutral, $+1$, or $+3$ by substituting one or both of the lysines with glutamic acid or by insertion of an additional lysine residue had no effect on cleavage and expression of pili (23). Most single mutations in the conserved hydrophobic domain also had no effect. Double and triple mutations were required to abolish prepilin processing and pilus biogenesis.

The apparent flexibility of the $+1$ position in the TFPP precursor cleavage site suggests that TFPPs are able to catalyze the reaction efficiently even when the substrate has a less than optimal structure. Kinetic studies on the *P. aeruginosa* TFPP PilD demonstrated that the cleavage of prepilin takes place with a K_m of 650 μM and k_{cat} of 180 min^{-1}, when both substrates and enzyme were part of enriched bacterial membranes (81). The high efficiency of the reaction is in part due to the high turnover rate measured in the substrate interaction with the enzyme (the high k_{cat}), and the relatively low affinity for substrate–enzyme binding. Further kinetic experiments examined the rates of PilD catalysis of unnatural substrates obtained by site-directed mutagenesis of the $+1$ phenylalanine codon. The results obtained demonstrated clearly that changes at this position do influence the catalytic rate, in spite of the fact that altered pilins are completely processed in growing bacteria with identical piliated phenotypes. Substitution of the phenylalanine with methionine, serine, or cysteine most strongly influenced the turnover rate without altering the affinity of the enzyme for the substrates. Interestingly, PilD processing of prepilin from *N. gonorrhoeae* demonstrated a higher k_{cat} than did *P. aeruginosa* prepilin, resulting in a more rapid processing. Substitution of the $+1$ phenylalanine with asparagine resulted in an overall prepilin cleavage rate (as measured by the catalytic efficiency ratio k_{cat}/K_m) that is comparable to that of wild-type prepilin. However, the actual measured turnover value (k_{cat}) for this substrate was considerably lower, suggesting that peptide bond catalysis was slower than seen on the wild-type substrate. This decrease appears to be compensated for by the apparent higher affinity of the enzyme for this substrate, as measured by the approximately 8-fold lower K_m than that measured for the wild-type substrate (81).

The kinetics of substrate catalysis where the substrate cleavage recognition sequence has been altered suggests that it may be possible to design enzyme inhibitors consisting of substrates with a sequence that is optimal for high affinity for TFPPs but with a low turnover rate. Such substrates could server two purposes. One would be to decouple the protease and N-methyltransferase activities, which could be used *in vivo* or *in vitro* to determine the role of substrate methylation in type IV pilus biogenesis or type II protein secretion. The second would be as potential therapeutic agents to attenuate pathogens during an infection. Of course other hurdles will have to be overcome in order for such a strategy to be effective, including delivery of the peptides to the site of infection and translocation into the cytoplasm of the targeted bacteria.

IX. Future Directions

The identification of type IV prepilin peptidases and their myriad substrates from numerous bacteria, and the characterization of their roles in extracellular protein secretion, adherence to eukaryotic cells, and uptake of DNA, clearly demonstrates their importance for bacterial function and pathogenesis. However, although we currently understand much about the enzyme in terms of catalytic mechanisms and substrates, several unanswered questions still remain. For example, we know that biogenesis of type IV pili and the assembly of the type II extracellular secretion machinery requires processing of specific substrates by the bifunctional TFPPs, but it is unknown why this two-step posttranslational modification is necessary. Removal of the leader peptide and subsequent N-methylation may be necessary prerequisites for entry into the assembly process, whether it be polymerization of pilin monomers into pili or maturation of the extracellular secretion pathway. However, if true, it is difficult to reconcile this hypothesis with data that suggest pilus assembly and protein secretion proceed in bacteria where TFPP-mediated N-methylation has been abolished or at least significantly decreased. Alternatively, the role of the leader peptides of TFPP substrates may be to prevent spontaneous polymerization of protein subunits until they reach their correct location in the cell membrane. Again, with this explanation it is difficult to assign a role for N-methylation of substrates after leader peptide cleavage.

The ability of TFPP homologs to sequentially carry out two unrelated enzyme reactions does present an interesting topic for studies on protease and methyltransferase catalytic mechanisms. TFPPs now appear to be a heretofore uncharacterized class of bilobed aspartic acid proteases, forming a unique group of proteases with a very specific set of substrates. The N-methyltransferase function also appears to be unique, only taking place on TFPP substrates where a leader peptide has been first removed by enzymatic cleavage. A more detailed analysis of TFPP *in vivo* functions is needed. These studies will require the isolation of defined mutants defective in either the protease or *N*-methyltransferase activity, or the identification of highly specific inhibitors of each of the functions. Such reagents are necessary before a complete understanding of the biological role of TFPPs and their substrates can be determined.

ACKNOWLEDGMENTS

We acknowledge funding from NIH grant AI21451 (SL) and the National Marine Fisheries Service, NOAA (MS).

REFERENCES

1. O'Toole, G. A., and Kolter, R. (1998). *Mol. Microbiol.* **30,** 295.
2. Kim, S. R., and Komano, T. (1997). *J. Bacteriol.* **179,** 3594.
3. Semmler, A. B., Whitchurch, C. B., and Mattick, J. S. (1999). *Microbiology* **145,** 2863.
4. Wall, D., and Kaiser, D. (1999). *Mol. Microbiol.* **32,** 1.
5. Pugsley, A. P. (1993). *Microbiol. Rev.* **57,** 50.
6. Nunn, D. (1999). *Trends Cell Biol.* **9,** 402.
7. Sauvonnet, N., Vignon, G., Pugsley, A. P., and Gounon, P. (2000). *EMBO J.* **19,** 2221.
8. Nunn, D. N., and Lory, S. (1991). *Proc. Natl. Acad. Sci. U.S.A.* **88,** 3281.
9. Strom, M. S., Nunn, D. N., and Lory, S. (1993). *Proc. Natl. Acad. Sci. U.S.A.* **90,** 2404.
10. Hu, N. T., Lee, P. F., and Chen, C. (1995). *Mol. Microbiol.* **18,** 769.
11. Dupuy, B., and Pugsley, A. P. (1994). *J. Bacteriol.* **176,** 1323.
12. Dupuy, B., Taha, M. K., Possot, O., Marchal, C., and Pugsley, A. P. (1992). *Mol. Microbiol.* **6,** 1887.
13. Zhang, H. Z., Lory, S., and Donnenberg, M. S. (1994). *J. Bacteriol.* **176,** 6885.
14. Nunn, D. N., and Lory, S. (1992). *Proc. Natl. Acad. Sci. U.S.A.* **89,** 47.
15. Paranjpye, R. N., Lara, J. C., Pepe, J. C., Pepe, C. M., and Strom, M. S. (1998). *Infect. Immun.* **66,** 5659.
16. Pepe, C. M., Eklund, M. W., and Strom, M. S. (1996). *Mol. Microbiol.* **19,** 857.
17. Strom, M. S., Nunn, D. N., and Lory, S. (1994). *Meth. Enzymol.* **235,** 527.
18. Froholm, L. O., and Sletten, K. (1977). *FEBS Lett.* **73,** 29.
19. Hermodson, M. A., Chen, K. C. S., and Buchanan, T. M. (1978). *Biochemistry* **17,** 442.
20. Sastry, P. A., Pearlstone, J. R., Smillie, L. B., and Paranchych, W. (1983). *FEBS Lett.* **151,** 253.
21. Strom, M. S., and Lory, S. (1987). *J. Bacteriol.* **169,** 3181.
22. Strom, M. S., and Lory, S. (1993). *Annu. Rev. Microbiol.* **47,** 565.
23. Strom, M. S., and Lory, S. (1991). *J. Biol. Chem.* **266,** 1656.
24. Roine, E., Raineri, D. M., Romantschuk, M., Wilson, M., and Nunn, D. N. (1998). *Mol. Plant Microbe Interact.* **11,** 1048.
25. Graupner, S., Frey, V., Hashemi, R., Lorenz, M. G., Brandes, G., and Wackernagel, W. (2000). *J. Bacteriol.* **182,** 2184.
26. Watson, A. A., Mattick, J. S., and Alm, R. A. (1996). *Gene* **175,** 143.

27. Elleman, T. C., Hoyne, P. A., and Lepper, A. W. (1990). *Infect. Immun.* **58,** 1678.
28. Beard, M. K., Mattick, J. S., Moore, L. J., Mott, M. R., Marrs, C. F., and Egerton, J. R. (1990). *J. Bacteriol.* **172,** 2601.
29. McNamara, B. P., and Donnenberg, M. S. (2000). *Microbiology* **146,** 719.
30. Stock, A., Clarke, S., Clarke, C., and Stock, J. (1987). *FEBS Lett.* **220,** 8.
31. Russell, M. A., and Darzins, A. (1994). *Mol. Microbiol.* **13,** 973.
32. Alm, R. A., and Mattick, J. S. (1995). *Mol. Microbiol.* **16,** 485.
33. Alm, R. A., and Mattick, J. S. (1996). *J. Bacteriol.* **178,** 3809.
34. Yoshida, T., Furuya, N., Ishikura, M., Isobe, T., Haino-Fukushima, K., Ogawa, T., and Komano, T. (1998). *J. Bacteriol.* **180,** 2842.
35. Lory, S. (1994). *In* "Signal Peptidases" (G. von Heijne, ed.). R. G. Landes Company, p. 31.
36. Bleves, S., Voulhoux, R., Michel, G., Lazdunski, A., Tommassen, J., and Filloux, A. (1998). *Mol. Microbiol.* **27,** 31.
37. Strom, M. S., Nunn, D., and Lory, S. (1991). *J. Bacteriol.* **173,** 1175.
38. Marsh, J. W., and Taylor, R. K. (1998). *Mol. Microbiol.* **29,** 1481.
39. Bally, M., Filloux, A., Akrim, M., Ball, G., Lazdunski, A., and Tommassen, J. (1992). *Mol. Microbiol.* **6,** 1121.
40. Lu, H. M., Motley, S. T., and Lory, S. (1997). *Mol. Microbiol.* **25,** 247.
41. Albano, M., Breitling, R., and Dubnau, D. A. (1989). *J. Bacteriol.* **171,** 5386.
42. Chung, Y. S., and Dubnau, D. (1998). *J. Bacteriol.* **180,** 41.
43. Chung, Y. S., Breidt, F., and Dubnau, D. (1998). *Mol. Microbiol.* **29,** 905.
44. Fussenegger, M., Rudel, T., Barten, R., Ryll, R., and Meyer, T. F. (1997). *Gene* **192,** 125.
45. Koomey, M. (1998). *APMIS Suppl.* **84,** 56.
46. Stone, B. J., and Kwaik, Y. A. (1999). *J. Bacteriol.* **181,** 1395.
47. Busch, S., Rosenplanter, C., and Averhoff, B. (1999). *Appl. Environ. Microbiol.* **65,** 4568.
48. Reyss, I., and Pugsley, A. P. (1990). *Mol. Gen. Genet.* **222,** 176.
49. Nunn, D., Bergman, S., and Lory, S. (1990). *J. Bacteriol.* **172,** 2911.
50. Pugsley, A. P., and Dupuy, B. (1992). *Mol. Microbiol.* **6,** 751.
51. Pugsley, A. P., and Reyss, I. (1990). *Mol. Microbiol.* **4,** 365.
52. Mohan, S., Aghion, J., Guillen, N., and Dubnau, D. (1989). *J. Bacteriol.* **171,** 6043.
53. Lindeberg, M., and Collmer, A. (1992). *J. Bacteriol.* **174,** 7385.
54. He, S. Y., Lindeberg, M., Chatterjee, A. K., and Collmer, A. (1991). *Proc. Natl. Acad. Sci. U.S.A.* **88,** 1079.
55. Reeves, P. J., Whitcombe, D., Wharam, S., Gibson, M., Allison, G., Bunce, N., Barallon, R., Douglas, P., Mulholland, V., Stevens, S., Walker, D., and Salmond, G. P. C. (1993). *Mol. Microbiol.* **8,** 443.
56. Lauer, P., Albertson, N. H., and Koomey, M. (1993). *Mol. Microbiol.* **8,** 357.
57. de Groot, A., Heijnen, I., de Cock, H., Filloux, A., and Tommassen, J. (1994). *J. Bacteriol.* **176,** 642.
58. de Groot, A., Krijger, J. J., Filloux, A., and Tommassen, J. (1996). *Mol. Gen. Genet.* **250,** 491.
59. Kaufman, M. R., Seyer, J. M., and Taylor, R. K. (1991). *Genes Dev.* **5,** 1834.
60. Fullner, K. J., and Mekalanos, J. J. (1999). *Infect. Immun.* **67,** 1393.
61. Sandkvist, M., Michel, L. O., Hough, L. P., Morales, V. M., Bagdasarian, M., Koomey, M., and DiRita, V. J. (1997). *J. Bacteriol.* **179,** 6994.
62. Karaolis, D. K., Somara, S., Maneval, D. R., Jr., Johnson, J. A., and Kaper, J. B. (1999). *Nature* **399,** 375.
63. Kirov, S. M., Barnett, T. C., Pepe, C. M., Strom, M. S., and Albert, M. J. (2000). *Infect. Immun.* **68,** 4040.
64. Felsenstein, J. (1993). PHYLIP (Phylogeny Inference Package). Department of Genetics, University of Washington, Seattle, WA.

65. Fitch, W. M., and Margoliash, E. (1967). *Science* **155,** 279.
66. Lory, S., and Strom, M. S. (1997). *Gene* **192,** 117.
67. Reeves, P. J., Douglas, P., and Salmond, G. P. (1994). *Mol. Microbiol.* **12,** 445.
68. Strom, M. S. (1992). "Analysis of determinants of pilus biogenesis in *Pseudomonas aeruginosa*" (Thesis, Microbiology), University of Washington, Seattle.
69. Nunn, D. N., and Lory, S. (1993). *J. Bacteriol.* **175,** 4375.
70. Pasloske, B. L., and Paranchych, W. (1988). *Mol. Microbiol.* **2,** 489.
71. Macdonald, D. L., Pasloske, B. L., and Paranchych, W. (1993). *Can. J. Microbiol.* **39,** 500.
72. Pugsley, A. P. (1993). *Mol. Microbiol.* **9,** 295.
73. Pepe, J. C., and Lory, S. (1998). *J. Biol. Chem.* **273,** 19120.
74. Bradley, D. E. (1972). *J. Gen. Microbiol.* **72,** 303.
75. Rawlings, N. D., and Barrett, A. J. (2000). *Nucleic Acids Res.* **28,** 323.
76. Strom, M. S., Bergman, P., and Lory, S. (1993). *J. Biol. Chem.* **268,** 15788.
77. LaPointe, C. F., and Taylor, R. K. (2000). *J. Biol. Chem.* **275,** 1502.
78. Jonsson, A. B., Pfeifer, J., and Normark, S. (1992). *Proc. Natl. Acad. Sci. U.S.A.* **89,** 3204.
79. Koomey, M., Bergstrom, S., Blake, M., and Swanson, J. (1991). *Mol. Microbiol.* **5,** 279.
80. Haas, R., Schwarz, H., and Meyer, T. F. (1987). *Proc. Natl. Acad. Sci. U.S.A.* **84,** 9079.
81. Strom, M. S., and Lory, S. (1992). *J. Bacteriol.* **174,** 7345.

Section II

Proprocessing

7

The Prohormone Convertases and Precursor Processing in Protein Biosynthesis

DONALD F. STEINER

Howard Hughes Medical Institute and Department of Biochemistry and Molecular Biology
The University of Chicago
Chicago, Illinois 60637

THE ENZYMES, Vol. XXII
Copyright © 2001 by Academic Press

I. Introduction and Historical Perspective

It is now widely appreciated that selective proteolysis of precursor proteins plays an essential role in many vital cellular processes, including embryonic development (1–5), gene expression (6), cell cycle progression (7), apoptosis (8), and intracellular protein topogenesis (9), as well as in endocrine regulation and neural function (10, 11). Recent work has revealed specialized families of proteolytic enzymes in many intracellular compartments and on cell surfaces (3, 6–8, 10, 11). Many of these systems play important roles in both normal and pathologic processes. In addition, there are the well-studied lysosomal and proteosomal proteolytic systems that also can regulate cellular balance via controlled degradation of key enzymes and/or transcription factors, as cells are reprogrammed in response to changing needs (12, 13).

The discovery of prohormones and their intracellular processing provided the first evidence that selective proteolysis plays a biosynthetic role within cells as an integral part of the posttranslational modification, intracellular transport, and sorting of newly synthesized protein molecules destined for export from the cell or to the plasma membrane. This set it apart from traditional zymogen activation, a mechanism for generating active enzymes from stable, inactive, usually secreted forms that had been known since the 19th century. Although proproteins usually also have attenuated biological activity, this is only one facet of the many functions they fulfill in enabling cells to produce product proteins or peptides efficiently and either to store them for later secretion in large quantities on demand (via dense core granules) or to release them rapidly in an unregulated manner (via small constitutive vesicles).

Knowledge of the existence of proproteins and of mechanisms for their intracellular processing began with the discovery of proinsulin and from early studies on its biosynthesis, conversion to insulin, and storage within the beta cell as insulin and C-peptide (14–16). This work and subsequent discoveries of other secreted precursor proteins, such as proparathyroid hormone (17), the ACTH/endorphin precursor proopiomelanocortin (18), proglucagon (19), proalbumin (20), and promellitin (21), in the early 1970s, indicated the greater generality of this mechanism, setting the stage for the later explosive developments in intracellular proteolysis.

A. PROINSULIN: THE FIRST BIOSYNTHETIC PRECURSOR PROTEIN

The discovery of proinsulin grew out of efforts to understand how the small two-chain insulin molecule (22) is assembled. Early evidence suggested that the two chains might be synthesized separately and then combined; in 1960 it was shown that insulin could be fully reduced into its constituent A and

B chains, which could then be recombined to active molecules by simple air oxidation to reform the disulfide bridges (23). This observation spurred several groups of synthetic organic chemists to attempt the synthesis of insulin (24), and this goal was achieved in the early 1960s (25–27). By the mid 1960s, yields of up to 50% or more were reported using directed chain recombination methods (28). It thus seemed likely that the biological mechanism of synthesis could operate similarly through chain combination, possibly via reversal of an enzymatic reaction characterized by Tomizawa in 1962 (29), the so-called gluthathione-insulin transhydrogenase, an enzyme that catalyzed insulin reduction by glutathione. This enzyme ultimately proved to be identical to protein disulfide isomerase, or PDI, which is now known to catalyze protein sulfhydryl oxidation in the endoplasmic reticulum (ER) during protein folding.

To examine the mechanisms of insulin biosynthesis required a robust system that would permit insulin, or any intermediates, that might exist in its biosynthesis to be readily identified. Although in the 1950s numerous attempts were made to study insulin biosynthesis using whole pancreas, it was difficult even to demonstrate the incorporation of labeled amino acids into the hormone in such systems because of the great excess of pancreatic exocrine enzymes (30). Earlier attempts to use a nonisotopic approach to detect a hypothetical single-chain form of insulin also failed to reveal any evidence for such a proinsulin, since the major form of the hormone stored in the pancreatic gland was mature insulin (31). This enabled pharmaceutical companies such as Eli Lilly, Inc., and Novo A/G to efficiently extract animal (beef, pork) glands to prepare large quantities of insulin for therapeutic use in the treatment of diabetics.

1. *Studies of Insulin Biosynthesis in Human Insulinomas*

The key to finding proinsulin was the use of highly enriched beta cells as a system to study insulin biosynthesis. In the mid 1960s, the only suitable sources of pure β cells were rarely occurring β cell tumors, or "insulinomas," in human subjects, or the large single islets (Brockman Bodies) of teleost fishes (32, 33). When a human insulinoma became available in the University of Chicago Hospitals in the fall of 1965 after its surgical excision from the pancreas of a patient who suffered from severe hypoglycemic episodes, my laboratory was fortunate to obtain a fresh sample of it (14). This small (approximately 1 cm) tumor mass consisted mainly of β cells and turned out to contain several *milligrams* of stored insulin. Tumor slices were labeled for 2.5 hr with medium containing either radioactive leucine or phenylalanine in separate incubations. Since these two amino acids are distributed very differently in insulin (Fig. 1A), it was hoped that they would aid in characterizing any intermediates of biosynthesis such as insulin chains or larger precursor molecules that might be detectable.

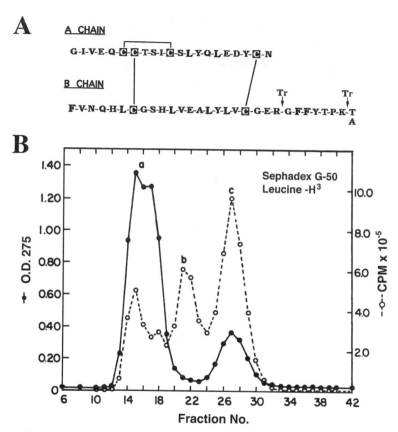

FIG. 1. The experimental design for studying insulin biosynthesis in a human islet cell tumor that led to the identification of proinsulin (*14*). (A) Structure of human insulin highlighting the placement of leucine (L) and phenylalanine (F) residues. Note that Phe is present only in the B chain, at the N terminus, and in a peptide that can be released by trypsin (Tr), whereas Leu is internal in both the A and B chain in a ratio of 1:2. These tritiated amino acids were incubated at 37° separately with slices from the tumor for 2.5 hr prior to extraction and analysis. Note also that the C-terminal residue of the bovine and porcine B chain is Ala (A) rather than Thr. (B) Gel filtration profile of leucine labeled material from the preceding experiment. Peaks designated a, b, and c are, respectively: a, high molecular weight proteins; b, proinsulin peak (MW 9–10 kDa); c, insulin monomer (~6 kDa). (The C-peptide was not recovered by the extraction method used.) Lower panel reproduced from (*14*).

To extract the labeled material, acid–ethanol was used to solubilize all insulin-related peptides. The resultant extracts contained large amounts of radioactivity and were then fractionated over columns of Sephadex G-50 in 1 *M* acetic acid (*14*). This separated the radioactivity into three peaks, which were each recovered and used for characterization; the first peak (component *a*) eluted in the void volume and consisted of much higher molecular weight proteins with associated spectrophotometrically measurable protein (Fig. 1B). Following this was a distinct peak of material of intermediate molecular size (9–10 kDa), which had little associated optical density at 275 μm, followed by a larger radioactive peak having a distinct associated OD_{275} peak. The latter peak, termed component c, was identified as insulin (M.W. $\sim$6000), which clearly was a major component of this well-differentiated tumor.

The earlier-eluting intermediate-sized peak (component b) had an apparent molecular weight of about 9–10 kDa and was of particular interest because few, if any, proteins of that size were known to exist at that time. Its very high specific radioactivity, much higher than that of the insulin peak, was compatible with the possibility that it might be a precursor. Its relationship to insulin was verified by demonstration of its cross-reactivity with insulin antisera. When aliquots of component b were incubated with trypsin, it was quantitatively converted to a form that coeluted with insulin with no other evident products (*14*). By using graded amounts of trypsin it was possible to show that conversion required very low levels of trypsin. Further incubations of component b with higher levels of trypsin demonstrated that the expected phenylalanine-containing heptapeptide fragment of the insulin B chain could be generated, along with desoctapeptide insulin, a form lacking the C-terminal eight amino acids of the B chain (*14, 33*) (Fig. 1). This provided further evidence that component b contained insulin as part of a higher molecular weight molecule.

2. *Component b (Proinsulin) Is a Single-Chain Protein*

The next important objective was to determine whether component b was a single polypeptide chain. However, after sulfitolysis or formic acid oxidation to open its putative disulfide bonds, the b component adhered to Sephadex G-50 columns, presumably because it was now disordered. This problem was overcome by using either 8 *M* urea or 50% acetic acid as a more chaotropic solvent for gel filtration and resulted in the demonstration that the oxidized b component did not release free insulin A or B chains, but rather remained as a single higher molecular weight component in contrast to the behavior of insulin under the same conditions (*14, 33*). This led to the conclusion that component b indeed was a single-chain form of insulin.

Since component b had been labeled with either phenylalanine or leucine, it was possible to answer yet another important question about the single-chain component, namely whether it began with the B chain or the A chain of insulin. Phenylalanine is the N-terminal amino acid of the B chain while the A chain begins with glycine (Fig. 1A). A microanalysis using the Sanger fluorodinitrobenzene technique to tag the N-terminal amino acids in both Phe- and Leu-labeled component b and insulin (component c) derived from the tumor was carried out on small amounts of the highly radioactive material

(*14*). The results with Phe-labeled component *b* clearly demonstrated that it contained an N-terminal phenylalanine residue, whereas the leucine labeled material did not yield a labeled dinitrophenyl derivative. Pretreatment of component *b* with trypsin before analysis did not change the relative amount of labeled N-terminal phenylalanine released, which amounted to approximately one-third of the total radioactivity of the molecule, consistent with the presence of only three phenylalanines in component *b*, as in insulin. From these findings it was concluded that component *b* was a single-chain precursor of insulin that began with the B chain and ended with the A chain of insulin (*14*). The latter conclusion was possible because tryptic digestion did not alter the electrophoretic mobility of the A chain–like component derived from component *b*, suggesting that it lacked a C-terminal basic residue and therefore must lie at the C terminus of the single chain.

However, despite the exciting nature of these findings, it was not yet clear that component *b* preceded insulin in biosynthesis. To conclusively demonstrate a precursor–product relationship would require pulse-chase experiments on more tumor material. In early 1966 we obtained a second insulinoma from a patient at the University of Iowa Hospital. We studied this tumor material there in the laboratories of helpful colleagues (*15*). Tumor slices were pulse-labeled for various times and then chase-incubated for relatively short periods on the assumption that conversion of the precursor to insulin must be a relatively rapid process. However, the chase times chosen, 20 and 40 min, turned out to be too short to demonstrate significant conversion to insulin of labeled component *b*, which was prominent in this tumor, whereas insulin was present at low levels. Thus, although the results supported a precursor role for component *b*, we had obtained a less well-differentiated tumor with impaired ability to convert proinsulin to insulin, a not uncommon feature of insulinomas.

3. *Rat Islet Biosynthetic Studies Demonstrate a Precursor–Product Relationship*

In late 1966 the first reports appeared of the isolation of islets of Langerhans from the rodent pancreas by an enzymatic digestion procedure devised by Moskolewski in Poland and perfected by Paul Lacy and co-workers at Washington University in St. Louis (*34, 35*). We undertook to isolate rat islets with rather limited success at first, but soon were able to routinely obtain adequate numbers of pure islets and carry out pulse-chase experiments with these, using various labeled amino acids. These experiments provided the definitive evidence that component *b* was indeed a precursor of insulin, and on the strength of these results, we now designated it *proinsulin* (*15*). However, a particularly striking aspect of our pulse-chase findings was the relatively slow rate of conversion of proinsulin to insulin (half-life of approximately 1 hr) and the surprising fact that no labeled insulin was detectable until 25–30 min after the beginning of the chase (*16*). It thus became obvious that proinsulin conversion was a late phenomenon relative to the biosynthesis of the peptide chain, and this along with other evidence derived from the effects of inhibitors led us to propose in 1970 that conversion of proinsulin to insulin occurred in the Golgi cisternae or in the early phases of granule maturation (*36*). Our findings and those of Howell and Taylor (*37*) showed that newly synthesized insulin was not released in significant amounts in response to glucose until 2 or 3 hr after its biosynthesis. Based on these results and emerging knowledge of the organization of the secretory pathway (*38*), we envisioned an orderly intracellular progression, e.g., biosynthesis in the RER, transfer to the Golgi complex, packaging of prohormone and convertase into secretory vesicles, and maturation of these vesicles in the cytosol via proteolysis, ultimately with acquisition of competence for secretion. This model was confirmed by subsequent elegant studies of Orci and colleagues (*39*) (Fig. 2). More details on these and other early experiments on insulin biosynthesis are available in Refs. *30, 33, 36, 40* and *41*.

4. *Purification of Proinsulin from Commercial Insulin Preparations*

Although we now had evidence that proinsulin was a precursor of insulin and was a single-chain protein that was about 1.5 times the size of insulin, we had no clear idea as to the nature of the additional peptide material between the B and A chains. We were particularly puzzled by the lack of any apparent product generated during its conversion to an insulin-like component by trypsin. To further examine this issue and get at its primary structure, we needed to purify larger amounts of proinsulin. Since it was unlikely to be a major component in pancreatic extracts, we looked for other sources. We examined various commercial preparations of insulin in the hopes that proinsulin might be present as a contaminant, and this indeed proved to be the case (*14, 16*). Crystalline preparations of insulin, especially those that had not been subjected to repeated recrystallization, contained 1–2% of proinsulin-like material (relative to insulin), as determined by gel filtration. However, component *b* from crystalline insulin preparations was not pure single-chain proinsulin, but contained a large proportion of two-chain intermediate cleavage products, as well as a significant proportion of a dimeric form of insulin, a chemical artifact generated by acid catalyzed isopeptide bond formation between insulin molecules during the isolation procedure (*42*). With the helpful cooperation of scientists at Novo Laboratories, beginning in 1967, we were able to obtain gram amounts of component *b* from gel filtration of commercial bovine insulin preparations, and with this as starting material we undertook the further purification of this fraction to obtain pure proinsulin and intermediate forms. By combining prior gel filtration with column chromatography on cationic and ionic cellulose columns in

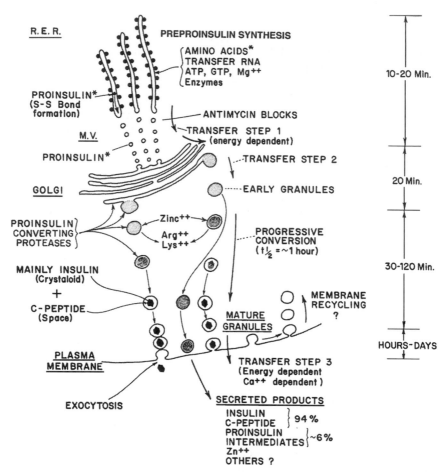

FIG. 2. Model for the subcellular organization of the insulin biosynthetic machinery of the β cell. Synthesis of preproinsulin occurs in the RER and results in translocation of the nascent proinsulin chain into the ER with cleavage of the signal peptide. After folding, proinsulin is transferred to the Golgi apparatus, where it moves from *cis* to *trans* leaflets. It is then packaged into secretory vesicles along with convertases (PC1/PC3, PC2 and CPE), which act as the granule pH drops to ~5.0. Insulin and C-peptide are retained in the granules and released together. Approximate times for each stage are shown on the right side. Reproduced with modification from (*40*).

7 *M* urea-containing buffers, we were able to develop a method that resolved intact proinsulin, intermediate components, and insulin dimers effectively (*42*). Working independently with porcine material at the Lilly Laboratories, Ronald Chance developed a somewhat similar method for purification of porcine proinsulin (*43*).

5. *Reduced Proinsulin Refolds to Regain Native Structure More Efficiently Than Insulin*

By 1968, we had obtained sufficient amounts of intact bovine proinsulin to determine its N-terminal residues by dansylation and to demonstrate that it begins with the phenylalanine of the B chain of insulin (and lacks a free A-chain N terminus) as expected from our results on the labeled human proinsulin from the insulinomas (*42*). Using this purified material we were also able to demonstrate that proinsulin, after complete reduction in 8 *M* urea, readily refolded to reconstitute the native disulfide bond structure of insulin when diluted into a suitable medium and exposed to air. Efficiencies of air reoxidation of up to 60% were achieved with proinsulin, in striking contrast to very meager yields with insulin chains (*44*). These results clearly supported our initial view that proinsulin must exist in order to solve the problems of chain combination in the biosynthesis of the hormone.

We then began efforts to sequence bovine proinsulin (*33*). However, before our structural analysis was completed, the primary structure of porcine proinsulin was reported by Chance *et al.* at Lilly Laboratories, confirming the proposed single-chain structure and revealing the existence of paired basic residues at the B and A chain junctions, explaining the ability of trypsin to rapidly release insulin-like material from it (*43*). The structure of bovine proinsulin was very similar to that of the porcine prohormone, except for a number of differences in the connecting peptide, including a three-residue deletion (*45*).

6. *Identification of the Proinsulin C-Peptide as a Cosecreted Product*

In studies on insulin biosynthesis in the rat, Jeffrey Clark, a graduate student in the laboratory, observed that polyacrylamide gel electrophoresis of the insulin-containing peak from rat islets labeled with radioactive amino acids revealed a rapidly migrating labeled component that contained proline and leucine, but not phenylalanine (*46*). Based on its compositional similarity to the bovine and porcine connecting segments, we surmised that this material was the rat connecting segments derived from the conversion of proinsulin to insulin. We had failed to detect these peptides in our earlier experiments because they comigrated with insulin on gel filtration. This stimulated us to try a similar electrophoretic approach to identifying the connecting segment in pancreatic extracts. Aliquots of insulin-containing fractions prepared by gel filtration of acid ethanol extracts of bovine or human pancreas were electrophoresed on paper in 30% formic acid and the electrophoretograms developed either with the standard Pauly stain or with ninhydrin. The Pauly stain (which reacts only with histidine- and tyrosine-containing peptides) gave only a single spot that was easily identified as insulin. However, ninhydrin revealed the presence in similar amounts of another peptide component with a slower anodal mobility than insulin. Compositional analysis revealed it to be the entire proinsulin connecting segment lacking the four basic amino acids (47). This natural product was dubbed the *proinsulin C-peptide*. Studies with rat islets showed that labeled insulin and C-peptide were generated and released in equivalent amounts from islets *in vitro* and cosecreted upon stimulation (48), suggesting that both peptides were stored together in the insulin-containing granules.

7. Structural Studies on C-Peptide

Once we had detected the C-peptide in pancreatic extracts, it was possible to readily purify it because it was present in relatively much larger amounts than proinsulin, which was present at levels of only 1–2% of total insulin. We first purified the C-peptide on a larger scale from bovine pancreas (49). In characterizing it, we prepared chymotryptic digests and found that these gave identical fingerprint patterns to those of the corresponding peptide derived from bovine proinsulin. We then undertook its primary structural elucidation to establish that it was identical in sequence. To do this, we developed a microadaptation of the manual Edmon degradation procedure, and by this means were able to sequence the entire 26-residue bovine C-peptide using just a few milligrams of peptide. We compared our sequence with that for the connecting segment of bovine proinsulin, then in progress in our collaborators' laboratory, and confirmed that the pancreatic C-peptide included the entire proinsulin connecting segment except for the basic amino acids linking it to the insulin chains (45, 49). We then went on to elucidate the structures of the major proinsulin intermediate cleavage components and showed that these consisted of material that was cleaved either at the B-chain–C-peptide junction or at the C-peptide–A-chain junction and lacked the pairs of basic residues (45). In subsequent studies, we determined the sequences of the C-peptides of the rat, dog, horse, and sheep (50, 51).

8. Structure of Human C-Peptide and Proinsulin

The primary structure the human C-peptide was similarly carried out by micro-Edmon degradation. This required extracting C-peptide from large numbers of human pancreas specimens at autopsy in the University of Chicago Pathology Department. This groundwork was carried out by Sooja Cho, an excellent technician who went several times each week to the morgue to collect pancreases for more than a year. This was necessary because postmortem autolysis greatly lowered yields, but we were eventually able to obtain enough purified intact human C-peptide to determine its amino acid sequence unequivocally (52). From the structure of the 31-residue human C-peptide, we could infer the structure of human proinsulin. However, to confirm this, we purified a small amount of proinsulin from several crystalline human insulin samples obtained from the Novo Research Laboratories in Copenhagen (53). We did not have enough human proinsulin for a complete amino acid sequence analysis with the methods then available, so we instead obtained its amino acid composition (52). This clearly showed that in addition to the amino acids present in human insulin and our human C-peptide sequence, there were only four additional amino acids—three arginines and one lysine. These results indicated that human proinsulin was identical in structure at the C-peptide B-chain and A-chain junctions to bovine and porcine proinsulin, both of which had Arg-Arg at the B-chain junction and Lys-Arg at the A-chain junction. We therefore proposed the structure of human proinsulin shown in Fig. 3, and this was ultimately confirmed by the cloning of the human insulin cDNA (54).

The availability of the human C-peptide enabled Arthur Rubenstein, a young physician who had joined my laboratory group in 1968, to prepare antibodies and develop a radioimmunoassay for the human C-peptide and proinsulin (53, 55). With these and other tools, he was soon able to demonstrate that both proinsulin and C-peptide are normally secreted in man, circulate in the blood, and also appear in the urine (56), and these observations led to a series of important developments in the clinical application of basic knowledge on proinsulin and C-peptide secretion in the diagnosis and treatment of diabetes and other metabolic disorders (57). This clinically important area will not be discussed in detail here, but has been reviewed (58, 59).

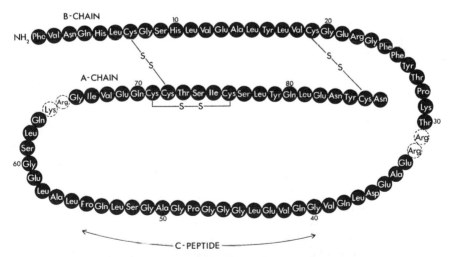

FIG. 3. Covalent structure of human proinsulin based on the structures of human insulin and C-peptide and compositional data on human proinsulin, indicating the presence of conserved Arg and Lys residues, as in bovine and porcine proinsulins. Reproduced from (52).

B.   MECHANISM OF PROTEOLYTIC CONVERSION OF PROINSULIN

1.   *Studies with Isolated Secretory Granules from Rat Islets*

The fact that pancreatic insulin and C-peptide both lacked the pairs of basic amino acids at the cleavage sites indicated that the cellular mechanism for conversion of proinsulin to insulin must result in the excision of these basic amino acids. We knew that trypsin could rapidly excise the C-peptide, but trypsin treatment left residual basic amino acids at the C termini of both the B chain and the C-peptide. To investigate the conversion mechanism further, we developed methods for isolating secretory granules from isolated islets of Langerhans that had been pulse-labeled with radioactive leucine and then chase incubated for a sufficient period of time to allow the newly synthesized proinsulin to be transferred into the maturing secretory granules (*60*). The isolated secretory granules were then incubated *in vitro* under various conditions to preserve their integrity, and we were able to demonstrate that conversion of proinsulin to insulin proceeded nicely in this isolated granule system. Disruption of the granules inhibited processing, but externally added protease inhibitors failed to inhibit conversion (*61*). We then tried pulse-labeling with ^{3}H-arginine so that we could specifically follow the fate of these residues. We were particularly interested in determining whether these basic amino acids were liberated as a pair or as single amino acids during conversion *in vitro*. To do this we developed a thin-layer chromatographic procedure for separating lysine and arginine. With this we could show that conversion led to the release of free lysine and arginine without any evidence for the production of the basic dipeptides (*61*). It thus seemed very likely that a carboxypeptidase B–like activity was involved in the conversion of proinsulin to insulin.

2.   *Trypsin and Carboxypeptidase B as a Model System for Proinsulin Processing to Insulin*

While efforts to identify converting enzymes in the isolated granule fractions continued, we also undertook studies with purified trypsin and carboxypeptidase B to see whether these enzymes in various proportions might reproduce the *in vivo* pattern of cleavage of proinsulin to insulin (*62*). We already knew from our results and those of Ronald Chance that tryptic digestion of proinsulin led to cleavage initially at the Arg-Arg and Lys-Arg sites to liberate C-peptide Lys-Arg and diarginyl insulin, which has two additional arginine residues at the end of the B chain. Diarginyl insulin is highly susceptible to further tryptic cleavage at lysine-B29, which results in the liberation of the tripeptide Ala-Arg-Arg and conversion of the diarginyl insulin to des-alanyl insulin, lacking alanine-B30 (see Fig. 1A). It seemed likely that if carboxypeptidase B were present during tryptic digestion, it would rapidly remove the two arginines exposed by tryptic cleavage after Arg-32, and this should prevent tryptic cleavage at Lys-29, as it was already known from work of Wang and Carpenter (63) that the Lys–Ala bond at B29–B30 in native insulin is not very susceptible to tryptic digestion. Indeed, with insulin, tryptic cleavage occurred after the B22 arginine before it occurred after lysine B29, generating only des-octapeptide insulin, whereas trypsin readily generated desalanyl insulin from proinsulin.

When various combinations of trypsin and carboxypeptidase B were tested for conversion of proinsulin, this tactic worked well immediately, generating the natural products, i.e., C-peptide lacking the basic amino acids, and native insulin with an intact B30 alanine at the end of the B chain. This gratifying result was rapidly confirmed in a number of experiments, and measurements were also made of the release of the basic amino acids; the products were carefully characterized (62). The yields were very high, well over 95% as judged by polyacrylamide electrophoresis. A similar method is used today to efficiently convert biosynthetic human proinsulin to insulin in the production of insulin for diabetes therapy at the Lilly Company.

It thus was clear that a combination of trypsin-like and carboxypeptidase B–like activities constituted a reasonable model for the mechanism acting in the beta cell in the conversion of proinsulin to insulin. Accordingly, a visiting scientist from Germany, Hartmut Zühlke, undertook experiments to detect carboxypeptidase B-like activity in islet extracts and secretory granule preparations, and was able to demonstrate the presence of an activity that was similar in its specificity to carboxypeptidase B, but had a lower, more acidic pH optimum (64). We did not pursue the purification of this islet carboxypeptidase B-like activity because we were committed to tracking down the more elusive endoproteolytic activity, which seemed to disappear once the secretory granules were opened. In 1982 Lloyd Fricker, working at Johns Hopkins in the laboratory of Sol Snyder, discovered a similar carboxypeptidase B-like activity in brain and went on to isolate and characterize it (65). It turned out to be carboxypeptidase E (CPE), a homolog of the exocrine pancreatic enzyme, carboxypeptidase B. CPE occurs in the islets and other neuroendocrine tissues (66) and is a major processing carboxypeptidase in some other tissues as well.

C.   THE EMERGENCE OF MULTIFUNCTIONAL PRECURSORS (POLYPROTEINS)

In the early 1970s evidence began to accumulate indicating that precursor processing was not restricted to complex two-chain molecules such as insulin. The first evidence came from studies of albumin biosynthesis, which indicated the existence of a short-lived precursor in the biosynthesis of this constitutively released serum protein (20). Proalbumin turned out to have a short N-terminal propeptide of six residues with a dibasic cleavage site ($R\cdot R\downarrow$) similar to that of proinsulin (*67*). Almost simultaneously, evidence was found by David Cohn and his associates (*68*) and by John Potts's group at the Massachusetts General Hospital (*69*) that the 84-residue parathyroid hormone (PTH), a single-chain protein lacking disulfide bridges, is also derived from a precursor with an N-terminal hexapeptide extension linked via a pair of basic residues to PTH. Further studies indicated that both proalbumin and proPTH were processed posttranslationally before release. In both instances the propeptide was clearly playing some role other than aiding in the folding of the mature protein.

In my laboratory, Howard Tager undertook a search for a proglucagon, which resulted in the isolation from crystalline pancreatic glucagon of a possible proglucagon fragment. It consisted of glucagon with a C-terminal extension of eight amino acids linked to the hormone via a Lys-Arg doublet (*19*) and was subsequently isolated from intestine and called oxyntomodulin (*70*). Cristoph Patzelt later on showed that proglucagon is an 18 kDa protein processed in islets to release glucagon and a 10 kDa fragment called major proglucagon fragment (MPGF) (*71*).

The demonstration by Rubenstein *et al.* (*55, 56*) that some proinsulin circulated in the blood along with insulin and was excreted in the urine prompted investigators to examine the heterogeneity of circulating forms of other peptide hormones using radioimmunoassay techniques, and these studies demonstrated the existence of larger immunoreactive forms in many endocrine systems, including gastrointestinal hormones, such as gastrin, and pituitary adrenocorticotropin (ACTH), as well as others (*72*). However, although these methods indicated size heterogeneity in circulating forms, they did not definitively establish the existence of biosynthetic precursors.

It had also been known since the early 1960s that the 39-residue ACTH molecule contained α-MSH (*N*-acetyl $ACTH_{1-13}$) within it followed by the sequence Lys-Lys-Arg-Arg, but the isolation of CLIP (corticotropin-like intermediate lobe peptide; $ACTH_{18-39}$) by Scott *et al.* (*73*) strengthened a possible precursor product relationship of ACTH and α-MSH. The discovery of the opioid peptides, Leu and Met enkephalin, and β endorphin in brain and pituitary gland and their identity with the C-terminal 31-residue region of β lipotropin (β LPH) (*74*), along with evidence that β MSH also was embedded within the γ lipotropin sequence, provided further tantalizing hints of a possible precursor–product relationship (*75*). Although it was evident that all of these forms coexisted in pituitary corticotrophs, as well as in pituitary intermediate lobe melanotrophs (*76*), it was not until Mains *et al.* (*77*) and Roberts and Herbert (*78*) in 1977 demonstrated the coexistence of the epitopes of β LPH/β endorphin and ACTH/α-MSH in a single large protein of 31 kDa that the existence of the first polyprotein precursor, proopiomelanocortin (POMC), was established (Fig. 4).

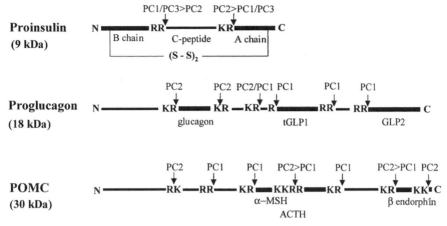

Fig. 4. Structural organization of three well-studied peptide hormone precursors—proinsulin, proglucagon, and proopiomelanocortin (POMC). The convertase(s) identified as responsible for processing at each cleavage site are indicated. Under normal conditions, proglucagon is differentially processed in the pancreatic islet alpha cells and the intestinal mucosal L cells. In the alpha cells, PC2 acting alone releases only active glucagon, whereas in the intestinal L cells, PC3 is the predominant convertase and acts to release only active forms of glucagon-like peptides GLP-1 and GLP-2. Similarly, in the pituitary anterior lobe corticotrophs, PC3 is the predominant convertase, and its action results in the release of intact ACTH and small amounts of beta endorphin. In the pituitary middle lobe, both PC2 and PC3 are expressed, resulting in the further processing of ACTH to α-MSH with increased production of beta endorphin. Thus, PC2 and PC3 have an absolute specificity for some sites in precursor proteins, whereas at others, both enzymes may act, as indicated above. Reproduced from (*111*); tGLP1 is GLP1$_{(7-37)}$.

In subsequent studies, two other opioid peptide precursors were identified, which both contained multiple copies of enkephalin-related sequences. These were proenkephalin and prodynorphin, both large polyproteins (*79*). In 1980, the longstanding mystery surrounding the mechanism of biosynthesis of the posterior pituitary peptide hormones was resolved by biosynthetic studies showing that both oxytocin and vasopressin were derived from precursors that contained their corresponding binding proteins, the neurophysins (*80*), as hypothesized earlier by Sachs *et al.* (*81*). Then, in the late 1970s and early 1980s, with the development of molecular cloning techniques, the structures of the precursors of many of the remaining known neuroendocrine regulatory peptides were rapidly revealed, either confirming prior biosynthetic evidence, or anticipating it by revealing novel precursor molecules and potential processing sites (*79*). Although many polyprotein precursors contain identical duplications of bioactive sequences, e.g., the precursors of TRH and FMRFamide (*82, 83*), others carry related sequences with either similar or differing functions. A prominent example in the latter category is POMC, which contains peptides with either ACTH/MSH activity, or unrelated peptides with opioid activity. Another is proglucagon, which contains glucagon and two homologous glucagon-like peptides (GLP1 and GLP2), which have divergent or even antagonistic functions (*84*) (Fig. 4). Thus glucagon is a glycogenolytic/hyperglycemic hormone and is released from the alpha cells of the pancreatic islets, whereas GLP1 released from intestinal L cells tends to oppose this action by stimulating insulin secretion from the pancreatic β cells in response to orally ingested meals or glucose (*85*). Both POMC and proglucagon undergo differential processing, dependent on their sites of production, to release differing mixtures of bioactive products (see Section II,B). This tissue-specific processing can thus generate great diversity from a limited number of precursor forms.

D.   DISCOVERY OF THE SUBTILISIN/KEXIN-LIKE PROPROTEIN CONVERTASES

Over the nearly 20-year period from the early 1970s until 1990, we and others pursued a number of experimental strategies in efforts to identify the converting endoproteases. Most of these either failed or gave misleading results. However, throughout this period, it seemed likely that the endoproteolytic enzymes would turn out to be trypsin-like serine proteases for several reasons. First of all, trypsin had proved to be an excellent model enzyme in the conversion of proinsulin to insulin, and second, although the lysosomal cathepsins, particularly cathepsin B, seemed to have some preference for synthetic substrates with basic residues, we were unable to demonstrate any activity of purified cathepsin B on proinsulin (unpublished data). Moreover, lysosomes are essentially degradative organelles, and it thus seemed unlikely that any lysosomal proteases would be involved in a highly specific biosynthetic process such as the proteolytic conversion of prohormones (see Ref. 30). The discovery of a mutant alpha 1 antitrypsin (Pittsburgh) associated with defective processing of precursors of blood coagulation cascade enzymes also pointed toward serine proteases as more likely candidates (*86*).

1.   *Identification of the Yeast Processing Endoproteinase, Kexin*

In the early 1980s, Jeremy Thorner began to study the mechanisms of production of the alpha mating factor in *Saccharomyces cerevisiae*, which he and others had shown was derived from a larger proprotein through processing at paired basic residues (*87*). By studying yeast strains defective in mating due to lack of processing of pro alpha-factor, he and his associates were able to identify three proteolytic activities that were essential for the release of active alpha-factor. These included a trypsin-like endoprotease, a carboxypeptidase B-like exopeptidase, and an aminopeptidase (*87, 88*). The endoprotease was cloned by Julius *et al.* (*87*) and called kex2p, or kexin, denoting its involvement in the production of killer toxin, another yeast peptide produced from a precursor protein.

When the amino acid sequence of kexin was published in 1988 (89), it was notable that, although it was indeed a serine protease, its catalytic module was clearly related to that of the bacterial subtilisins rather than the trypsin family. This was a great surprise in that large numbers of trypsin-like enzymes were known to occur in eukaryotes, while the subtilisin-like fold seemed to be confined to bacteria. It was thus not clear at first that the converting enzymes in higher eukaryotes would be related to kexin. Kexin was a large type 1 integral membrane protein, about 100 kDa in size, and it was dependent on calcium for activity (90) (Fig. 5). It also could be readily covalently tagged by a modified chloromethyl ketone that was available in my laboratory in the early 1980s (90). Elliot Shaw had kindly provided us with a very useful tripeptidyl chloromethyl ketone inhibitor, Ala-Lys-Arg-CH$_2$-Cl. Kevin Docherty modified this compound by adding ^{125}I-tyrosine to its amino terminus to generate ^{125}I-Tyr-Ala-Lys-Arg-CH$_2$-Cl at high specific activity. We initially had used this reagent in attempts to detect candidate convertase activities in isolated secretory granule lysates from rat islets and found a 31 kDa component that reacted strongly. This turned out to be an enzymatically active form of procathepsin B, which for a time we thought might be involved in proinsulin processing (91, 92). However, the addition of inhibitors of cathepsins, such as E64, paradoxically, stimulated the proinsulin-converting activity of lysed islet granule fractions (D. F. Steiner, unpublished data). This finding suggested that lysosomal cathepsins were degrading the convertases in our impure granule preparations and indicated that we were on the wrong track in our pursuit of a modified form of cathepsin B as a convertase candidate. However, when [^{125}I]chloromethylketone-labeled islet granule lysates were run out on PAGE and radioautographed, we were unable to detect any labeled components in the size range of kexin (100 kDa). Indeed, the only other band visible on many of our autoradiographs was at about 60–65 kDa. Since our chloromethyl ketone reacted with BSA to some extent (probably via reactive thiols), we made the unfortunate assumption that this band was an artifact. In fact, it most likely contained the as yet undiscovered convertases PC2 and PC1/PC3 (see later discussion).

2. Discovery of the Prohormone Convertases PC2 and PC1/PC3

In the late 1980s, Steven Smeekens in my laboratory attempted to identify cDNAs encoding prohormone convertases from insulin-producing cell lines and/or islets by complementation of kex2 defective yeast strains obtained from Reed Wickner (NIH). This approach, which often works for functionally highly conserved proteins from yeast to *Drosophila* to man, failed to

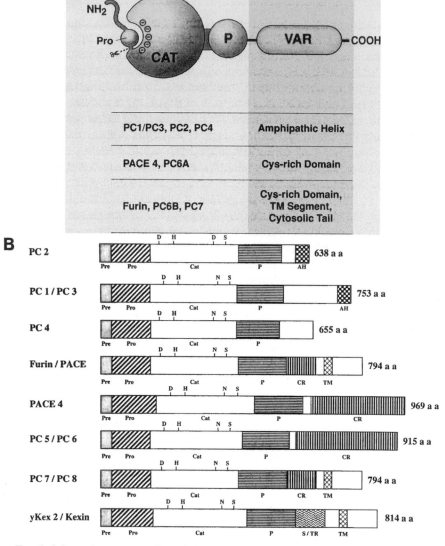

FIG. 5. Schematic representation of the structural features of the mammalian family of subtilisin-like proprotein convertases (SPCs). (A) All seven members have well-conserved signal peptides, proregions (Pro), catalytic domains (CAT), and P domains (P), but differ in their carboxyl terminal domains (VAR), as indicated. Autocatalytic cleavage and release of the prodomain results in activation (see text for details). Modeling studies predict that the P domain folds to form an eight-stranded beta barrel that interacts with the catalytic domain through a hydrophobic patch (*109*). (B) Comparison of the structures of the convertases and kexin. Note: Only PC6A structure is shown here. Symbols as above and (AH) amphipathic helix, (CR) cysteine-rich repeats, (TM) transmembrane domain, (S/TR) serine/threonine-rich domain. A note on alternative terminology: Furin = SPC1/PACE; PC2 = SPC2; PC1/PC3 = SPC3; PACE4 = SPC4; PC4 = SPC5; PC6 = SPC6; PC7 = SPC7/PC8/LPC (see Ref. *102*).

elicit any positive results over a frustrating 3-year period (*41*). Meanwhile, in England, John Hutton and his colleagues succeeded in identifying convertases in insulin secretory granules isolated from a transplantable rat insulinoma. These tumors provided an excellent source of much larger amounts of material than could be achieved with isolated islets (*93*). They described two calcium-dependent proteolytic activities from granules that cleaved selectively at one or the other dibasic sites in proinsulin.

When the catalytic domain of kexin was found to be related to subtilisins (*89*), Steven hypothesized that the amino acid sequences of the mammalian convertases might be highly conserved in the regions surrounding the active site catalytic residues and so designed PCR primers based on the sequences surrounding the catalytic Asp and Ser residues in kexin and subtilisin. When these were used to amplify cDNAs from a human insulinoma cDNA library, several bands were obtained. Two of these, denoted PC2 and PC3, turned out to be related to kexin/subtilisin and were clearly convertase candidates (Fig. 5). A full-length clone of PC2 indicated a protein considerably smaller than kexin and lacking a transmembrane domain near the C terminus (*94*). A full length cDNA for the PC3 partial cDNA was found in a mouse AtT20 cDNA library and this convertase turned out to be similar to PC2 in its organization (*95*). Using a similar PCR strategy Nabil Seidah and Michel Chrétien in Montreal also identified mouse PC2 and PC3, which they called PC2 and PC1, respectively (*96, 97*). These findings were rapidly confirmed by a number of other laboratories (*98*).

3. *Identification of Furin*

While preparing our findings on PC2 for publication in 1989, we noticed that a partial cDNA for another probable convertase had turned up in the protein database. The encoded protein, termed furin by the group in the Netherlands that identified it, was initially thought to be a receptor because it contained a cysteine-rich domain followed by a hydrophobic transmembrane domain and a C-terminal tail (*99, 100*). Although its putative catalytic domain was incomplete, the furin protein was more similar to kexin in overall structure than was PC2. Moreover, our PC2 sequence indicated that its oxyanion residue was not asparagine, as in all known subtilisins and in kexin, but instead was aspartic acid (*94*). Our concerns were reinforced when Fuller *et al.* reported that a full-length furin clone, which they had obtained, clearly encoded a convertase with conservation of the catalytic residues and other kexin-like features, including the C-terminal TM domain (Fig. 5) (*101*). Nonetheless, the structures of both PC2 and PC3 were consistent with their specialized role as untethered enzymes acting in neuroendocrine secretory granules. And now we know that the Asp substitution of the PC2 oxyanion residue is highly conserved in all PC2s throughout metazoan evolution and is consistent with its function as a convertase in the acidic environment of the secretory granule, where the aspartyl side-chain carboxyl will be protonated, and hence capable of forming a hydrogen bond with the carbonyl oxygen of the scissile residue during catalysis (*102*).

II. The Proprotein Convertase Family

A. GENERAL FEATURES

The subtilisin-like proprotein convertases (SPCs), or more simply PCs, comprise a subfamily of the subtilases and are widely distributed in eukaryotes (*102, 103*). These specialized proteases are all calcium dependent and contain additional downstream domains that play roles in their stability, regulation and subcellular localization. Seven members of this family have now been identified and characterized in mammals (Fig. 5). Although a three-dimensional structure is not yet available, their catalytic modules have been modeled by homology with the X-ray structures of subtilisins (*104, 105*). Like subtilisins, these proteases become active by autocatalytic cleavage of an N-terminal propeptide, which is also required for folding of the proenzymes (*103, 106*). A downstream domain of about 150 amino acids, called the P or homo B domain (Fig. 5) (*90, 102, 107*), is also required for folding and activity. This domain plays a regulatory role, influencing both calcium dependency and pH optimum (*108*). Molecular modeling predicts that the P domain may consist of an eight-stranded beta barrel that interacts with the catalytic domain through a surface hydrophobic patch on the side of the catalytic domain opposite the substrate binding groove (*109*). This buttress-like configuration may be required to overcome the destabilizing influence of the strongly negative surface charge in the region surrounding the catalytic site. The variable C-terminal regions of the PCs (Fig. 5) are less conserved and appear mainly to play a role in their subcellular localization, but in some instances may inhibit or modify their activity (*110*).

The mammalian SPCs are designed to function in either the regulated or constitutive branches of the secretory pathway (*111*; see Chapter 9, this volume). The prohormone convertases PC2 and PC1/PC3 are the major convertases expressed in the neuroendocrine system and brain and are responsible for processing prohormone and neuropeptide precursors within the dense-core vesicles of the regulated secretory pathway (*102, 103;* see Chapter 11, this volume). PC4, which is expressed in the testis (*112*), and an isoform of PC6 that lacks a TM domain, PC6A, are also members of this group (*113*). As reviewed in Section II,B, the differential expression of PC2 and PC1/PC3 in various endocrine cells and brain centers gives rise to varied mixtures of bioactive peptide products, having divergent or opposing activities, sometimes derived from the same or related precursors. Both the transcription and translation of PC2 and PC1/PC3 mRNA is regulated in neuroendocrine cells by glucose, second messengers, and other factors (*102, 114–116*), paralleling similar regulatory changes in prohormone expression. However, the developmental appearance of the prohormones and their convertases is not closely synchronized (*117, 118*).

The other major convertase family branch includes furin, PACE4, PC6B, and the more recently discovered PC7 (*111*) (Fig. 5). These convertases all process precursors in the constitutive pathway in most tissues, including the neuroendocrine system, liver, kidney, gut, and brain (*113*). They are localized in the TGN and the small clathrin-clad secretory vesicles of the constitutive pathway (*103, 107, 119, 120* ; see Chapter 8, this volume). Because of alternative splicing, multiple forms of some of these convertases occur, as in the case of PACE4 (*121*) and PC5/PC6 (*113*). These enzymes are more similar to the yeast subtilisin-related convertase, kexin, which also is localized in the TGN by a TM/cytosolic tail and functions analogously (*122*). The convertase genes exhibit extensive similarity in intron/exon structure (*102, 103*), implying their origin in early metazoans from an ancestral kexin/subtilisin-like protease via gene duplication and divergence into TGN-localized and nonanchored secretory granule-localized forms (*103, 123*).

The classical motifs for processing by the prohormone convertases are K•R↓ or R•R↓ (*102, 103, 107*). However upstream basic residues at the P4 position may also enhance cleavage of some substrates such as proinsulin (*105, 124, 125*). Furin preferentially recognizes the motif R•X•K/R•R↓, but also is known to cleave R•X•X•R↓ sites in some precursors (*107, 126*). Further upstream basic residues at the P6 position may also contribute to cleavage activity. Endoproteolytic cleavage is almost invariably followed by exoproteolytic removal of the exposed C-terminal basic residues by carboxypeptidase E (CPE) in neuroendocrine and a few other tissues (*65, 66*), or by other more recently discovered carboxypeptidases, including CPD (*127, 128*), CPZ (*129, 130*), and/or CPM (*131*). For further information see Chapter 15, this volume.

B. Processing of Neuroendocrine Precursor Proteins

An enlarging body of evidence supports a key role for PC2 and PC3 in the processing of prohormones and neuroendocrine precursors. The earliest experiments used vaccinia virus vectors to coexpress kexin, PC2, PC3, or furin with various prohormones, usually in nonendocrine cell lines lacking a regulated secretory pathway (*132*). Using this approach, proopiomelanocortin (POMC) was shown to be processed by PC3 to ACTH, in a pattern similar to that seen in the anterior pituitary corticotroph, whereas inclusion of PC2 with PC3 gave rise to more complete processing to alpha-MSH, characteristic of the middle or intermediate pituitary lobe (*133, 134*) (Fig. 4). These results were consistent with the observed high level of expression of PC3 in anterior pituitary corticotrophs, whereas both PC2 and PC3 are expressed in the middle lobe (*135*). The anterior pituitary corticotroph cell line AtT20 also expresses PC3 at high levels with little or no PC2 and gives rise mainly to ACTH and small amounts of β endorphin. The addition of PC2 by stable transfection changes this pattern, altering it to that of the intermediate lobe (*136*). Thus, although both furin and kexin are also capable of processing POMC extensively when expressed at high levels in nonendocrine cells (*132, 133*), the role of furin in normal POMC processing in neuroendocrine cells appears to be minor.

Proinsulin requires both PC2 and PC3 for normal processing in the pancreatic β cells (Fig. 4). Early experiments with vaccinia virus vectors indicated that PC3 cleaves more actively at the B-chain–C-peptide junction, while PC2 cleaves the C-peptide–A-chain junction more readily (*137*). Furin cleaved at both sites in rat proinsulin 1, which has a P4 basic residue at both cleavage sites, unlike most other mammalian proinsulins (*137*). Both PC2 and PC3 are expressed at high levels in β cells, consistent with those findings and with the earlier studies of Hutton and co-workers, who had identified two calcium-dependent proinsulin cleavage enzymes with properties similar to PC2 and PC3 in insulinoma secretory granules (*93, 98*). The intermediate generated by cleavage by PC3 at the B chain junction (split 32,33 proinsulin) is a better substrate for PC2 than is intact proinsulin (*138*). Thus proinsulin processing is believed to begin with cleavage at the B chain site by PC3 followed by cleavage at the A chain site by PC2 (Fig. 6). However, either enzyme, when present at high levels, is capable of cleaving proinsulin completely to insulin, i.e., AtT20 cells expressing mainly PC3 efficiently process proinsulin to insulin, as do αTC1,6 cells, which express only PC2 at high levels (unpublished results). Nonetheless, mice lacking PC2 (*139*) and a human subject lacking PC3 (*140*) both exhibit moderate to severe defects in proinsulin processing.

Proglucagon, like POMC, is a complex precursor comprised of glucagon and two glucagon-like peptide sequences (GLP 1 and 2), which are processed differently in the pancreatic islets and intestinal L cells (*84, 85*) (Fig. 4). The islet alpha cells express only high levels of PC2 (*102, 117, 141*), as does the alpha cell line αTC1,6 (*142*), and only glucagon is released from proglucagon, whereas both GLP sequences are retained in a large inactive C-terminal fragment, denoted the major proglucagon fragment, or MPGF. On the other hand, the intestinal L cells release both GLP1 and GLP2, while glucagon remains bound in a larger inactive fragment called glicentin (*85, 143*). Evidence accumulated to date strongly supports the likelihood that PC3 is

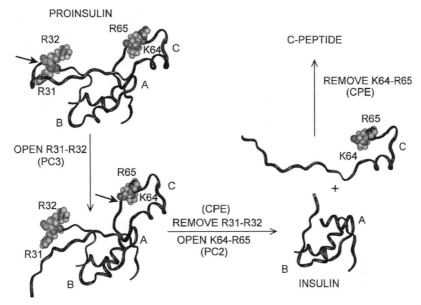

Fig. 6. Proposed schemes for conversion of proinsulin to insulin based on evidence, cited in text, that PC1/PC3 cleavage at the B-chain/C-peptide junction precedes cleavage by PC2 at the C-peptide/A-chain junction. The conformation for proinsulin was modeled based on a structural orientation determined to be optimal for interaction with the convertases (*186*). Note that formation of the enzyme substrate complex at the A chain junction requires unfolding of the N-terminal A-chain alpha helix to allow the isoleucine side chain at A2 to move from the insulin core (in contact with Tyr-A19) to the S2' subsite of PC2 (see Fig. 3 in Ref. 186). (Figure courtesy of G. Lipkind.)

responsible for the L cell phenotype, including cleavage of GLP1 at position 6 (a single arginine) to produce an active truncated form of GLP1, i.e., GLP1$_{7-37}$ (*143, 144*). Although it has been suggested that other enzymes are required in addition to PC2 to excise glucagon (*145, 146*), especially for cleavage at the N terminus, it seems likely that the failure to demonstrate PC2 cleavage at this site may have been due to lack of sufficient amounts of active PC2 (*147*). Mice lacking PC2 activity due to targeted gene disruption (*139*) are unable to cleave either N- or C-terminal to glucagon in proglucagon, leaving the proglucagon essentially intact until secreted in the islet alpha cells (M. Furuta, A. Zhou, G. Webb, and D. F. Steiner, unpublished results).

Other precursors that have been examined include proenkephalin, prodynorphin, propancreatic polypeptide, and prosomatostatin. The last is processed in the islets to liberate somatostatin 28 (SS-28), possibly by furin or another ubiquitous convertase (*148*), and then by PC2 to generate somatostatin 14, the major form found in the CNS and in the D cells of the islets of Langerhans (*139*). In the islet G cells, the pancreatic polypeptide precursor is also processed by PC2 to the mature peptide (A. Zhou and D. F. Steiner, unpublished results). PC2 plays a major role in processing both proenkephalin (*149*) and prodynorphin (*150, 151*) to active products. In its absence, the level of opioid peptides in the CNS is greatly reduced (*150, 151*).

III. Effects of Mutations and/or Disruptions in Convertase Genes

A. FURIN AND SPC4 (PACE4) NULL MUTATIONS ARE EMBRYONIC LETHALS

Furin is expressed very early in development and is involved in the processing of many growth factors and/or their receptors (*5*). The introduction of a null mutation by homologous recombination results in early embryonic lethality, probably attributable to defects in maturation of several bone morphogenetic protein (BMP) activities (*4*). Developmental studies indicate that null embryos develop normally until day 8.5, but then fail to undergo normal axial rotation. Consequent disruption of the development of many systems, but especially of the heart and vascular systems, results in embryonic death between days 10.5 and 11.5. SPC4 (PACE4) also is required early in embryogenesis for the processing of several developmental factors, including *nodal* (*5*). Null embryos exhibit situs ambiguous, left pulmonary isomerism and other cardiovascular defects, and craniofacial malformations; they die during gestation by day E15.5 (*5*).

B. PC4 NULL MICE ARE INFERTILE

PC4 is similar in overall structure and properties to PC2 and PC1/PC3, but is restricted in its expression in rodents to the male gonad in the spermatocytes and round spermatids (*112*). Disruption of the gene encoding PC4 severely impairs fertility in homozygous mutant males, but does not affect spermatogenesis (*152*). The fertility of PC4 null spermatozoa is reduced and fertilized ova fail to develop. The exact nature of the requirement for PC4 in production of fertile and developmentally competent spermatozoa has not been elucidated.

C. MULTIPLE EFFECTS OF A PC2 INACTIVATING MUTATION

Mice lacking active PC2, because of deletion of the activation site and surrounding sequences encoded in exon 3, survive and reproduce, but with reduced litter size, and a slightly subnormal growth rate (*139*). The mutant proenzyme is made, but does not reach the distal secretory pathway in either brain or islet cells. Homozygous PC2 null mice exhibit a complex polyendocrine phenotype, whereas heterozygotes are normal. Pancreatic proinsulin levels are increased to 35–40% of total insulin-like material (normal levels are <5%), and these are the source of elevated circulating proinsulin. Biosynthetic pulse-chase experiments with isolated islets indicate a significant block in proinsulin conversion in (−/−) mice; half-times for the conversion of both mouse proinsulins I and II (rats and mice both have two insulin genes) are prolonged approximately threefold (125). In keeping with the scheme for proinsulin conversion shown in Fig. 6, increased amounts of des-31,32 proinsulin, an intermediate cleaved at the B chain–C-peptide junction, the preferential site of action of PC3, are generated. These results indicate that PC2 accounts for processing of approximately one-third of proinsulin and that consequently PC3 is the more important convertase in insulin production.

Despite the defective processing of circulating proinsulin, PC2 null mice do not develop diabetes. Instead, their blood sugar level is chronically below normal and the rise in response to glucose is also reduced, because of the lack of circulating active glucagon (139). As discussed in the preceding section, only PC2 is present in the islet alpha cells, and it has been shown to act alone to generate the characteristic alpha cell pattern of processing of the multifunctional proglucagon molecule, resulting in the selective release of only glucagon (142, 143, 147) (see Fig. 4). Accordingly, mature glucagon is not detectable in the plasma, whereas large amounts of proglucagon and some partially processed larger forms are present in the alpha cells and the circulation (139). The chronic hypoglycemia confirms that glucagon plays a major role in physiology as a tonic antagonist of insulin.

Prosomatostatin also is not processed to somatostatin 14 in the islet D cells (139), or in the brain in PC2$^{-/-}$ mice (G. Chiu and D. F. Steiner, unpublished results). Similarly, propancreatic polypeptide (proP-P) processing is inhibited in the G cells of the islets, as indicated by increased numbers of PP-positive cells and elevated levels of proP-P (see Section II,B). Lack of PC2 in the beta cells also impairs islet amyloid polypeptide (IAPP/amylin) production (153). The metabolic consequences of the lack of islet somatostatin 14 and pancreatic polypeptide have not been assessed. Lack of IAPP/amylin may tend to reduce peripheral insulin resistance (154) and hence the already low requirement for insulin would be further reduced. The PC2$^{-/-}$ islets are enlarged and show marked hypertrophy and hyperplasia of the alpha, delta, and gamma cells in the periphery, while the central beta cell mass appears to be diminished significantly (139). Hyperplasia in the alpha, delta, and gamma cell populations presumably represents an attempt to compensate for the lack of their normal processed hormonal products. Studies have indicated that long-term administration of glucagon to PC2 null mice results in normalization of blood sugar levels and reversal of α cell hyperplasia (G. Webb and D. F. Steiner, unpublished data). These findings demonstrate that the lack of glucagon in the PC2$^{-/-}$ animals accounts for the two major metabolic phenotypes and further, dramatically illustrates the existence of dynamic feedback mechanisms that sense and regulate the relative size of the islet cell compartments via changes in cell proliferation and/or apoptosis.

In addition to the marked changes in the hormones of the islets of Langerhans, the $PC2^{-/-}$ mice also have multiple defects in neuropeptide production that have yet to be fully characterized. Studies have shown marked reductions in NEI (neuropeptide-EI) derived from promelanin concentrating hormone (155), pro-CCK (cholecystokinin) derived peptides (156), opioid peptides, including enkephalins from proenkephalins (149) and dynorphins from prodynorphin in brain (150, 151), as well as in α-MSH, which, in the intermediate lobe of the pituitary, is generated by the conjoint action of PC2 and PC3 on proopiomelanocortin (POMC) (see Fig. 4) (A. Zhou and D. F. Steiner, unpublished data).

D.   LACK OF NE PROTEIN 7B2 EXPRESSION MIMICS PC2 DISRUPTION

Interestingly, a 7B2 knockout mouse that lacks active PC2 because of the failure of proPC2 to undergo normal activation in the absence of 7B2 (111; see Chapter 11, this volume) also has the same intermediate lobe defect in processing of ACTH to α-MSH, as do the PC2 nulls, but develops severe Cushing's syndrome (adrenocortical hyperfunction), leading to death within 5–7 weeks of birth (157). The lack of 7B2, or the presence of inactive proPC2 (or degradation products thereof), evidently leads to increased secretion of ACTH from the intermediate lobe, leading to hyperadrenocorticalism and marked suppression of the normal hypothalamic, anterior pituitary, adrenal axis of regulation of glucocorticoid production (157). Further studies of this interesting syndrome may elucidate novel regulatory mechanisms of neurointermediate lobe secretion.

E.   PC1/PC3 DEFICIENCY IN MAN

An adult human subject with severe obesity and hyperproinsulinemia (140) has been found to be a compound heterozygote for inactivating PC3 gene mutations (158). This patient has multiple endocrine deficits arising from an absolute or severe lack of active PC3. These include markedly elevated proinsulin and elevated ACTH precursors (POMC and intermediates) in the plasma (140); PC3 is the major convertase that cleaves ACTH from POMC in the anterior pituitary corticotrophs (Section II,B) (see Fig. 4). Increased levels of des-64,65 proinsulin intermediates were found with the elevated intact proinsulin, indicating PC2 action. The absence of des-31,32 proinsulin (Fig. 6), which is normally present in significant amounts in human plasma, would be expected from a lack of PC3-mediated cleavage at the B-chain–C-peptide junction. The absence of detectable insulin in the blood confirms the major role played by PC3 in proinsulin processing. A similar severe defect in proinsulin processing occurs in PC3 ($-/-$) mice (X. Zhu, R. Carroll, and D. F. Steiner, unpublished results).

The striking obesity of this subject, which began at an early age, may result from defective processing of neuropeptides involved in hypothalamic regulation of food intake and/or metabolism (*159*). This subject also suffers from hypogonadotropic hypogonadism (*140*), leading to the suggestion that PC3 may be involved in the processing of precursors of pituitary gonadotropin releasing hormone (GRH).

F. CARBOXYPEPTIDASE E (CPE) DEFICIENCY SYNDROME

Another genetic model of severe obesity with hyperglycemia is the CPE^{fat} mouse (*160*). These mice lack the processing carboxypeptidase, CPE, because of a missense mutation that leads to misfolding and degradation of CPE. Homozygous mice have markedly reduced CPE-like activity in both the pituitary and islets of Langerhans, but less severely in brain, because of the presence of other carboxypeptidases, such as CPD (*127*) and CPZ (*129*), which share the role of CPE in some regions. Affected mice have elevated levels of immunoreactive insulin from early life, but without insulin resistance, despite obesity similar to that seen in the obese hyperglycemic mouse (ob/ob). The ob/ob and diabetic (db/db) mice lack either leptin or the leptin receptor, respectively, which accounts for their obesity (*161*).

The circulating immunoreactive insulin-like material in CPE^{fat} homozygotes is mainly proinsulin; pancreatic extracts contain about 40–50% proinsulin and increased numbers of immature secretory granules are seen by electron microscopy of islet beta cells (*160*). HPLC analysis revealed elevated levels of arginine-extended forms of each of the two normal mouse insulins (I and II). These intermediates are generated by convertase action and are normally converted rapidly to insulin by CPE. The rapid buildup of such C-terminally extended forms within secretory granules may inhibit the endoproteases (see Refs. *151, 162, 163*). In this regard a 29 kDa neuroendocrine peptide (proSAAS) has been identified that contains an inhibitor sequence that is highly selective for PC3, and thus may inhibit PC3 processing more severely than that of the other convertases (*164*). CPE may normally relieve this inhibition by removing C-terminal basic residues from the processed inhibitory peptide from proSAAS. Defects in processing of hypothalamic neuropeptides involved in satiety may account for the marked obesity, whereas maturation of POMC to ACTH may be impaired in the anterior pituitary and could explain the mild hypoadrenocorticism of these mice (*160*). All of these endocrine defects in the CPE^{fat} mice resemble those in the human subject with PC3 deficiency described above.

An alternative mechanism for the elevated secretion of POMC and proinsulin in CPE^{fat} mice has been proposed by Loh and co-workers. By demonstrating binding of several neuroendocrine precursors to CPE, they have inferred that CPE is a receptor for sorting precursors into the regulated secretory pathway and have postulated that missorting in CPE^{fat} mice accounts for elevated precursor secretion (*165, 166*). Studies directly measuring the efficiency of proinsulin sorting in islets of CPE^{fat} mice do not support this hypothesis (*167*), nor does the finding of large amounts of proinsulin and POMC in abundant secretory granules in islet and pituitary cells, respectively, of the CPE^{fat} mice (*160*).

IV. Are There Other Convertases?

As we have seen, the subtilisin-like proprotein convertase family consists of at least seven members in mammals plus a number of alternatively spliced forms. Comparative studies indicate the existence of PC2, PC3, and a furin-like enzyme, most similar to PC6, in amphioxus (*168*). PC2 and its activator protein 7B2 are highly conserved in insects, worms, and mollusks (*169, 170–172*), whereas a PC3-like protein has been described in hydra (*173*). There clearly are two conserved branches of the convertase family, one operating primarily in the unregulated, or constitutive, secretory pathway and the other in the dense core granules of the regulated secretory pathway. But although a large body of data strongly supports the notion that the SPC family plays a central role in the processing of a large number of precursor proteins, a spectrum of other enzymes have been identified that also participate in specialized aspects of processing. Of particular interest is a recently discovered ER/Golgi-localized protease that has a subtilisin-like catalytic domain, but is only distantly related to the SPCs. It cleaves at sites such as R•S•V•L↓ (*6*) or R•G•L•T↓ (*174*), as well as others (*175*). One of its substrates is the precursor of the sterol regulatory element binding protein and it thus plays an important role in regulating cholesterol metabolism (*6*). It also cleaves other substrates, including the brain-derived neurotrophic factor precursor (*174*). It represents the first member of a novel subfamily of convertase-like enzymes, as similar cleavage sites with neutral or hydrophobic P1 residues have been noted in prorelaxin and several other precursor proteins (*103*).

A membrane-bound aspartyl enzyme in yeast, called Yap3 or Yapsin-1, can complement kexin deficiency in yeast when overexpressed, restoring cleavage of alpha mating factor (*176*). This enzyme has also been shown to be responsible for cleavage in yeast cells of the somatostatin precursor protein at a single basic residue to generate somatostatin-28 (SS-28), a larger active form that contains somatostatin-14 (SS-14) as its C-terminal region (*177, 178*). Yapsin-1 cleaves at single basic residues, usually arginines, which occur in a number of precursors (*176, 178*). However, in some instances, these monobasic sites can be cleaved by convertases like furin or PC3, especially when upstream basic residues are present at the P4 or P6 positions. It is unclear as yet whether a true mammalian homolog of Yapsin-1 truly exists, but two aspartyl proteases have been identified (*179–183*), one of which is the β secretase responsible for cleavage at the β site in the amyloid precursor to generate the characteristic amyloid β peptide found in the brains of Alzheimer patients.

Efforts to identify other types of processing proteases, such as thiol or metalloproteases (reviewed in *184, 185*), are promising, but have not yet yielded definitive genetic evidence of their participation in normal neuroendocrine precursor processing.

V. Conclusions and Future Glimpses

From its humble beginnings 35 years ago, the field of protein precursor processing has steadily enlarged to encompass a great diversity of proteolytic systems within and on the surface of cells. These act in response to various specific internal or external signals, or within defined subcellular compartments, to generate important regulatory peptides, enzymes, transcription factors, signaling molecules, or structural components. These in turn regulate many vital processes. The reports in this volume deal with subsets of these systems that primarily function within the secretory pathway.

The identification of the mammalian subtilisin/kexin-like proprotein convertase (SPC) family represents a major achievement of the past decade. The recognition that various members of this protease family act in either the *trans*Golgi network in many cells or in the clathrin-clad vesicles or nascent secretory granules of the regulated secretory pathway in peptidergic neurones and neuroendocrine cells has demonstrated how subtle variations in a basic motif for processing integrates with other sorting signals to permit selective cleavage of a rich variety of precursor proteins as they travel to their respective destinations.

The convertases have been studied in considerable detail and much has been learned regarding their structural organization, pH and calcium dependence, autoactivation mechanisms, substrate selectivity, evolution, and many other important aspects, as described in the chapters that follow. However, we still lack the precise three-dimensional structural description for these proteases that is essential for a complete understanding of their activation, catalytic mechanisms, and substrate selectivity. Their similarity to subtilisin is a tantalizing feature, but the essential presence of an additional domain, the downstream P domain, adds to the complexity. The specialized aspects of the activation of PC2 through its interaction with the neuroendocrine protein 7B2 also challenge the imagination, despite the significant achievements in analysis of the details of this interaction (*187*). The discovery of another 7B2-like protein, proSAAS, which also bears a strongly inhibitory sequence, as does 7B2, raises the interesting question as to whether other members of the SPC family interact with specialized helper/inhibitor proteins during their synthesis, activation and/or function (*188*). The interactions between the processing endoproteases and exoproteases, such as CPE, CPD, and CPZ, also appear to have important regulatory aspects that need to be more fully elucidated.

Finally, the search for additional processing enzymes continues with the successful identification of additional subtilisin-like, as well as aspartyl, proteases in the secretory pathway that have novel substrate specificities and functional roles. It has become increasingly clear that processing enzymes can also participate in pathologic processes, as evidenced by the cleavages of sites in amyloid precursor proteins to generate both normal or mutant peptides with a high potential to aggregate into insoluble β sheet structures, contributing to such diseases as Alzheimer's (*179–183, 189, 190*) or familial British dementia (*191*). Recent studies also indicate that furin and related convertases may also participate in the degradation of some of their misfolded substrates, e.g., mutant human insulin receptor precursors (*192*). Defects in processing enzymes or their activators, on the other hand, can give rise to complex developmental, endocrine, and metabolic syndromes, as described herein. Disorders in SPC tissue-specific expression are a more subtle, but potentially equally important, factor in pathology, especially in the makeup of tumor cells (*193, 194*). Clearly the future holds many opportunities for further fruitful investigation in the field of intracellular precursor processing.

ACKNOWLEDGMENTS

I am indebted to many students, postdocs, and colleagues who have contributed to the work from my laboratory discussed in this review. In particular, I would like to mention Philip Oyer, Jeffrey Clark, Wolfgang Kemmler, Dennis Cunningham, Franco Melani, Arthur Rubenstein, Simon Pilkis, Howard Tager, James Peterson, Shu Jin Chan, Susan Terris, Sooja Cho Nehrlich, Ray Carroll, Ole Madsen, Sture Falkmer, Stefan Emdin, Christoph Patzelt, Åke Lernmark, Cecelia Hoffman, Simon Kwok, Kevin Docherty, John Hutton, David Nielsen, Michael Welsh, Susumu Seino, Kishio Nanjo, Steve Smeekens, Graeme Bell, Kenneth Polonsky, Steve Duguay, Shinya Ohagi, Machi Furuta, Shinya Nagamatsu, Tadashi Hanabusa, Hisako Ohgawara, Yves Rouillé, Grigory Lipkind, Sean Martin, An Zhou, Gene Webb, Joe Bass, Iris Lindberg, Per Westermark, and Gunilla Westermark. I thank Rosie Ricks for expert assistance in preparing this manuscript. Work from the author's laboratory has been supported by NIH grants DK 13914 and DK20595 and by the Howard Hughes Medical Institute.

REFERENCES

1. Cui, Y., Jean, F., Thomas, G., and Christian, J. L. (1998). *The EMBO Journal* **17**, 4735.
2. Dubois, C. M., Laprise, M.-H., Blanchette, F., Gentry, L. E., and Leduc, R. (1995). *J. Biol. Chem.* **270**, 10618.
3. Peschon, J., Slack, J. L., Reddy, P., Stocking, K. L., Sunnarborg, S. W., Lee, D. C., Russell, W. E., Castner, B. J., Johnson, R. S., Fitzner, J. N., Boyce, R. W., Nelson, N., Kozlosky, C. J., Wolfson, M. F., Rauch, C. T., Cerretti, D. P., Paxton, R. J., March, C. J., and Black, R. A. (1998). *Science* **282**, 1281.
4. Roebroek, A. J., Umans, L., Pauli, I. G., Robertson, E. J., van Leuven, F., Van de Ven, W. J., and Constam, D. B. (1998). *Development* **125**, 4863.
5. Constam, D. B., and Robertson, E. J. (2000). *Genes Devel.* **14**, 1146.
6. Sakai, J., Rawson, R. B., Espenshade, P. J., Cheng, D., Seegmiller, A. C., Goldstein, J. L., and Brown, M. S. (1998). *Molec. Cell* **2**, 505.
7. Wilkinson, K. (1995). *Annual Rev. Nutr.* **15**, 161.
8. Thornberry, N. A., and Lazebnik, Y. (1998). *Science* **281**, 1312.
9. Martoglio, B., and Dobberstein, B. (1998). *Trends in Cell Biology* **8**, 410.
10. Steiner, D. F. (1998). *Curr. Opin. Chem. Biol.* **2**, 31.
11. Seidah, N., Chrétien, M., and Day, R. (1994). *Biochimie (Paris)* **76**, 197.
12. Karin, M., and Ben-Neriah, Y. (2000). *Ann. Rev. Immunol.* **18**, 621.
13. Dace, A., Zhao, L., Park, K. S., Furuno, R., Takamura, N., Nakanishi, M., West, B. L., Hanover, J. A., and Cheng, S.-Y. (2000). *Proc. Natl. Acad. Sci. USA* **97**, 8985.
14. Steiner, D. F., and Oyer, P. E. (1967). *Proc. Natl. Acad. Sci. USA* **57**, 473.
15. Steiner, D., Cunningham, D., Spigelman, L., and Aten, B. (1967). *Science* **157**, 697.
16. Steiner, D. F. (1967). *Trans. NY Acad. Sci.* **30**, 60.
17. Habener, J., Kemper, B., Potts, J., Jr., and Rich, A. (1972). *Science* **178**, 630.
18. Eipper, B. A., and Mains, R. E. (1980). *Endocrine Reviews* **1**, 1.
19. Tager, H., and Steiner, D. (1973). *Proc. Natl. Acad. Sci. USA* **70**, 2321.
20. Judah, J., Gamble, M., and Steadman, J. (1973). *Biochem. J.* **134**, 1083.
21. Suchanek, G., Kreil, G., and Hermodson, M. (1978). *Proc. Natl. Acad. Sci. USA* **75**, 701.
22. Sanger, F. (1959). *Science* **129**, 1340.
23. Dixon, G., and Wardlaw, A. (1960). *Nature* **188**, 721.
24. Zahn, H. (2000). *J. Peptide Sci.* **6**, 1.
25. Meienhofer, J., Schnabel, E., Bremer, H., Brinkhoff, O., Zabel, R., Sroka, W., Klostermeyer, H., Brandenburg, D., Okuda, T., and Zahn, H. (1963). *Z. Naturforsch.* **18b**, 1120.
26. Katsoyannis, P., Fukuda, K., Tometsko, A., Suzuki, K., and Tilak, M. (1964). *J. Am. Chem. Soc.* **86**, 930.
27. Kung, Y., Du, Y., Huang, W., Chen, C., Ke, L., Hu, S., Jiang, R., Chang, W., Cheng, L., Li, H., Wang, Y., Loh, T., Chi, A., Li, C., Shi, P., Yich, Y., Tang, K., and Hsing, C. (1965). *Scientia Sinica* **14**, 1710.
28. Katsoyannis, P. G., and Tometsko, A. (1966). *Proc. Natl. Acad. Sci. USA* **55**, 1554.
29. Tomizawa, H. (1962). *J. Biol. Chem.* **237**, 3393.
30. Steiner, D. F., Kemmler, W., Clark, J. L., Oyer, P. E., and Rubenstein, A. (1972). *In* "Handbook of Physiology—Endocrinology I" (D. F. Steiner and N. Freinkel, eds.), p. 175. Williams & Wilkins, Baltimore.
31. Wang, S.-S., and Carpenter, F. H. (1965). *J. Biol. Chem.* **240**, 1619.
32. Humbel, R. (1965). *Proc. Natl. Acad. Sci. USA* **53**, 853.
33. Steiner, D. F., Clark, J. L., Nolan, C., Rubenstein, A. H., Margoliash, E., Aten, B., and Oyer, P. (1969). *In* "Recent Progress in Hormone Research," Vol. 25, p. 207.
34. Moskalewski, S. (1965). *Gen. Comp. Endocrinol.* **5**, 342.

35. Lacy, P., and Kostianovsky, M. (1967). *Diabetes* **16,** 35.
36. Steiner, D., Clark, J., Nolan, C., Rubenstein, A., Margoliash, E., Melani, F., and Oyer, P. (1970). *In* "The pathogenesis of diabetes mellitus. Nobel symposium 13" (E. Cerasi and R. Luft, eds.), p. 57. Almqvist and Wiksell, Stockholm.
37. Howell, S., and Taylor, K. (1967). *Biochem. J.* **102,** 922.
38. Palade, G. (1975). *Science* **189,** 347.
39. Orci, L., Ravazzola, M., Amherdt, M., Madsen, O., Vassalli, J., and Perrelet, A. (1985). *Cell* **42,** 671.
40. Steiner, D., Kemmler, W., Tager, H., and Peterson, J. (1974). *Fed. Proc.* **33,** 2105.
41. Steiner, D. (1991). *In* "Peptide biosynthesis and processing" (L. Fricker, ed.), p. 1. CRC Press Inc., Boca Raton.
42. Steiner, D. F., Hallund, O., Rubenstein, A. H., Cho, S., and Bayliss., S. (1968). *Diabetes* **17,** 725.
43. Chance, R., Ellis, R., and Bromer, W. (1968). *Science* **161,** 165.
44. Steiner, D. F., and Clark, J. L. (1968). *Proc. Natl. Acad. Sci. USA* **60,** 622.
45. Nolan, C., Margoliash, E., Peterson, J., and Steiner, D. (1971). *J. Biol. Chem.* **246,** 2780.
46. Clark, J., and Steiner, D. (1969). *Proc. Natl. Acad. Sci. USA* **62,** 278.
47. Clark, J., Cho, S., Rubenstein, A., and Steiner, D. (1969). *Biochem. Biophys. Res. Comm.* **35,** 456.
48. Rubenstein, A., Clark, J., Melani, F., and Steiner, D. (1969). *Nature* **224,** 697.
49. Steiner, D., Cho, S., Oyer, P., Terris, S., Peterson, J., and Rubenstein, A. (1971). *J. Biol. Chem.* **246,** 1365.
50. Peterson, J., Nehrlich, S., Oyer, P., and Steiner, D. (1972). *J. Biol. Chem.* **247,** 4866.
51. Tager, H., and Steiner, D. (1972). *J. Biol. Chem.* **247,** 7936.
52. Oyer, P., Cho, S., Peterson, J., and Steiner, D. (1971). *J. Biol. Chem.* **246,** 1365.
53. Rubenstein, A., and Steiner, D. (1970). *In* "Early Diabetes" (R. Camerini-Davalos and R. Levine, eds.), p. 159. Academic Press, New York.
54. Bell, G. I., Swain, W. F., Pictet, R., Cordell, B., Goodman, H. M., and Rutter, W. J. (1979). *Nature* **282,** 525.
55. Melani, F., Rubenstein, A., and Steiner, D. (1970). *J. Clin. Invest.* **49,** 497.
56. Rubenstein, A., Cho, S., and Steiner, D. (1968). *Lancet* June 22, 1353.
57. Rubenstein, A., Steiner, D., Horwitz, D., Mako, M., Block, M., Starr, J., Kuzuya, H., and Melani, F. (1977). *Rec. Prog. Horm. Res.* **33,** 435.
58. Polonsky, K. S., and O'Meara, N. M. (1995). *In* "Endocrinology" (L. J. DeGroot, ed.), Vol. 2, p. 1354. W. B. Saunders, Philadelphia.
59. Hartling, S. G. (1999). *Danish Medical Bulletin* **46,** 291.
60. Kemmler, W., and Steiner, D. (1970). *Biochem. Biophys. Res. Com.* **41,** 1223.
61. Kemmler, W., Steiner, D. F., and Borg, J. (1973). *J. Biol. Chem.* **248,** 4544.
62. Kemmler, W., Peterson, J. D., and Steiner, D. F. (1971). *J. Biol. Chem.* **246,** 6786.
63. Wang, S.-S., and Carpenter, F. H. (1969). *J. Biol. Chem.* **244,** 5537.
64. Zühlke, H., Steiner, D., Lernmark, Å., and Lipsey, C. (1976). *In* "Polypeptide hormones: molecular and cellular aspects," p. 183. Elsevier/Excerpta Medica, North-Holland.
65. Fricker, L., Evans, C., Esch, F., and Herbert, E. (1986). *Nature* **323,** 461.
66. Davidson, H., and Hutton, J. (1987). *Biochem. J.* **245,** 575.
67. Russell, J., and Geller, D. M. (1975). *J. Biol. Chem.* **250,** 3409.
68. Cohn, D., MacGregor, R. R., Chu, L. L., Kimmel, J. R., and Hamilton, J. W. (1972). *Proc. Natl. Acad. Sci. USA* **69,** 1521.
69. Potts, Jr., J. T., Kronenberg, H. M., and Rosenblatt, M. (1982). *Adv. Prot. Chem.* **35,** 323.
70. Thim, L., and Moody, A. (1981). *Peptides* **2**(Suppl. 2), 37.
71. Pazelt, C., Tager, H., Carroll, R., and Steiner, D. (1979). *Nature* **282,** 260.

72. Yalow, R. S. (1974). *In* "Recent Progress in Hormone Research" (R. O. Greep, ed.), Vol. 30, p. 597. Academic Press, New York.
73. Scott, A., Ratcliff, J., Rees, L., and Landon, J. (1973). *Nature New Biol.* **244,** 65.
74. Hughes, J., Smith, T., Kosterlitz, H., Fothergill, L., Morgan, B., and Morris, H. (1975). *Nature* **258,** 577.
75. Chrétien, M., and Li, C. (1967). *Can. J. Biochem.* **45,** 1163.
76. Smith, A. I., and Funder, J. W. (1988). *Endocrine Reviews* **9,** 159.
77. Mains, R., Eipper, B., and Ling, N. (1977). *Proc. Natl. Acad. Sci. USA* **74,** 3014.
78. Roberts, J., and Herbert, E. (1977). *Proc. Natl. Acad. Sci. USA* **74,** 4826.
79. Douglass, J., Civelli, O., and Herbert, E. (1984). *In* "Annual Review of Biochemistry" (C. C. Richardson, P. D. Boyer, and A. Meister, eds.), Vol. 53, p. 665. Annual Reviews Inc., Palo Alto.
80. Brownstein, M., Russell, J., and Gainer, H. (1980). *Science* **207,** 373.
81. Sachs, H., Fawcett, P., Takabatake, Y., and Portanova, R. (1969). *In* "Recent Progress in Hormone Research," Vol. 25, p. 447. Academic Press, New York.
82. Nillni, E. A., and Sevarino, K. A. (1999). *Endocrine Reviews* **20,** 599.
83. Grimmelikhuijzen, C., Leviev, I., and Carstensen, K. (1996). *Int. Rev. Cytol.* **167,** 37.
84. Orskov, C. (1992). *Diabetologia* **35,** 701.
85. Holst, J. J., Bersani, M., Johnsen, A. H., Kofod, H., Hartmann, B., and Orshov, C. (1994). *J. Biol. Chem.* **269,** 18827.
86. Carrell, R., Jeppsson, J.-O., Laurell, C.-B., Brennan, S., Owen, M., Vaughn, L., and Bosell, D. (1982). *Nature (London)* **246,** 329.
87. Julius, D., Brake, A., Blair, L., Kunisawa, R., and Thorner, J. (1984). *Cell* **37,** 1075.
88. Fuller, R., Sterne, R., and Thorner, J. (1988). *Ann. Rev. Physiol.* **50,** 345.
89. Mizuno, K., Nakamura, T., Ohshima, T., Tanaka, S., and Matsuo, M. (1988). *Biochem. Biophys. Res. Commun.* **156,** 246.
90. Fuller, R., Brake, A., and Thorner, J. (1989). *Proc. Natl. Acad. Sci. USA* **86,** 1434.
91. Docherty, K., Carroll, R., and Steiner, D. (1982). *Proc. Natl. Acad. Sci. USA* **79,** 4613.
92. Docherty, K., Hutton, J., and Steiner, D. (1984). *J. Biol. Chem.* **259,** 6041.
93. Davidson, H., Rhodes, C., and Hutton, J. (1988). *Nature* **333,** 93.
94. Smeekens, S. P., and Steiner, D. F. (1990). *J. Biol. Chem.* **265,** 2997.
95. Smeekens, S. P., Avruch, A. S., LaMendola, J., Chan, S. J., and Steiner, D. F. (1991). *Proc. Natl. Acad. Sci.* **88,** 340.
96. Seidah, N., Gaspar, L., Mion, P., Marcinkiewicz, M., Mbikay, M., and Chretien, M. (1990). *DNA and Cell Biology* **9,** 415.
97. Seidah, N., Marcinkiewicz, M., Benjannet, S., Gaspar, L., Beaubien, G., Mattei, M., Lazure, C., Mbikay, M., and Chretien, M. (1991). *Molec. Endocrinol.* **5,** 111.
98. Steiner, D. F., Smeekens, S. P., Ohagi, S., and Chan, S. J. (1992). *J. Biol.Chem.* **267,** 23435.
99. Roebroek, A., Schalken, J., Leunissen, J., Onnekink, C., Bloemers, H., and Van de Ven, W. (1986). *EMBO J.* **5,** 2197.
100. Schalken, J. A., Roebroek, A. J., Oomen, P. P., Wagenaar, S. S., and Debruyne, F. M. (1987). *J. Clin. Invest.* **80,** 1545.
101. Fuller, R., Brake, A., and Thorner, J. (1989). *Science* **246,** 482.
102. Rouillé, Y., Duguay, S., Lund, K., Furuta, M., Gong, Q., Lipkind, G., Oliva, Jr., A., Chan, S., and Steiner, D. (1995). *Frontiers in Neuroendocrinology* **16,** 322.
103. Seidah, N., Mbikay, M., Marcinkiewicz, M., and Chrétien, M. (1998). *In* "Proteolytic and cellular mechanisms in prohormone processing" (V. Hook, ed.), p. 49. RG Landes, Georgetown.
104. Siezen, R. J., de Vos, W. M., Leunissen, J. A., and Dijkstra, B. W. (1991). *Protein Engineering* **4,** 719.

105. Lipkind, G., Gong, Q., and Steiner, D. F. (1995). *J. Biol. Chem.* **270,** 13277.
106. Fu, X., Inouye, M., and Shinde, U. (2000). *J. Biol. Chem.* **275,** 16871.
107. Nakayama, K. (1997). *Biochem. J.* **327,** 625.
108. Zhou, A., Martin, S., Lipkind, G., LaMendola, J., and Steiner, D. (1998). *J. Biol. Chem.* **273,** 11107.
109. Lipkind, G., Zhou, A., and Steiner, D. (1998). *Proc. Natl. Acad. Sci. USA* **95,** 7310.
110. Muller, L., and Lindberg, I. (2000). *In* "Progress in Nucleic Acid Research and Molecular Biology," Vol. 63, p. 69. Academic Press.
111. Zhou, A., Webb, G., Zhu, X., and Steiner, D. F. (1999). *J. Biol. Chem.* **274,** 20745.
112. Mbikay, M., Raffin-Sanson, M.-L., Tadros, H., Sirois, F., Seidah, N., and Chretien, M. (1994). *Genomics* **20,** 231.
113. De Bie, I., Marcinkiewicz, M., Malide, D., Lazure, C., Nakayama, K., Bendayan, M., and Seidah, N. G. (1996). *J. Cell Biol.* **135,** 1261.
114. Schuppin, G. T., and Rhodes, C. J. (1996). *Biochem. J.* **313,** 259.
115. Mania-Farnell, B. L., Botros, I., Day, R., and Davis, T. P. (1996). *Peptides* **17,** 47.
116. Jansen, E., Ayoubi, T. A., Meulemans, S. M., and Van de Ven, W. J. (1997). *Biochem. J.* **15,** 69.
117. Marcinkiewicz, M., Ramla, D., Seidah, N. G., and Chretien, M. (1994). *Endocrinology* **135,** 1651.
118. Holling, T. M., van Herp, F., Durston, A. J., and Martens, G. J. (2000). *Molec. Brain Res.* **75,** 70.
119. Molloy, S., Anderson, E., Jean, F., and Thomas, G. (1999). *Trends in Cell Biol.* **9,** 28.
120. Denault, J.-B., and Leduc, R. (1996). *FEBS Lett.* **379,** 113.
121. Zhong, M., Benjannett, S., Lazure, C., Munzer, S., and Seidah, N. G. (1996). *FEBS Lett.* **396,** 31.
122. Brickner, J. H., and Fuller, R. S. (1997). *J. Cell Biol.* **139,** 23.
123. Seidah, N. G., Day, R., Marcinkiewicz, M., and Chrétien, M. (1998). *Ann. N. Y. Acad. Sci.* **839,** 9.
124. Sizonenko, S. V., and Halban, P. A. (1991). *Biochem. J.* **278,** 621.
125. Furuta, M., Carroll, R., Martin, S., Swift, H., Ravazzola, M., Orci, L., and Steiner, D. (1998). *J. Biol. Chem.* **273,** 3431.
126. Molloy, S. S., Bresnahan, P. A., Leppla, S. H., Klimpel, K. R., and Thomas, G. (1992). *J. Biol. Chem.* **267,** 16396.
127. Varlamov, O., and Fricker, L. D. (1998). *J. Cell Science* **111,** 877.
128. Dong, W., Fricker, L. D., and Day, R. (1999). *Neuroscience* **89,** 1301.
129. Song, L., and Fricker, L. (1997). *J. Biol. Chem.* **272,** 10543.
130. Xin, X., Day, R., Dong, W., Lei, Y., and Fricker, L. D. (1998). *DNA Cell Biol.* **17,** 311.
131. McGwire, G., and Skidgel, R. (1995). *J. Biol. Chem.* **270,** 17154.
132. Thomas, G., Thorne, B. A., Thomas, L., Allen, R. G., Hruby, D. E., Fuller, R., and Thorner, J. (1988). *Science* **241,** 226.
133. Thomas, L., Leduc, R., Thorne, B., Smeekens, S. P., Steiner, D. F., and Thomas, G. (1991). *Proc. Natl. Acad. Sci. USA* **88,** 5297.
134. Benjannet, S., Rondeau, N., Day, R., Chrétien, M., and Seidah, N. G. (1991). *Proc. Natl. Acad. Sci. USA* **88,** 3564.
135. Day, R., Schaefer, M. K.-H., Watson, S. J., Chrétien, M., and Seidah, N. G. (1992). *Molec. Endocrinol.* **6,** 485.
136. Zhou, A., and Mains, R. E. (1994). *J. Biol. Chem.* **269,** 17440.
137. Smeekens, S. P., Montag, A. G., Thomas, G., Albiges-Rizo, C., Carroll, R., Benig, M., Phillips, L. A., Martin, S., Ohagi, S., Gardner, P., Swift, H. H., and Steiner, D. F. (1992). *Proc. Natl. Acad. Sci. USA* **89,** 8822.
138. Rhodes, C., Lincoln, B., and Shoelson, S. (1992). *J. Biol. Chem.* **267,** 22719.

139. Furuta, M., Yano, H., Zhou, A., Rouille, Y., Holst, J., Carroll, R., Ravazzola, M., Orci, L., Furuta, H., and Steiner, D. (1997). *Proc. Natl. Acad. Sci. USA* **94**, 6646.
140. O'Rahilly, S., Gray, H., Humphreys, P., Krook, A., Polonsky, K., White, A., Gibson, S., Taylor, K., and Carr, C. (1995). *New Eng. J. Med.* **333**, 1386.
141. Blache, P., Le-Nguyen, D., Boegner-Lemoine, C., Cohen-Solal, A., Bataille, D., and Kervran, A. (1994). *FEBS Lett.* **344**, 65.
142. Rouillé, Y., Westermark, G., Martin, S. K., and Steiner, D. F. (1994). *Proc. Natl. Acad. Sci. USA* **91**, 3242.
143. Rouillé, Y., Martin, S., and Steiner, D. (1995). *J. Biol. Chem.* **270**, 26488.
144. Rouillé, Y., Kantengwa, S., Irminger, J.-C., and Halban, P. A. (1997). *J. Biol. Chem.* **272**, 32810.
145. Dhanvantari, S., Seidah, N. G., and Brubaker, P. L. (1996). *Molec. Endocrinol.* **10**, 342.
146. Rothenberg, M. E., Eilertson, C. D., Klein, K., Mackin, R. B., and Noe, B. D. (1996). *Molec. Endocrinol.* **10**, 331.
147. Rouillé, Y., Bianchi, M., Irminger, J.-C., and Halban, P. (1997). *FEBS Lett.* **413**, 119.
148. Galanopoulou, A., Seidah, N. G., and Patel, Y. (1995). *Biochem. J.* **309**, 33.
149. Johanning, K., Juliano, M. A., Juliano, L., Lazure, C., Lamango, N. S., Steiner, D. F., and Lindberg, I. (1998). *J. Biol. Chem.* **273**, 22672.
150. Berman, Y., Mzhavia, M., Polonskaia, A., Furuta, M., Steiner, D., Pintar, J., and Devi, L. A. (2000). *J. Neurochem.* **75**, 1763.
151. Day, R., Lazure, C., Basak, A., Boudreault, A., Limperis, P., and Dong, W. (1998). *J. Biol. Chem.* **273**, 829.
152. Mbikay, M., Tadros, H., Ishida, N., Lerner, C., De Lamirande, E., Chen, E., El-Alfy, M., Clermont, Y., Seidah, N., Chretien, M., Gagnon, C., and Simpson, E. (1997). *Proc. Natl. Acad. Sci. USA* **94**, 6842.
153. Wang, J., Xu, J., Finnerty, J., Furuta, M., Steiner, D. F., and Verchere, C. B. (2001). *Diabetes* **50**, 534.
154. Gebre-Medhin, S., Mulder, H., Pekny, M., Westermark, G., Törnell, J., Westermark, P., Sundler, F., Ahrén, B., and Betsholtz, C. (1998). *Biochem. Biophys. Res. Commun.* **250**, 271.
155. Viale, A., Ortola, C., Hervieu, G., Furuta, M., Barbero, P., Steiner, D. F., Seidah, N. G., and Nahon, J.-L. (1999). *J. Biol. Chem.* **274**, 6536.
156. Vishnuvardhan, D., Connolly, K., Cain, B., and Beinfield, M. C. (2000). *Biochem. Biophys. Res. Commun.* **273**, 188.
157. Westphal, C., Muller, L., Zhou, A., Zhu, X., Bonner-Weir, S., Steiner, D., Lindberg, I., and Leder, P. (1999). *Cell* **96**, 689.
158. Jackson, R., Creemers, J., Ohagi, S., Raffin-Sanson, M.-L., Sanders, L., Montague, C., Hutton, J., and O'Rahilly, S. (1997). *Nature Genetics* **16**, 303.
159. Kalra, S. P., Dube, M. G., Pu, S., Xu, B., Horvath, T. L., and Kalra, P. (1999). *Endocrine Reviews* **20**, 68.
160. Naggert, J., Fricker, L., Varlamov, O., Nishina, P., Rouille, Y., Steiner, D., Carroll, R., Paigen, B., and Leiter, E. (1995). *Nature Genetics* **10**, 135.
161. Friedman, J. M., and Halaas, J. L. (1998). *Nature* **395**, 763.
162. Rovere, C., Viale, A., Nahon, J.-L., and Kitabgi, P. (1996). *Endocrinology* **137**, 2954.
163. Udupi, V., Gomez, P., Song, L., Varlamov, O., Reed, J. T., Leiter, E. H., Fricker, L. D., and Greeley, G. H. J. (1997). *Endocrinology* **138**, 1959.
164. Fricker, L. D., McKinzie, A. A., Sun, J., Curran, E., Qian, Y., Yan, L., Patterson, S. D., Courchesne, P. L., Richards, B., Levin, N., Mzhavia, N., Devi, L. A., and Douglass, J. (2000). *J. Neurosci.* **20**, 639.
165. Cool, D., Normant, E., Shen, F.-S., Chen, H.-C., Pannell, L., Zhang, Y., and Loh, Y. (1997). *Cell* **88**, 73.

166. Shen, F.-S., and Loh, Y. (1997). *Proc. Natl. Acad. Sci. USA* **94,** 5314.
167. Irminger, J.-C., Verchere, C. B., Meyer, K., and Halban, P. A. (1997). *J. Biol. Chem.* **272,** 27532.
168. Oliva, Jr., A. A., Chan, S. J., and Steiner, D. F. (2000). *Biochim. Biophys. Acta* **1477,** 338.
169. Siekhaus, D. E., and Fuller, R. S. (1999). *J. Neurosci.* **19,** 6942.
170. Mentrup, B., Londershausen, M., Spindler, K.-D., and Weidemann, W. (1999). *Insect Molec. Biol.* **8,** 305.
171. Hwang, J. R., Siekhaus, D. E., Fuller, R. S., Taghert, P. H., and Lindberg, I. (2000). *J. Biol. Chem.* **275,** 17886.
172. Thacker, C., and Rose, A. M. (2000). *BioEssays* **22,** 545.
173. Chan, S. J., Oliva, Jr., A. A., LaMendola, J., Grens, A., Bode, H., and Steiner, D. F. (1992). *Proc. Natl. Acad. Sci. USA* **89,** 6678.
174. Seidah, N. G., Mowla, S. J., Hamelin, J., Mamarbachi, A. M., Benjannet, S., Touré, B. B., Basak, A., Munzer, J. S., Marcinkiewicz, J., Zhong, M., Barale, J.-C., Lazure, C., Murphy, R. A., Chrétien, M., and Marcinkiewicz, M. (1999). *Proc. Natl. Acad. Sci. USA* **96,** 1321.
175. Touré, B., Munzer, J., Basak, A., Benjannet, S., Rochemont, J., Lasure, C., Chrétien, M., and Seidah, N. (2000). *J. Biol. Chem.* **275,** 2348.
176. Egel-Mitani, M., Flygenring, H., and Hansen, M. (1990). *Yeast* **6,** 127.
177. Bourbonnais, Y., Ash, J., Dingle, M., and Thomas, D. (1993). *EMBO J.* **12,** 285.
178. Cawley, N. X., Noe, B. B., and Loh, Y. P. (1993). *FEBS Lett.* **332,** 273.
179. Vassar, R., Bennett, B. D., Babu-Khan, S., Kahn, S., Mendiaz, E. A., Denis, P., Teplow, D. B., Ross, S., Amarante, P., Loeloff, R., Luo, Y., Fisher, S., Fuller, J., Edenson, S., Lile, J., Jarosinski, M. A., Biere, A. L., Curran, E., Burgess, T., Louis, J.-C., Collins, F., Treanor, J., Rogers, G., and Citron, M. (1999). *Science* **286,** 735.
180. Yan, R., Bienkowski, M., Shuck, M., Miao, H., Tory, M., Pauley, A., Brashier, J., Stratman, N., Mathews, W., Buhl, A., Carter, D., Tomaselli, A., Parodi, L., Heinrikson, R., and Gurney, M. (1999). *Nature* **402,** 533.
181. Lin, X., Koelsch, G., Wu, S., Downs, D., Dashti, A., and Tang, J. (2000). *Proc. Natl. Acad. Sci. USA* **97,** 1456.
182. Hussain, I., Powell, D., Howlett, D., Tew, D., Meek, T., Chapman, C., Gloger, I., Murphy, K., Southan, C., Ryan, D., Smith, T., Simmons, D., Walsh, F., Dingwall, C., and Christie, G. (1999). *Mol. Cell Neurosci.* **14,** 419.
183. Acquati, F., Accarino, M., Nucci, C., Fumagalli, P., Jovine, L., Ottolenghi, S., and Taramelli, R. (2000). *FEBS Lett.* **468,** 59.
184. Hook, V. Y. (1998). *In* "Handbook of proteolytic enzymes" (A. J. Barrett, ed.), p. 779. Academic Press, New York.
185. Loh, Y., and Cawley, N. (1995). *Methods Enzymol.* **248,** 136.
186. Lipkind, G., and Steiner, D. (1999). *Biochem.* **38,** 890.
187. Apletalina, E. V., Muller, L., and Lindberg, I. (2000). *J. Biol. Chem.* **275,** 14667.
188. Qian, Y., Devi, L. A., Mzhavia, N., Munzer, S., Seidah, N. G., and Fricker, L. D. (2000). *J. Biol. Chem.* **275,** 23596.
189. Haass, C., and DeStrooper, B. (1999). *Science* **286,** 916.
190. Lopez-Perez, E., Seidah, N. G., and Checler, F. (1999). *J. Neurochem.* **73,** 2056.
191. Kim, S.-H., Wang, R., Gordon, D. J., Bass, J., Steiner, D. F., Lynn, D. G., Thinakaran, G., Meredith, S. C., and Sisodia, S. S. (1999). *Nature Neurosci.* **2,** 984.
192. Bass, J., Turck, C., Rouard, M., and Steiner, D. F. (2000). *Proc. Natl. Acad. Sci. USA* **97,** 11905.
193. Bassi, D. E., Mahloogi, H., and Klein-Szanto, A. J. (2000). *Molec. Carcinogenesis* **28,** 63.
194. Sawada, Y., Kameya, T., Aizawa, T., Izumi, T., and Takeuchi, T. (2000). *Endocrine Pathology* **11,** 31.

8

Furin

SEAN S. MOLLOY • GARY THOMAS

Vollum Institute, Oregon Health Sciences University
Portland, Oregon 97201

I. Introduction

The observation that the mature, two-chain form of insulin is produced from an inactive proinsulin precursor molecule during its biosynthesis and transport provided some of the first evidence of activation by proteolytic processing (*1*). Identification of the cleavage sites in proinsulin (-Lys-Arg- and -Arg-Arg-) (*2*) and those of the γ-lipotropin(LPH)/β-melanocyte stimulating hormone junction within β-LPH (*3*) further defined pairs or clusters of basic residues as common sites for this proteolytic processing event. Since these initial studies, an ever-growing number of

THE ENZYMES, Vol. XXII
Copyright © 2001 by Academic Press
All rights of reproduction in any form reserved.

bioactive proteins have been found to be initially synthesized as inactive proproteins that become active on limited proteolysis at sites flanked by basic residues within various compartments of the secretory/endocytic pathway [see (4) for recent review]. Examples include growth factors, hormones, neuropeptides, plasma proteins, adhesion factors, and receptors, as well as bacterial toxins and viral envelope proteins. The finding that the egg-laying hormone of the marine mollusk *Aplysia* was similarly generated by processing of a larger precursor protein at sites defined by paired basic residues (5) indicated that the process of proteolytic proprotein activation was shared by a diverse cross section of metazoans. Indeed, the conservation of basic residue-directed processing between yeast and humans eventually provided the mechanism for identification of furin as the first bona-fide mammalian endoprotease.

The broad range and biological importance of proproteins activated by processing at multiple basic sites prompted considerable effort toward identification of the endoprotease(s) responsible. However, biochemical approaches to purifying specific proprotein processing activities generally proved unsuccessful. This can be explained, at least in part, by the low expression levels observed for some of the proprotein convertases (PCs) as well as the overlapping proteolytic profiles of contaminating enzymes and lack of sensitive and specific assays. The first significant breakthrough in the identification of the PCs came from analyses of *Saccharomyces cerevisiae* strains defective in the processing of α-mating factor and killer toxin. The product of the defective KEX2 gene in these yeast, kex2p or kexin, proved to be an endoprotease with structural similarity to the bacterial subtilisn family of proteases. The role of kexin in the processing of α-mating factor and killer toxin was established by genetic complementation in yeast (6), and sequence analysis showed significant similarity to the catalytic domain of the subtilisins, including the conservation of the critical triad of active site residues—Asp, His, Ser (6, 7). These findings became of particular interest to the field of mammalian protein processing when the ability of kexin to cleave proalbumin *in vitro* (8) and process the proopiomelanocortin precursor (POMC) (9) was reported. Coexpression of kexin and POMC resulted in the production of mature β-LPH, γ-LPH, and β-endorphin, indicating that the enzyme could recognize and correctly cleave multiple basic residue sites within the prohormone molecule (9). These data strongly suggested that the endogenous mammalian PCs were catalyticaly and structurally related to kexin and the subtilisins. The subsequent identification of furin and six additional members of the mammalian proprotein convertase (PC) family [PC1/PC3 (*10–13*) PC2 (*10, 14*) PACE4 (*15*), PC4 (*16, 17*), PC5/PC6 (*18, 19*), and PC7/PC8/LPC/SPC7 (*20–23*)] has clarified the role of endoproteases in the activation of proproteins *in vivo*. However, the discovery of multiple enzymes with potentially overlapping function has raised questions concerning redundancy and specialization within the PC family. Recent discoveries regarding furin (the focus of this review) and the other PCs have helped address these issues as well as provide insight into the mechanisms of intracellular protein trafficking.

II. Identification/Expression

The expected homology between kexin and the PCs led to the recognition of furin as the first PC and ultimately to the cloning of several additional family members (see Fig. 1) [reviewed in (24) and (4)]. The gene encoding furin was first identified as an open reading frame adjacent to the fes/fps

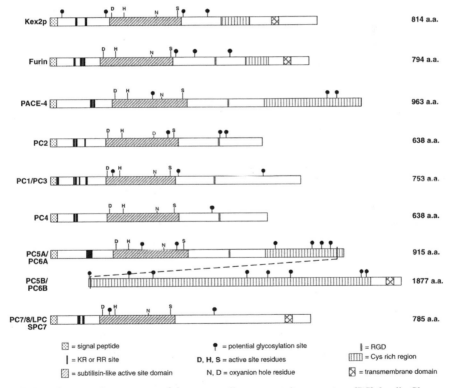

Fig. 1. Comparative structure of the mammalian proprotein convertase (PC) family. Shown are schematic representations of the seven identified members of the proprotein convertase family as well as the structurally and enzymatically related yeast kexin protease (KeX2p). See text for details.

protooncogene [fur: fes/fps upstream region (25, 26)]. The *fur* gene product, furin, has sequence similarity to both kexin and the subtilisins (as predicted) (27) and proved capable of correctly processing precursors for proteins such as pro β-NGF (28), von Willebrand factor (29, 30), albumin (31), and complement C3 (31) in initial coexpression experiments. A combination of Northern analyses and *in situ* hybridization studies subsequently showed that furin mRNA (predominantly a 4 kb species) was expressed ubiquitously in virtually every tissue and cell type examined (32, 33). However, developmental studies indicate that significant changes in distribution and expression pattern do occur (34). In rat, furin is detectable at embryonic day 7 (e7) relatively uniformly expressed in both endoderm and mesoderm. At e10, higher expression is observed in the heart and liver primordia and by mid to late gestation furin is widely detected in peripheral tissues. This developmental expression pattern is similar to yet distinct from the other ubiquitously expressed PCs [PACE4, PC5/PC6 (35) and PC7 (22)] and strikingly different from the neuroendocrine specific patterns of PC2 and PC1/PC3 (36, 37) and testis-specific expression of PC4 (38). The widespread but differential expression of furin is consistent with the proposal that the *fur* gene is regulated by at least three distinct promoters, two of which have characteristics indicative of "housekeeping" gene function (lack of TATA or CAAT boxes and GC-rich) while the third resembles those associated with regulated function (39). Broad and yet regulatable expression of furin is consistent with its role in the processing of substrates involved in homeostasis in adults, key developmental steps during embryogenesis, and physiological changes associated with disease (as discussed later). For example, TGFβ (a furin substrate; see later discussion) appears to regulate the expression of furin in an apparent feedback loop (40) and furin and pro-β-NGF expression are coregulated in damaged sciatic nerve (41). Furin has subsequently been cloned in organisms as divergent as insects [*Drosophila* (42, 43) and *Aedes aegypti* (44)], nematodes [*Caenorhabditis elegans* (45), and *Dirofilaria immitis* (46)], and *Aplysia* (47), consistent with the conservation of proprotein processing throughout evolution. In contrast to the mammalian gene, these nonvertebrate forms of furin can show significant alternative splicing (46), suggesting that some of the roles of the multiple PCs in mammals may be filled by furin variants in these organisms. Despite the discovery of furin in such a wide range of organisms, it may not represent the most "primitive" form of PC based on the apparent absence of a homolog in the protochordate amphioxius [reviewed in (48)].

Members of the PC family have been divided into as many as four classes (4) based primarily on their expression patterns and subcellular localization (see later discussion). One class would consist of furin and PC7, which function in the constitutive secretory pathway; a second of the neuroendocrine specific PC1/PC3 and PC2 (and potentially the soluble A isoforms of PC5/PC6); a third of PACE4 and PC5/PC6, which are expressed in both endocrine and nonendocrine cells and could function in the constitutive and regulated pathways; and a fourth group of the germ-cell-specific PC4. This categorization reflects some general characteristics of PC expression and localization; however, detailed studies of the cell biology of furin as described later indicate that the roles of these enzymes *in vivo* are complex and often interrelated.

III. Structure

Furin (human) is a 794 residue Type I transmembrane protein with a large lumenal segment, including signal peptide, pro-, catalytic, P or "homoB," and Cys-rich domains, a transmembrane domain, and a 56 aa cytoplasmic domain (see Fig. 1). This topology is most similar to that of PC7 and the B splice variant of PC5/PC6 (hereafter called PC6B), which are the only other transmembrane PCs and also have Cys-rich regions. The overall structure of the lumenal portion of furin, however, is shared by all PCs with the highest level of sequence homology (54 to 70% identity to furin) occurring within the catalytic domains and including the Asp, His, Ser catalytic triad as well as the oxyanion hole Asn (Asp in PC2). In addition to the signal peptide necessary for delivery to the secretory pathway, furin and the rest of the PC family have a prodomain defined by the signal peptide cleavage site (N terminus) and a conserved paired basic sequence (C terminus) upstream of the catalytic domain. The prodomain is essential (*49–51*) and plays a critical role in the biosynthesis and regulation of the endproteases (see Section VI). Following the catalytic domain, the PCs share a well conserved P or homoB domain (33 to 53% identity to furin) that is indispensable for enzyme activity. The C-terminal border of the P domain is thought to be at or near a conserved Thr (residue 573) that marks the point of sequence divergence among the PCs. This is consistent with reports showing that furin constructs truncated at Glu-583 (*52*) or Thr-573 (*53*) are still active. Studies using chimeric PC constructs have shown that heterologous P domains could support the production of active endoproteases (*54*). However, the resultant chimeras had altered enzymatic properties, including a shift to more basic pH optima and atypical calcium requirements (see Section IV). These data are consistent with the idea that P domains of the PCs are crucial for obtaining the proper conformation of the catalytic domain and contribute to the divergent pH and calcium optima within the family. In addition to the overall sequence identity, each PC prodomain contains an Arg-Gly-Asp (RGD) integrin binding motif (RGS in PC7) that has an as-yet unidentified function but, at least in the case of PC1/PC3, appears to be essential for normal activity (55). PC1/PC3 constructs with point mutations within the RGD sequence displayed more rapid ER degradation as well as inefficient conversion to lower molecular weight (more active) forms and sorting into secretory granules. The possibility of furin interacting with integrin is of interest in the context of the endoprotease's potential role in extracellular matrix regulation (e.g., activation of matrix metalloproteases and collagens). The transmembrane domain does not appear to affect either the catalytic properties (56, 57) or intracellular trafficking of furin (58) and may only act to provide a membrane anchor and correct topology for recognition of the sorting information contained within the enzyme's cytoplasmic domain. By contrast, the cytoplasmic domain of furin has been shown to control the movement of furin through the secretory and endocytic pathways and has proved to be a seminal model for understanding regulated protein trafficking within mammalian cells (see Section VII).

IV. Biochemical Characteristics

Although the low expression level and transmembrane topology of furin made characterization difficult, expression of truncated soluble furin constructs resulted in a detailed description of the enzyme's biochemical properties reported simultaneously by the Thomas and Nakayama laboratories (56, 57). The sequence and apparent structural similarities between the catalytic domains of furin, kexin, and the subtilisins indicated that the mammalian enzyme is a serine protease. Interestingly, however, the concentrations of serine-protease inhibitors such as PMSF and diisopropyl fluorophosphate (DIFP) necessary for inhibition of furin are very high (see Table I). Furin is, however, effectively inhibited by heavy metal and mercuric salts that typically target cysteine proteases. Although the basis for the relative insensitivity of furin to serine-protease inhibitors is not yet understood, the effectiveness of the cysteine-directed inhibitors has been attributed to the presence of an unpaired Cys (residue 178) proximal to the active site His (59). Furin has a broad pH optimum with greater than 50% activity between pH 6 and 8. The protease is also calcium-dependent, requiring concentrations of approximately 1 mM for full activity with a $K_{0.5}$ of 200 μM. This calcium dependence explains the observed inhibition of furin by the chelators EDTA and EGTA and appears to be relatively selective based on the ineffectiveness of other divalents to substitute *in vitro* (56). Modeling studies that predict the three-dimensional structure of the core of the furin catalytic domain based on alignment with that of bacterial subtilisins (59, 60) indicate at least two calcium binding pockets, CA1 and CA2, with

TABLE I

BIOCHEMICAL CHARACTERISTICS OF FURIN

Inhibitor	Concentration (mM)	Activity (% remaining)
Inhibitor profile		
Pepstatin	1.0	100[a]
Leupeptin	1.0	67,[a] 75.4[b]
E-64	1.0	96[a]
PMSF	2.0	63.8[b]
	3.0	97[a]
	10.0	19[a]
pAPMSF	2.0	92.5[b]
	3.0	97[a]
	10.0	37[a]
DIFP	1.0	99[a]
	2.0	54.8[b]
	30.0	0[a]
Iodoacetamide	2.0	72.5[b]
	3.0	103[a]
	10.0	68[a]
N-Ethylmaleimide	2.0	70.3[b]
DTT	1.0	14[a]
pCMBS	2.0	0.2[b]
HgCl$_2$	1.0	0[a]
	2.0	0.1[b]
ZnCl$_2$	1.0	0[a]
	2.0	0.2[b]
o-Phenanthroline	1.0	82[a]
α_1-AT PITT	2.0×10^{-3}	50[c]
α_1-AT PDX	0.6×10^{-6}	50[c]

Divalent cation requirements	Concentration (mM)	Activity
EDTA	2.0	0[a]
	5.0	0.3[b]
EGTA	2.0	0[a]
EGTA + BaCl$_2$	2.0 each	1[a]
EGTA + MgCl$_2$	2.0 each	0[a]
EGTA + MnCl$_2$	2.0 each	70[a]
EGTA + CaCl$_2$	2.0 each	80[a]
MnCl$_2$ (no CaCl$_2$)	1.0	0[a]

[a] Values from Molloy *et al.* (*56*).
[b] Values from Hatsuzawa *et al.* (*57*).
[c] Values from Anderson *et al.* (*129*).

predicted high and medium affinities, respectively. However, the lack of a crystal structure for furin has prevented a clear understanding of the role of calcium in regulating activity. The relative inability of other divalent cations to substitue for calcium (see Table I) indicates potentially strict structural requirements. Although the other members of the PC family share several biochemical characteristics with furin, there are distinctions that relate to their respective roles *in vivo*. For example, the neuroendocrine specific endoproteases PC2 and PC1/PC3 require higher calcium concentrations for optical activity (10 to 50 mM) and have narrow, more acidic pH optima (5.0 to 6.5) (*61–65*). These differences are consistent with the function of PC2 and PC1/PC3 in processing of neuropeptides and hormones within the specialized acidic milieu of secretory granules of the regulated secretory pathway. By contrast, the relatively permissive activity requirements of furin reflect its role in the processing of a much wider range of substrates in multiple cellular compartments with varied physiochemical properties (see also Section VII).

V. Furin Cleavage Site Requirements

The observation that furin was able to correctly cleave pro-β-NGF to produce bioactive NGF (*28*), as well as the precursors for von Willebrand factor (*29, 30*), albumin (*31*), and complement C3 (*31*), in coexpression studies indicated that the endoprotease recognized sites defined by basic residues as predicted by its homology to kexin. The initial biochemical characterizations of furin (see above) supported these findings by demonstrating that the protease preferentially cleaved synthetic peptides with P1 (first position upstream of the scissile bond) Arg. Furthermore, in studies by the Thomas laboratory (*56, 66*), purified furin was shown to accurately cleave a protein precursor, anthrax toxin protective antigen (PA), at a multiple basic site of the type -Arg-X-Lys/Arg-Arg(P1)- (where X represents any residue). By digesting PA molecules containing point mutations in the cleavage site with limiting amounts of furin *in vitro*, this group showed that furin recognized the native -Arg-X-Lys/Arg-Arg(P1)- site most efficiently but would still process molecules with only P1 and P4 Arg residues. Although, as expected, Arg at the P1 position was essential, the P4 Arg was also critical for cleavage, whereas the basic residue in the P2 position was not required but greatly enhanced processing efficiency ($\sim$10-fold). Based on the observations with PA and fluorogenic peptides *in vitro*, -Arg-X-Lys/Arg-Arg(P1)- was proposed as the optimal cleavage site with -Arg-X-X-Arg(P1)- representing the minimal motif required for efficient processing. The analyses with purified furin supported studies by Nakayama's group that used a set of prorenin constructs

containing cleavage site point mutations to begin examining the specificity
of furin in coexpression studies (67). These coexpression experiments also
indicated a preference for the -Arg-X-Lys/Arg-Arg(P1)- motif and underscored the importance of a P4 Arg for efficient processing by furin. A more
extensive analysis of mutant prorenin processing *in vitro* and *in vivo* led the
Nakayama group to formulate a set of sequence rules proposed to govern
cleavage by furin (68–70). These sequence rules extend the initial furin cleavage site definition by introducing the importance of a basic residue at the P6
position in addition to P2 and P4, and suggesting that Arg is required at two
out of three of these upstream positions for efficient processing. Furthermore, their studies indicated that residues with hydrophobic aliphatic side
chains are not tolerated in the P'1 position (P'# indicating position downstream of the scissile bond). Data obtained from studies of proalbumin processing were also consistent with the prorenin results (71, 72). Substitution
of either the P1 or P2 Arg residues within the cleavage site of proalbumin
[-Arg-Gly-Val-Phe-Arg-Arg(P1)-] blocked processing, as did introduction
of Val in the P'1 position, consistent with the disruptive nature of hydrophobic aliphatic amino acids. Interestingly, the proalbumin cleavage site lacks
the P4 Arg of an optimal furin consensus site and is efficiently cleaved at
acidic pH (5.5 to 6.0). This concept of pH dependent cleavage has potential
relevance for both the compartment-specific recognition of substrates and
the activation pathway of furin itself (see Section VI).

Several additional studies have added further support to the initial observations regarding recognition sequences, including analyses of furin cleavage
site preferences by phage display (73) and by detailed kinetic experiments
using efficiently cleaved fluorogenic substrates (74). In the later study, furin
was compared directly to kexin and shown to have several key distinctions
that fit the cleavage site rules generated from analyses of protein substrates.
Although interaction at the P1 position contributes equally to catalysis by
both enzymes, furin showed a 10-fold lower dependence on the P2 position
than kexin. By contrast, furin was 10 to 100 times more affected than kexin
by residues in the P4 position. These analyses also showed that the positive
charge at the P6 position will contribute 1.4 kcal/mol to catalysis by furin
independently of the P4 residue, and that favorable residues at P2 and P6 can
compensate for less favorable ones at P1 and P4. Together, these data indicate that the furin recognition motif is complicated and may include multiple
contact sites within the enzyme's catalytic pocket. The importance of multiple basic residues for efficient cleavage prompted the proposal that the furin
substrate-binding pocket contains complementary negatively charged amino
acids to facilitate high-affinity binding (69, 70). A modeling study in which
the catalytic domain of furin was reconstructed based upon the crystal structures of subtilisin BPN' thermitase supported this idea by predicting negative

charges at S1, S2, and S4 subsites within the putative binding pocket (59, 60). Mutation of these predicted contact sites was found to change the specificity of furin (52), suggesting that they could potentially act to stabilize substrate binding by charge pairing. The potential for S1, S2, and S4 subsites to dictate specificity has also been independently demonstrated in a study where the normal nonpolar residues in subtilisin BPN' were replaced with Asp. This altered subtilisin—called furilisin—displayed a new preference for cleavage after the furin recognition motif -Arg-X-Lys/Arg-Arg- (75, 76). Although the modeling and mutational studies conducted to date argue strongly for charge pairing in the furin–substrate binding interaction, the lack of a crystal structure for the enzyme precludes any definitive determinations.

VI. Furin Biosynthesis

The biosynthesis of furin has been the focus of considerable attention and has provided a model of ordered, compartment-specific enzyme activation that appears to apply to other members of the PC family (Fig. 2). In addition, these analyses have revealed interesting links between the activation pathway and regulation of protein traffic in the secretory pathway. In initial studies, furin expression by transient transfection resulted in production of two principal membrane-associated translation products of approximately 90 and 96 kDa that did not represent alternative glycosylation states of the enzyme (28, 29). Further studies by the Thomas laboratory (77) demonstrated that these principal molecular weight forms represented the full-length translation product (without signal sequence) and an N-terminally truncated form of furin arising from proteolytic processing at an internal -Arg-Thr-Lys-Arg_{107}(P1)- site (numbering based on the full open reading frame of human furin) that defines the C-terminal border of the prodomain. Analyses of point mutants that either eliminated the P1 or P4 arginines at the prodomain cleavage site or eliminated the essential Asp-153 of the catalytic triad further showed that conversion of the 96 kDa profurin molecule to the 90 kDa form required both active furin and an intact (P4 Arg containing) consensus processing site, indicating that this biosynthetic step is autocatalytic (77). The observation that overexpression of active furin failed to rescue the processing of the active site and cleavage site mutants further argued that the autocatalysis is intramolecular. Comparison of the activity of the various constructs toward exogenous substrate in vitro demonstrated that prodomain cleavage is required for activation of furin. These findings were supported by the observation that mutation of residues at the P2 and P4 positions within the prodomain cleavage site disrupted furin maturation and activity in coexpression experiments (49). Subsequent studies

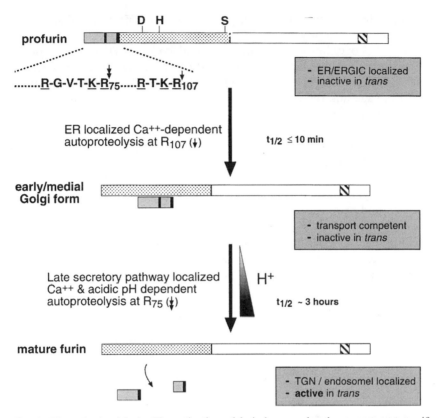

Fig. 2. Biosynthesis of furin. The activation of furin by an ordered, compartment-specific series of autoproteolytic steps is diagrammed. See text for details.

by W. J. M. Van de Ven and co-workers found identical results for furin constructs mutated at any of the three positions of the catalytic triad residues (Asp, His, or Ser) as well as the oxyanion hole asparagine, thereby supporting the conclusion that the maturation of furin is an autocatalytic intramolecular process (52).

Additional studies of furin localization and processing have demonstrated that the autocatalytic cleavage of the prodomain is necessary but not sufficient to produce active furin, and that this processing step regulates the efficient export of furin from the ER/cis-Golgi. In a more detailed characterization of furin biosynthesis and localization (78), the Thomas laboratory demonstrated that the initial cleavage of profurin at Arg-107 to remove the prodomain occurred rapidly ($t_{1/2} < 10$ min) while the enzyme was still endoglycosidase H sensitive, and was unaffected by brefeldin A treatment, suggesting an ER/cis-Golgi localized processing step. Furthermore, these studies indicated that an inactive point mutant of furin in which Asp-153 was replaced with Ala not only failed to undergo proregion cleavage as described above, but also remained endoglycosidase H sensitive and had an immunofluorescence localization pattern indicative of the ER/ERGIC (ER–Golgi intermediate compartment). Both the ER localization of proregion cleavage and the requirement for processing as a prerequisite for efficient transport from the ER/ERGIC were confirmed in subsequent studies by the laboratories of W. Garten (79) and W. J. M. Van de Ven (80). These later studies examined additional inactive furin constructs and also found that the ER retention of uncleaved furin appeared saturable, suggesting interaction of profurin with chaperones or similar quality-control binding proteins. In the Thomas laboratory study (78), the observation that ER-retained profurin constructs with intact catalytic sites (e.g., those with mutated proregion cleavage sites lacking P1 or P4 Arg) were inactive was explored further using a truncated soluble form of furin either with or without the addition of a C-terminal KDEL-ER retention signal. This analysis showed that furin that underwent normal proregion cleavage was inactive toward substrates in *trans* if retained in the ER, even when assayed under optimal cleavage conditions *in vitro*. An analogous construct was previously reported to be inactive in coexpression studies with vonWillebrand factor precursor; however, the role of enzyme/substrate compartmentalization had not been explored (49). The finding that proregion cleavage at Arg-107 was necessary but not sufficient for activation of furin indicated that additional activation steps required transport to late secretory pathway compartments.

The importance of late secretory pathway localization to furin activation was explored further by the Thomas laboratory, who reconstituted the activation of ER-retained furin *in vitro* and described a role for the proregion in regulating this activation step (81). These studies demonstrated activation of an ER-retained furin construct by incubation under conditions resembling those found in the late secretory pathway [i.e., acidic pH (6.0) in the presence of free calcium]. This pH/calcium dependent activation step could be mimicked by limited trypsinization, suggesting a secondary cleavage event. The *in vitro* activation of furin was correlated with a loss of propeptide association, as determined by coimmunoprecipitation, and the appearance of a smaller propetide fragment. Mass spectrometric analysis of the furin propeptide fragment indicated that it resulted from a second cleavage occurring at Arg-75. The importance of cleavage at this internal P6 Arg-containing proregion site [-$\underline{Arg}$-Gly-Val-Thr-$\underline{Lys}$-$\underline{Arg}_{75}$(P1)-] was demonstrated by mutation of either Arg_{70} or Arg_{75} to Ala—both of which block pH/calcium dependent activation. The pH-dependent cleavage of the furin proregion at this P6 Arg-containing site is reminiscent of the more efficient processing of proalbumin under acidic conditions (71, 72). The requirement for proregion dissociation and degradation in furin activation was explained by the observation that the proregion itself is a highly effective inhibitor of the mature protease with a $K_{0.5}$ of 14 nM. The ability of furin propeptide to function as a selective slow tight binding inhibitor of furin both *in vitro* and *in vivo* when expressed in *trans* has been established independently by work in the laboratory of N. Seidah. These observations led to a model of the multistep compartment specific activation of furin (see Fig. 2) whereby the initial preproenzyme (w/signal peptide) undergoes essentially instantaneous cleavage to release the signal peptide and form profurin. Within the neutral pH environment of the ER the non-transport-competent profurin molecules undergo a second autoproteolytic cleavage at Arg-107 to release the propeptide. The liberated propeptide, however, remains associated with the mature furin domain, acting as an intramolecular inhibitor to prevent ectopic processing of other secretory pathway proteins. This furin/proregion complex is competent for transport from the ER, and upon delivery to late secretory compartments with acidic pH (i.e., the TGN/endosomes) it undergoes a second autoproteolytic processing event within the proregion at Arg-75. This secondary cleavage results in release of the proregion fragments from furin and disinhibition of the enzyme.

Additional coimmunoprecipitation and immunofluorescence colocalization analyses (*51*) demonstrated that, as predicted in this model, the liberated furin propeptide remains associated with mature furin until transport to the TGN. Further analyses of furin constructs with P1 or P6 mutations within the internal proregion cleavage site indicate that efficient export of furin from the TGN, however, does not require processing at this second site. Consistent with the importance of both P1 and P4 residues for processing, *in vitro* dilution assays indicate that the internal proregion cleavage is also an autocatalytic and largely intramolecular event. Furthermore, kinetic analyses using peptides representing the primary (Arg-107) and secondary (Arg-75) proregion cleavage sites indicate that furin can process the initial proregion cleavage site efficiently at neutral pH, whereas acidic conditions are required for efficient processing of the second site. This differential pH sensitivity appears to be due to the presence of a P6 instead of a P4 Arg, since substitution of the P4 Val with Arg in the secondary site peptide eliminates the effect. The presence of an ionizable His near this second site may also influence the pH dependence of its processing *in vivo*. This tightly controlled process of furin proregion dissociation and cleavage is in stark contrast to the apparently unregulated degradation of the proregions of the bacterial subtilisins. The importance of the ordered, pH regulated furin activation process was demonstrated *in vivo* by mutation of the internal proregion cleavage site to introduce a P4 Arg (Val$_{72}$ to Arg), thereby promoting cleavage at neutral pH and presumably allowing activation within the ER. On expression in cells, however, the Val-$_{72}$ to Arg construct failed to become active and was not transported out of the ER. The overall importance of the proregion itself to the proper activation of furin has been independently demonstrated using a construct in which the signal peptide sequence was directly linked to mature furin. This proregion deleted construct fails to exit the ER and is inactive. Activity can be rescued, at least in part, by coexpression of the proregion, suggesting that it can function *in trans*. It is not yet clear how the proregion affects the activation of furin beyond its function as an intramolecular inhibitor. However, one possibility is that the initial proregion cleavage site, which fits the furin optimal recognition motif, functions as a nucleation site to facilitate proper folding of the furin catalytic pocket. In this case, mutation of the upstream secondary (non-P4 Arg) cleavage site to produce a second optimal motif may competitively interfere with the nucleation/folding process. This potential facilitory function is consistent with the description of protease proregions as intramolecular chaperones or foldases based on studies of the subtilisins (*82*).

VII. Intracellular Localization and Trafficking of Furin

The characterization of furin localization and trafficking not only has provided insight into the role of the protease *in vivo,* but also has led to the identification of novel targeting signals and components of the cellular sorting machinery. These advances have produced a detailed model of trafficking in the TGN/endosomal system that has helped in understanding processes as diverse as phosphorylation-dependent protein sorting and down regulation of surface proteins by HIV. In initial expression studies furin was described as having a Golgi-like localization based on its steady-state distribution determined by immunofluorescence analysis (*28*). In more detailed studies, G. Thomas and co-workers used a combination of immunofluorescence and immunoelectron microscopy to show that furin was largely found in the TGN, where it colocalized with TGN38 (*78*). In addition, antibody uptake experiments revealed that furin actively cycled between the TGN and plasma membrane. Furthermore, analyses of truncation mutant showed that the 56 amino acid cytoplasmic domain of furin (see Fig. 3) was required for its localization to the TGN and recycling from the plasma membrane. Subsequent experiments by the laboratory of S. Munro using chimeric constructs demonstrated that the cytoplasmic domain of furin was sufficient to convey TGN localization and recycling from the cell surface (*58*). In addition, these studies showed that localization and recycling of a furin chimera was disrupted by chloroquine, a lysosomotropic drug, suggesting that sorting of furin in endosomes is essential for maintaining normal localization of the protein.

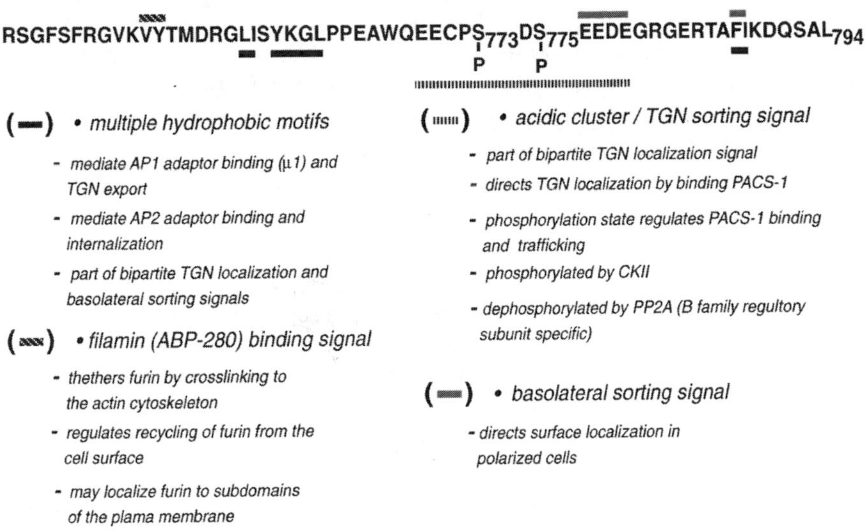

RSGFSFRGVKVYTMDRGLISYKGLPPEAWQEECPS$_{773}$DS$_{775}$EEDEGRGERTAFIKDQSAL$_{794}$

P   P

(▬) • multiple hydrophobic motifs

- mediate AP1 adaptor binding (μ1) and TGN export
- mediate AP2 adaptor binding and internalization
- part of bipartite TGN localization and basolateral sorting signals

(∞∞) • filamin (ABP-280) binding signal

- thethers furin by crosslinking to the actin cytoskeleton
- regulates recycling of furin from the cell surface
- may localize furin to subdomains of the plama membrane

(⏸) • acidic cluster / TGN sorting signal

- part of bipartite TGN localization signal
- directs TGN localization by binding PACS-1
- phosphorylation state regulates PACS-1 binding and trafficking
- phosphorylated by CKII
- dephosphorylated by PP2A (B family regultory subunit specific)

(▬) • basolateral sorting signal

- directs surface localization in polarized cells

FIG. 3. Sorting signals in the furin cytoplasmic domain (cd). The sequence of the human furin cd is shown with its multiple intracellular sorting motifs indicated. See text for details.

A subsequent paper by the laboratory of J. Bonifacino confirmed the TGN localization of furin and chimeric proteins containing the furin cytoplasmic domain (cd) (83). This later study also argued for the presence of lumenal domain signals that direct furin to a late endosome/lysosomal degradation pathway, presumably by virtue of protein aggregation in the TGN (84). The importance of such a lysosomal pathway in the biology of furin, however, is not yet clear and somewhat conflicts with a variety of studies indicating that furin is primarily lost from cells via a discrete TGN localized proteolytic cleavage leading to secretion or "shedding" of the active lumenal domain (79, 85, 29). Time course immunofluorescence studies of furin routing in the endocytic pathway by the laboratory of F. Maxfield indicate that following internalization, furin is delivered from early sorting endosomes to late endosomes from which it is efficiently retrieved to the TGN by a single-pass mechanism (86). These results suggest that the normal furin trafficking pathway can intersect with the lysosomal pathway, but that it is not necessarily a prelude degradation.

The initial studies demonstrating TGN localization and recycling of furin were verified and extended in a series of papers that further defined the trafficking itinerary of furin as well as the sorting signals and machinery responsible for various transport steps. Using a combination of cytoplasmic domain truncations, deletions, and point mutations, the laboratory of W. Garten uncovered two predominant targeting signals within the furin cd (*87*). These studies indicated that a stretch of sequence including a cluster of acidic residues (CPSDSEEDEG) was the key determinant of TGN localization while the tyrosine-based motif, YKGL, functioned as a retrieval signal to direct furin recycling from the cell surface (see Fig. 3). Independent studies by the laboratories of J. Bonafacino and G. Thomas also identified these same regions of the furin cd as essential sorting determinants using an analogous mutational strategy (*88, 89*). The initial study by W. Garten and co-workers argued that the acidic region alone was sufficient to promote TGN localization. However, this interpretation was complicated by the presence of a previously unidentified internalization signal, LI (*90*), present in conjunction with the acidic region. As described below, later studies indicate that both efficient targeting to the TGN and retrieval from the cell surface require a cooperative effect of the acidic region and a separate hydrophobic internalization motif. Similarly, studies by the laboratory of W. Hunziker have shown that the polarized sorting of furin to the basolateral surface in MDCK cells is a function of the EEDE residues within the acidic cluster in conjunction with a phenylalanine–isolucine motif (*91*). The importance of bipartite signals to the routing of furin in the TGN endosomal system is reminiscent of previous analyses of LDL receptor trafficking (*92*). This cooperative effect of acidic and hydrophobic motifs can be ascribed to their interaction with components of the cytoplasmic sorting machinery, as discussed below.

Additional analyses by the laboratories of K. Nakayama and G. Thomas revealed the importance of two serine residues that constitute CK2 phosphorylation sites within the furin cd acidic region ($S_{773}DS_{775}EEDE$). Nakayama and co-workers showed that mutation of these serines to alanines resulted in disruption of TGN localization of furin and that the furin cd could be phosphorylated *in vitro* by CK2 (*93*). Studies by the Thomas laboratory using phosphate labeling and mass spectromery showed that CK2 phosphorylated furin at S_{773} and S_{775} both *in vitro* and *in vivo* (*89*). Furthermore, their analyses showed that substitution of these serines with aspartic acid residues to mimic their phosphorylated state led to an accumulation of recycling furin in early endocytic structures. The same effect was observed with native furin in the presence of a protein phosphatase inhibitor, tautomycin, suggesting that dephosphorylation of the cd regulated the efficient sorting of furin from endosomes to the TGN *in viro* (*89*). In subsequent studies, Thomas and co-workers have identified specific protein phosphatase 2A isoforms containing B family regulatory subunits as the furin phosphatase *in vivo* and have also demonstrated that furin with the phosphomimic aspartic acid substitutions is recycled to the plasma membrane with kinetics similar to those of the transferrin receptor, whereas a nonphosphorylatable construct is delivered to later secretory compartments (*94*). Studies by the laboratory of W. Garten have confirmed the importance of S_{773} and S_{775} for the endosomal sorting of furin. These analyses, however, suggest that phosphorylated furin is more efficiently retrieved to the TGN than is the dephosphorylated form (95). One reason for this apparent discrepancy could be the resialylation assay used to measure TGN recycling in the later study. The efficient resialylation of protein delivered to the TGN by the resident sialyltransferase enzyme may require a significant dwell time in this compartment. Based on the model of phosphorylation-dependent furin trafficking proposed by the Thomas laboratory (96) (see Fig. 4), the nonphosphorylatable construct would be preferentially delivered from endosomes to the TGN local cycling loop (see below) but would not remain there, and hence would be inefficiently resialylated. By contrast, the phosphomimic construct would be preferentially recycled from endosomes to the plasma membrane; however, any of this construct that was delivered to the TGN (perhaps due to saturation of the phosphorylation-dependent sorting machinery) would be effectively maintained and hence efficiently resialylated.

The observation that the phosphorylation state of the furin cd regulates the sorting of the protein predicted that some component(s) of the cytoplasmic sorting machinery would interact specifically with phosphorylated furin. A yeast two-hybrid screen conducted by the Thomas laboratory discovered a novel phosphorylation specific furin cd binding protein, PACS-1, which functions as a connector protein linking the cds of cargo molecules to the sorting machinery (97). Mutational studies have defined independent cargo and adaptor protein binding sites that allow PACS-1 to facilitate the formation of ternary complexes that connect the cargo to the clathrin-dependent sorting machinery (98). Depletion and dominant negative expression have shown that PACS-1 is necessary to maintain the TGN localization of furin as well as that of additional TGN/endosomal proteins with acidic cluster sorting motifs, including the mannose 6-phosphate receptor (M6PR), CPD, and HIV-1 nef (97–99). Analyses of TGN budding show that the region of the furin cd necessary for interaction with PACS-1 has no effect on export from this compartment. Instead, the tyrosine-based YKGL motif previously identified as a key determinant of endocytosis appears to direct efficient TGN budding, suggesting that PACS-1 promotes TGN localization by an efficient retrieval pathway. Efficient retrieval/recycling of phosphorylated furin to the TGN from immediate post TGN/endosomal compartments by PACS-1 is supported by studies of CPD trafficking (L. Fricker, personal communication). These results suggested a model in which furin is retained in one of two local cycling loops by the combined function of CK2 and PACS-1—one loop is represented by the TGN and post-TGN/endosomal compartments while the other consists of the plasma membrane and early endosomes (see Fig. 4). By contrast, dephosphorylation of furin by PP2A allows the enzyme to move between these two local cycling loops.

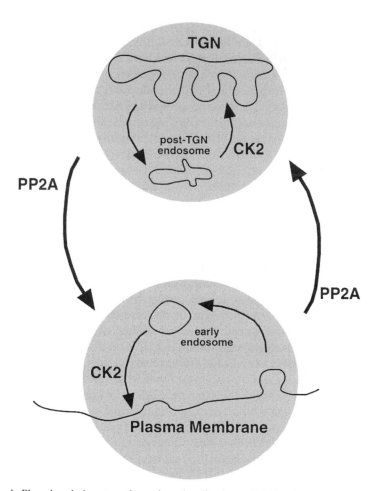

Fɪɢ. 4. Phosphorylation-state-dependent localization of furin. The function of CK2/phosphorylation-dependent retrieval in the TGN and plasma membrane local cycling loops is diagrammed along with the PP2A/dephosphorylation-dependent sorting between loops. See text for details.

Studies by the laboratory of W. Garten have provided additional insight into the interplay of hydrophobic and acidic motifs in furin trafficking. Using *in vitro* binding and permeablized cell analyses, these studies show that the YKGL motif is the major determinant of interaction between the furin cd and the AP-1Golgi/TGN adaptor complex and that this interaction is enhanced by CK2 phosphorylation of the acidic cluster (*100*). Furthermore, the tyrosine motif supports direct binding to the μ1 subunit of the AP-1 complex. Whereas the alternate LI internalization motif can support

$\mu 1$ binding at a lower level, the phenylalanine motif involved in basolateral targeting only binds the intact AP-1 complex. This rather complicated interplay of motifs is also observed for the interaction of the furin cd with the AP-2 adaptor complex that regulates clathrin-dependent endocytosis (95). In these studies W. Garten and co-workers showed that both the YKGL and LI motifs support binding to the $\mu 2$ subunit of AP-2 and that this binding can be enhanced by CK2 phosphorylation or Ser to Asp substitutions within the acidic cluster motif. By contrast, the phenylalanine motif had little influence on binding. These studies support the idea that hydrophobic motifs drive the efficient budding of furin into transport vesicles derived from the TGN (AP-1) or plasma membrane (AP-2) by direct interaction with the adaptor complexes while PACS-1 binding regulates the retrieval step (Fig. 5, see color plate). The apparent effect of CK2 phosphorylation on direct adaptor binding suggests the scenario may be more complicated. However, the presence of PACS-1 in extracts used as the source of adaptor could account for the enhanced binding observed in these assays with CK2 phosphorylation.

In addition to the prominent acidic cluster and "classical" hydrophobic motifs described above, a 19-residue stretch of membrane proximal amino acids including a key Val and Tyr can regulate the cell surface trafficking of furin. A two-hybrid screen and mutational analysis demonstrated that this segment of the furin cd bound specifically to filamin (ABP-280) (101). Filamin is a component of the actin cytoskeleton that cross-links actin fibrils and tethers membrane proteins at the cell surface by virtue of its multiple protein binding domains. The importance of filamin to the trafficking of furin was demonstrated using a filiman-deficient cell line that showed enhanced furin internalization and disrupted endosomal sorting. These data suggest that the amount and distribution of cell surface furin could be controlled by regulating the interaction of furin with filamin and hence the actin cytoskeleton. Such a surface tethering effect could provide a mechanism for regulating the cleavage of extracellular furin substrates (e.g., anthrax toxin PA and proaerolysin; see later discussion).

Together, the analyses of furin trafficking have provided an understanding of how the enzyme becomes available to process such a wide range of proproteins in multiple cellular compartments. In addition, these studies have greatly enhanced our knowledge of both the signals and machinery used to direct protein sorting in the TGN/endosomal system.

VIII. Furin Substrates

Although studies defining the furin recognition motif have produced a good understanding of the endoprotease's cleavage site preferences, a more complicated and vexing question has been to establish which proprotcins containing consensus furin sites represent actual substrates of the enzyme *in vivo*. This has proved to be a difficult question for several reasons, including the often overlapping sequence preferences and expression patterns of the various PCs as well as the ever increasing list of potential substrates containing multiple basic processing sites [see Table II and reference (*4*) for a recent review]. Furin, PACE4, PC7, and PC5/PC6 are all widely expressed but have distinct patterns. Often, however, subsets of these enzymes can be found in the same tissues and/or cell types. Furthermore, although these enzymes display subtle differences in their cleavage site preferences, these differences are insufficient to be predictive of cleavage by a particular PC. This scenario is further complicated by the observation that cleavage site recognition can be significantly altered by its context within a protein. Proproteins expressed exclusively in neuroendocrine cells and routed through the regulated secretory pathway are generally thought to be processed by PC2 and/or PC1/PC3. The determination that non-neuroendocrine-specific PCs, including furin (*102*) and PC5/PC6 (*103*), can at least transiently associate with regulated pathway compartments indicates that PCs normally connected with constitutive pathway processing could also contribute to the maturation of regulated pathway proteins.

Despite these complicating factors, progress has been made in determining bona-fide furin substrates using a combination of approaches. The availability of furin-deficient cell lines has provided one means of exploring the role of furin in the processing of identified proproteins. The human colon carcinoma cell line, LoVo, has been shown to have inactivating mutations in both furin alleles (a frame shift or a point mutation in a conserved Trp within the P domain) (*104, 105*) and fails to process endogenously expressed insulin proreceptor or hepatocyte growth factor receptor (*106*). Expression of functional furin restored the processing of both receptors, consistent with their proposed processing by furin *in vivo* (*107*). In addition, LoVo cells have been used to assess the processing of exogenously expressed proproteins in a furin-deficient background. Some of these proproteins, including Newcastle disease virus envelope glycoprotein Fo (*108*), avian influenza virus hemagglutinin (HA) (*109*), diphtheria toxin (*110*), transforming growth factor β (*111*), measles virus Fo (*112*), and human cytomegalovirus gB (*113*), reportedly have a relatively strict dependence on furin expression based on the lack of processing in LoVo and rescue by reexpression of active furin. Other putative furin substrates, however, show residual processing in LoVo and/or rescue by additional proteases [e.g., integrin α chains (*114*) and HIVgp160 (*108*), suggesting that multiple PCs may contribute to the processing of some substrates *in vivo* (see also Section X)]. Similar results have been observed with a second furin-deficient cell line obtained

TABLE II

Proposed Furin Substrates[a]

	P6	P4	P2	P1	↓	P'1	P'2	Site(s) of processing
				Endogenous Substrates				
Serum proteins								
Proalbumin	R	G	V	F	R	R	D	A	Biosynthetic pathway (bp)
Pro-factor IX	L	N	R	P	K	R	Y	N	bp
Pro-protein C	R	S	H	L	K	R	D	T	bp
Pro-von Willebrand Factor	S	H	R	S	K	R	S	L	bp
Hormones and growth factors								
Pro-β-nerve growth factor	T	H	R	S	K	R	S	S	bp
Pro-BNP	T	L	R	A	P	R	S	P	bp
Pro-parathyroid hormone	K	S	V	K	K	R	S	V	bp
Pro-semaphorin D (PCS 1)	K	R	R	T	R	R	Q	D	bp
Pro-TGF β1	S	S	R	H	R	R	A	L	bp
MIC-1	G	R	R	R	A	R	A	R	bp
Cell surface receptors								
Insulin pro-receptor	P	S	R	K	R	R	S	L	bp
Notch1 receptor	G	G	R	Q	R	R	E	L	bp
Scatter factor receptor	E	K	R	K	K	R	S	T	bp
Vitamin B$_{12}$ pro-receptor	L	Q	R	Q	K	R	S	I	bp
Misc.								
Pro-7B2	Q	R	R	K	R	R	S	V	bp
BIR-L demential rel. pep.	R	G	I	Q	K	R	E	A	bp/?
Extracellular matrix and signalling proteina								
BMP-1	R	S	R	S	R	R	A	A	bp/?
BMP-4 precursor	R	R	R	A	K	R	S	P	bp/?
C. elegans rol-6	S	N	R	V	R	R	Q	Q	?
C. elegans sqt-1	S	K	R	V	R	R	Q	Y	?
Human MT-MMP1	N	V	R	R	K	R	Y	A	bp
Integrin α3 (shown), 5,6 & v	P	Q	R	R	R	R	Q	L	bp/?
Profibrillin	R	G	R	K	R	R	S	T	bp/extracellular
Stromelysin-3	R	N	R	Q	K	R	F	V	bp
Xenopus laevis XMMP	K	I	R	R	K	R	F	L	?
ZP1	I	A	R	R	R	R	S	S	?
ZP2	S	L	R	S	K	R	E	A	?
ZP3α	A	A	R	R	R	R	S	S	?
TACE/ADAM17	V	E	R	V	K	R	R	A	bp/?
Muc2 (intestinal mucin)	S	R	R	A	R	R	S	P	bp/?
Disintegrin MDC9	L	L	R	R	R	R	A	V	bp/?
				Pathogenic Substrates				
Bacterial toxins								
Anthrax toxin PA	N	S	R	K	K	R	S	T	Cell surface
Clostridium septicum α-tox.	K	R	R	G	K	R	S	V	Endosomes
Diphtheria toxin	G	N	R	V	R	R	S	V	Endosomes
Proaerolysin	K	V	R	R	A	R	S	V	Cell surface
Exotoxin A	R	H	R	Q	P	R	G	W	Endosomes
Shiga toxin	A	S	R	V	A	R	M	A	Endosomes
Viral coat proteins								
Avian influenza HA (H5N1)	R	R	R	K	K	R	G	L	bp
Boma disease virus	L	K	R	R	R	R	D	T	bp
Cytomegalovirus gB	T	H	R	T	R	R	S	T	bp
Ebola zalre GP	G	R	R	T	R	R	E	A	bp
Epstein–Barr virus gB	L	R	R	R	R	R	D	A	bp
HIV-1 gp160	V	Q	R	E	K	R	A	V	bp
Infect, bronchitis virus E2	T	R	R	F	R	R	T	V	bp
Japan B encephalitis M	S	K	R	S	R	R	S	V	bp
Measles virus F$_o$	S	R	R	H	K	R	F	A	bp
Respiratory-sync4. virus F	K	K	R	K	R	R	F	L	bp
Rous sarcoma virus env	G	I	R	R	K	R	S	Vi	bp
Yellow fever virus M	S	G	R	S	R	R	S	V	bp

[a] P6 through P'2 indicate residue position before and after the identified precursor cleavage site (arrow). Site of processing indicates the probable location of the processing event *in vivo*. Question marks indicate cases that are undetermined.

by J. Moehring and co-workers using ethane methane sulfonate mutagenesis of Chinese hamster ovary (CHO) cells and selection for resistance to *Pseudomonas* exotoxin A (PEA) and Sindbis virus (SbV) (*115, 116*). Sequence characterization demonstrated that the RPE.40 cells, like the LoVo, had inactivating mutations in both furin alleles, thereby providing an independent furin minus cell model (*117*). Reexpression studies demonstrated that active furin can restore the processing defects in RPE.40 cells as well as support the processing of additional coexpressed proproteins (*118*). Expression of PACE4, however, can also restore SbV PE2 envelope glycoprotein and low-density lipoprotein-like receptor related protein (LRP) processing, suggesting the potential for overlap between PACE4 and furin (*119*). By contrast, PEA sensitivity was nearly unaffected, indicating that the toxin may be cleaved exclusively by furin.

The data obtained with LoVo and RPE.40 cells have been useful in identifying what appear to be exclusive furin substrates (e.g., PEA) as well as proproteins that could be substrates for multiple PCs *in vivo*. The apparent overlap between PACE4 and furin in a coexpression paradigm would not be surprising considering that PACE4 is purported to have a consensus recognition sequence, -Arg-X-Lys/Arg-Arg-, similar to that of furin—the difference being the relative ability of furin to accommodate P6 arginines. Interestingly, inactivation of furin in RPE.40 cells is sufficient to block processing of SbV PE2 and LRP even though PACE4 is endogenously expressed in these cells, suggesting that PACE4 expressed by gene transfer is either more active or differentially localized in the cells, perhaps because of its level of overexpression (*119*) (see also Section VII). Because of caveats such as these, no single analysis method has proved definitive in assigning PC/substrate pairings. The most convincing data has come from studies combining *in vitro* processing analyses with coexpression experiments. Examples of such studies include an analysis of influenza HA processing and a recent comparison of integrin α chain cleavage. In the case of HA processing, the ability of cells and endogenous Golgi associated protease to process a battery of HA cleavage site mutants was compared to that of limiting concentrations of furin *in vitro* (*120*). The results showed that furin activity exactly matched *in vivo* and Golgi-associated protease processing, whereas overexpression of furin with the substrates resulted in more permissive cleavages. In the case of α-integrin processing, furin, PC5A, and PACE4 were able to efficiently cleave the proprotein *in vitro*, whereas only furin and PC5A were effective in promoting cleavage in LoVo cells (*114*). These later data demonstrate that *in vitro* digests alone can be potentially misleading because of cleavage resulting from changes in substrate conformation or inappropriate enzyme/substrate interaction (i.e., loss of intracellular compartmentalization).

IX. Furin Inhibitors

Another approach that has proved effective in establishing the role of furin in the processing of particular proproteins is the design and application of selective inhibitors. Because of their importance in multiple biological processes, the PC family has been the focus of considerable effort in attempts to generate inhibitors that are both effective and selective. Early studies used short acylated peptides based on the furin consensus site that incorporate a reactive chloromethylketone (CMK) group at the C terminus. This strategy produced a series of stoichiometric PC inhibitors that were effective at nanomolar concentration *in vitro* and micromolar concentrations in cell culture (*121*). However, the reactive nature of the CMK moiety renders these compounds relatively toxic, limiting their usefulness *in vivo*. More importantly, it has proved difficult to engineer sufficient specificity into these short peptide inhibitors to target individual PCs, which is not surprising considering the overlap in cleavage site specificities. In a direct comparison (*113*), the peptide Dec-Arg-Val-Lys-Arg-CH_2Cl inhibited all the PCs (with the exception of PC4, which was not examined) with a k_i ranging from 0.12 nM for PC7 to 3.6 nM for PACE4. Despite these limitations, the CMK-peptide inhibitors have been used to implicate furin in the processing of several proproteins, including HIV gp160, influenza HA, HCMV gB, and NDV Fo (*121–124*). Variations on the peptide strategy include the introduction of a methylene group into the scissile bond between the P1 and P′1 residues to produce a reversible nanomolar inhibitor (*125*) and substitution of unnatural amino acids in the P′1 position (*126*). None of these peptide approaches, however, have yet proved both effective and selective for inhibition of furin *in vivo* without cytotoxicity.

As an alternative to these peptide-based strategies, other groups have attempted to alter naturally occurring protease inhibitors such as ovomucoid third domain (*127*), α_2-macroglobulin (*128*), and α_1-antitrypsin (*129, 130*) to target furin. In the case of ovomucoid, replacement of the reactive-center -Ala-Cys-Thr-Leu(P1)- sequence with a minimal furin recognition motif, -**Arg**-Cys-Lys- **Arg**(P1)-, produces a submicromolar inhibitor (*128*). In a similar approach, the laboratory of G. Thomas changed the reactive-site loop of the naturally occurring serpin (serine protease inhibitor), α_1-antitrypsin, from -Ala-Ile-Pro-Met(P1)-, which normally targets elastase *in vivo*, to the minimal furin site -**Arg**-Ile-Pro-**Arg**(P1)-, thereby generating a subnanomolar PC inhibitor designated α_1-antitrypsin Portland or α_1-PDX (*129, 130*). The feasibility of using α_1-antitrypsin as a scaffold for the design of a furin inhibitor was based on the naturally occurring variant called α_1-antitrypsin Pittsburgh (α_1-PIT) in which the Met at the P1 position within the reactive site loop has been changed to Arg. This point mutation completely shifted the specificity of α_1-antitrypsin from elastase to thrombin, hence inducing a severe clotting disorder in the affected patient. Early studies showed that at α_1-PIT could inhibit proalbumin cleavage by kexin (8) and an endogenous liver activity *in vitro* (131) at high concentrations. Furthermore, overexpression of α_1-PIT by transfection resulted in inhibition of complement C3 and proalbumin processing in cell culture (132), suggesting that when coexpressed at high levels in the secretory pathway α_1-PIT could affect the activity of PCs. As anticipated based on the cleavage site preferences of furin, the introduction of a second Arg at the P4 position to create α_1-PDX shifted the specificity of the serpin from thrombin to furin (129), which it inhibited with a k_i of 0.6 nM (130). α_1-PDX also formed an SDS-stable complex with furin, characteristic of the "suicide substrate" mechanism of enzyme inhibition by serpins. Furthermore, kinetic analyses showed that α_1-PDX was functioning as a slow tight binding inhibitor of furin and had a stoichiometry of inhibition (SI) of 2 (equivalent to the average number of serpin molecules needed to inhibit one molecule of protease). The low SI value indicates that α_1-PDX inactivates furin very efficiently. A quantitative comparison of the various PCs, showed that α_1-PDX was by far most efficient at inhibiting furin, with only one other PC, PC6B, having a nanomolar k_i (2.3) (130). In addition to a 4-fold difference in k_i, the stoichiometry of inhibition of PC6B by α_1-PDX was also 4-fold higher, suggesting that this PC could be targeted significantly less efficiently than furin by α_1-PDX *in vivo*. The only other PC that was significantly inhibited by α_1-PDX *in vitro* was PC1/3 with a k_i of 0.26 mM and an SI value of 40 (130).

Some subsequent studies have indicated that α_1-PDX can inhibit PCs other than furin and PC6B. However, these were generally coexpression paradigms [e.g., (133)] in which the concentration of α_1-PDX in the secretory pathway was not controlled and likely to be very high compared to the effective concentrations measured *in vitro*. One exception is a report that PACE4 immunologically isolated from the conditioned medium of transiently transfected COS cells is inhibited in a dose-dependent manner by α_1-PDX (134). However, the effective concentration of α_1-PDX appears to be quite high in these studies ($K_{0.5} = 1 \ \mu M$) and k_i and SI values were not determined. Similarly, α_1-PDX has been reported to inhibit the HIV gp160 directed processing of PC1/PC3, PACE4 and PC5/PC6 as well as furin *in vitro,* but no kinetic analyses were performed (135). Together, these data suggest that α_1-PDX can be used to inhibit PCs in addition to furin and possibly PC5/PC6 if both enzyme and inhibitor can be sufficiently concentrated to drive the reaction. However, the kinetic analyses reported to date clearly show that α_1-PDX displays significant intrinsic specificity toward furin when used at limiting enzyme and inhibitor concentrations. Recent studies also demonstrate that of the membrane-anchored PCs (furin, PC5B/PC6B, and PC7), only expression of furin led to the internalization and accumulation of extracellularly applied α_1-PDX (113), indicating that the kinetic data represent real differences in the ability of the inhibitor to target PCs. Despite possible caveats about its specificity, α_1-PDX has proved to be a very effective tool for investigating the role of PC-dependent processing in vivo (see Section X).

The generation of even more selective PC inhibitors will likely require more detailed information regarding the interaction of current inhibitors and substrates with the catalytic pockets of the enzymes (i.e., cocrystallization studies). An interesting potential template for the design of selective inhibitors are the proregions of the PCs themselves. In independent studies (81, 50), the proregions of furin, PC1/PC3, and PC7 have been shown to function as effective and selective inhibitors of their cognate enzymes both in vitro and in vivo. Although the C-terminal Arg of the liberated propeptides appears to be essential for inhibition, synthetic peptides derived from this portion of the proregions were significantly less effective and acted as simple competitive inhibitors (50) rather than slow tight binding inhibitors [as observed for the intact proregions (81)]. This observation suggests that the properties of the propeptides that make them effective inhibitors of the PCs in vivo may not translate to simple peptide chemistry. Alternatively, naturally occurring endogenous furin-specific inhibitors may yet be discovered. To date the most potent naturally occurring inhibitor of furin is the ovalbumin-type serpin, proteinase inhibitor-8 (PI8), with a k_i of 53.8 pM (136). It is not yet known if the reactive site sequence of PI8 [-Arg-Asn-Ser-Arg(P1)-] can be modified to increase the effectiveness and selectivity of this serpin toward furin. The absence of a signal peptide to target PI8 to the secretory pathway, however, complicates the serpin's potential role as a furin inhibitor in vivo.

X. Role of Furin in Vivo

The importance of furin activity to biological systems has been demonstrated in several different ways. The first indication of the critical role of furin or furin-like proteolytic activity in vivo came from analyses of protein sequences found in nonprocessed proproteins associated with various clinical disorders. Numerous instances of hemophilia B have been linked to point mutations within the normal -Arg-Pro-Lys-Arg(P1)-cleavage site sequence of pro-Factor IX that lead to accumulation of the precursor in serum [reviewed in (137)]. Many of these mutations are substitutions eliminating the basic residues at the P1, P2, or P4 position, including the initially described P4 Arg-to-Gln mutation responsible for hemophilia B Oxford-3 (138). Similarly, mutations in the cleavage site [-Arg-Gly-Val-Phe-Arg-Arg(P1)-] of proalbumin have been documented in multiple clinical cases in which the serum protein remains unprocessed [e.g., proalbumin Christchurch (139)]. These proalbumin mutations include substitutions of the P1 and P2 basic residues, as well as the introduction of a Val at P'1 [proalbumin Blenheim]. These mutant forms of proalbumin cannot be processed by furin (71, 140), consistent with the importance of an intact furin consensus site and the idea that hydrophobic aliphatic residues in the P'1 position are unfavorable for cleavage.

Cleavage site point mutations in additional nonserum proproteins have also been linked to clinical conditions. In one case of a familial form of extreme insulin resistance, the insulin proreceptor [-Arg-Pro-Ser-Arg-Lys-Arg-Arg(P1)-] was discovered to have undergone a mutation converting the P1 Arg to Ser that disrupted processing (141, 142). In more recent studies, a patient displaying characteristics of Marfan syndrome was found to have a point mutation in one fibrillin allele that eliminated the P6 Arg in the profibrillin cleavage site [-Arg-Gly-Arg-Lys-Arg-Arg(P1)-] (143), thereby blocking the processing step required for formation of microfibrils and deposition in the extracellular matrix (144). Other studies have shown that a form of familial British dementia (FBD) is linked to a mutation in the BRI gene that alters the processing of its encoded precursor protein by furin, generating ABri peptides that accumulate in the amyloid deposits characteristic of the disease (145). Unlike the previous examples, the mutation occurs downstream of the cleavage site and eliminates the normal stop codon to produce an elongated precursor, BRI-L. Coexpression analyses in CHO and furin deficient RPE.40 cells showed that BRI-L was processed to ABri peptides by furin and that this processing was more efficient than that observed for the native BRI. This link between furin and FBD is especially intriguing considering the recent observation that the Alzheimer's β-secretase, BACE, is a proprotein that is activated by a furin-like convertase during its biosynthesis (146).

A more direct indication of the role of furin has been provided by the generation of transgenic knockout mice by A. J. M. Roebroek and D. Constam (147). The inactivation of furin in these mice creates an embryonic lethal phenotype with death occurring at or near day 10.5 postcoitus. The arrested embryos display a well-defined phenotype with defects in ventral closure, heart-looping morphogenesis, and axial rotation consistent with normal areas of furin expression. Additional analyses by D. Constam and E. J. Robertson found that these defects correlated with a loss of asymmetry in the expression of the TGFβ family member lefty-2 and the transcription factor pitx2 (*148*). The use of chimeric embryos (consisting of tissues of diverse genetic constitution) also showed that furin activity is required in multiple mesodermal and endodermal tissues for normal development. Disruption of these various cell fate and morphogenic steps in the absence of furin is consistent with the protease's proposed role in activating TGFβ and the related proteins BMP-4 and nodal (*111, 149, 150*). The noted similarity of the turning and closure defects of the furin knockout mouse to those seen with embryos lacking the downstream regulator of BMP-1 receptor activation, Smad5, suggested that furin may play a key role in this signaling pathway as well (*148*). Interestingly, the observation that loss of function mutations in some of these genes can produce more dramatic defects than the furin knockout suggests that additional PCs could be involved in their activation (*151*). However, the fact that the furin knockout mouse is indeed an embryonic lethal demonstrates that other PC family members are unable to completely compensate for the loss of furin activity *in vivo*, despite the similarity in their cleavage site recognition sequences.

An alternative approach to the disruption of furin function *in vivo* has been used by the laboratory of J. Christian to explore the role of furin-like processing in developing *Xenopus* embryos (*149, 152*). In these studies, α_1-PDX was introduced into various regions of the developing embryos by the targeted injection of synthetic mRNA. When injected near the ventral midline of 4- or 8-cell embryos, α_1-PDX mRNA altered the fate of the ventral mesodermal cells, causing them to adopt dorsal characteristics. Similarly, the α_1-PDX mRNA caused a neuralization of ectodermal cells when injected into animal caps (*149*). In both cases, the α_1-PDX-induced phenotype mimicked that seen for disruption of the BMP signaling pathway in the affected cells (e.g., by expression of truncated dominant negative receptor). In addition, bypassing the BMP-4 receptor activation step by expression of intracellular transducers of this pathway blocked the effects of α_1-PDX. These data indicate a role for an α_1-PDX sensitive protease (e.g., furin or PC6B) in activation of the BMP-4 receptor and/or ligand. Consistent with such a role, coinjection of α_1-PDX and BMP-4 precursor into oocytes resulted in a complete block in processing of the precursor to lower molecular weight active forms. As a correlate to the cellular studies, the proBMP-4 was incubated with furin, PC6B, PC7, and PACE-4 *in vitro* to assess the relative ability of the constitutive pathway PCs to correctly process the precursor. The data from these *in vitro* digests show that although all four PCs can cleave proBMP-4, only the activity of furin and PC6B was sensitive to α_1-PDX. Together, these results indicate that the developmental defects observed in α_1-PDX mRNA injected embryos arise from the inhibition of furin and/or PC6B dependent processing of the BMP-4 precursor. In addition, analysis of the cleavage products generated from proBMP-4 *in vitro* revealed the presence of two P4-P1 Arg cleavage sites that were differentially recognized by the PCs. The primary site within proBMP-4, -Arg-Ser-Lys-Arg(P1)-, represents the border between mature BMP-4 and the prodomain. The secondary site, -Arg-Ile-Ser-Arg(P1)-, is located approximately 30 residues upstream within the prodomain. Whereas furin and PACE4 can cleave at both sites, PC1/PC3 and PC7 only cleaved at the primary site. Further metabolic labeling pulse-chase analyses have shown that these sites are sequentially cleaved *in vivo* as well as *in vitro* (*152*). Point mutations within the secondary cleavage site also showed that although processing at this site was not a prerequisite for production of active BMP-4, either blocking cleavage or enhancing cleavage (by introduction of a P2 Arg) increased the range of expression of BMP-4 target genes in embryonic ectoderm. By contrast, the secondary cleavage site mutants had no effect on BMP-4 target gene expression in mesodermal tissues. Thus, sequential cleavage of proBMP-4 by furin and/or PACE4 controls the range of BMP-4 signaling *in vivo* in a tissue-specific manner. The presence of conserved primary and secondary cleavage sites in both proBMP-4 and proBMP-2 from *Xenopus,* chicken, and humans, as well as in decapentaplegic (DPP) of *Drosophila,* suggests that regulation of function by ordered sequential PC-dependent processing may be widely used *in vivo*.

The data obtained to date using both gene disruption and α_1-PDX mRNA injection strategies indicate that furin activity is essential for maintaining cell signaling pathways required for embryogenesis, including aspects of morphogenesis and cell fate determination. The continued expression of furin in adult tissues and demonstrated function in the processing of a host of proproteins (see Section VIII) indicates that this PC is also essential to the homeostasis of adult organisms. The lethality of the current germline knockouts, however, makes these additional functions difficult to assess. Ultimately, the targeted tissue specific inhibition of furin (e.g., expression of α_1-PDX by tissue-specific developmentally regulated promoters) may provide insight into the apparent multiple roles of furin in the adult.

XI. Furin and Pathogenesis

A host of studies have linked furin or furin-like activity to the maturation of both viral and bacterial pathogen proteins. Initially, this link was based on the presence of multiple basic processing sites in the precursors of several viral envelope glycoproteins and bacterial protoxins (see Table II). The subsequent cloning of furin and development of inhibitors has provided strong support for the role of this PC in the pathogenicity of both viruses and bacteria. The first data indicating a direct role for furin in the activation of bacterial toxins was obtained from *in vitro* digests using purified furin and a battery of anthrax protective antigen (PA) mutants. PA is the toxin component that, upon binding to a cell surface receptor and cleavage by furin, allows translocation of the active toxin from endsomes into the cytoplasm. These studies showed that the processing efficiency of the PA proteins by furin directly correlated with their toxicity *in vivo* (*56, 66*). In addition to indicating that furin activated PA *in vivo*, these analyses also provided the first evidence for the importance of furin in processing substrates at the cell surface and in endosomes. Subsequent studies using *in vitro* digestions, inhibitors, and furin deficient cells (RPE.40 and LoVo) have also demonstrated a role for furin in the processing and activation of diphtheria toxin (*110*), *Pseudomonas* exotoxin A (PEA) (*153*), proaerolysin (*154*), *Clostridium septicum* α-toxin (*155*), and Shiga toxin (*156*). In the case of PEA, the dose–response curve is shifted several logs between the furin-deficient RPE.40 cells and their parental line, an effect that is reversed by expression of exogenous active furin (*116*). Similarly, inhibition of cellular furin by incubation with extracellularly applied α_1-PDX completely blocks toxicity of PEA on a human melanoma cell line (*130*).

An even more compelling connection can be made between furin and the pathogenicity of viruses of several different classifications. Prominent envelope glycoproteins from many viruses are synthesized as proproteins that are processed to their mature active form during their transport through the host cell secretory pathway and incorporation into virions. As with the bacterial toxins, these cleavage sites are generally defined by basic residues at the P1 position. Initial studies comparing the cleavage efficiency of avian influenza hemagglutinin (HA) from various viral strains showed that the specificity of purified furin was indistinguishable from the protease activity associated with HA processing *in vivo* (*120*). Additional analyses using inhibitors and furin-deficient cells provided further evidence for the role of furin in processing envelope protein precursors. Both CMK peptide inhibitors and α_1-PDX are able to block the processing of the gp160 envelope glycoprotein (*129, 122*). Similar results have been obtained for measles virus (*112*) and human cytomegalovirus (*130*) where inhibition of furin by α_1-PDX blocks the processing of the envelope glycoproteins Fo and gB, respectively, as well as infectivity of the viruses. The presence of furin sites in numerous other glycoproteins (see Table II) suggests that the protease is critical to the production of infectious particles of additional viruses.

Indeed, comparison of envelope protein sequences from multiple viral strains indicates that efficient cleavage by furin is a key determinant of virulence and even host species *in vivo*. In the case of Ebola virus, its lethality is directly correlated with the cleavage site sequence of GP (*157*). In the apparently nonpathogenic Ebola Reston strain, GP lacks a consensus furin motif at its cleavage site [-Lys-Gln-Lys-Arg(P1)-]. By contrast, the highly pathogenic Ebola Zaire and Ebola Ivory Coast strains, which cause fulminant hemorrhagic fever resulting a 90% lethality rate, contain consensus furin sequences at their GP cleavage sites [-Arg-Thr/Ala-Arg-Arg(P1)- and -Arg-Lys-Arg-Arg(P1)-, respectively]. Similarly, sequence analysis of HA genes from 16 isolates of a fatal strain of avian influenza A (H5N1) revealed a consistent alteration in the viral genomes whereby a second consensus furin site at the HA1/HA2 junction was generated [-Arg-Lys-Lys-Arg(P1)- to -Arg-Glu-Arg-Arg-Arg-Lys-Lys-Arg(P1)-] (*158*). These observations linking furin processing to pathogenicity are supported by the finding that cleavage efficiency of the Fo fusion proteins of NDV by furin and/or PC6 *in vitro* closely parallels the virulence of the parental strain of virus (*159*). Furthermore, the studies of viral and bacterial pathogenesis suggest that furin is a potential target for the development of broad-based therapeutics, a possibility bolstered by the initial findings with peptide inhibitors and α_1-PDX.

XII. Conclusion

The discovery of furin as the first member of the PC family of kexin/subtilisin-like mammalian endoproteases prompted an intense and wide-ranging investigation into the biochemistry and cell biology of these enzymes. Furin was found to be a calcium-dependent serine endoprotease with an atypical inhibitor profile and wide pH range. Cleavage site analyses show a strong preference for substrates with the -Arg-X-Lys/Arg-Arg(P1)-motif; however, additional sites lacking P2 and/or P4 basic residues and containing P6 residues can also be efficiently processed *in vitro* and *in vivo*. These cleavage site preferences strongly suggest an active site having multiple acidic residues within the binding pocket, an observation that fits predictions of furin catalytic domain structure based on computer modeling. Furin was found to be synthesized as a zymogen that undergoes an ordered compartment-specific series of activation steps during its biosynthesis. These steps include an initial autocatalytic cleavage of the proregion within the ER as a prerequisite for export, followed by transport to the TGN where a second cleavage of the proregion occurs coincident with dissociation of this intramolecular inhibitor from the mature enzyme. Studies of furin trafficking demonstrated that the enzyme was concentrated in the TGN by a retrieval-based mechanism that involves CK2-dependent phosphorylation of an acidic region in the furin cd and interaction with a novel component of the cellular sorting machinery, PACS-1. Additional furin sorting steps include PP2A-dependent dephosphorylation and transport to and from the cell periphery as well as cell surface sorting via interaction with filamin/ABP-280. These trafficking analyses have provided the basis for new models of protein sorting and transport within the TGN/endosomal system and also demonstrated the presence of furin in multiple cellular processing compartments, consistent with the enzyme's role in processing a wide range of proproteins *in vivo*. The combination of germline knockout and targeted inhibition strategies have shown that furin activity is essential in several aspects of early development, indicating a key role in regulating cellular signaling pathways. Based on its continued expression in the adult and its broad substrate range, furin also appears to function in homeostatic processes such as blood clotting cascades and sugar metabolism. In addition, the pathogenesis of several bacteria and viruses is tightly coupled to furin-dependent processing of their respective toxins or envelope glycoproteins, demonstrating the importance of furin to acute and chronic disease states. The documented role of furin in such a wide range of biological processes, combined with the ever-increasing list of furin substrates, suggests that this member of the PC family will continue to be the focus of intensive research addressing both the biochemical and cell biological properties of the protease.

ACKNOWLEDGMENTS

Financial support for some of the work described in this review was provided by NIH grants DK 37274 and AI 49793.

REFERENCES

1. Steiner, D. F., Cunningham, D., Spigelman, L., and Aten, B. (1997). *Science* **157**, 697.
2. Chance, R. E., Ellis, R. M., and Bromer, W. W. (1968). *Science* **161**, 165.
3. Chretien, M., and Li, C. H. (1967). *Can. J. Biochem.* **45**, 1163.
4. Seidah, N. G., and Chretien, M. (1999). *Brain Res.* **848**, 45.
5. Newcomb, R., Fisher, J. M., and Scheller, R. H. (1988). *J. Biol. Chem.* **263**, 12514.
6. Fuller, R. S., Sterne, R. E., and Thorner, J. (1988). *Annu. Rev. Physiol.* **50**, 345.
7. Mizuno, K., Nakamura, T., Ohshima, T., Tanaka, S., and Matsuo, H. (1988). *Biochem. Biophys. Res. Commun.* **156**, 246.
8. Bathurst, I. C., Brennan, S. O., Carrell, R. W., Cousens, L. S., Brake, A. J., and Barr, P. J. (1987). *Science* **235**, 348.

9. Thomas, G., Thorne, B. A., Thomas, L., Allen, R. G., Hruby, D. E., Fuller, R., and Thorner, J. (1988). *Science* **241,** 226.
10. Seidah, N. G., Gaspar, L., Mion, P., Marcinkiewicz, M., Mbikay, M., and Chretien, M. (1990). *DNA Cell Biol.* **9,** 415.
11. Seidah, N. G., Marcinkiewicz, M., Benjannet, S., Gaspar, L., Beaubien, G., Mattei, M. G., Lazure, C., Mbikay, M., and Chretien, M. (1991). *Mol. Endocrinol.* **5,** 111.
12. Smeekens, S. P., Avruch, A. S., LaMendola, J., Chan, S. J., and Steiner, D. F. (1991). *Proc. Natl. Acad. Sci. USA* **88,** 340.
13. Nakayama, K., Hosaka, M., Hatsuzawa, K., and Murakami, K. (1991). *J. Biochem.* (*Tokyo*) **109,** 803.
14. Smeekens, S. P., and Steiner, D. F. (1990). *J. Biol. Chem.* **265,** 2997.
15. Kiefer, M. C., Tucker, J. E., Joh, R., Landsberg, K. E., Saltman, D., and Barr, P. J. (1991). *DNA Cell Biol.* **10,** 757.
16. Nakayama, K., Kim, W. S., Torii, S., Hosaka, M., Nakagawa, T., Ikemizu, J., Baba, T., and Murakami, K. (1992). *J. Biol. Chem.* **267,** 5897.
17. Seidah, N. G., Day, R., Hamelin, J., Gaspar, A., Collard, M. W., and Chretien, M. (1992). *Mol. Endocrinol.* **6,** 1559.
18. Nakagawa, T., Hosaka, M., Torii, S., Watanabe, T., Murakami, K., and Nakayama, K. (1993). *J. Biochem.* (*Tokyo*) **113,** 132.
19. Lusson, J., Vieau, D., Hamelin, J., Day, R., Chretien, M., and Seidah, N. G. (1993). *Proc. Natl. Acad. Sci. USA* **90,** 6691.
20. Meerabux, J., Yaspo, M. L., Roebroek, A. J., Van de Ven, W. J., Lister, T. A., and Young, B. D. (1996). *Cancer Res.* **56,** 448.
21. Bruzzaniti, A., Goodge, K., Jay, P., Taviaux, S. A., Lam, M. H., Berta, P., Martin, T. J., Moseley, J. M., and Gillespie, M. T. (1996). *Biochem. J.* **314,** 727.
22. Seidah, N. G., Hamelin, J., Mamarbachi, M., Dong, W., Tardos, H., Mbikay, M., Chretien, M., and Day, R. (1996). *Proc. Natl. Acad. Sci. USA* **93,** 3388.
23. Constam, D. B., Calfon, M., and Robertson, E. J. (1996). *J. Cell. Biol.* **134,** 181.
24. Steiner, D. F. (1998). *Curr. Opin. Chem. Biol.* **2,** 31.
25. Roebroek, A. J., Schalken, J. A., Bussemakers, M. J., van Heerikhuizen, H., Onnekink, C., Debruyne, F. M., Bloemers, H. P., and Van de Ven, W. J. (1986). *Mol. Biol. Rep.* **11,** 117.
26. Roebroek, A. J., Schalken, J. A., Leunissen, J. A., Onnekink, C., Bloemers, H. P., and Van de Ven, W. J. (1986). *EMBO J.* **5,** 2197.
27. Fuller, R. S., Brake, A. J., and Thorner, J. (1989). *Science* **246,** 482.
28. Bresnahan, P. A., Leduc, R., Thomas, L., Thorner, J., Gibson, H. L., Brake, A. J., Barr, P. J., and Thomas, G. (1990). *J. Cell Biol.* **111,** 2851.
29. Wise, R. J., Barr, P. J., Wong, P. A., Kiefer, M. C., Brake, A. J., and Kaufman, R. J. (1990). *Proc. Natl. Acad. Sci. USA* **87,** 9378.
30. van de Ven, W. J., Voorberg, J., Fontijn, R., Pannekoek, H., van den Ouweland, A. M., van Duijnhoven, H. L., Roebroek, A. J., and Siezen, R. J. (1990). *Mol. Biol. Rep.* **14,** 265.
31. Misumi, Y., Oda, K., Fujiwara, T., Takami, N., Tashiro, K., and Ikehara, Y. (1991). *J. Biol. Chem.* **266,** 16954.
32. Schalken, J. A., Roebroek, A. J., Oomen, P. P., Wagenaar, S. S., Debruyne, F. M., Bloemers, H. P., and Van de Ven, W. J. (1987). *J. Clin. Invest.* **80,** 1545.
33. Hatsuzawa, K., Hosaka, M., Nakagawa, T., Nagase, M., Shoda, A., Murakami, K., and Nakayama, K. (1990). *J. Biol. Chem.* **265,** 22075.
34. Zheng, M., Streck, R. D., Scott, R. E., Seidah, N. G., and Pintar, J. E. (1994). *J. Neurosci.* **14,** 4656.
35. Zheng, M., Seidah, N. G., and Pintar, J. E. (1997). *Dev. Biol.* **181,** 268.

36. Marcinkiewicz, M., Day, R., Seidah, N. G., and Chretien, M. (1993). *Proc. Natl. Acad. Sci. USA* **90,** 4922.

37. Marcinkiewicz, M., Ramla, D., Seidah, N. G., and Chretien, M. (1994). *Endocrinology* **135,** 1651.

38. Torii, S., Yamagishi, T., Murakami, K., and Nakayama, K. (1993). *FEBS Lett.* **316,** 12.

39. Ayoubi, T. A., Creemers, J. W., Roebroek, A. J., and Van de Ven, W. J. (1994). *J. Biol. Chem.* **269,** 9298.

40. Blanchette, F., Day, R., Dong, W., Laprise, M. H., and Dubois, C. M. (1997). *J. Clin. Invest.* **99,** 1974.

41. Marcinkiewicz, M., Marcinkiewicz, J., Chen, A., Leclaire, F., Chretien, M., and Richardson, P. (1999). *J. Comp. Neurol.* **403,** 471.

42. Hayflick, J. S., Wolfgang, W. J., Forte, M. A., and Thomas, G. (1992). *J. Neurosci.* **12,** 705.

43. Roebroek, A. J., Creemers, J. W., Pauli, I. G., Kurzik-Dumke, U., Rentrop, M., Gateff, E. A., Leunissen, J. A., and Van de Ven, W. J. (1992). *J. Biol. Chem.* **267,** 17208.

44. Chen, J. S., and Raikhel, A. S. (1996). *Proc. Natl. Acad. Sci. USA* **93,** 6186.

45. Gomez-Saladin, E., Luebke, A. E., Wilson, D. L., and Dickerson, I. M. (1997). *DNA Cell Biol.* **16,** 663.

46. Jin, J., Poole, C. B., Slatko, B. E., and McReynolds, L. A. (1999). *Gene* **237,** 161.

47. Chun, J. Y., Korner, J., Kreiner, T., Scheller, R. H., and Axel, R. (1994). *Neuron* **12,** 831.

48. Oliva, A. A., Jr., Chan, S. J., and Steiner, D. F. (2000). *Biochim. Biophys. Acta* **1477,** 338.

49. Rehemtulla, A., Dorner, A. J., and Kaufman, R. J. (1992). *Proc. Natl. Acad. Sci. USA* **89,** 8235.

50. Zhong, M., Munzer, J. S., Basak, A., Benjannet, S., Mowla, S. J., Decroly, E., Chretien, M., and Seidah, N. G. (1999). *J. Biol. Chem.* **274,** 33913.

51. Anderson, E. D., Jean, F., and Thomas, G. (submitted).

52. Creemers, J. W., Siezen, R. J., Roebroek, A. J., Ayoubi, T. A., Huylebroeck, D., and Van de Ven, W. J. (1993). *J. Biol. Chem.* **268,** 21826.

53. Creemers, J. W., Usac, E. F., Bright, N. A., Van de Loo, J. W., Jansen, E., Van de Ven, W. J. M., and Hutton, J. C. (1996). *J. Biol. Chem.* **271,** 25284.

54. Zhou, A., Martin, S., Lipkind, G., LaMendola, J., and Steiner, D. F. (1998). *J. Biol. Chem.* **273,** 11107.

55. Lusson, J., Benjannet, S., Hamelin, J., Savaria, D., Chretien, M., and Seidah, N. G. (1997). *Biochem. J.* **326,** 737.

56. Molloy, S. S., Bresnahan, P. A., Leppla, S. H., Klimpel, K. R., and Thomas, G. (1992). *J. Biol. Chem.* **267,** 16396.

57. Hatsuzawa, K., Nagahama, M., Takahashi, S., Takada, K., Murakami, K., and Nakayama, K. (1992). *J. Biol. Chem.* **267,** 16094.

58. Chapman, R. E., and Munro, S. (1994). *EMBO J.* **13,** 2305.

59. Siezen, R. J., Creemers, J. W., and Van de Ven, W. J. (1994). *Eur. J. Biochem.* **222,** 255.

60. Siezen, R. J. (1996). *Adv. Exp. Med. Biol.* **379,** 63.

61. Zhou, Y., and Lindberg, I. (1993). *J. Biol. Chem.* **268,** 5615.

62. Jean, F., Basak, A., Rondeau, N., Benjannet, S., Hendy, G. N., Seidah, N. G., Chretien, M., and Lazure, C. (1993). *Biochem. J.* **292,** 891.

63. Rufaut, N. W., Brennan, S. O., Hakes, D. J., Dixon, J. E., and Birch, N. P. (1993). *J. Biol. Chem.* **268,** 20291.

64. Lamango, N. S., Zhu, X., and Lindberg, I. (1996). *Arch. Biochem. Biophys.* **330,** 238.

65. Lindberg, I., van den Hurk, W. H., Bui, C., and Batie, C. J. (1995). *Biochemistry* **34,** 5486.

66. Klimpel, K. R., Molloy, S. S., Thomas, G., and Leppla, S. H. (1992). *Proc. Natl. Acad. Sci. USA* **89,** 10277.

67. Hosaka, M., Nagahama, M., Kim, W. S., Watanabe, T., Hatsuzawa, K., Ikcmizu, J., Murakami, K., and Nakayama, K. (1991). *J. Biol. Chem.* **266,** 12127.
68. Watanabe, T., Nakagawa, T., Ikemizu, J., Nagahama, M., Murakami, K., and Nakayama, K. (1992). *J. Biol. Chem.* **267,** 8270.
69. Watanabe, T., Murakami, K., and Nakayama, K. (1993). *FEBS Lett.* **320,** 215.
70. Takahashi, S., Hatsuzawa, K., Watanabe, T., Murakami, K., and Nakayama, K. (1994). *J. Biochem.* (*Tokyo*) **116,** 47.
71. Brennan, S. O., and Nakayama, K. (1994). *FEBS Lett.* **347,** 80.
72. Brennan, S. O., and Nakayama, K. (1994). *FEBS Lett.* **338,** 147.
73. Matthews, D. J., Goodman, L. J., Gorman, C. M., and Wells, J. A. (1994). *Protein Sci.* **3,** 1197.
74. Krysan, D. J., Rockwell, N. C., and Fuller, R. S. (1999). *J. Biol. Chem.* **274,** 23229–74.
75. Ballinger, M. D., Tom, J., and Wells, J. A. (1995). *Biochemistry* **34,** 13312.
76. Ballinger, M. D., Tom, J., and Wells, J. A. (1996). *Biochemistry* **35,** 13579.
77. Leduc, R., Molloy, S. S., Thorne, B. A., and Thomas, G. (1992). *J. Biol. Chem.* **267,** 14304.
78. Molloy, S. S., Thomas, L., VanSlyke, J. K., Stenberg, P. E., and Thomas, G. (1994). *EMBO J.* **13,** 18.
79. Vey, M., Schafer, W., Berghofer, S., Klenk, H. D., and Garten, W. (1994). *J. Cell. Biol.* **127,** 1829.
80. Creemers, J. W., Vey, M., Schafer, W., Ayoubi, T. A., Roebroek, A. J., Klenk, H. D., Garten, W., and Van de Ven, W. J. (1995). *J. Biol. Chem.* **270,** 2695.
81. Anderson, E. D., VanSlyke, J. K., Thulin, C. D., Jean, F., and Thomas, G. (1997). *EMBO J.* **16,** 1508.
82. Fu, X., Inouye, M., and Shinde, U. (2000). *J. Biol. Chem.* **275,** 16871.
83. Bosshart, H., Humphrey, J., Deignan, E., Davidson, J., Drazba, J., Yuan, L., Oorschot, V., Peters, P. J., and Bonifacino, J. S. (1994). *J. Cell. Biol.* **126,** 1157.
84. Wolins, N., Bosshart, H., Kuster, H., and Bonifacino, J. S. (1997). *J. Cell. Biol.* **139,** 1735.
85. Vidricaire, G., Denault, J. B., and Leduc, R. (1993). *Biochem. Biophys. Res. Commun.* **195,** 1011.
86. Mallet, W. G., and Maxfield, F. R. (1999). *J. Cell. Biol.* **146,** 345.
87. Schafer, W., Stroh, A., Berghofer, S., Seiler, J., Vey, M., Kruse, M. L., Kern, H. F., Klenk, H. D., and Garten, W. (1995). *EMBO J.* **14,** 2424.
88. Voorhees, P., Deignan, E., van Donselaar, E., Humphrey, J., Marks, M. S., Peters, P. J., and Bonifacino, J. S. (1995). *EMBO J.* **14,** 4961.
89. Jones, B. G., Thomas, L., Molloy, S. S., Thulin, C. D., Fry, M. D., Walsh, K. A., and Thomas, G. (1995). *EMBO J.* **14,** 5869.
90. Stroh, A., Schafer, W., Berghofer, S., Eickmann, M., Teuchert, M., Burger, I., Klenk, H. D., and Garten, W. (1999). *Eur. J. Cell. Biol.* **78,** 151.
91. Simmen, T., Nobile, M., Bonifacino, J. S., and Hunziker, W. (1999). *Mol. Cell. Biol.* **19,** 3136.
92. Matter, K., Yamamoto, E. M., and Mellman, I. (1994). *J. Cell. Biol.* **126,** 991.
93. Takahashi, S., Nakagawa, T., Banno, T., Watanabe, T., Murakami, K., and Nakayama, K. (1995). *J. Biol. Chem.* **270,** 28397.
94. Molloy, S. S., Thomas, L., Kamibayashi, C., Mumby, M. C., and Thomas, G. (1998). *J. Cell. Biol.* **142,** 1399.
95. Teuchert, M., Berghofer, S., Klenk, H. D., and Garten, W. (1999). *J. Biol. Chem.* **274,** 36781.
96. Molloy, S. S., Anderson, E. D., Jean, F., and Thomas, G. (1999). *Trends Cell. Biol.* **9,** 28.
97. Wan, L., Molloy, S. S., Thomas, L., Liu, G., Xiang, Y., Rybak, S. L., and Thomas, G. (1998). *Cell* **94,** 205.

98. Crump, C. M., Xiang, Y., Thomas, L., Austin, C., Tooze, S. A., and Thomas, G. (2001). *EMBO J.* **20,** 2191.

99. Piguet, V., Wan, L., Borel, C., Mangasarian, A., Demaurex, N., Thomas, G., and Trono, D. (2000). *Nature Cell Biol.* **2,** 163.

100. Teuchert, M., Schafer, W., Berghofer, S., Hoflack, B., Klenk, H. D., and Garten, W. (1999). *J. Biol. Chem.* **274,** 8199.

101. Liu, G., Thomas, L., Warren, R. A., Enns, C. A., Cunningham, C. C., Hartwig, J. H., and Thomas, G. (1997). *J. Cell. Biol.* **139,** 1719.

102. Dittie, A. S., Thomas, L., Thomas, G., and Tooze, S. A. (1997). *EMBO J.* **16,** 4859.

103. De Bie, I., Marcinkiewicz, M., Malide, D., Lazure, C., Nakayama, K., Bendayan, M., and Seidah, N. G. (1996). *J. Cell. Biol.* **135,** 1261.

104. Takahashi, S., Kasai, K., Hatsuzawa, K., Kitamura, N., Misumi, Y., Ikehara, Y., Murakami, K., and Nakayama, K. (1993). *Biochem. Biophys. Res. Commun.* **195,** 1019.

105. Takahashi, S., Nakagawa, T., Kasai, K., Banno, T., Duguay, S. J., Van de Ven, W. J., Murakami, K., and Nakayama, K. (1995). *J. Biol. Chem.* **270,** 26565.

106. Mondino, A., Giordano, S., and Comoglio, P. M. (1991). *Mol. Cell. Biol.* **11,** 6084.

107. Komada, M., Hatsuzawa, K., Shibamoto, S., Ito, F., Nakayama, K., and Kitamura, N. (1993). *FEBS Lett.* **328,** 25.

108. Ohnishi, Y., Shioda, T., Nakayama, K., Iwata, S., Gotoh, B., Hamaguchi, M., and Nagai, Y. (1994). *J. Virol.* **68,** 4075.

109. Horimoto, T., Nakayama, K., Smeekens, S. P., and Kawaoka, Y. (1994). *J. Virol.* **68,** 6074.

110. Tsuneoka, M., Nakayama, K., Hatsuzawa, K., Komada, M., Kitamura, N., and Mekada, E. (1993). *J. Biol. Chem.* **268,** 26461.

111. Dubois, C. M., Laprise, M. H., Blanchette, F., Gentry, L. E., and Leduc, R. (1995). *J. Biol. Chem.* **270,** 10618.

112. Watanabe, M., Hirano, A., Stenglein, S., Nelson, J., Thomas, G., and Wong, T. C. (1995). *J. Virol.* **69,** 3206.

113. Jean, F., Thomas, L., Molloy, S. S., Liu, G., Jarvis, M. A., Nelson, J. A., and Thomas, G. (2000). *Proc. Natl. Acad. Sci. USA* **97,** 2864.

114. Lissitzky, J. C., Luis, J., Munzer, J. S., Benjannet, S., Parat, F., Chretien, M., Marvaldi, J., and Seidah, N. G. (2000). *Biochem. J.* **346 Pt 1,** 133.

115. Watson, D. G., Moehring, J. M., and Moehring, T. J. (1991). *J. Virol.* **65,** 2332.

116. Inocencio, N. M., Moehring, J. M., and Moehring, T. J. (1993). *J. Virol.* **67,** 593.

117. Spence, M. J., Sucic, J. F., Foley, B. T., and Moehring, T. J. (1995). *Somat. Cell Mol. Genet.* **21,** 1.

118. Moehring, J. M., Inocencio, N. M., Robertson, B. J., and Moehring, T. J. (1993). *J. Biol. Chem.* **268,** 2590.

119. Sucic, J. F., Moehring, J. M., Inocencio, N. M., Luchini, J. W., and Moehring, T. J. (1999). *Biochem. J.* **339,** 639.

120. Walker, J. A., Molloy, S. S., Thomas, G., Sakaguchi, T., Yoshida, T., Chambers, T. M., and Kawaoka, Y. (1994). *J. Virol.* **68,** 1213.

121. Stieneke-Grober, A., Vey, M., Angliker, H., Shaw, E., Thomas, G., Roberts, C., Klenk, H. D., and Garten, W. (1992). *EMBO J.* **11,** 2407.

122. Hallenberger, S., Bosch, V., Angliker, H., Shaw, E., Klenk, H. D., and Garten, W. (1992). *Nature* **360,** 358.

123. Vey, M., Schafer, W., Reis, B., Ohuchi, R., Britt, W., Garten, W., Klenk, H. D., and Radsak, K. (1995). *Virology* **206,** 746.

124. Ortmann, D., Ohuchi, M., Angliker, H., Shaw, E., Garten, W., and Klenk, H. D. (1994). *J. Virol.* **68,** 2772.

125. Angliker, H. (1995). *J. Med. Chem.* **38,** 4014.

126. Basak, A., Schmidt, C., Ismail, A. A., Seidah, N. G., Chretien, M., and Lazure, C. (1995). *Int. J. Pept. Protein. Res.* **46**, 228.
127. Lu, W., Zhang, W., Molloy, S. S., Thomas, G., Ryan, K., Chiang, Y., Anderson, S., and Laskowski, M., Jr. (1993). *J. Biol. Chem.* **268**, 14583.
128. Van Rompaey, L., Ayoubi, T., Van De Ven, W., and Marynen, P. (1997). *Biochem. J.* **326**, 507.
129. Anderson, E. D., Thomas, L., Hayflick, J. S., and Thomas, G. (1993). *J. Biol. Chem.* **268**, 24887.
130. Jean, F., Stella, K., Thomas, L., Liu, G., Xiang, Y., Reason, A. J., and Thomas, G. (1998). *Proc. Natl. Acad. Sci. USA* **95**, 7293.
131. Brennan, S. O., and Peach, R. J. (1988). *FEBS Lett.* **229**, 167.
132. Misumi, Y., Ohkubo, K., Sohda, M., Takami, N., Oda, K., and Ikehara, Y. (1990). *Biochem. Biophys. Res. Commun.* **171**, 236.
133. Benjannet, S., Savaria, D., Laslop, A., Munzer, J. S., Chretien, M., Marcinkiewicz, M., and Seidah, N. G. (1997). *J. Biol. Chem.* **272**, 26210.
134. Tsuji, A., Hashimoto, E., Ikoma, T., Taniguchi, T., Mori, K., Nagahama, M., and Matsuda, Y. (1999). *J. Biochem. (Tokyo)* **126**, 591.
135. Decroly, E., Wouters, S., Di Bello, C., Lazure, C., Ruysschaert, J. M., and Seidah, N. G. (1996). *J. Biol. Chem.* **271**, 30442.
136. Dahlen, J. R., Jean, F., Thomas, G., Foster, D. C., and Kisiel, W. (1998). *J. Biol. Chem.* **273**, 1851.
137. Giannelli, F., Green, P. M., High, K. A., Lozier, J. N., Lillicrap, D. P., Ludwig, M., Olek, K., Reitsma, P. H., Goossens, M., Yoshioka, A., Sommer, J., and Brownlee, G. G. (1990). *Nucleic Acids Res.* **18**, 4053.
138. Bentley, A. K., Rees, D. J., Rizza, C., and Brownlee, G. G. (1986). *Cell* **45**, 343.
139. Brennan, S. O., and Carrell, R. W. (1978). *Nature* **274**, 908.
140. Oda, K., Ikeda, M., Tsuji, E., Sohda, M., Takami, N., Misumi, Y., and Ikehara, Y. (1991). *Biochem. Biophys. Res. Commun.* **179**, 1181.
141. Yoshimasa, Y., Seino, S., Whittaker, J., Kakehi, T., Kosaki, A., Kuzuya, H., Imura, H., Bell, G. I., and Steiner, D. F. (1988). *Science* **240**, 784.
142. Kobayashi, M., Sasaoka, T., Takata, Y., Ishibashi, O., Sugibayashi, M., Shigeta, Y., Hisatomi, A., Nakamura, E., Tamaki, M., and Teraoka, H. (1988). *Biochem. Biophys. Res. Commun.* **153**, 657.
143. Milewicz, D. M., Grossfield, J., Cao, S. N., Kielty, C., Covitz, W., and Jewett, T. (1995). *J. Clin. Invest.* **95**, 2373.
144. Raghunath, M., Putnam, E. A., Ritty, T., Hamstra, D., Park, E. S., Tschodrich-Rotter, M., Peters, R., Rehemtulla, A., and Milewicz, D. M. (1999). *J. Cell. Sci.* **112**, 1093.
145. Kim, S. H., Wang, R., Gordon, D. J., Bass, J., Steiner, D. F., Lynn, D. G., Thinakaran, G., Meredith, S. C., and Sisodia, S. S. (1999). *Nature Neurosci.* **2**, 984.
146. Bennett, B. D., Denis, P., Haniu, M., Teplow, D. B., Kahn, S., Louis, J. C., Citron, M., and Vassar, R. (2000). *J. Biol. Chem.* **275**, 37712.
147. Roebroek, A. J., Umans, L., Pauli, I. G., Robertson, E. J., van Leuven, F., Van de Ven, W. J., and Constam, D. B. (1998). *Development* **125**, 4863.
148. Constam, D. B., and Robertson, E. J. (2000). *Development* **127**, 245.
149. Cui, Y., Jean, F., Thomas, G., and Christian, J. L. (1998). *EMBO J.* **17**, 4735.
150. Constam, D. B., and Robertson, E. J. (1999). *J. Cell. Biol.* **144**, 139.
151. Constam, D. B., and Robertson, E. J. (2000). *Genes Dev.* **14**, 1146.
152. Cui, Y., and Christian, J. L. (in preparation).
153. Inocencio, N. M., Moehring, J. M., and Moehring, T. J. (1994). *J. Biol. Chem.* **269**, 31831.

154. Abrami, L., Fivaz, M., Decroly, E., Seidah, N. G., Jean, F., Thomas, G., Leppla, S. H., Buckley, J. T., and van der Goot, F. G. (1998). *J. Biol. Chem.* **273,** 32656.
155. Gordon, V. M., Benz, R., Fujii, K., Leppla, S. H., and Tweten, R. K. (1997). *Infect. Immun.* **65,** 4130.
156. Garred, O., van Deurs, B., and Sandvig, K. (1995). *J. Biol. Chem.* **270,** 10817.
157. Volchkov, V. E., Feldmann, H., Volchkova, V. A., and Klenk, H. D. (1998). *Proc. Natl. Acad. Sci. USA* **95,** 5762.
158. Subbarao, K., and Shaw, M. W. (2000). *Rev. Med. Virol.* **10,** 337.
159. Fujii, Y., Sakaguchi, T., Kiyotani, K., and Yoshida, T. (1999). *Microbiol. Immunol.* **43,** 133.

9

Cellular Limited Proteolysis of Precursor Proteins and Peptides

NABIL G. SEIDAH

Laboratory of Biochemical Neuroendocrinology
Clinical Research Institute of Montreal
Montreal, Quebec H2W 1R7, Canada

I. Processing of Precursors at Single and Dibasic Residues and at Nonbasic Sites

Cellular communication occurs at both short (paracrine, juxtacrine) or long (endocrine, exocrine) distances. The effectors of this dialogue are proteins or peptides that act as transmitters, ligands, or receptors of the signals. Evolution led to the elaboration of complex signaling molecules that,

237

THE ENZYMES, Vol. XXII
Copyright © 2001 by Academic Press

in addition to variations in amino acid (aa) sequences, are further modulated by posttranslational modifications, thereby extending their diversity and flexibility. Some of these modifications include variable carbohydrate attachments (both N- and O-glycosylations), sulfation (of Tyr and carbohydrate moieties), phosphorylation (at Tyr, Ser, or Thr), disulfide bridges, carboxy-terminal amidation and prenylation (geranylation and farnesylation), and limited proteolysis. The last involves cleavage of relatively inactive precursor proteins (proproteins) to generate bioactive entities.

Processing of proproteins into bioactive polypeptides is an ancient phenomenon that must have been established more than 600 million years ago, because it occurs in unicellular organisms such as hydra and in both prokaryotic and eukaryotic phyla. A number of posttranslational processing enzymes participate in the shaping of the final form of the bioactive moieties. One of the major events occurring following the translation of proteins is their limited proteolysis at either single or pairs of basic residues. This phenomenon has been observed in diverse proteins including polypeptide hormones, cell-surface proteins such as receptors, growth factors, signaling and adhesion molecules, transcription factors, enzymes, and even in viral glycoproteins and bacterial toxins. The enzymes responsible for such processing, which commonly occurs within the secretory pathway, i.e., in the endoplasmic reticulum (ER), Golgi apparatus (usually the *trans*-Golgi network, TGN), secretory vesicles or granules and at the cell surface, have been identified and several review articles have dealt with the events that led to their discovery (*1–10*).

Seven proprotein convertases are presently known, namely, PC1 (also called PC3), PC2, furin (also called PACE), PC4, PACE4, PC5 (also called PC6) and PC7 (also called SPC7, PC8, or LPC) (*1–8*). Of these, only furin, and the isoforms PC5-B and PACE4-E are type-I membrane-bound proteins (Fig. 1, see color plate). These enzymes were shown to process various precursors at both single and pairs of basic residues, usually within the consensus sequence $[\mathbf{R/K}]$-X_n-$[\mathbf{R/K}]\downarrow$, where the number of spacer amino acids $X_n = 0$, 2, 4, or 6. Although the tissue distribution of each PC was found to be unique, in each cell of the organism at least two PCs colocalize. This suggests that not all cellular processing events are performed by a single PC, but rather that a cascade of cleavage events occurs, each accomplished by one or more PCs. A typical example is the processing of proopiomelanocortin (POMC) where the production of ACTH and β-LPH is performed by PC1 and PC2 further cleaves these intermediates into products leading to the formation of α-MSH and β-endorphin (*11*). Another example involves the processing of the PC2-specific binding protein pro7B2, which is cleaved first by furin in the TGN (*12*) and then by PC2 itself in immature secretory granules (*13*).

TABLE I

PROCESSING OF PRECURSORS AT NONBASIC RESIDUES[a]

Precursor protein	Cleavage site sequence																
	P8	P7	P6	P5	P4	P3	P2	P1		P1'	P2'	P3'	P4'	P5'	P6'	P7'	P8'
(h) proBDNF	**Lys**-Ala-Gly-Ser-**Arg**-Gly-**Leu**-Thr								→	Ser-Leu-Ala-Asp-Thr-Phe-Glu-His							
(r) proBDNF	**Lys**-Ala-Gly-Ser-**Arg**-Gly-**Leu**-Thr								→	Thr-Thr-Ser-Leu-Ala-Asp-Thr-Phe							
(r) pro-Somatostatin (Antrin)	Asp-Pro-**Arg**-Leu-**Arg**-Gln-**Phe**-Leu								→	Gln-Lys-Ser-Leu-Ala-Ala-Ala-Thr							
(h) SREBP-2	Ser-Gly-Ser-Gly-**Arg**-Ser-**Val**-Leu								→	Ser-Phe-Glu-Ser-Gly-Ser-Gly-Gly							
(h) SREBP-1a	**His**-Ser-Pro-Gly-**Arg**-Asn-**Val**-Leu								→	Gly-Thr-Glu-Ser-Arg-Asp-Gly-Pro							
(h) ATF6	Ala-Asn-Gln-Arg-**Arg**-His-**Leu**-Leu								→	Gly-Phe-Ser-Ala-Lys-Glu-Ala-Gln							
(r) pro-Relaxin (B-chain)	Ala-Ser-Val-Gly-**Arg**-Leu-**Ala**-Leu								→	Ser-Gln-Glu-Glu-Pro-Ala-Pro-Leu							
(h) pro-CCK (CCK5)	**Arg**-Ile-Ser-Asp-**Arg**-Asp-**Tyr**-Met								→	Gly-Trp-Met-Asp-Phe-Gly-Arg-Arg							
(b) Chromogranin A (309↓310)	Met-Ala-**Arg**-Ala-Pro-Gln-**Val**-Leu								→	Phe-Arg-Gly-Gly-Lys-Ser-Gly-Glu							
(r) pro-Renin	**Lys**-Ser-Ser-Phe-Thr-Asn-**Val**-Thr								→	Ser-Pro-Val-Leu-Thr-Asn-Tyr							
(h) PDGF-B	Ile-Val-Thr-Pro-**Arg**-Pro-**Val**-Thr								→	Arg-Ser-Pro-Gly-Thr-Ser-Arg-Glu							
(h) proSKI-1/S1P: site B	**Arg**-Lys-Val-Phe-**Arg**-Ser-**Leu**-Lys								→	Tyr-Ala-Glu-Ser-Asp-Pro-Thr-Val							
site C	**Arg**-His-Ser-Ser-**Arg**-Arg-**Leu**-Leu								→	Arg-Ala-Ile-Pro-Arg-Gln-Val-Ala							

[a] The precursors listed in this table include those of human and rat proBDNF, rat pro-Somatostatin to produce an antrin-like peptide, human sterol responsive element binding proteins 1a and 2, rat relaxin, human cholecystokinin (CCK), bovine chromogranin A, rat pro-Renin, human platelet derived growth factor B (PDGF-B), and proSKI-1/S1P, all of which involve cleavage at either a hydrophobic or small residue and exhibit the presence of a basic residues **(R/K/H)** at P4, and/or P6 or P8.

Aside from cleavage of proproteins at single and pairs of basic residues, it has been realized since the early 1970s that cellular processing of precursors also can occur at sites that are not occupied by basic residues. Selected examples of such processing events are shown in Table I, in which we focused on precursors that exhibit the presence of basic residue at P4, P6, or P8 and a hydrophobic one at P2. The search for the enzyme(s) responsible for such cleavages has recently led to the identification of a novel subtilase known either as SKI-1 (*14, 15*) or S1P (*16, 17*). In our laboratory, SKI-1 was shown to process pro-brain derived neurotrophic factor (proBDNF) at the noncanonical site **Arg**-Gly-**Leu**-Thr↓Ser-Leu (*14*), and in collaboration with the laboratory of Yogesh Patel (McGill University) we also demonstrated that it can process pro-somatostatin at the antrin site: **Arg**-Leu-**Arg**-Gln-**Phe**-Leu↓Gln-Lys (*18*). In addition, we (*15*) and Espenshade *et al.* (*17*) showed that proSKI-1 undergoes autocatalytic activation in the endoplasmic reticulum (ER) and early Golgi compartment through sequential cleavages at **Arg**-Ser-**Leu**-Lys↓Tyr-Ala and **Arg**-Arg-**Leu**-Leu↓Arg-Ala. From these results and mutagenesis studies (*19, 20*), it can be surmised that SKI-1 processes precursors exhibiting the consensus motif: **[R/K]**-X_n-**[hydrophobic]**-Z↓, where Z is any amino acid except Pro, Glu, Asp, or Cys and the spacer $X_n = 1, 3,$ or 5. Predicted examples of such processing are shown in Table I, which includes growth factors such as PDGF-B. Future studies should define in more details the extent of involvement of SKI-1/S1P in these and other processes.

II. The Regulation of Processing within the Constitutive and Regulated Secretory Pathways

Analysis of the tissue expression and intracellular localization of the PCs and SKI-1/S1P as well as their processing of various precursors revealed a hierarchy in the order at which these convertases cleave their cognate substrates. The question that arises is how does each enzyme selectively cleave its substrate at, and only at, specific loci within the cell? Intuitively, one could think that aside from their inherent cleavage selectivity, these enzymes must also be under stringent control elements ensuring that processing only occurs at a specific time and place within the cellular environment. A number of factors regulate this ordered process, the first of which is that convertases require prior removal of their inhibitory prosegment before activation (*1–10*). This operation involves one (primary) or two (primary and secondary) cleavages followed by carboxypeptidase trimming of the exposed basic residues (Fig. 2). Figure 3, summarizes the intracellular sites where such autocatalytic activation occurs for each PC. Thus, whereas all convertases lose their signal

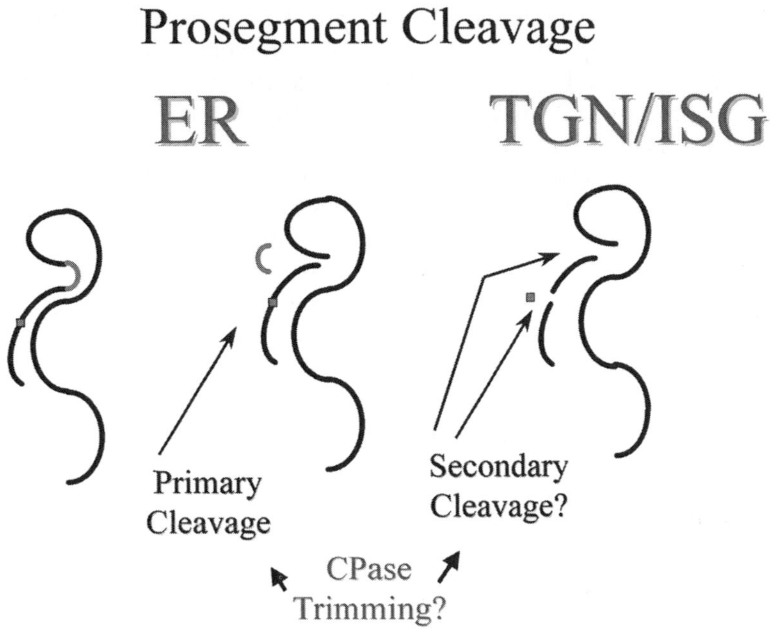

Prosegment Cleavage

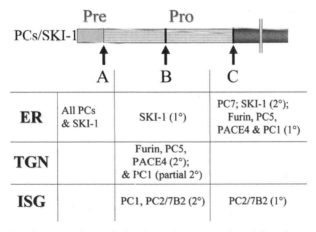

FIG. 2. Proposed pathway of autoactivation of pro-convertases. This involves one or two cleavages, each followed by carboxypeptidase D or E trimming of the exposed C-terminal basic residues.

	A	B	C
ER	All PCs & SKI-1	SKI-1 (1°)	PC7; SKI-1 (2°); Furin, PC5, PACE4 & PC1 (1°)
TGN		Furin, PC5, PACE4 (2°); & PC1 (partial 2°)	
ISG		PC1, PC2/7B2 (2°)	PC2/7B2 (1°)

FIG. 3. Identification of the intracellular sites where processing of the primary (1°) and secondary (2°) cleavages of the prosegments of PCs and SKI-1 occur. ER = endoplasmic reticulum; TGN = *trans*-Golgi network; ISG = immature secretory granules. See text for description of A, B, and C.

peptide (site A cleavage) in the ER, only proSKI-1 is processed at both sites B and C in this compartment. This result is in agreement with the fact that SKI-1 is active in the ER (*14–17*). With the exception of PC2, all other PCs undergo a primary (1°) cleavage in the ER and a secondary (2°) one either in the Golgi/TGN (furin, PACE4, PC5) or in immature secretory granules (ISG) (PC1 and PC2). The convertases PC7 (*21*), and possibly PC4, exhibit only a 1° cleavage and no 2° processing within their prodomain. In the case of PC2, prosegment cleavage at both sites occur in ISG.

Analysis of the biosynthesis of furin (*22*), PC1 and PC2 (*23*), PACE4 (*24*), PC5 (*25*), PC7 (*26, 27*), and SKI-1 (*14–17*) revealed that these convertases are first synthesized as zymogens that undergo autocatalytic cleavage of their N-terminal inhibitory prosegment, which seems to act both as a chaperone and an inhibitor of its cognate enzyme. Indeed overexpression of the prosegments of furin and PC7 as independent domains confirmed their inhibitory potency (*15, 28*) and revealed that the C terminus of the prosegment contains the critical inhibitory elements (*28*). Nuclear magnetic resonance analysis of this inhibitory region in PC7 revealed it to contain an α helix stabilized by salt bridges and H bonds (*29*). Interestingly, expression of some of these prosegments in *trans* to enzymes in which the prosegments have been deleted (Δ-pro) did not allow production of active proteins in mammalian cells (*22*). In contrast, in yeast *Saccharomyces cerevisae* such an experiment with Kex2p (known as kexin) led to an active protease (*30*). However, coexpression of either of the mammalian PC-prosegments with kexin-Δ-pro did not result in active kexin, implying that the prosegment of kexin is specific for this enzyme and that it cannot be replaced by that of foreign PCs (*30*). On the other hand, swapping of the prodomains of mammalian PCs demonstrated that the prosegment of PC1 can replace that of PACE4 and that the furin prosegment can replace that of either PC1 and to a lesser extent that of PC2 (*31*). However, the prodomains of PC1 and PC2 are not interchangeable and the prodomain of PC2 cannot replace that of furin (*22*). These results further emphasize the uniqueness of PC2, which is the only convertase undergoing prosegment removal late along the secretory pathway (TGN/ISG), that has an Asp instead of the usual Asn at the site of the oxyanion hole, and that requires the participation of a specific binding protein 7B2 (*32–36*) for its efficient zymogen activation. Nevertheless, these domain-swap data suggested that some of the prosegments are interchangeable and hence could act as inhibitors of more than one PC. Finally, synthetic peptides mimicking the C-terminal segment of the prodomain of furin and PC7 have been shown to be very potent tight binding inhibitors of these enzymes (*28, 29*).

Following autocatalytic removal of the inhibitory prosegment, the second control element would be the trafficking of these enzymes to different intracellular organelles. Cellular localization experiments revealed that SKI-1 is sorted to the *cis/medial* Golgi (*14, 17*), thereby suggesting that it is poised to process its cognate precursors before any other PC (Fig. 4). The next in line along the secretory pathway is the isoform PC5/6-B (Fig. 1), which has been shown to localize to an earlier, brefeldin A (BFA) collapsible, saccule of the Golgi apparatus (*37*) (Fig. 4). This interesting observation rationalizes an earlier one whereby PC5/6-A, and not PC5/6-B, is the best convertase of α-chain integrins (*38*), possibly because the TGN wherein PC5/6-A is concentrated in cells devoid of secretory granules (SGs, Fig. 4) (*25*) may be a more favorable environment for the efficient processing of the α-chain within the $\alpha\beta$ heterodimer. In the TGN, various laboratories demonstrated that furin, PC7, and PACE4 concentrate. In addition, furin and PC7 circulate to the cell surface and are retrieved back to the TGN by mechanisms controlled by their cytosolic tails [for review, see (*9*)]. It is thus believed that

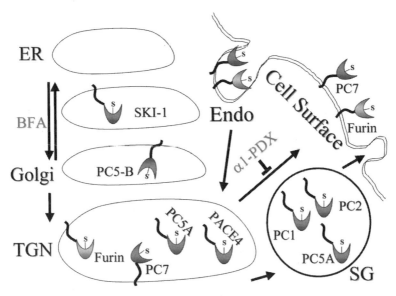

Fig. 4. Cellular localization of the PCs and SKI-1. Thus both SKI-1 and PC5-B are localized in Golgi saccules that fuse with the ER upon treatment of the cells with brefeldin A (BFA). Furin, PACE4, PC5-A, and PC7 are concentrated in the TGN and furin and PC7 cycle to the cell surface (constitutive secretory pathway) and are retrieved back to the TGN via endosomes (Endo). The serpin α1-PDX block the processing within the constitutive secretory pathway. In contrast, PC1, PC2, and some PC5-A enter the regulated secretory pathway and are concentrated in secretory granules (SG).

SKI-1, furin, PC7, PACE4, and PC5-B and, in some cases, PC5-A are the major mammalian convertases acting within the constitutive secretory pathway and/or at the cell surface or within early endosomes (Fig. 4) (9). Precursors processed in this pathway include most growth factors (1–8), transcription factors entering the secretory pathway, e.g., SREBPs (16, 17), receptors such as Notch (39), adhesion molecules and surface glycoproteins (8, 9, 38). All these processes can be inhibited by the overexpression of specific inhibitors such as the modified serpin α1-PDX (40, 41) (Fig. 4) or the prosegments (28).

Most polypeptide hormone processing occurs within the regulated secretory pathway. This generates bioactive proteins and peptides that are secreted from mature SGs following specific stimuli, e.g., following cellular exposure to secretagogues. The only enzymes entering the regulated secretory pathway and concentrating in mature SGs are PC1, PC2 (42), and PC5-A (25) (Fig. 4). Although their sorting to SGs physically separates them form the other PCs, it is becoming apparent that, at least for some convertases, e.g., PC1 and PC2, further controls are in place, thereby regulating their activities. For PC2, the binding protein/inhibitor 7B2 (32) is necessary both for productive folding of the enzyme (33–35) and for temporarily inhibiting its activity via its C-terminal 31 aa segment (36). In the case of PC1, two inhibitory ploys are in place. The first is provided by its C-terminal inhibitory segment, as deduced from the cleavage of human prorenin to renin by PC1 (43). This segment also contains two α helices that act as sorting domains directing the protein toward SGs (44). In addition, a novel inhibitor of PC1, called proSAAS, has been identified (45). The PC1-inhibitory domain within this granin-like protein has been localized (46) and synthetic peptides mimicking this process were characterized (46, 47).

III.   Convertases' Gene Structures, Evolution, and Knockout

A phylogenetic analysis of the catalytic domains of the PCs and SKI-1 revealed that the latter is the most ancient mammalian convertase, which is closer to the pyrolysin subfamily of subtilases than to that which includes the PCs and yeast kexin (8). In an effort to further understand the function and evolution of these genes, we first compared their chromosomal localization (Table II). It is seen that except for furin and PACE4 each convertase is on a different chromosome in both human and mouse. Interestingly, PC4 is not only found on chromosome 19 (48), but mining the database revealed that an identical gene is also on chromosome 1 (Table II). Barring an error of assignment, the usefulness and consequences of such gene duplication is not yet understood.

TABLE II

PC Genes: Chromosomal Localization, Structure, and Phenotypic Consequences of Their Inactivation[a]

| PC | Chromosomal localization | | | | Gene structure (species) | | Morbid phenotype (species) |
| | Human (h) | | Mouse (m) | | | | |
	Locus Symbol	Chr (c, s, l)	Locus Symbol	Chr (c, s, l)	Size (kb)	No of Exons	
PC1	PCSK1	5q (c)	Pcsk1	13 (c, l)	35 (h), 42 (m)	14 (h), 14 (m)	Obesity, hypogonadotropic hypogonadism, hypercortisolism (h)
PC2	PCSK2	20p (c)	Pcks2	2 (c, l)	>140 (h)	11 (h)	Slow growth rate, hyperplasia of pancreatic alpha and delta cells (m)
Furin	PCSK3	15q (c)	Pcsk3	7 (c, l)	10 (h)	15 (h)	Embryonic dysmorphisms and lethality (m)
PC4	PCSK4	19 and 1 (s)	Pcsk4	10 (l)	9.5 (m)	14 (m)	Reduced male fertility, preimplantation embryonic lethality (m)
PC5	PCSK5	9q (c, s)	Pcsk5	19 (l)	nd	nd	nd
PACE4	PCSK6	15q (c, s)	Pcsk6	7 (l)	>250 (h)	25 (h)	Cranofacial abnormalities, CNS defects[b]
PC7	PCSK7	11q (c, s)	Pcsk7	9 (l)	13 (m)	9 (m)	Viable, no visible phenotype yet[b]
SKI-1	SKI-1	16 (c, s)	SKI-1	8 (l)	>65 (h)	24 (h)	nd

[a] Reference on PC gene structure and on morbid phenotypes of PC mutations are presented in the text. This has not yet been determined (nd) for PC5 and SKI-1. References on chromosomal localizations were reviewed by Mbikay et al. (48). Localization was by cytogenetic (c) methods (including in situ hybridization), by analysis of somatic (s) cell hybrids, or by linkage (l) analysis in human (h) or mouse (m).

[b] D. S. Constam (Personal communication).

Alignment of the seven members of the mammalian (mouse) family of precursor convertases

```
                1                                                   50                                              100
mPACE4    ~..praadva rgagaagrhg lpplaIrpwr ..wlllaIp aacsaI...p pprpvytnhw avqv.lgg.p gaAdrvAaah Gylnlqqign lddyynfyhs
mPC5      ~......... ~......... ~.mdwd wgnrcsrpgr rdlicvlall agcllp...v crtrvytnhw avki.agg.f aeAdrlAsky Gfinvgqiga lkdyynfyhs
mfurin    ~......... ~......... ~.melrs wllwvvaaag avvllaadaq gq.kiftntw avhi.pgg.p avAdrvAqkh Gfhnlgqlfg d..yyhfwhr
mPC4-A    ~......... ~......... ~.mrpsq telwlgltlt laillavrwas agaplyvsw avrv.tkg.y qeAerlArkf Gfvnlgqlfp ddqyfhlrhr
mPC2      ~......... ~.......megc gsqwkaagf. .ifcvmvfas aerpvftnhf lvelhkdg.e eeArqvAaeh Gfg.vrklpf aeglyhfyhn
mPC1      ~......... ~.meqrgw tiqctaff cvwcalnsvk akrq.fvnew aael.pgg.q eaAsaiAeel Gydligqigs ienhylfkhk
mPC7      mpkgrqkvph idahlgipic lwielaiffl vpqvmglsea gqldiqtgg lswavhldsl egerkeeslt qqAdavAqaa Glvnagrige iqghylfvgp

            101                          1°                                   150                                    200
mPACE4    ktfkrstlss rgphtfIrmd pqvkwlqqge vkrrvkrq.. .arsdslyfn DfiwsnmWym hctdknsrcr s.emnvqaaw krgyTGknvv vtilDDGier  ●
mPC5      rtikrsvlss rgthsfisme pkvewlqqgv vkkrtkrdyd lshaqstyfn DfkwpsmWym hcsdnthpcq s.dmniegaW krgyTGkniv vtilDDGier
mfurin    avtkrslsph rprhsrlqre pqvkwleqqv akrrakrdvy qept...... DfkfpqqWyl ....sgvtq r.dinvkeaW agfTGhgiv vsilDDGiek
mPC4-A    gvaqgsltph wghrlrlkkd pkvrwfeqqt lrrrvkrslv .vpt...... DfwfskqWym .....nkeiq q.dinllkaW nqglTGrgvv isilDDGiek
mPC2      glakakrrs lhhkrqlerd pikxmalqge gfdrkkrgyr dineidinmn DplftkqWyl fntqgadgtp gldlnveaaW elgyTGkgvt igimDDGidy
mPC1      shprsrsa lhitkIrsdd drvtwaeqqy ekerskrsvq kdsaldl.fn DPmwnqqWyl qdtrmtaalp kldlhvipvW ekgiTGkgvv itvlDDGlew
mPC7      tghrqameve ...amrqg aeavlarhea vrwhseqtll krakrsihfn DPkypqqWhl ....nnrrsp grdinvtgvw ernvTGrgvt vvvvDDGvet

            201                                           250                                                     300
mPACE4    nhpDlapNYd syaSyDvngn DydPaPryda sneNkHGTRC AGEVaasanN syClvGlAyn akigGiRmLD .gdvTDvvEA kslglrpnyi dIYSaSWGPd
mPC5      thpDlmqNYd alaScDvngn DldPmPryda sneNkHGTRC AGEVaatanN shCtvGlAfn akigGvRmLD .gdvTDmvEA ksvsynpqhv hIYSaSWGPd
mfurin    nhpDlagNYd pgaSfDvndq DpdPqPrytq mndNtHGTRC AGEVaavanN gvCgvGvAyn arigGvRmLD .gevTDavEA rslglnpnhi hIYSaSWGPe
mPC4-A    dhpDlwaNYd plaSyDfndy DpdPqPrytp ndeNrHGTRC AGEVsatanN gfCgaGvAfn arigGvRmLD .gaiTDlvEA qslslqpqhi hIYSaSWGPe
mPC2      lhpDlayNYn adaSyDfssn DpyPyPrytd dwfNsHGTRC AGEVsaaasN niCgvGvAyn skvaGiRmLD qpfmTDilEA ssishmpqli dIYSaSWGPt
mPC1      nhtDiyaNYd peaSyDfndn DhdPfPrydl tneNkHGTRC AGEiamqanN hkCgvGvAyn skvgGiRmLD .givTDaiEA ssigfnpghv dIYSaSWGPn
mPC7      hrqDiapNYs pegSyDlnsn DpdPmPhpde engNhHGTRC engNhHGTRC AGEiaavpnN sfCavGvAyg sriaGiRvLD .gplTDsmEA vafnhhyqin dIYScSWGPd
```

FIG. 5.

```
        301                                                     350                                                     400
mPACE4  DdGkTVdGPg rLakqAfeyG ikkGRqGlGs ifvwASGnGG regDhCscDG YtnSiYTisv ssttenGhkp wYlEecaStl attYSsGafy erk..ivttD
mPC5    DdGkTVdGPa pLtrqAfenG vrmGRrGlGs vfvwASGnGG rskDhCscDG YtnSiYTisi sstaesGkkp wYlEecSStl attYSsGesy dkk..iittD
mfurin  DdGkTVdGPa rLaeeAffrG vsgqGRgGlGs ifvwASGnGG rehDsCncDG YtnSlYT.si ssatqfGnvp wYsEaCsStl attYSsGnqn ekq..ivttD
mPC4-A  DdGrTVdGPg lLtqeAfrrG vtkGRqGlGt lfiwASGnGG lhyDnCncDG YtnSihTlsv gsttrgGrvp wYsEaCStf tttfSsGvvt dpq..ivttD
mPC2    DnGkTVdGPr eLtlqAmadG vnkGRqGkGs iyvwASGdGG sy.DdcncDG YasSmwTisi nsaindGrta lYdEsCsStl astfSnGrkr npeagvattD
mPC1    DdGkTVeGPg rLaqkAfeyG vkqGRqGkGs ifvwASGnGG rqgDnCdcDG YtdSiYTisi sasqqGlsp wYaEkCsStl atsySsGdyt dqr..itsaD
mPC7    DdGkTVdGPh qLgkaAlqhG vmaGRqGfGs ifvwASGnGG qhnDnCnyDG YanSlYIvti gavdeeGrmp fYaEeCaSml avtfSgGdkm lrsivttdwD
```

```
        401                                                     450                                                     500
mPACE4  Lrq..rCtdg HtGTSvsAPm vAGiiALaLe annqLTWRDv QHllVktsrp ahl..kasdW kvNgaChkvs hlyGfGlvdA ealVleA..r kWtavpsqhv
mPC5    Lrq..rCtdn HtGTSasAPm aAGiiALaLe anpfLTWRDv QHviVrtsra ghl..nandW ktNaaGfkvs hlyGfGlmdA eamVmeA..e kWttvpqqhv
rrfurin Lrq..kCtes HtGTSasAPl aAGiiAltLe arknLTWRDm QHlvVqtskp ahl..naddW atNgvGrkvs hsyGyGlldA gamVaIA..q nWttvapqrk
mPC4-A  Lhh..qCtdk HtGTSasAPl aAGmiALaLe anplLTWRDl QHlvVrasrp aql..caedW riNgvGrqvs hhyGyGlidA gllVdlA..r vWlptkpqkk
mPC2    Lyg..nCTlr HsGTSaaAPe aAGvfALaLe anldLTWRDm QHlvVltsk: nqlhdevhqW rNgvGlefn hlfGyGyldA gamVkmA..k dWktvperfh
mPC1    Lhh..dCtet HtGTSasAPl aAGifALaLe anpnLTWRDm QHlvVwtsey dplasn.pgW kkNgaGlmvn srfGfGllnA kalVdlAdpr tWrnvpekke
mPC7    LqkgtgCTeg HtGTSaaAPl aAGmiAlmLq vrpclTWRDv QHliVfta.. iqyedhhadW ltNeaGfshs hqHGfGllnA wrlVnaA..k iWtsvp.yla
```

```
        501                                                     550                                                     600
mPACE4  cvat.adkrp rsipivqvlr ttaltnacad hsdqr.vvyL EHVvvris:s hpRRGdIqih lispsGtksg llakRl.lDf sneGftnWeF mtvhcWgEka
mPC5    cves.tdrqi ktirpnsavr siykasgcsd npnhh.vnyL EHVvvritt hpRRGdLaiy ltspsGtrsq llanRl.fDh smeGfknWeF mtihcWgEra
mfurin  cive.ilvep kdigkrlevr ka..vtaclg epnh..itrL EHVqarltls ynRRGdLaih lispmGtrst llaaRp.hDy sadGfndWaF mtthsWdEdp
mPC4-A  cair.vvhtp tpilprmlvp kn..vtacsd gsrrlirsL EHVqvqlsls ysRRGdIeif ltspmGtrst lvaiRp.lDi sqgGynnWiF msthyWdEdp
mPC2    c.vggsvqnp ekipptgklv ltlktnaceg ken..fvryL EHVqavitvn atRRGdlnin mtspmGtksi llsrRprdDd skvGfdkWpF mtthtWgEda
mPC1    cvvkdnnfep ra:kangevi veiptraceg qen..aiksL EHVqfeatie ysRRGdLnvt ltsavGtstv llaeRer.Dt spnGfknWdF msvhtWgEnp
mPC7    syvspmlken kav.prsphs levlwnvsrt dlemsglktL EHVavtvsit hpRRGsLelk lfcpsGmmsl igapR.smDs dpnGfnaWtF stvrcWgEra
```

FIG. 5. (*continued*)

```
601                                                650                                                700

mPACE4   eGewtlevqd  ipsqvrnpek  qGklkewsLi  LyGtaehpyr  tfsshqsrsr  mlelsvpeqe  ppkaagqppq  aetpedeeey  tqvchpecgd  kgcdgpnadq
mPC5     aGdwvLevyd  tpsqlrnfkt  pGklkewsLv  LyGtsvqpys  ptnefpkver  frysrvedpt  ddyga.....  ........edy  agpcdpecse  vgcdgpqpdh
mfurin   aGewvLeien  t....seann  yGtltkftLv  LyGtapegls  ...tppessg  cktltssq..  ...acvvcee  gyslhqkscv  qhcppgfipq  vldthysten
mPC4-A   qGlwtLqien  k....gyyfn  tGtlyyytLl  LyGtaedmta  rpqapqvtsr  aracvqrd..  ...teglcqe  shs..plsil  aglclissqq  wwwlyshpqq
mPC2     rGtwtLel.g  ...fvgsapq  kGllkewtLm  LhGtqsapyi  dqvvrdyqs.  klamskkqel  eeeldeaver  slqsilrkn~
mPC1     vGtwtLkitd  ...msgrmqn  eGrivnwkLi  LhGtssqpeh  mkqprvytsy  ntvqndirgv  ekmvnvvekr  ptqkslngnl  lvpknsssn   vegrrdeqvq
mPC7     rGvyrLvird  vgd..eplq   mGilqqwqLt  LyGsmwspv.  ..dikdrqsl  lesamsgkyl  hdgftlpcpp  glkipeedgy  t.itpnlikt  lvlvgcfsvf


701                                                750                                                800

mPACE4   clncvhfslg  nsktnrkcvs  ecplgyfgda  aarrcrrchk  goetctgrsp  aqclscrrgf  yhhqetntcv  tlcpaglyad  esqrlcirch  pscqkcvdep
mPC5     csdclhyyyk  lknntricvs  scppghy.ha  dkkrcrkcap  ncesofgshg  dqclsckygy  fineetsscv  tqcpdgsyed  ikknvcgkcs  enckacigf.
mfurin   dveiirasvc  tpchascatc  qgpaptdcls  cpshasldpv  eqtcsrqsqs  sresrpqqqp  palrpeveme  prlqaglash  lpevlaglsc  liivlifgiv
mPC4-A   pvteggasch  ppvtpaaaa~
mPC2     ~
mPC1     gtpskamlrl  lqsafsknal  skqspksps   aklisipyesf  yealeklnkp  sklegsedsl  ysdyvdvfyn  tkpykhrddr  llqalmdiln  een~~~~~~~
mPC7     wtiyymlevc  lsqrnkasth  gcrkgccpwa  prrqnskdag  talesmplcs  skdldgvdse  hgdcttassf  lapellgead  wslsqnskse  ldcpphqppd


801                                                850                                                900

mPACE4   ekctvckegf  slargscipd  cepgtyfdse  lvkcqechht  crtcvqpsre  echncaksfh  fqdwkcvpac  gegfypeemp  glphkvcrrc  eenclscegs
mPC5     hncteckggl  slqsrcsvt   cedqqfngh   ..dcapchrf  catcsgagad  gcinctegyv  meegrcvqsc  svsyyldhss  eggyksckrc  dnscltcngp
mfurin   flflhrcsgf  sfrgmkvytm  drglisvykgl  ppeawqeecp  sdseedegrg  ertafikdgs  al~~~~~~~
mPC4-A   ~
mPC2     ~
mPC1     ~
mPC7     llqgkic~~~
```

Fig. 5. (*continued*)

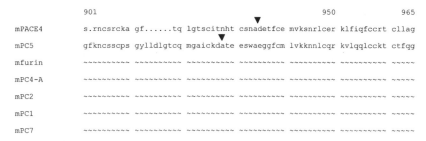

Fig. 5. Alignment of PCs and intron positions. Symbols: ▼ intron position; ●, active site residues in the catalytic domain; ■, oxyanion hole Asn (Asp); ✦, zymogen activation site. The PC5 (mining the data bases) and PACE4 (49) intron/exon positions are based on their corresponding human genomic sequences.

In Fig. 5, we compare the intron/exon boundaries of the genes coding for mouse PC1, PC2, furin, PC4, PACE4, PC5 (obtained from mining recent databases), and PC7. It is seen that except for PC7, within the prodomain and catalytic subunit (containing the active sites D, H, and S of the catalytic triad; ●, Fig. 5), all intron/exon boundaries are very similar. Interestingly, whereas the prodomains of all PCs are interrupted by two introns (separating the 1° and 2° cleavage sites), the whole prodomain of PC7 is within a single exon, befitting the absence of a 2° cleavage site in this enzyme (see above). The largest differences occur following the Ser of the active site (residue 415 in Fig. 5), including the P- and C-terminal domains. The high conservation of intron/exon boundaries is in line with the hypothesis that these genes evolved from an ancestral one through gene duplication and insertion/deletions. It is notable that except for the absence of one exon in PC5 in an area containing a gap (close to residue 670 in Fig. 5), PC5 and PACE4 have a very similar intron/exon organization throughout their sequences, including their Cys-rich domains.

In the case of SKI-1, mining the database of the human genome allowed us to define the intron–exon boundaries within the SKI-1 gene (Figs. 6 and 7). The size of the human SKI-1 gene is expected to be larger than 67 kb with 24 exons. Here also, the prodomain primary (**RSLK↓**) and secondary (**RRLL↓**) cleavage sites are found on two separate exons, namely exons III and IV (Figs. 6, 7). Within the catalytic subunit, SKI-1 exhibits intron–exon boundaries closest to those of PC7 (compare Figs. 5 and 7). For example, in both genes the distinct exons containing the oxyanion hole Asn and the Ser of the active site are further separated by an extra exon, whereas in all other PCs these two exons are juxtaposed. However, following the exon of the active site Ser, the genomic organization of SKI-1 is more complex than that of the other PC, exhibiting the presence of 13 more exons (Fig. 7), whereas the most and least complex PC genes, namely PACE4

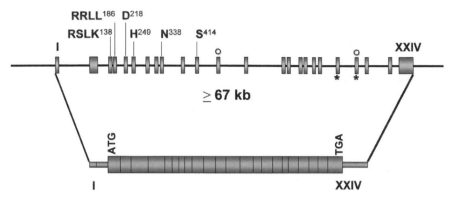

FIG. 6. Genomic organization of the human *SKI-1* gene. Intron I is in the 5' UTR of the mRNA. The gene is predicted to be ≥67 kb in size. Alternatively spliced exons XII and XXI (°) and exons XX and XXI coding for the growth factor cytokine receptor motif (*) are emphasized.

(*49*) and *PC2* (*50*), contain 12 and 2 more exons, respectively (Fig. 5). It is interesting to note that the growth factor/cytokine receptor motif found in SKI-1 (*14*) is split between two separate exons (exons XX and XXI in Fig. 6). Finally, the isolation of various mRNA forms of SKI-1 (*14*) can now be readily explained by alternative splicing of the heteronuclear mRNA resulting from the primary transcript of the *SKI-1* gene (exons XII and XXI in Fig. 6).

In Fig. 7, we also show the alignment of human SKI-1 with its rat (r), mouse (m), Chinese hamster (ch), and *Drosophila melanogaster* (dm) orthologs. The extent of the absolutely conserved consensus sequence reveals that this is a highly conserved protein. In addition, the dmSKI-1 shows the presence of insertions and deletions in the segment following the catalytic subunit and is presumably a soluble form of SKI-1, since it does not exhibit the presence of a transmembrane segment. Furthermore, we also noticed that the dmSKI-1 contains a 53 aa insertion 12 residues before the growth factor/cytokine receptor motif found in all SKI-1 orthologues (underlined in Fig. 7).

Analysis of PC-gene knockout in mice [reviewed in (*7, 8*)] revealed that furin and PC1 (−/−) mice are embryonic lethals, whereas PC2, PACE4, PC4, and PC7 nulls are not (Table II). The result in mice for PC1 (−/−) is in sharp contrast to the reported viable PC1 double point mutant human patient (*51*) and raises three possibilities: that there is a fundamental difference between humans and mice, that the patient may have other mutations compensating for the "presumed" inactivity of endogenous PC1, or simply that the mutant PC1 in one allele may still be partially active. So far, no PC5 or SKI-1 null mice have been produced. It is expected that constitutive inactivation of SKI-1 (−/−) will result in a lethal phenotype, since this will deplete the cells from

```
          1                                            50                                  100
hski-1    MKLVNIWLLL LVVLLCGKKH LGDRLEKKSF EKAPC?GCSH LTLKVEFSST VVEYEYIVAF NGYFTAKARN SFISSALKSS EVDNWRIIPR NNPSSDYPSD
rski-1    MKLVNIWLLL LVVLLCGKKH LGDRLGKKAF EKAPC?SCSH LTLKVEFSST VVEYEYIVAF NGYFTAKARN SFISSALKSS EVDNWRIIPR NNPSSDYPSD
chski-1   MKLININWLLL LVVLLCGKKH LGDRLGKKAF EKASCPSCSH LTLKVEFSST VVEYEYIVAF NGYFTAKARN SFISSALKSS EVENWRIIPR NNPSSDYPSD
mski-1    MKLVSTWLLV LVVLLCGKRJ LGDRLGTRAL EKAPCPSCSH LTLKVEFSST VVEYEYIVAF HSKYFAPVRE SYIAAKLLGS NVTNWRIVPR LNLAWQYPSD
dmski-1   ~~~~~~~~~ ~~~~~~~~~ ~~~MNVFTF LFIISAICS. ..LDA.FKTA VVPNEFIVHF ----L---F-- -I---A-R-S -V-NWRI-PR -N----YPSD
Consensus ---------- ---------- ---------- ---------- ---------- VV--E-IV-F ---------- ---------- ---------- --------

          101                                          150                                 200
hski-1    FEVIQIKEKQ KAG...LLT LEDHPNIKRV TPQRKVFRSL KYAESDPTVP CNETRWSQKW QSSRPLRRAS LSLGSGFWHA TGRHSSRRLL RAIPRQVAQT
rski-1    FEVIQIKEKQ KAG...LLT LEDHPNIKRV TPQRKVFRSL KFAESDPIVP CNETRWSQKW QSSRPLRRAS LSLGSGFWHA TGRHSSRRLL RAIPRQVAQT
chski-1   FEVIQIKEKQ KAG...LLT LEDHPNIKRV TPQRKVFRSL KFAESDPIVP CNETRWSQKW QSSRPLRRAS LSLGSGFWHA TGRHSSRRLL RAIPRQVAQT
mski-1    FEVIQIKEKQ KAG...LLT LEDHPNIKRV TPQRKVFRSL KFAESNPIVP CNETRWSQKW QSSRPLKRAS LSLGSGFWHA TGRHSSRRLL RAIPRQVAQT
dmski-1   FDILRVCDGY ESSEFFIER LQTHPSVKAV VPQRVRRIL NY........ .......DAY SNLTYIHR.. ........HP QGVLRNRNEN NDRHRQLCSV
Consensus F--------- ---------- L-HP--K-V -PQR-V-R-L ---------- ---------- ------R-- -------H-- ----G----R- ----RQ----

          201                                          250                                 300
hski-1    LQADVLWQMG YTGANVRVAV FDTGLSEKHP HFKNVKERTN WTNERTLDDG LGHGTFVAGV IASMRECQGF APDAELHIFR VETNNQVSYT SWFLDAFNYA
rski-1    LQADVLWQMG YTGANVRVAV FDTGLSEKHP HFKNVKERTN WTNERTLDDG LGHGTFVAGV IASMRECQGF APDAELHIFR VETNNQVSYT SWFLDAFNYA
chski-1   LQADVLWQMG YTGANVRVAV FDTGLSEKHP HFKNVKERTN WTNERTLDDG LGHGTFVAGV IASMRECQGF APDAELHIFR VETNNQVSYT SWFLDAFNYA
mski-1    LQADVLWQMG YTGANVRVAV FDTGLSEKHP HFKNVKERTN WTNERTLDDG LGHGTFVAGV IASMRECQGF APDAELHIFR VETNNQVSYT SWFLDAFNYA
dmski-1   LHANILWKLG ITGKGVKVAI FDTGLTKNHP HFRNVKERTN WTNEKSLDDR VSHGTFVAG. .......... .......... .....VSYT SWFLDAFNYA
Consensus L-A--LW--G -TG-V-VA- FDTGL---HP 3F-N-KERTN WTNE--LDD- --HGTFVAG- ---------- ---------- ----VSYT SWFLDAFNYA

          301                                          350                                 400
hski-1    ILKKIDVLNL SIGGPDFMDH PFVDKVWELT ANNVIMVSAI GNDGPLYGTL NNPADQMDVI GVGGIDFEDN IARFSSRGMT TWELPGGYGR MKPDIVTYGA
rski-1    ILKKMDVLNL SIGGPDFMDH PFVDKVWELT ANNVIMVSAI GNDGPLYGTL NNPADQMDVI GVGGIDFEDN IARFSSRGMT TWELPGGYGR VKPDIVTYGA
chski-1   ILKKIDVLNL SIGGPDFMDH PFVDKVWELT ANNVIMVSAI GNDGPLYGTL NNPADQNDVI GVGGIDFEDN IARFSSRGMT TWELPGGYGR VKPDIVTYGA
mski-1    ILKKMDVLNL SIGGPDFMDH PFVDKVWELT ANNVIMISAA GNDGPLYGTL NNPADQMDVI GVGGIDFEDN IAKFSSRGMT TWELPGGYGR VKPDIVTYGA
dmski-1   IYRKINIINL SIGGPDFMDS PFVEKVLELS ANNVIMISAA GNDGPLYGTL NNPGDQSDVV GVGGIQFDDK ........... TWELPLGYGR MGLDIVTYGS
Consensus I--K---LNL SIGGPDFMD- PFV-KV-EL- ANNVIM-SA- GNDGPLYGTL NNP-DQ-DV- GVGGI-F-D- IA-FSSRGMT TWELF-GYGR --DIVTYG-

          401                                          450                                 500
hski-1    GVRGSGVKGG CRALSGTSVA SPVVAGAVTL LVS.TVQKRE LVNPASMKQA LIASARRLPG VNMFEQGHGK LDLLRAYQIL NSYKFQASLS PSYIDLTECP
rski-1    GVRGSGVKGG CRALSGTSVA SPVVAGAVTL LVS.TVQKRE LVNPASVKQA LIASARRLPG VNMFEQGHGK LDLLRAYQIL SSYKPQASLS PSYIDLTECP
chski-1   GVRGSGVKGG CRALSGTSVA SPVVAGAVTL LVS.TVQKRE LVNPASVKQA LIASARRLPG VNMFEQGHGK LDLLRAYQIL SSYKPQASLS PSYIDLTECP
mski-1    GVRGSGVKGG CRALSGTSVS SPVVAGAVTL LVS.TVQKRE LVNPASVKQA LIASARRLPG VNMFEQGHGK LDLLRAYQIL SSYKPQASLS PSYIDLTECP
dmski-1   QVEGSDVRKG CRRLSGTSVS SPVVAGAAAL LISGAFQKID YINPASLKQV LIEGAEKLPH YNMFEQGAGK LNLLKSMQLL LSYKPKITLI PAYLDFTQ.N
Ccnsensus -V-GS-V--G CR-LSGTSV- SPVVAGR--L -L-S---QK-- -NPAS-KQ- LI--A--LP- -NMFEQG-GK L-LL---Q-L -SYKP---L P-Y-D-T--
```

FIG. 7.

```
          501                                                                      550
hski-1    YMWPYCSQPI YYGGMPTVVN VTILNGMGVT GRIVDKPDWQ PYLPQNGDNI EVAFSYSSVL WPWSGYLAIS ISVTKKAASW EGIAQGHVMI TVASPAETES
rski-1    YMWPYCSQPI YYGGMPTIVN VTILNGMGVT GRIVDKPEWR PYLPQNGDNI EVAFSYSSVL WPWSGYLAIS ISVTKKAASW EGIAQGHIMI TVASPAETEL
chski-1   YMWPYCSQPI YYGGMPTIVN VTILNGMGVT GRIVDKPEWR PYLPQNGDNI EVAFSYSSVL WPWSGYLAIS ISVTKKAASW EGIAQGHIMI TVASPAETEA
mski-1    YMWPYCSQPL YYGSSVAIAN VTILNGISVT SHIVGIPKWI PDEENQSQFL QVSAQVSPIV WPWTGWMSVF IAVKKEGENF EGVCKGSITL VLESFKQT..
dmski-1   YMWPY-SQP- YYG------N VTILNG--VT --IV-P-W-- P------G-- -V----S--- WPW-G----- I-V-K----- EG----G--- -----S----T---
Consensus

          601                                                                      650
hski-1    KNGAEQTSTV KLPIKVKIIP TPPRSKRVLW DQYHNLRYPP GYFPRDNLRM KNDPLDWNGD HIHTNFRDMY QHLRSMGYFV EVLGAPFTCF DASQYGTLLM
rski-1    KNGAEHTSTV KLPIKVKIIP TPPRSKRVLW DQYHNLRYPP GYFPRDNLRM KNDPLDWNGD HVHTNFRDMY QHLRSMGYFV EVLGAPFTCF DATQYGTLLM
chski-1   KNGAEHTSTV KLPIKVKIIP TPPRSKRVLW DQYHNLRYPP GYFPRDNLRM KNDPLDWNGD HVHTNFRDMY QHLRSMGYFV EVLGAPFTCF DATQYGTLLM
mski-1    HSGAEHTSTV KLPIKVKIIP KPPRNKRIIW DQYHSLRYPP RYIPRDDLKV KLDPLDWRAD HIHTNFRDMY THLRNVGYYI DVLREPFTCF NASDYGALLI
dmski-1   .TNETHVTEV DFPLTIKVTP -PPR-KR-LW DQYH-LRYPP -Y-PRD-L-- K-DPLDW--D H-HTNFRDMY -HLR--GY-- -VL--PFTCF -A--YG-LL-
Consensus --------V- EI--L--V-

          701                                                                800  800
hski-1    VDSEEEYFPE EIAKLRRDV. DNGLSLVIFS DWYNTSVMRK VKFYDENTRQ WWMPDTGGAN IPALNELLSV WNMGFSDGLY EGEFTLANHD MYYASGCSIA
rski-1    VDSEEEYFPE EIAKLRRDV. DNGLSLVVFS DWYNTSVMRK VKFYDENTRQ WWMPDTGGAN VPALNELLSV WNMGFSDGLY EGEFALANHD MYYASGCSIA
chski-1   VDSEEEYFPE EIAKLRRDV. DNGLSLVIFS DWYNTSVMRK VKFYDENTRQ WWMPDTGGAN IPALNELLSV WNMGFSDGLY EGEFALANHD MYYASGCSIA
mski-1    VDSEEEYFPE EIAKLRRDV. DNGLSLVIFS DWYNTTVMKK IKFFDENTRQ WWMPDTGGAN IPALNDLLKP WNMGFSDGLY EGEFVLANHD MYYASGCSIA
dmski-1   VDPERGFGDE EINALQENVY KRGLNVVVFG DWYNT-VM-K -KF-DENTRQ WW-PDTGGAN -PALN-LL-- FGIAEGDFVG EGHFKLGDHS MYYASGATIV
Consensus VD-E-----E EI--L---V- ---GL--V-F- DWYNT-VM-K                           -PALN-LL-- ---F-D--- EG-F-L--H- MYYASG--I-

          801                                                                      900
hski-1    KFP.EDGVVI TQT.FKDQGL EVLKQET..A VVE.NVPILG LYQ...... .........IPAE .......... .........G
rski-1    RFP.EDGVVI TQT.FKDQGL EVLKQET..A VVD.NVPILG LYQ...... .........IPAE .......... .........G
chski-1   KFP.EDGVVI TQT.FKDQGL EVLKQET..A VVE.NVPILG LYQ...... .........IPAE .......... .........G
mski-1    KFP.EDGVVI TQT.FKDQGL EVLKQET..A VVE.NVPILG LYQ...... .........IPSE .......... .........G
dmski-1   KFPMNPGDII VGTKLNDQGL SIINSKTPSK VAKLDVPIFG MFQTKANSIQ SNEEIVVNAE SNLAEAIPTD YSTFKNRVLL LRTKQRSISF AKSNNHETKN
Consensus -FP--G--I --T---DQGL ------------ V----VPI-G -----------------IP-- ----------- ------------

          901                                                                      1000
hski-1    GGRIVLYGDS NCLDDSHRQK DCFWLLDALL QYTSYGVTPP SLSHSGNRQR PPSGAGSVTP ERMEGNHLHR YSKVLEAHLG DPKPRPLPAC PRLSWAKPQP
rski-1    GGRIVLYGDS NCLDDSHRQK DCFWLLDALL QYTSYGVTPP SLSHSGNRQR PPSGAGLAPP ERMEGNHLHR YSKVLEAHLG DPKPRPLPAC PHLSWAKPQP
chski-1   GGRIVLYGDS NCLDDSHRQK DCFWLLDALL QYTSYGVNPP SLSHSGNRQR PPSGAGLAPP ERMEGNHLHR YSKVLEAHLG DPKPRPLPAC PHLSWAKPQP
mski-1    GGRIVLYGDS NCLDDSHRQK DCFWLLDALL QYTSYGVTPP SLSHSGNRQR PPSGAGLAPP ERMEGNHLHR YSKVLEAHLG DPKPRPLPAC PHLSWAKPQP
dmski-1   EGRIAVYGDS NCLDSTHLEK ACYWLLLITFL DFAINSHKSS LLQNL----- -L-------- ---------- ---------- ---------- ----------
Consensus -GRI--YGDS NCLD--H--K -C-W-LL--------------L---------------- ----------- ---------- ---------- ----------

          1001                                                                     1100
hski-1    LNETAPSNLW KHQKLLSIDL DKVVLPNFRS NRPQVRPLSP GESGAWDIPG VGQTIPVFAF LGAMVVLAFF VVQINKAKSR PKRRKPRVKR
rski-1    LNETAPSNLW KHQKLLSIDL DKVVLPNFRS NRPQVRPLSP GESGAWDIPG VGQTIPVFAF LGAMVALAFF VVQISKAKSR PKRRRPRAKR
chski-1   LNETAPSNLW KHQKLLSIDL DKVVLPNFRS NRPQVRPLSP GESGAWDIPG VGQTIPVFAF LGAMVALAFF VVQISKAKSR PKRRRPRAKR
mski-1    LNETAPSNLW KHQKLLSIDL DKVVLPNFRS NRPQVRPLSP GESGAWDIPG VGQTIPVFAF LGAMVALAFF VVQISKAKSR PKRRRPRAKR
dmski-1
Consensus -------------------- --------- ---------- ---------- ---------- ---------- ----------
```

FIG. 7. (continued)

```
                    1101        1115
       hski-1       PQLMQQVHPP  KTPSV
       rski-1       PQLAQQAHPA  RTPSV
       chski-1      PQLTQQTHPP  RTPSV
       mski-1       PQLAQQAHPA  RTPSV
       Consensus    ----------  -----
```

FIG. 7. Orthologs of SKI-1 and intron–exon junctions. Symbols: ▼ intron position; ●, active site residues in the catalytic domain; ■, oxyanion hole Asn; ↓, zymogen activation sites. The intron–exon junctions were obtained from the human genome database. The species of aligned SKI-1 are human (h), rat (r), Chinese hamster (ch), mouse (m), and *Drosophila melanogaster* (dm).

all active forms of SREBP-1 and SREBP-2 as well as prevent the processing of other precursor substrates of SKI-1. Indeed, just the inactivation of one of SKI-1/S1P's substrates, namely SREBP-1 $(-/-)$, results in 50–85% of the homozygous $(-/-)$ mice dying *in utero* at embryonic day 11 (52). The 15–50% surviving knockout mice appear normal at birth and throughout life, probably because of a compensatory two- to threefold increase in the amount of mature SREBP-2 in liver nuclei (52). Therefore, in the cases of furin, PC1, and SKI-1 it will be necessary to obtain conditional knockout mice in order to better define their functions in adult tissues. In conclusion, the availability of PC-null mice enriches our understanding of their biological functions and provides useful models of human pathologies, ultimately leading to the identification of their physiological substrates. In addition, crossing these animals to each other may further extend our understanding of the degree of functional redundancy that these enzymes may exhibit.

IV. Proprotein Convertases and the Processing of α- and β-Secretases Implicated in the Etiology of Alzheimer Disease

Alzheimer disease (AD) is a progressive degenerative disorder of the brain characterized by mental deterioration, memory loss, confusion, and disorientation. Among the cellular mechanisms contributing to this pathology are two types of fibrous protein deposition in the brain: intracellular neurofibrillary tangles composed of polymerized tau protein, and abundant extracellular fibrils composed largely of β-amyloid [for reviews, see (53, 54)]. Beta-amyloid, also known as Aβ, arises from proteolytic processing of the β-amyloid precursor protein (βAPP) at the β- and γ-secretase cleavage sites. The cellular toxicity and amyloid-forming capacity of the two major forms of Aβ (Aβ_{40} and especially Aβ_{42}) have been well documented (53, 54).

An alternative anti-amyloidogenic cleavage site performed by α-secretase is located within the Aβ peptide sequence of βAPP and thus precludes formation of intact insoluble Aβ. Cleavage by α-secretase within the

[HisHisGlnLys↓LeuVal] sequence of βAPP is the major physiological route of maturation. In several recent reports, metalloproteinases such as ADAM9, 10, and 17 were shown to be involved in the α-secretase cleavage of βAPP (55–57). Enzymes within this family are typically synthesized as inactive zymogens that subsequently undergo prodomain cleavage and activation in the *trans*-Golgi network (TGN). To date, several of the ADAMs have been shown to be activated in a nonautocatalytic manner by other enzymes such as the PCs (8). Thus, it is conceivable that such enzymes may participate in a cascade leading to the activation of α-secretase. In support of this proposal, we have demonstrated that inhibition of PC-like enzymes in HK293 cells by the α1-antitrypsin serpin variant α1-PDX (58) blocks the α-secretase cleavage of the Swedish mutant βAPP$_{sw}$ (59). Correspondingly, overexpression of a PC (e.g., PC7) increases α-secretase activity. Of the above-mentioned candidate α-secretases, our ontogeny and tissue-expression analyses suggest that, in adult human and/or mouse brain neurons, ADAM10 is a more plausible α-secretase than ADAM17 (60).

The amyloidogenic pathway of βAPP processing begins with β-secretase. This enzyme(s) generates the N terminus of Aβ by cleaving βAPP within the GluValLysMet↓AspAla sequence, or by cleaving the Swedish mutant βAPP$_{sw}$ within the GluVal<u>AsnLeu</u>↓AspAla sequence. Five different groups have simultaneously reported the isolation and initial characterization of two novel human aspartyl proteinases, BACE (61–65) and its closely related homolog BACE2 (64, 65). Both BACE and BACE2 are type-I membrane-bound proteins with a prodomain that at least for BACE (62) is rapidly cleaved intracellularly. Our developmental analysis of the comparative tissue expression of mouse BACE and BACE2 suggested that BACE, but not BACE2, is a good candidate β-secretase in the brain (60).

Until very recently, little else was known about the mechanism of zymogen processing of BACE, including whether its activation is autocatalytic or carried out by other enzymes. Data derived from BACE overexpressed in bacteria (65) have suggested that zymogen processing of the prosegment's $\mathbf{R}_{42}$LP$\mathbf{R}_{45}$↓ site, which is reminiscent of PC-cleavage sites (8), is not autocatalytic; rather, it is effected by another proteinase(s). We have demonstrated that furin and to a lesser extent PC5 are the major proprotein convertases responsible for the conversion of proBACE into BACE within the *trans*-Golgi network (66). Although removal of the 24 aa prosegment is required in order for BACE to achieve its maximal catalytic activity *in vitro*, we also have provided evidence that proBACE is active in the ER and can produce significant quantities of the β-secretase product C99 (66). Therefore, both α- and β-secretases are processed by PCs and their zymogen activation seems to occur at or near the TGN.

V. Conclusions

The identification of a family of serine proteinases of the subtilisin/kexin type (PCs) and proof of their function(s) as intracellular dibasic processing enzymes resolved a long-standing quest for such convertases. The cumulative knowledge acquired over the past 10 years revealed that a wide array of secretory precursors are processed intracellularly by a limited set of convertases. Other processing enzyme(s) cleaving at nonbasic residues are yet to be identified. The discovery of SKI-1 has revealed new cleavage specificities and cellular processing compartments distinct from those of PCs, e.g., the ER and medial Golgi, where the latter are inactive, emphasizing the complexity of the regulatory steps needed for high fidelity and efficient processing. Structure–function and cellular localization and inhibitor studies revealed that complex, well-orchestrated events regulate the temporal expression and zymogen activation, tissue and intracellular localization, and fine substrate specificity of the proprotein convertases. The implication of convertases in pathologies is fast becoming a reality, and it is believed that they should be seriously considered as novel drug targets.

VI. Summary

Limited proteolysis at single and pairs of basic residues has now been recognized to be a fundamental process in all eukaryotes. The major enzymes that catalyze this reaction are serine proteinases related to yeast kexin and bacterial subtilisins and are known as the proprotein convertases (PCs). So far, seven human dibasic subtilases have been identified: furin, PC1/3, PC2, PC4, PACE4, PC5/6, and PC7/8. Together, and often in combination, these enzymes ensure that tissue-specific processing of precursors occurs in a regulated and timely fashion. The genetic organization of these enzymes was found to be similar, with PC7 being the most distant member of the family. Gene knockouts have now been achieved for furin, PC1, PC2, PC4, PACE4, and PC7. The null phenotypes are most severe for furin $(-/-)$ and PC1 $(-/-)$, which result in embryonic lethals. Much milder consequences are observed for PC2, PC4, PACE4, and PC7 nulls. Aside from these kexin-like subtilases, a novel pyrolysin-like proteinase known as SKI-1/SIP has been discovered. This enzyme is responsible for the processing of precursors containing basic and hydrophobic residues at P4 and P2, respectively, with a relatively relaxed acceptance of P1 amino acids. SKI-1/S1P is now recognized to be the major luminal processing enzyme of sterol regulatory element binding proteins, thereby regulating cholesterol and fatty acid biosynthesis. It is also critical for the N-terminal processing of brain-derived neurotrophic factor and somatostatin, resulting in products that are sorted to the regulated secretory pathway of endocrine and neural cells. Homology searches revealed that SKI-1 is a highly conserved vertebrate and invertebrate enzyme. Therefore, together PCs and SKI-1 ensure that the limited processing and activation of various proproteins would occur in a timely and regulated fashion thereby allowing efficient homeostatic cellular communication and signaling.

ACKNOWLEDGMENTS

This work was supported by Medical Research Council program grant PG-11474 and operating grant MOP-36496, and by the Protein Engineering Network of Centres Excellence (PENCE) Program supported by the Government of Canada. The secretarial assistance of Mrs. Sylvie Emond was highly appreciated.

REFERENCES

1. Seidah, N. G. (1995). *In* "Intramolecular Chaperones and Protein Folding" (U. Shinde and M. Inouye, eds.), pp. 181–203. R. G. Landes Co., Austin, TX.
2. Rouillé, Y., Duguay, S. J., Lund, K., Furuta, M., Gong, Q. M., Lipkind, G., Oliva, A. A., Chan, S. J., and Steiner, D. F. (1995). *Front. Neuroendocrinol.* **16,** 322.
3. Seidah, N. G., and Chrétien, M. (1997). *Curr. Opin. Biotech.* **8,** 602.
4. Nakayama, K. (1997). *Biochem. J.* **327,** 625.
5. Seidah, N. G., Day, R., Marcinkiewicz, M., and Chrétien, M. (1998). *Ann. N.Y. Acad. Sci.* **839,** 9.
6. Seidah, N. G., Mbikay, M., Marcinkiewicz, M., and Chrétien, M. (1998). *In* "Proteolytic and Cellular Mechanisms in Prohormone and Neuropeptide Precursor Processing" (V. Y. H. Hook, ed.), pp. 49–76. R. G. Landes Co., Georgetown, TX.
7. Zhou, A., Webb, G., Zhu, X., and Steiner, D. F. (1999). *J. Biol. Chem.* **274,** 20745.
8. Seidah, N. G., and Chrétien, M. (1999). *Brain Res.* **848,** 45.
9. Molloy, S. S., Anderson, E. D., Jean, F., and Thomas, G. (1999). *Trends Cell. Biol.* **9,** 28.
10. Thacker, C., and Rose, A. M. (2000). *Bioessays* **22,** 545.
11. Benjannet, S., Rondeau, N., Day, R., Chrétien, M., and Seidah, N. G. (1991). *Proc. Natl. Acad. Sci. USA* **88,** 3564.
12. Paquet, L., Bergeron, F., Seidah, N. G., Chrétien, M., Mbikay, M., and Lazure, C. (1994). *J. Biol. Chem.* **269,** 19279.
13. Zhu, X., Rouille, Y., Lamango, N. S., Steiner, D. F., and Lindberg, I. (1996). *Proc. Nat. Acad. Sci. USA* **93,** 4919.
14. Seidah, N. G., Mowla, S. J., Hamelin, J., Mamarbachi, A. M., Touré, B. B., Benjannet, S., Basak, A., Munzer, J. S., Zhong, M., Marcinkiewicz, J., Barale, J.-C., Lazure, C., Murphy, R. A., Chrétien, M., and Marcinkiewicz, M. (1999). *Proc. Natl. Acad. Sci. USA* **96,** 1321.
15. Touré, B. B., Basak, A., Munzer, J. S., Benjannet, S., Rochemont, J., Lazure, C., Chrétien, M., and Seidah, N. G. (2000). *J. Biol. Chem.* **275,** 2349.
16. Sakai, J., Rawson, R. B., Espenshade, P. J., Cheng, D., Seegmiller, A. C., Goldstein, J. L., and Brown, M. S. (1998). *Mol. Cell.* **2,** 505.

17. Espenshade, P. J., Cheng, D., Goldstein, J. L., and Brown, M. S. (1999). *J. Biol. Chem.* **274,** 22795.
18. Mouchantaf, R., Seidah, N. G., and Patel, Y. C. (2000). *Neuroscience Annual Meeting,* New Orleans, November 4–9, Abstract #8766.
19. Duncan, E. A., Brown, M. S., Goldstein, J. L., and Sakai, J. (1997). *J. Biol. Chem.* **272,** 12778.
20. Prat, A., Mowla, J. S., Basak, A., Chrétien, M., Murphy, R. A., and Seidah, N. G. (2000). *Neuroscience Annual Meeting,* New Orleans, November 4–9, Abstract #504.
21. Munzer, J. S., Benjannet, S., Chrétien, M., and Seidah, N. G. (2000). Submitted.
22. Rehemtulla, A., Dorner, A. J., and Kaufman, R. J. (1992). *Proc. Natl. Acad. Sci. USA* **89,** 8235.
23. Benjannet, S., Rondeau, N., Paquet, L., Boudreault, A., Lazure, C., Chrétien, M., and Seidah, N. G. (1993). *Biochem. J.* **294,** 735.
24. Nagahama, M., Taniguchi, T., Hashimoto, E., Imamaki, A., Mori, K., Tsuji, A., and Matsuda, Y. (1998). *FEBS Lett.* **434,** 155.
25. De Bie, I., Marcinkiewicz, M., Malide, D., Lazure, C., Nakayama, K., Bendayan, M., and Seidah, N. G. (1996). *J. Cell Biol.* **135,** 1261.
26. Munzer, J. S., Basak, A., Zhong, M., Mamarbachi, M., Hamelin, J., Savaria, D., Lazure, C., Benjannet, S., Chrétien, M., and Seidah, N. G. (1997). *J. Biol. Chem.* **272,** 19672.
27. van de Loo, J. W., Creemers, J. W., Bright, N. A., Young, B. D., Roebroek, A. J., and Van de Ven, W. J. (1999). *J. Biol. Chem.* **272,** 27116.
28. Zhong, M., Munzer, J. S., Basak, A., Benjannet, S., Mowla, S. J., Decroly, E., Chrétien, M., and Seidah, N. G. (1999). *J. Biol. Chem.* **274,** 33913.
29. Bhattacharjya, S., Xu, P., Zhong, M., Chrétien, M., Seidah, N. G., and Ni, F. (2000). *Biochemistry* **39,** 2868.
30. Lesage, G., Prat, A., Lacombe, J., Thomas, D. Y., Seidah, N. G., and Boileau, G. (2000). *Mol. Biol. Cell.* **11,** 1947.
31. Zhou, A., Paquet, L., and Mains, R. E. (1995). *J. Biol. Chem.* **270,** 1509.
32. Seidah, N. G., Hsi, K. L., De Serres, G., Rochemont, J., Hamelin, J., Antakly, T., Cantin, M., and Chrétien, M. (1983). *Arch. Biochem. Biophys.* **225,** 525.
33. Benjannet, S., Savaria, D., Chrétien, M., and Seidah, N. G. (1995). *J. Neurochem.* **64,** 2303.
34. Braks, J. A., and Martens, G. J. M. (1994). *Cell* **78,** 263.
35. Muller, L., Zhu, X. R., and Lindberg, I. (1997). *J. Cell Biol.* **139,** 625.
36. Zhu, X., Rouillé, Y., Lamango, N. S., Steiner, D. F., and Lindberg, I. (1996). *Proc. Natl. Acad. Sci. USA* **93,** 4919.
37. Xiang, Y., Molloy, S. S., Thomas, L., and Thomas, G. (2000). *Mol. Biol. Cell* **11,** 1257.
38. Lissitzky, J. C., Luis, J., Munzer, J. S., Benjannet, S., Parat, F., Chrétien, M., Marvaldi, J., and Seidah, N. G. (2000). *Biochem. J.* **346,** 133.
39. Logeat, F., Bessia, C., Brou, C., LeBail, O., Jarriault, S., Seidah, N. G., and Israel, A. (1998). *Proc. Natl. Acad. Sci. USA* **95,** 8108.
40. Anderson, E. D., Thomas, L., Hayflick, J. S., and Thomas, G. (1993). *J. Biol. Chem.* **268,** 24887.
41. Benjannet, S., Savaria, D., Laslop, A., Chrétien, M., Marcinkiewicz, M., and Seidah, N. G. (1997). *J. Biol. Chem.* **272,** 26210.
42. Malide, D., Seidah, N. G., Chrétien, M., and Bendayan, M. (1995). *J. Histochem. Cytochem.* **43,** 11.
43. Jutras, I., Seidah, N. G., Reudelhuber, T. L., and Brechler, V. (1997). *J. Biol. Chem.* **272,** 15184.
44. Jutras, I., Seidah, N. G., and Reudelhuber, T. L. (2000). *J. Biol. Chem.* **275,** 40337.
45. Fricker, L. D., McKinzie, A. A., Sun, J., Curran, E., Qian, Y., Yan, L., Patterson, S. D., Courchesne, P. L., Richards, B., Levin, N., Mzhavia, N., Devi, L. A., and Douglass, J. (2000). *J. Neurosci.* **20,** 639.

46. Qian, Y., Devi, L. A., Mzhavia, N., Munzer, S., Seidah, N. G., and Fricker, L. D. (2000). *J. Biol. Chem.* **275,** 23596.

47. Cameron, A., Fortenberry, Y., and Lindberg, I. (2000). *FEBS Lett.* **473,** 135.

48. Mbikay, M., Seidah, N. G., Chrétien, M., and Simpson, E. M. (1995). *Genomics* **26,** 123.

49. Tsuji, A., Hine, C., Tamai, Y., Yonemoto, K., Mori, K., Yoshida, S., Bando, M., Sakai, E., Mori, K., Akamatsu, T., and Matsuda, Y. (1997). *J. Biochem.* (*Tokyo*) **122,** 438.

50. Ohagi, S., LaMendola, J., LeBeau, M. M., Espinosa, R., Takeda, J., Smeekens, S. P., Chan, S. J., and Steiner, D. F. (1992). *Proc. Natl. Acad. Sci. USA* **89,** 4977.

51. Jackson, R. S., Creemers, J. W. M., Ohagi, S., Raffin-Sanson, M.-L., Sanders, L., Montague, C. T., Hutton, J. C., and O'Rahilly, S. (1997). *Nature Genet.* **16,** 303.

52. Shimano, H., Shimomura, I., Hammer, R. E., Herz, J., Goldstein, J. L., Brown, M. S., and Horton, J. D. (1997). *J. Clin. Invest.* **100,** 2115.

53. Checler, F. (1995). *J. Neurochem.* **65,** 1431.

54. Näslund, J., Haroutunian, V., Mohs, R., Davis, K. L., Davies, P., Greengard, P., and Buxbaum, J. D. (2000). *JAMA* **283,** 1571.

55. Buxbaum, J. D., Liu, K.-N., Luo, Y., Slack, J. L., Stocking, K. L., Peshon, J. J., Johnson, R. S., Castner, B. J., Cerretti, D. P., and Black, R. A. (1998). *J. Biol. Chem.* **273,** 27765.

56. Lammich, S., Kojro, E., Postina, R., Gilbert, S., Pfeiffer, R., Jasionowski, M., Haass, C., and Fahrenhol, F. (1999). *Proc. Natl. Acad. Sci. USA* **96,** 3922.

57. Koike, H., Tomioka, S., Sorimachi, H., Saido, T. C., Maruyama, K., Okuyama, A., Fujisawa-Sehara, A., Ohno, S., Suzuki, K., and Ishiura, S. (1999). *Biochem. J.* **343,** 371.

58. Anderson, E. D., Thomas, L., Hayflick, J. S., and Thomas, G. (1993). *J. Biol. Chem.* **268,** 24887.

59. Lopez-Perez, E., Seidah, N. G., and Checler, F. (1999). *J. Neurochem.* **73,** 2056.

60. Marcinkiewicz, M., and, Seidah, N.G. (2000). *J. Neurochem.* **75,** 2133.

61. Vassar, R., Bennett, B. D., Babu-Khan, S., Kahn, S., Mendiaz, E. A., Denis, P., Teplow, D. B., Ross, S., Amarante, P., Loeloff, R., Luo, Y., Fisher, S., Fuller, J., Edenson, S., Lile, J., Jarosinski, M. A., Biere, A. L., Curran, E., Burges, T., Louis, J.-C., Collins, F., Treanor, J., Rogers, G., and Citron, M. (1999). *Science* **286,** 735.

62. Hussain, I., Powell, D., Howwlett, D. R., Tew, D. G., Meek, T. D., Chapman, C., Gloger, I. S., Murphy, K. E., Southan, C. D., Ryan, D. M., Smith, T. S., Simmons, D. L., Walsh, F. S., Dingwall, C., and Christie, G. (1999). *Mol. Cell. Neurosci.* **14,** 419.

63. Sinha, S., Anderson, J. P., Barbour, R., Basi, G. S., Caccavello, R., Davis, D., Doan, M., Dovey, H. F., Frigon, N., Hong, J., Jacobson-Croak, K., Jewett, N., Keim, P., Knops, J., Lieber-burg, I., Power, M., Tan, H., Tatsuno, G., Tung, J., Schenk, D., Seubert, P., Suomensaari, S. M., Wang, S., Walker, D., John, V., Zhao, J., McConlogue, L., and John, V. (1999). *Nature* **402,** 533.

64. Yan, R., Bienkowski, M., Shuck, M. E., Miao, H., Tory, M. C., Pauley, A. M., Brashier, J. R., Stratman, N. C., Mathews, W. R., Buhl, A. E., Carter, D. B., Tomasselli, A. G., Parodi, L. A., Heirikson, R. L., and Gurney, M. E. (1999). *Nature* **402,** 533.

65. Lin, X., Koelsch, G., Wu, S., Downs, D., Dashti, A., and Tang, J. (2000). *Proc. Natl. Acad. Sci. USA* **97,** 1456.

66. Benjannet, S., Elagoz, A., Wickham, L., Mamarbachi, A., Munzer, J.-S., Basak, A., Lazure, C., Cromlish, J. A., Sisodia, S., Checler, F., Chretien, M., and Seidah, N. G. (2001). *J. Biol. Chem.* **276,** 10879.

10

Yeast Kex2 Protease

NATHAN C. ROCKWELL[1] • ROBERT S. FULLER

Department of Biological Chemistry
University of Michigan Medical School
Ann Arbor, Michigan

[1]Current address: Department of Molecular and Cell Biology, University of California, Berkeley, California.

THE ENZYMES, Vol. XXII

I. General Introduction

A. OVERVIEW

Kex2 protease (kexin, EC 3.4.21.61) represents a prototype for the family of subtilisin-related serine proteases involved in endoproteolytic processing of proprotein substrates in late compartments—the *trans*-Golgi network (TGN) and secretory granules—of the eukaryotic secretory pathway. The processing events catalyzed by these enzymes typically involve cleavage carboxyl to clusters of Lys and Arg residues. Precursors that undergo such processing include prohormones and neuropeptide precursors that are expressed only in neuroendocrine and neural tissues, as well as a variety of proproteins secreted via the constitutive secretory pathway, including growth factors, growth factor receptors, vitamin K–dependent coagulation factors, and metalloprotease precursors. In the past decade, fundamental progress has been made in understanding the basic biochemistry of such processing reactions in mammals through the discovery of a family of seven distinct processing enzymes related to the prototype enzyme, yeast Kex2 protease, studies of which led directly to the identification of the mammalian enzymes [(*1*); see (*2*) for review]. Related processing enzymes have been identified in invertebrate metazoans, in protozoans, and in other fungi as well, though not in plants. This chapter will focus on the enzymological properties of Kex2 protease as the "founding member" of the eukaryotic proprotein processing enzyme family. However, as biochemical information has become increasingly integrated with information on biological function, the chapter will also review aspects of the physiology and cell biology of Kex2 as well.

B. IDENTIFICATION OF KEX2

In 1976, Wickner and Leibowitz identified mutations in two yeast (*Saccharomyces cerevisiae*) genes that blocked expression of the killer phenotype by preventing secretion of killer toxin (*3*). Mutations in *KEX2* (killer expression defective) also blocked production of the mating pheromone produced by α haploid yeast cells, α-factor (*4*). Characterization of the biosynthesis of α-factor and killer toxin, identification of intermediates in α-factor biosynthesis that accumulated in mutant strains (including *kex2* mutants), and the cloning and sequencing of the genes encoding α-factor and killer toxin precursors showed that both precursors underwent cleavage carboxyl to pairs of basic residues (Lys-Arg↓ and Arg-Arg↓) during their maturation [for review, see ref. (*5*)]. Biochemical analysis of *kex2* mutant cells and cloning, sequence analysis and expression studies of the *KEX2* gene demonstrated that *KEX2* encoded a transmembrane serine protease of the subtilisin family (*1, 7–9*). In standard yeast genetic nomenclature, the *Saccharomyces KEX2* gene product is termed Kex2p or Kex2 protein (or protease). Kex2 activity was designated protease YscF in one report (*10*). The name "kexin" was recommended in one case (*11*). Homologous enzymes carrying out similar processing reactions have been identified in other fungi, including *Kluyveromyces lactis, Yarrowia lipolytica,* and *Schizosaccharomyces pombe* (*12–14*). This chapter will utilize the common name, Kex2, in conformance with the original yeast nomenclature.

C. INITIAL CHARACTERIZATION OF KEX2 ENZYMATIC ACTIVITY

Resistance to "classical" serine protease inhibitors combined with sensitivity to heavy metals and thiol reagents led to the erroneous conclusion that Kex2 was a thiol protease (*6, 10, 15, 16*). Kex2 is resistant to PMSF, TPCK, and TLCK and although sensitive to DFP, requires >10 mM for inactivation. The enzyme is also inactivated by heavy metals, including Zn^{2+}, Cu^{2+}, and Hg^{2+}, by PCMPS, and by iodoacetate and iodoacetamide. However, homology to subtilisin, sensitivity to DFP, albeit only at high concentrations, and the effects of mutation of catalytic triad residues (His-213, Ser-385) and the oxanion hole (Asn-314) confirmed that Kex2 is a serine protease (*1, 8, 17, 18*). Purification of the full-length transmembrane form of Kex2 has not been achieved because of the susceptibility of the C-terminal cytosolic tail to artefactual proteolysis during purification (*8*). Characterization has instead focused on the secreted, soluble form of the enzyme generated by placing a translational stop codon downstream from the P-domain (*19, 20*). This form of the enzyme can be expressed in yeast at high levels and purified from culture medium to homogeneity with good yield (*19, 21*).

D. KEX2 AND THE HOMOLOGOUS PROCESSING ENZYMES

The sequence of the *KEX2* gene led directly to the identification of seven mammalian homologues through homology searches (*1*) and through PCR using degenerate primers/hybridization approaches (*22–29*). Kex2 and its seven distinct mammalian homologs contain regions of ~25% identity to the catalytic domain of subtilisin and possess a signal peptide and a prodomain that is removed by intramolecular proteolysis (*20, 30–32*). A Kex2 family of processing enzymes can be distinguished clearly from degradative subtilisins found in both eukaryotes and prokaryotes (*20, 33*). Distinctive features of Kex2 family members include their high degree of sequence specificity, their roles in specific processing as opposed to protein degradation, the presence of distinct patterns of conserved residues near active site residues, a high degree of overall conservation within the core subtilisin domain (~45 to 50% between the yeast and mammalian enzymes), and the conservation of an essential sequence of ~150 residues, the P-domain, that follows the subtilisin domain and is found in none of the degradative subtilisins (*20*). Beyond the P-domain, sequences of the enzymes diverge and appear to function in unique ways in the cell biology and localization of the enzymes.

II. Primary Structure of Kex2 Protease

A. THE PRODOMAIN AND AUTOACTIVATION

Like most if not all subtilases, Kex2 requires a segment N-terminal to its conserved catalytic domain for proper folding (*20, 33–36*). After folding is complete, this prodomain is removed from the mature protein through a rapid, intramolecular cleavage at Lys-Arg-110 that occurs prior to export of the enzyme from the endoplasmic reticulum (*19, 20, 30*). Mutant forms of Kex2 that lack catalytic activity and fail to undergo prodomain cleavage are retained in a stable form in the ER, suggesting that these are properly folded molecules (*20*). The prodomain of Kex2, like those of other processing proteases, contains two potential autoproteolytic cleavage sites, Lys-Arg-80 and Lys-Arg-109. One study suggests that these sites are interchangeable (*37*). However, the only active species of Kex2 protease that has been purified has undergone cleavage carboxyl to Lys-Arg-110 (*19*). A possible explanation for the conservation of these two sites is suggested by studies of maturation of the mammalian Kex2 homolog furin, which also contains two potential autoproteolytic sites. The furin prodomain is cleaved at the more carboxy-proximal site prior to exit from the ER. The prodomain remains associated with the enzyme after ER export and is only released upon cleavage at the more amino-proximal cleavage site in an acidic post-ER compartment (37–39).

The N-terminal sequence of the mature Kex2 protease is normally generated by the removal of two dipeptides by the Ste13 dipeptidyl aminopeptidase that follow Lys-Arg-110; however, this trimming event is not required for activity (19, 20). Although the Kex2 pro-domain has multiple Kex2 cleavage sites, it is not yet known whether a similar process of delayed activation is present.

B.   THE CATALYTIC DOMAIN

C-terminal to the proregion, the Kex2 catalytic domain has extensive homology to the degradative proteases of the subtilisin family (33). Processing proteases such as Kex2 are therefore members of the subtilase superfamily of serine proteases and other acyl-transfer enzymes (33, 34, 40). These enzymes employ the classical serine protease catalytic triad of Ser, His, and Asp (41, 42).

The catalytic domain of Kex2 (and related processing proteases) contains several insertion loops and other characteristic deviations from the well-known subtilisin sequence (34, 38). Several of these insertions are in regions thought to be in involved in substrate recognition, which has complicated attempts to model the structure of the processing proteases (40, 43). Kex2 and other processing proteases also have a conserved Cys residue 4 positions C-terminal to the catalytic His, a situation also found in certain degradative subtilases such as proteinase K (34). The Kex2 catalytic domain requires Ca^{2+} for activity and can be reversibly inactivated by Ca^{2+} chelation (8). Analysis of Ca^{2+}-binding by other members of the subtilase family indicates that bound Ca^{2+} is structural and that Ca^{2+} removal can result in reduced stability and in conformational changes that affect activity (44, 45).

C.   THE CATALYTIC TRIAD AND OXYANION HOLE

The subtilisin-type catalytic triad is conserved in all known processing proteases of Kex2 family (34, 40). It is known to be required for function, and a great deal of kinetic evidence supports the idea that the classical serine protease mechanism is utilized by these enzymes (see below). Like other subtilases, Kex2 also has a fourth conserved catalytic residue, an Asn, which is key for transition state stabilization (18, 46, 47). This residue and a backbone amide hydrogen donate hydrogen bonds to the carbonyl oxygen of the scissile bond, stabilizing the buildup of negative charge and oxyanion character on this atom during nucleophilic attack (46, 47). In Kex2, substitution of Asp for this Asn results in a greatly reduced catalytic rate (18). Surprisingly, one metazoan Kex2 homolog, PC2, has a conserved Asp at this position in place of Asn (23). It is not yet clear what significance this substitution has for PC2 function *in vivo*.

D.   THE P-DOMAIN

C-terminal to the catalytic domain, there is an additional region termed the "P-domain" that is conserved in the processing proteases of the Kex2/furin family (20). The P-domain is required for proper biogenesis of the enzyme, in that deletion of even the last evolutionarily conserved residue of the domain results in abrogation of intramolecular cleavage of the prodomain (20). The proprotein fails to exit the ER. It is not yet known whether the P-domain is dispensable for activity after the enzyme has folded; however, chimeric proteins that mix P-domains and catalytic domains among the various metazoan homologs of Kex2 show slight variations in pH optimum, Ca^{2+} affinity, and substrate specificity (48). Modeling studies based on alignments of a large collection of P-domain sequences suggest that the P-domain forms an eight-stranded β-barrel that may pack against a hydrophobic patch on a face of the subtilisin domain that is oblique to the catalytic cleft (49).

E.   THE C-TAIL

Following the P-domain, Kex2 has a single transmembrane segment followed by a 115-residue cytosolic extension, the "C-tail." This domain is required for proper sorting of Kex2 to and net retention of Kex2 in a late Golgi compartment, the yeast equivalent of the *trans*-Golgi network (TGN). Deletion of the tail or mutation of the sorting signals within it results in the rapid transport to and degradation of Kex2 in the yeast vacuole (50). The multiple sorting signals identified in this region of the protein and their role in localization of Kex2 (50, 51) will be discussed in greater detail in Section V of this chapter.

III.   Phenotypes Associated with *kex2* Mutations

A.   PHENOTYPES ASSOCIATED WITH KNOWN SUBSTRATES

The *KEX2* gene was first identified in a screen for genes required for the production of the *Saccharomyces cerevisiae* M_1 killer toxin (3, 4). Production of the mature toxin, an α/β disulfide-linked heterodimer from the single-chain preprotoxin, requires processing reminiscent of proinsulin maturation, involving removal of a connecting peptide (δ) between the α and β chains by cleavage at Arg-Arg and Lys-Arg processing sites (5, 6). *kex2* strains are unable to convert pro–killer toxin into mature killer toxin because the endoproteolytic events required in maturation do not take place (5, 6, 52).

Additional phenotypes of *kex2* mutant strains have also been reported. Most notably, strains of the α mating type that lack functional Kex2 protease are sterile. This is because maturation of the precursor of the α mating pheromone, pro-α-factor, requires Kex2-dependent cleavage at multiple Lys-Arg sites (5, 6). The six Kex2 processing sites within the two genes encoding pro-α-factor along with sites in pro–killer toxin provided the most insightful early information about Kex2 specificity (5).

B. Phenotypes Not Associated with Known Substrates

A number of other phenotypes are associated with loss-of-function mutations in the *KEX2* gene but are not yet linked to known substrates of Kex2 protease. *kex2* homozygous mutant diploid cells are sporulation defective (4). *kex2* mutant cells are cold sensitive, undergoing growth arrest at or below 20°C on rich media (53, 54). Alterations in cell volume, shape, and chitin distribution suggest that aberrant cell wall deposition occurs at nonpermissive temperature. *kex2* mutants are also hypersensitive to a number of drugs, including quinidine (55), hygromycin B, cyclosporin A, and FK506 (L. S. Kean, G. Green, and R. S. Fuller, unpublished data). The further observations that *kex2* mutations are suppressed by mutations in genes encoding subunits of RNA polymerase II (56) and synthetically lethal with mutations in transcription factor genes *SWI1* and *SNF5* (57) suggest that loss of Kex2-dependent processing results in a stress, perhaps involving cell wall defects, that signals alterations in transcription patterns within the yeast cell. This hypothesis has yet to be tested.

C. Other Genetic Interactions

Some additional information about the *in vivo* function of Kex2 protease has emerged from other genetic studies. For instance, screens for multicopy suppressors of phenotypes associated with the loss of Kex2 function resulted in the isolation of the first two members of a newly recognized family of glycosylphosphatidylinositol-linked, cell surface aspartyl proteases, the yapsins, which exhibit proteolytic activity toward basic residue cleavage sites (53, 54, 58). Overexpression of these proteases within the yeast secretory pathway can at least partially compensate for the loss of Kex2-dependent processing, suggesting that these enzymes arc also processing enzymes. This view was further substantiated by the synergistic phenotypic effects of combining mutations in the *YPS1* and *YPS2/MKC7* genes with a *kex2* mutation (*53*). These results also imply partial functional redundancy between Kex2 and the yapsins, suggesting that these enzymes process overlapping sets of substrates or substrates with redundant functions (*53*).

IV. Characterization of Kex2 Specificity *in Vivo*

A. Exogenous Substrates and Autoactivation

A number of *in vivo* results have provided additional information about Kex2 specificity. For instance, Kex2 can correctly cleave mammalian precursors having Arg at the P_4 residue (*59*), indicating that Kex2 is likely to tolerate basic residues at P_4 in addition to the recognized dibasic cleavage motif. Mutagenesis of the intramolecular cleavage sites at the junction of the Kex2 prodomain and catalytic domain has also been attempted, although in this case mutations at P_1 or P_2 did not produce significant effects on cleavage [(*37*); P. Gluschankoff and R. S. Fuller, unpublished data]. This is in contrast to a great deal of work *in vitro,* as well as work that has been done *in vivo* with a known Kex2 substrate, pro-α-factor (*60*). Taken together, these data indicate that the initial autoactivation of Kex2 is much less dependent on the nature of the substrate side chains and highlights the difference between this intramolecular cleavage after completion of folding and subsequent, bimolecular catalytic events.

B. Pro-α-factor Mutagenesis

The requirement of yeast mating for Kex2 cleavage of pro-α-factor has been exploited to examine the effects of P_2 substitutions on this substrate. Mutations that impede cleavage thus result in the production of less mature α-factor, resulting in a titrable mating defect. A quantitative mating assay has been employed to quantify these defects over a dynamic range of six orders of magnitude (*60*). The results indicate that basic residues are optimal at this position, with title discrimination between Lys and Arg. Other residues result in defects that are roughly correlated with side chain volume, with certain exceptions. Proline is well tolerated at this position, as expected on the basis of Kex2 cleavage of a Pro–Arg site in pro–killer toxin (*52*). Surprisingly, Glu produces only a modest mating defect. These data have been compared with *in vitro* results, and a rough correlation exists (see below).

V. Cellular Aspects of Kex2 Function

A. BIOSYNTHESIS AND LOCALIZATION OF KEX2 IN THE YEAST
    SECRETORY PATHWAY

Kex2 is a rare protein in yeast cells—less than one part in 10^5, equivalent to at most a few hundred molecules per cell (*8*)—yet at its wild-type level of expression, it can process an enormous volume of substrate molecules. The precision and efficiency exhibited *in vivo* by Kex2 is due to the cellular mechanisms that transiently concentrate the enzyme with its substrates in a "processing compartment." The organization of the processing compartment and the signals that localize Kex2 and (possibly) its substrates to the compartment are fundamental to understanding the molecular basis of processing specificity and provide an attractive model for cellular targeting events.

Indirect immunofluorescence using an antibody against the C-terminal, cytosolic tail (C-tail) of Kex2 identified a punctate, cytoplasmic compartment that corresponds to the yeast Golgi complex (*61*). This conclusion was further strengthened by colocalization of Kex2 with the Sec7 and Sec14 proteins, which are required for transport through and out of Golgi compartments (*61, 62*). Cleavage of pro-α-factor occurs in a compartment distal to that in which α-1,3-linked mannose is added to N-linked oligosaccharides, the last oligosaccharyl modification known to occur in yeast Golgi (*63*). As a result, Kex2 has become a definitive marker for late Golgi/TGN in yeast.

Biosynthetic studies have provided us with a detailed view of the posttranslational modifications of Kex2, its transport through the secretory pathway, its retention in a late Golgi compartment, and its ultimate degradation in the vacuole (*30, 64*). Signal peptide cleavage and both N-linked and O-linked glycosylation occur rapidly in the ER, resulting in formation of proKex2p ("I_1"), which undergoes a rapid ($t_{1/2} \sim 1$ min at $30°$C) intramolecular cleavage to produce species I_2 prior to exiting the ER (*20, 30, 64*). In the Golgi, I_2 undergoes rapid modification of both N- and O-linked carbohydrate ($t_{1/2}$ of ~ 2 min). For the remainder of its lifetime, mature Kex2p undergoes progressive modification of N-linked and O-linked oligosaccharide by addition of α-1,3-linked mannose, which most likely results from intermittent exposure of mature Kex2p to late Golgi glycosyl transferases by periodic cycling of the protein from a more distal compartment to a late Golgi compartment containing the transferases (*30, 64*). Kex2 is eventually degraded in the yeast vacuole (lysosome-equivalent). From the half-time of transport of Kex2p to the late Golgi (~ 3 min) and the half-life of the protein (~ 80 min), one can estimate that Kex2p must accumulate ~ 27-fold in the pro-α-factor processing compartment (*30, 64*).

B. LOCALIZATION SIGNALS IN THE KEX2 CYTOSOLIC TAIL

A signal in the first ~19 residues of the C-terminal cytosolic tail (C-tail) promotes net retention of the protein in the TGN (*50*). This TGN-localization signal or "TLS1" lies in the first 19 residues of the tail, residues 700–718 in Kex2p (*65*). Deletions that remove TLS1 or the entire C-tail or point mutations in TLS1 of Thr-712 (to Pro), Tyr-714 (to any of several residues), or Phe-715 (to Ala) abrogate TGN localization (*50, 65*). TLS1 mutant proteins reach the Golgi but are transported at an accelerated rate to the vacuole by a default pathway that does not traverse the plasma membrane (*50, 65*). This default vacuolar pathway, which is independent of signals in the cytosolic tail of Kex2p, is nevertheless dependent on clathrin heavy chain (*66*) and may define the major role of clathrin in localization of Kex2p (*67, 68*). Deletion of the C-tails of two other TGN transmembrane processing enzymes, Ste13 and Kex1, also results in transport to the vacuole by the default pathway (*69–71*). A TLS has been identified in the N-terminal cytosolic tail of Ste13p that contains a Phe-Gln-Phe motif (residues 85–87), similar to the Tyr-Glu-Phe sequence (residues 713–715) in the TLS1 of Kex2 (*70*). The aromatic-residue-based signals in Kex2p and Ste13p appear to function in retrieval of the proteins from the yeast equivalent of the late endosome, also known as the prevacuolar compartment (PVC) (*51, 72*). The C-tails of both proteins also contains second signals ("TLS2" in Kex2) that appear to function in TGN retention (*51, 72*).

C. LOCALIZATION BY TGN-ENDOSOMAL CYCLING

Net localization of Kex2p, Ste13p, and other membrane proteins to the TGN is thought to be achieved by a process of continual cycling between the TGN and the PVC, punctuated by periods of TGN retention. The cycling pathway appears to be similar or identical to the pathway followed by the vacuolar precursor sorting receptor, Vps10 protein (*73–75*). Several *VPS* genes required for vacuolar protein sorting are also important in localization of Kex2p and Ste13 (*51, 65, 76–78*). Isolation of genetic suppressors of mutations in TLS1 of Kex2 has led to identification of three additional genes involved in TLS1 and TLS2 function (*51, 65*).

One of these genes, *SOI1* (also known as *VPS13*), which encodes a 3144 amino acid, conserved protein that promotes the cycling of TGN transmembrane proteins, including Kex2 and Ste13, between the TGN and the late endosome, has been analyzed (*51*). Soil antagonizes the function of "retention" signals (TLS2 in Kex2p) in the TGN, promoting entry of proteins into transport vesicles targeted to the PVC. At the PVC, Soil promotes entry of proteins containing aromatic-residue containing retrieval signals (TLS1 in Kex2) into retrograde transport vesicles.

VI. Purification and Characterization of Kex2 Protease

A. ENGINEERING OVEREXPRESSION

The facile manipulation of the yeast genetic system aided development of Kex2 as a model system for the function of the processing proteases. It has proven possible to overexpress a truncated version of Kex2 lacking the C-tail [Kex2Δ613 or "ssKex2:"(19)]. Expression in yeast of a form of Kex2 with a translational termination codon following the P-domain resulted in a secreted, soluble form of the protease, "ssKex2" (19, 20). "Kex2" will refer to "ssKex2" in all subsequent discussion of the enzyme. This species is released into the culture medium, which is largely free of other proteins, providing an excellent starting point for purification (19, 20). Engineering expression of this form of Kex2 with a high-copy plasmid containing the *TDH3* promoter allows routine production of 5–10 mg Kex2 per liter of culture (19, 21, 79).

B. PURIFICATION

Although detailed protocols for purification of Kex2 (21) and furin (80) have been presented elsewhere, features of the purification scheme will be highlighted here that may be important in purification of other processing proteases. First, a yeast strain freshly transformed with the expression plasmid, pG5-Kex2Δ613, is used. Pep⁻ yeast strains, deficient in vacuolar proteases (81), have proven most suitable for expression. Overnight cultures of individual transformants are grown in a synthetic medium designed to stabilize the secreted enzyme ("1040 medium," buffered with Bistris). Cell-free supernatants are screened for activity and cultures with high levels are used as inocula for large-scale expression. Optimal expression is obtained by growing the cells in 5- to 10-ml volumes for 26 hr. Cell-free culture medium is diluted and Kex2 is recovered by stirring with Fast Flow Q Sepharose for 60 min at 4°C (all subsequent steps being performed in a cold room). Recovered resin is poured onto a column with a small bed of additional Q Sepharose, the column is washed, and Kex2 is eluted with a salt gradient. Peak fractions are pooled and concentrated by pressure dialysis using an Amicon cell (XM50 and YM30 membranes have both been used, but YM30 has proved more reliable). Concentrated material is purified to homogeneity by chromatography on a high-performance Mono Q or Resource Q column, and the resulting material is again pressure dialyzed, to a concentration of 5–10 μM.

It has not yet proved possible to express other processing proteases in yeast. However, purification of furin using a similarly truncated form expressed in a baculovirus expression system has been shown to give good yields of purified protein (*80, 82*). These purifications have both demonstrated that correct choice of buffer is essential: Kex2 is optimally stable in Bistris at neutral pH, but furin is unstable in Bistris and stable instead in MES or MOPS buffers. The acidic character of the processing proteases makes it likely that a first step such as that used for Kex2 purification will prove generally useful to concentrate a secreted form of the enzyme. Highly concentrated Kex2 exhibits autoproteolysis, which can be eliminated by keeping the enzyme concentration below 10 μM and by keeping the Ca^{2+} concentration high (1 mM $CaCl_2$ is standard for Kex2 assay, but concentrations of 10 mM or higher result in less active enzyme that can readily be reactivated by lowering the concentration).

C. Variations on the Established Purification

Conventional dialysis can be substituted for pressure dialysis, but the final material will be ~50% active because of autoproteolysis and will also be less concentrated. Bistris is the buffer of choice throughout the purification, as Kex2 is less stable in other buffers such as HEPES or Tris; additionally, $CaCl_2$ is required to maintain stable enzyme. Alternative buffers have proven suitable for the final concentration step if necessary. For instance, repeated cycles of pressure dialysis with an ethanolamine/acetic acid buffer in the presence of octylglucoside have proven successful (*83*). In this case $CaCl_2$ was omitted from the final buffer and dialysis was carried out for a finite number of cycles such that the final $CaCl_2$ concentration was approximately 10 nM. The enzyme can be frozen at any stage after the first column by snap-freezing in liquid nitrogen, although typically approximately 10% of the activity is lost with each cycle of freeze–thaw. Losses can be much higher in the absence of either glycerol or octylglucoside. Frozen stocks are prepared directly after the final concentration step, with the enzyme dialyzed against 40 mM Bistris pH 7.26, 100 mM NaCl, 10 mM $CaCl_2$, and 10% (w/v) glycerol.

D. Conditions for Kinetic Characterization

Most kinetic characterization of Kex2 has been carried out in a Bistris–HCl buffer system with 1 mM $CaCl_2$ and 0.1% (w/v) Triton X-100. For dilution of Kex2 prior to addition to a reaction, a hybrid buffer containing 40 mM Bistris, 100 mM NaCl, 1 mM $CaCl_2$, and 0.1% Triton X-100 has proved useful. The enzyme has also been shown to retain high specific activity in the presence of 30% glycerol in the above buffers. The final specific activity of purified Kex2 is typically 3.0×10^7 U/mg [a unit being defined as the amount of Kex2 sufficient to cleave 1 pmol/min of the commercially available fluorogenic peptide substrate boc-Gln-Arg-Arg-MCA: *(19)*]. Ca^{2+} is required to observe activity, but not for catalysis per se, a feature also seen in other members of the subtilase superfamily. Triton X-100 is useful for preventing the loss of Kex2 through adsorption or denaturation at low concentrations. The enzyme displays markedly different stability in different buffers *(21)*, with Bistris providing good overall stability. Other buffer systems that have been shown to give good results include ethanolamine/acetic acid (pH 7.0), sodium acetate (pH 6.0), and *N*-methylmorpholine/acetic acid (pH 7.0). The enzyme is unstable at pH> 7.5, but it has proven to be well behaved under slightly acidic conditions (pH 6).

E. GENERAL KINETIC BEHAVIOR

Characterization of purified Kex2 has primarily relied on two types of peptide substrates *(19, 79, 83, 84)*. The enzyme cleaves peptidyl-methylcoumarinamides [peptidyl-MCA substrates *(85)*] very efficiently, with k_{cat}/K_M values well in excess of 10^7 M^{-1} s^{-1} *(19, 79, 84, 86)*. Additionally, Kex2 has been shown to cleave internally quenched fluorescent peptide substrates or "IQ" substrates, in which the scissile bond is a true peptide bond *(87)*, with comparable k_{cat}/K_M values *(79)*, showing that peptidyl-MCA substrates serve as valid models for understanding Kex2 specificity and catalysis.

However, these two types of substrates differ in one important respect. Whereas peptidyl-MCA substrates exhibit normal Michaelis–Menten kinetics, IQ substrates typically exhibit substrate inhibition *(79)*. This behavior is not an artifact of the IQ system, because the same substrates are cleaved with normal kinetics by trypsin under identical conditions. While the behavior is well described by a model for substrate inhibition based on a second molecule of substrate interacting with an enzyme–substrate complex *(42, 79)*, the available data are insufficient to prove such a model definitively. Fitting the data to artificially constrained models indicates that neither k_{cat} nor K_M are well defined, and separate values of k_{cat} and K_M have therefore not been measured reliably for these substrates. However, using low substrate concentrations ($<<K_M$) and pseudo-first-order kinetic analysis, it has proven possible to determine k_{cat}/K_M reliably for IQ substrates and thereby compare information on Kex2 specificity with IQ substrates to information determined with peptidyl-MCA substrates. This was possible because the observed substrate inhibition is only significant when the concentration of enzyme–substrate complex or intermediate is itself significant (either the

Michaelis complex or the acylenzyme). Therefore, under pseudo-first-order conditions in which the vast majority of the enzyme is unbound, occupancy of the second substrate-binding site is kinetically insignificant and the reaction is well behaved. In the case of Kex2, substrate inhibition is independent of enzyme concentration and occurs in the presence of detergent, indicating that a proposed aggregation model (88) cannot explain the observed behavior of Kex2 with IQ substrates.

Both of these IQ and peptidyl-MCA substrates have been used to define the enzyme–substrate contacts that are important for catalysis. The equivalent results obtained with IQ and peptidyl-MCA substrates indicate that the nature of the C-terminal cleavage product is unimportant for substrate recognition by Kex2. Similarly, the nature of the P_3 side chain is also unimportant, although preference against Asp (but not Glu or Asn) at this position has been observed (79). Contacts beyond P_4 are also relatively unimportant for Kex2 catalysis (79, 84, 86). Thus, the enzyme has been shown to recognize cleavage sites by means of the P_4, P_2, and P_1 side chains, which are also the primary specificity determinants for a number of degradative subtilisins (89).

F. INHIBITION

The P_4, P_2, and P_1 positions have also been shown to be important for efficient Kex2 inhibition (82). Naturally occuring [α1-antitrypsin Pittsburg (90)] and engineered [eglin C (82)] protein protease inhibitors can inhibit Kex2 with K_I values below 1 nM in the best cases. In the case of eglin c, this work has involved engineering the P_4 and P_1 residues to provide a better consensus site for Kex2 binding and inhibition. The most potent of a series of peptidyl chloromethylketone inhibitors were obtained with favorable residues at P_4, P_2, and P_1 (91). In this case, the best inhibitors were able to bind sufficiently tightly to permit active-site titration, although a precise 2-fold excess of inhibitor was required, possibly because of racemization of the P_1 residue during synthesis.

The response of Kex2 to a variety of small-molecule inhibitors has also been reported (8, 15). Thiol reagents such as DTT can inactivate Kex2 at high concentrations, as can the mercury compound PCMPS (p-chloromercuriphenylsulfonate). In addition, heavy metals such as Zn^{2+} and Cu^{2+} also inactivate Kex2. It is thus likely that the Kex2 active site contains a free thiol, presumably Cys-217, which can react with such agents, thereby occluding the active site and inactivating the enzyme. Kex2 is also sensitive to chelators such as EDTA or EGTA, because of the loss of Ca^{2+} required for stability, and is inactivated by high concentrations of the electrophilic serine protease inhibitor diisopropylfluorophosphate [DFP (8)].

VII. Detailed Characterization of Specificity *in Vitro*

A. Specificity at P_4

It is known that many enzymes of the subtilase superfamily exhibit interactions with the substrate side chain at the P_4 position [for a review, see (*34*); model structures for processing proteases are discussed in references (*40, 43*)]. This interaction has been examined both by varying the substrate and by mutagenesis of the enzyme in both subtilisins and processing proteases (*89, 92–94*). In the processing enzyme furin, the importance of the P_4 residue in determining cleavage sites has long been recognized (*95*). However, the somewhat weaker apparent preference of Kex2 for P_4 residues (Table I) as compared to P_2 and P_1 led to the expectation that Kex2 did not recognize the P_4 sidechain. A preliminary study of Kex2 specificity seemed to bear this out, as tripeptidyl-MCA substrates were cleaved with comparable efficiency to pentapeptidyl-MCA substrates (*19*). Moreover, this study also examined cleavage of substrates having either Arg or Met at P_4, and it was shown that Kex2 did not exhibit a marked preference for either substrate. However, more recent work has demonstrated that specificity at P_4 is significant, with alternate modes of recognition of either aliphatic or basic residues.

TABLE I

Physiological Kex2 Cleavage Sites[a]

Cleavage site	Sequence
k1 killer toxin (44↓45)	--LLPR↓EA--
k1 killer toxin (149↓150)	--VARR↓DI--
k1 killer toxin (188↓189)	--YVKR↓SD--
k1 killer toxin (233↓234)	--VAKR↓YV--
MFα1 pro-α-factor (85↓86)	--LDKR↓EA--
MFα1 pro-α-factor (104↓105)[b]	--MYKR↓EA--
KEX2 prodomain (109↓110)	--LFKR↓LP--
Exo-1,3-β-glucanase (40↓41)	--NKKR↓YY--
MFα2 pro-α-factor (80↓81)	--LAKR↓EA--
MFα2 pro-α-factor (101↓102)	--MYKR↓EA--
S. cerevisiae HSP150 (71↓72)	--KAKR↓AA--

[a] Eleven known Kex2 cleavage sites from *S. cerevisiae* are listed. Residues surrounding the cleavage site are shown from P_4 to P_2'. The residue numbers of P_1 and P_1' are given, and cleavage sites are indicated by arrows. Reprinted with permission from (*84*). Copyright (1998) American Chemical Society.

[b] The same cleavage site is repeated at 125↓126 and 145↓146.

1. Recognition of Aliphatic P_4 Residues

During the examination of a series of IQ substrates, it was initially noted that cleavage of substrates with P_1 substitution of Lys or ornithine for the preferred Arg resulted in a much better substrate than expected based on the prior examination of commercially available tripeptidyl-MCA substrates (19, 79). This difference was shown to be caused by the presence in the IQ substrates of a favorable interaction with the P_4 side chain (in this case norleucine, Nle), because the effect could be duplicated with tetrapeptidyl-MCA substrates containing Nle at this position. However, a tetrapeptidyl-MCA substrate with a smaller P_4 side chain (Ala) did not show such stimulation of cleavage (84). This implicated the longer, more hydrophobic Nle side chain at P_4 as being responsible for stimulation of cleavage by 77-fold relative to P_4 Ala (Table II).

This observation was extended by the synthesis and characterization of substrates with P_4 side chains of intermediate length (norvaline and α-aminobutyric acid). The results obtained from this analysis indicated that a straight alkyl chain of at least three carbons was sufficient to provide stimulation. However, examination of a substrate with P_4 valine showed that it gave an intermediate result between these two substrates, indicating that the Kex2 P_4 pocket can accommodate branched-chain residues and gain partial benefits from the extra hydrophobic surface. It was also possible to

TABLE II

Kex2 Protease Possesses P_4 Specificity[a]

Sequence	k_{cat}/K_M $(M^{-1}\,s^{-1})$	Relative k_{cat}/K_M
AcβYKK↓MCA	9.2×10^4	1
AcπYKK↓MCA	7.8×10^4	0.85
AcεYKK↓MCA	5.7×10^3	0.062
AcAYKK↓MCA	1.2×10^3	0.013
AcVYKK↓MCA	1.4×10^4	0.15
AcFYKK↓MCA	3.7×10^4	0.40
AcχYKK↓MCA	1.3×10^5	1.4
AcRYKK↓MCA	1.2×10^5	1.3
AcÇYKK↓MCA	4.9×10^3	0.053
AcDYKK↓MCA	<250	<0.0027

[a] Cleavage sites are indicated by arrow (↓). $\varepsilon = \alpha$-aminobutyric acid, $\pi =$ norvaline, $\beta =$ norleucine, $\chi = (\beta$-cyclohexyl)alanine, and Ç = citrulline. Relative k_{cat}/K_M values are normalized to the substrates with norleucine at P_4 for each P_1 residue. Reprinted with permission from (84). Copyright (1998) American Chemical Society.

demonstrate that the P_4 pocket is able to accommodate quite bulky residues, as a substrate with cyclohexylalanine (Cha) at P_4 was cleaved with comparable efficiency to the substrate with P_4 Nle. However, a substrate with P_4 Phe was cleaved with a 3- to 4-fold defect relative to P_4 Cha, showing that aromatic residues are not accommodated quite as readily at this position, perhaps because of their more rigid character. This work thus showed that Kex2 is able to recognize aliphatic P_4 side chains and use energy from that recognition to drive catalysis at the level of k_{cat}/K_M.

2. Recognition of Basic P_4 Residues

The recognition of aliphatic P_4 residues by Kex2 presented an apparent contradiction with earlier work that had showed no difference between aliphatic and basic residues at this position. Additional substrates were examined to resolve this matter (Table II). It was found that P_4 Arg also gave substantial stimulation relative to P_4 Ala, suggesting that the enzyme either preferred flexible yet bulky P_4 residues or had dual specificity. However, the uncharged Arg analog citrulline resulted in an approximately 3-fold defect in cleavage relative to Arg when introduced at P_4, suggesting that the positive charge of the Arg side chain was indeed involved in substrate recognition. Moreover, a substrate containing an acidic P_4 residue, Asp, was cleaved so poorly that it was not possible to rule out the possibility that the observed activity was due to a trace contaminating protease. It is therefore clear that Kex2 is not simply selecting for bulky, flexible residues at this position, but rather that it can recognize either basic or aliphatic P_4 side chains.

B. Specificity at P_2

Examination of known Kex2 cleavage sites *in vivo* indicates a clear preference for basic residues, with only one known exception (Table I). Both Lys and Arg are known at this position, indicating that recognition most likely involves simple electrostatics. Unlike other aspects of Kex2 specificity, it has been possible to test this question using both *in vitro* characterization of model substrates and examination of pro-α-factor cleavage *in vivo* (*60, 79*).

1. Electrostatic Recognition of Basic P_2 Residues in Vitro

Substitutions at P_2 in either peptidyl-MCA substrates or IQ substrates were examined. In both cases, it was found that replacement of P_2 Lys with norleucine (effectively removing the ε-amino group from the Lys side chain) resulted in a substantial defect in k_{cat}/K_M, and effects were similar in the two types of substrates (Table III). Additionally, Kex2 cleavage of

TABLE III

P_2 SUBSTITUTIONS[a]

Sequence	k_{cat}/K_M (M^{-1} s^{-1})	Relative k_{cat}/K_M	Relative mating
A. IQ Substrates			
RJβYKR↓EAEABR	2.5×10^7	1	1
RJβYRR↓EAEABR	3.0×10^7	1.2	0.51
RJβYOR↓EAEABR	7.5×10^6	0.3	N/D
RJβYER↓EAEABR[*]	1.1×10^5	0.0044	0.088
RJβYβR↓EAEABR	1.4×10^6	0.056	N/D
RJβYFR↓EAEABR[*]	5.6×10^4	0.0022	6.3×10^{-6}
RJβYPR↓EAEABR	3.5×10^6	0.14	0.095
B. AMC Substrates			
AcSLNKR↓MCA	3.9×10^7	1	1
AcSLNDR↓MCA	1.4×10^4	3.6×10^{-4}	0.0027
AcSLNQR↓MCA	6.8×10^4	0.0017	0.0024
AcSLNLR↓MCA	5.0×10^4	0.0013	0.00043
AcSLNβR↓MCA	2.2×10^5	0.0056	N/D
AcSLNAR↓MCA	1.3×10^5	0.0033	0.035
AcSLNYR↓MCA	1.5×10^4	3.8×10^{-4}	0.00017

[a] Cleavage sites are indicated with an arrow (↓), and IQ substrate sequences are thus in the form -P_3 -P_2-P_1↓P_1'-P_2'-P_3'-. IQ substrate cleavage sites indicated with asterisks were determined by amino acid sequencing. AMC substrate sequences are of the form -P_3-P_2-P_1↓MCA, where MCA indicates the C-terminal methylcoumarinamide. Mating values were determined by quantitative mating assay. Data are from Refs. *60* and *79*. Adapted in part with permission from (*79*). Copyright (1997) American Chemical Society.

peptidyl-MCA substrates containing either norleucine or alanine at P_2 proceeded with k_{cat}/K_M values within 2-fold of each other, indicating that the energetic contribution of the aliphatic portion of the Lys side chain to P_2 recognition is negligible.

However, it was also noted that certain types of P_2 side chains resulted in k_{cat}/K_M values below those observed with alanine, a presumably "neutral" side chain that has neither specific contacts with enzyme nor the extra conformational flexibility of glycine. Phe, Tyr, and Asp P_2 residues consistently resulted in even less favorable cleavage. This suggested that these side chains were actually being excluded from the P_2 pocket, resulting in a rearrangement of the enzyme–substrate complex with consequent loss of hydrogen bonding to the substrate backbone (*79*). Thus, *in vitro* data suggest that the P_2 residue is recognized by virtue of its positive charge alone, while either negative or bulky residues are excluded from the pocket at additional energetic cost.

2. *In Vivo Studies Correlate with in Vitro Results*

The effect of P_2 substitutions on cleavage *in vivo* has also been examined using a genetic system involving the Kex2 substrate pro-α-factor, the precursor of a pheromone required for *Saccharomyces cerevisiae* cells of the α mating type to mate with **a** cells. Thus, Kex2 cleavage is required for cells to mate, allowing *in vivo* cleavage to be assayed by formation of diploid cells from haploid yeast with appropriate auxotrophies (*60*). Control experiments showed that the observed α-factor cleavage was Kex2-dependent and that substitutions at P_2 did not adversely affect precursor expression (*60*).

Quantitative mating data were then generated for pheromone precursors carrying all 20 possible P_2 residues (*60*). As shown in Table III and Fig. 1, the mating assay allowed examination of Kex2 efficiency over several orders of magnitude. The results obtained *in vivo* demonstrated a rough correlation with *in vitro* data, validating the two approaches. Moreover, several additional trends could be discerned from this analysis (*60*). First, a rough correlation between side-chain volume and cleavage efficiency could be seen, with basic residues showing significant deviation from this trend (as would be expected in the presence of specific, favorable enzyme–substrate contacts). Futhermore, it was noted that β-branched residues were cleaved somewhat better than might be expected on the basis of their size. Finally, it was noted that Glu was a surpisingly good residue at this position, comparable to Thr. However, none of these residues have yet been found in authentic Kex2

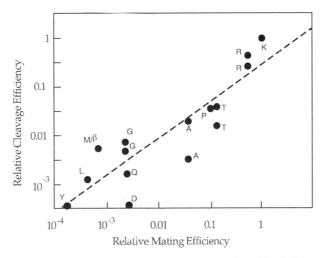

Fig. 1. Data on P_2 specificity from *in vivo* and *in vitro* studies with Kex2 protease plotted against each other. The rough correlation between these data suggests that *in vitro* studies should give relevant information about Kex2 behavior *in vivo* and demonstrate the flexibility of the yeast system. Figure taken from Ref. (*60*) (Figure 4A) with permission.

cleavage sites from yeast, suggesting that these residues can be viewed as establishing a level of cleavage that is insufficient *in vivo* though readily detectable *in vitro*. Similarly, the known, less pronounced defect in cleavage of Pro-Arg sites [which are known in authentic yeast cleavage sites (52)] places a lower bound on physiologically significant values of k_{cat}/K_M.

C. Specificity at P_1

All known *in vivo* Kex2 cleavage sites have Arg at P_1. Preliminary characterization of P_1 substitution of Arg for Lys in commercially available peptidyl-MCA substrates suggested that even this conservative substitution resulted in a massive k_{cat}/K_M defect of approximately four orders of magnitude (19). However, this substitution also involved the loss of a favorable interaction at P_4, as well as different residues at P_3 and different N-terminal protecting groups on the synthetic peptide substrates. More systematic studies have subsequently shed further light on the enzyme–substrate interaction at P_1.

1. *Exquisite Recognition of P_1 Arginine*

The effect of substitution of P_1 Arg to Lys or ornithine upon Kex2 cleavage was examined using IQ substrates (Table IV). Both substitutions gave very similar results, with a k_{cat}/K_M defect of approximately two orders of

TABLE IV

P_1 SUBSTITUTIONS[a]

Sequence	k_{cat}/K_M $(M^{-1}\,s^{-1})$	Relative k_{cat}/K_M
A. IQ Substrates		
RJβYKR↓EAEABR	2.5×10^7	1
RJβYKK↓EAEABR[*]	3.6×10^5	0.014
RJβYKO↓EAEABR	2.2×10^5	0.0088
RJβYRK↓EAEABR[*]	4.7×10^4	0.0019
B. AMC Substrates		
AcPβYKR↓MCA	3.4×10^7	1
AcPβYKK↓MCA	1.5×10^5	0.0044
AcPβYKÇ↓MCA	<500	1.5×10^{-5}

[a] Cleavage sites are indicated with an arrow (↓), and IQ substrate sequences are therefore in the form $-P_3-P_2-P_1 \downarrow P_1'-P_2'-P_3'-$. IQ substrate cleavage sites indicated with asterisks were determined by amino acid sequencing. AMC substrate sequences are of the form $-P_3-P_2-P_1 \downarrow MCA$, where MCA indicates the C-terminal methylcoumarinamide. Reprinted with permission from (79). Copyright (1997) American Chemical Society.

magnitude. This discrepancy raised the possibility that either the P_4 interaction present in these substrates or the presence of authentic amino acids C-terminal to the cleaved bond in this context was in some way able to compensate for the defect caused by the P_1 substitution.

A series of peptidyl-MCA substrates were synthesized to answer this question and provided unequivocal evidence that the observed discrepancy could be traced to the presence of a proper P_4 side chain in the IQ substrates (see below). However, the nature of the interactions involving the P_1 side chain was still not fully clear. Although lysine and ornithine maintain the positive charge of the Arg side chain, these residues have very different hydrogen bonding potential. Therefore, cleavage of an otherwise identical peptidyl-MCA substrate containing P_1 citrulline was examined. In this case, Kex2 cleavage of the substrate was exceptionally poor (Table IV). It was estimated that the positive charge on the Arg side chain at P_1 contributed approximately 6.8 kcal/mol to catalysis. Thus, Kex2 specificity at P_1 is extremely high, with Arg being recognized by both shape and charge (79). The observed defect with Lys is comparable to those seen with uncharged P_2 residues such as Ala or Thr, making it extremely unlikely that any Kex2 substrates with different P_1 residues will ever be found.

2. Interplay with Recognition of P_4

As discussed earlier, the original estimate of the P_1 defect associated with substitution of Lys for Arg proved incorrect because of the simultaneous presence of a P_4 effect. This implied a nonadditive relationship between the two subsites, which has been experimentally confirmed (84). Such interplay is a well-known feature of degradative subtilisins as well (96), although the reasons for it are not entirely clear. Indeed, the related processing protease furin provides a rare example of a situation in which two subsites (P_4 and P_6) seem independent of one another. Although attempts to engineer specificity in subtilases are ongoing (92–94, 97–102), to date the effects of these mutations on subsite interdependence have not been systematically examined. It may well prove that the extensive library of available substrates and the ready availability of large quantities of purified enzyme will make the Kex2 system amenable to such analysis.

VIII. Pre-Steady-State Characterization of Kex2 Protease

A. ACYLATION IS NOT RATE LIMITING FOR CLEAVAGE OF SYNTHETIC AMIDE SUBSTRATES

Serine proteases had long been thought to exhibit rate-limiting acylation in cleavage of amide substrates and rate-limiting deacylation in cleavage of

esters (*41, 103, 104*). Although more recent work has challenged this notion for enzymes of the trypsin/chymotrypsin family (*105*), this model has
been validated for degradative subtilisins because they exhibit a pronounced
leaving-group effect when equivalent esters and amides are examined, indicative of rate-limiting acylation for the amide (*103*). However, one of
the first studies of purified Kex2 reported an initial burst of formation of
C-terminal cleavage product in cleavage of a peptidyl-MCA fluorogenic
amide substrate (*19*). This result allowed convenient active-site titration of
Kex2, but it also indicated that acylation could not be rate-determining in
cleavage of this amide substrate, in contrast to the degradative enzymes of
the subtilisin family.

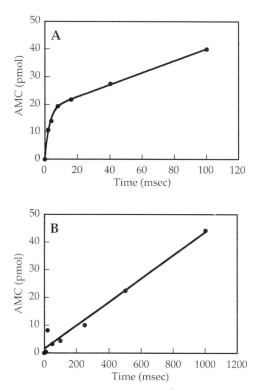

FIG. 2. Kex2 cleavage of Ac-Nle-Tyr-Lys-Arg↓MCA (A) and Ac-Nle-Tyr-Lys-Lys↓MCA (B)
was examined by rapid-quenched flow as described (*84*). The substrate containing a P_1 Arg
(the correct sequence by analogy to authentic cleavage sites) demonstrates an initial burst of
C-terminal cleavage product that is stoichiometric with enzyme (*19*). However, an otherwise
identical substrate with a substitution of Lys for Arg at P_1 does not display such a burst, indicative
of a change in rate-limiting step. Panel A adapted with permission from (*84*). Copyright (1998)
American Chemical Society.

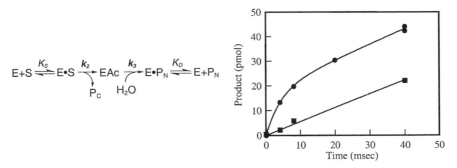

FIG. 3. (*Left*) A scheme of the serine protease mechanism. (*Right*) Pre-steady-state examination of Kex2 cleavage of Ac-Nle-Tyr-Lys-Arg↓MCA, with both the C-terminal product (circles) and the N-terminal product (squares) shown. The presence of burst kinetics in formation of the acylation product (C-terminal) but not the deacylation product (N-terminal) indicates that deacylation is likely to be rate-limiting for Kex2 cleavage of this substrate. Right hand panel adapted from (*83*). Copyright (2001) American Chemical Society.

B.   BURST KINETICS ARE ALSO SEEN WITH FURIN AND WITH PEPTIDE BONDS

More recent work has demonstrated that burst kinetics are also seen with furin (*80, 86*), indicating that this behavior may reflect a conserved feature of the processing proteases rather than a mere kinetic curiosity. Moreover, burst kinetics are not seen in Kex2 cleavage of otherwise identical substrates with substitution of Lys for Arg at P_1 [(*84*); Fig. 2], suggesting that rapid acylation may be an adaptation for specificity in the processing proteases. It has also been shown that Kex2 cleavage of an authentic peptide bond within an IQ substrate shows a similar burst with correct P_1 Arg but not with P_1 Lys, indicating that this is not an artifact caused by using substrates with an activated coumarin leaving group (*83*). Thus, the rate-limiting step for Kex2 cleavage of correct substrates was identified as some step subsequent to acylation, such as deacylation or release of the N-terminal cleavage product (Fig. 3).

C.   UTILIZATION OF ENZYME–SUBSTRATE CONTACTS DURING THE CATALYTIC CYCLE

Classical means of examining deacylation with Kex2 by the addition of exogenous nucleophiles (*42, 106, 107*) failed because of the instability of the enzyme in the required conditions. However, it has been shown that Kex2 deacylation is at or after the rate-limiting step, because there is no burst in the formation of the N-terminal cleavage product [(*83*) and Fig. 3]. The minimal interpretation of these results is that deacylation itself is rate limiting for Kex2 cleavage of correct sequences, although some conformational change after acylation but before deacylation cannot be ruled out on the basis of current data.

To determine the nature of the defect seen in Kex2 cleavage of substrates with P_1 Lys, equivalent peptidyl-methylcoumarin ester substrates [peptidyl-MCE substrates (108)] were synthesized. These substrates were shown to exhibit burst kinetics, indicating that acylation is rate limiting for Kex2 cleavage of substrates with Lys at P_1 (108). Thus, Kex2 protease specifically uses the energy from P_1 recognition to stabilize the acylation transition state and effect catalysis. However, although data sets under identical conditions have not yet been reported, it is clear that the deacylation rates for substrates with P_1 Lys will be very comparable to those for equivalent substrates with P_1 Arg (83, 108). Thus, Kex2 protease does not utilize the same interaction for stabilization of the deacylation transition state, but rather utilizes specific contacts with correct substrates at different points in the catalytic cycle. It is hoped that further experiments and structural results will shed more light on this interesting phenomenon.

IX. Kex2 as a Paradigm for Processing Protease Structure and Function

The detailed *in vitro* characterization of Kex2 protease has established a number of key results that together outline a number of features one might expect to find in other processing proteases. In other words, the characterization of Kex2 has established this enzyme as a paradigm for the processing proteases of the subtilisin superfamily. However, such a paradigm is not a dogmatic canon, but rather a framework of results and hypotheses against which other enzymes can be measured.

A. Other Fungal Proteases

Several close relatives of Kex2 are known from other fungi, including *krp1* from *Schizosaccharomyces pombe*, KEX1 from *Kluyveromyces lactis*, *KEX2* from *Candida albicans*, and XPR6 from *Yarrowia lipolytica* (12–14, 109). Additionally, other fungi such as *Pichia pastoris* have been shown to correctly process pro-alpha-factor fusion proteins at Kex2 cleavage sites (110), suggesting that similar enzymes may be responsible. However, none of these enzymes has yet been purified and characterized *in vitro*. By sequence homology, *K. lactis* KEX1, *C. albicans* KEX2, and *Y. lipolytica*

XPR6 are likely to be very similar to Kex2, with *krp1* being somewhat more distant. Interestingly, *krp1* is an essential gene in *S. pombe* (*14*), in contrast to Kex2 and other members of this family [furin is essential for embryonic viability in mice, but is not required for cell viability per se (*111, 112*)].

B. FURIN

Like Kex2, furin is localized to the constitutive secretory pathway, residing in such compartments as the TGN, the endosome, and the plasma membrane (*113*). In contrast to Kex2, furin has a requirement for basic residues at the P_4 position, with significant discrimination between Lys and Arg (*86, 95, 114, 115*). Additionally, the P_6 residue can also contribute toward catalysis, with simple electrostatic recognition of basic side chains providing a 10-fold increase in k_{cat}/K_M. Interestingly, this 10-fold increase is roughly constant in the presence of multiple P_4 residues, indicating that the P_6 interaction is independent of the P_4 subsite. In contrast to Kex2, furin cleavage is less sensitive to the nature of the P_2 residue than to that of the P_4 residue. The energetics of the $S_1–P_1$ interaction are less well understood than in Kex2, but it is clear that there is significant discrimination between Lys and Arg at this position.

Furin is also prone to substrate inhibition with hexapeptidyl-MCA substrates, although such inhibition is not seen with tetrapeptidyl-MCA substrates (*80, 86*). This complicates analysis of early work with furin, which neglects to take such behavior into account (*114, 115*). The presence of substrate inhibition makes it difficult to determine values for k_{cat} reliably, although pseudo-first-order kinetics behave as predicted and permit accurate determination of k_{cat}/K_M. The presence of a second substrate-binding site on the enzyme is also likely to explain the complicated behavior of furin inhibitors that mimic substrates (*115*), as these molecules may bind to either site and are essentially competing with substrate for both sites.

Furin exhibits burst kinetics with amide substrates (both tetrapeptidyl-MCA and hexapeptidyl-MCA), indicating that pre-steady-state behavior of this enzyme may well be similar to that of Kex2 (*80, 86*). However, the simple presence of burst kinetics does not rigorously imply that deacylation is rate limiting. At the current time, rate-limiting product release cannot be ruled out. An important test of the lessons learned with Kex2 will be the measurement of furin deacylation rates, either by classical means (*42, 107*) or the recently reported direct assay developed for use with Kex2 (*83*).

C.   PC2

PC2 is localized to the regulated secretory pathway, with most activity in compartments such as secretory granules. Surprisingly, the oxyanion hole Asn found in most subtilases is not conserved in PC2, which instead has a conserved Asp at this position. This substitution has been proposed to confer a more acidic pH optimum on the enzyme. However, attempts to active-site titrate PC2 with amide substrates were unsuccessful, raising the possibility that PC2 did not exhibit burst kinetics with such substrates. Indeed, burst kinetics were observed with an ester substrate (*108*), indicating that PC2 exhibits rate-limiting acylation. This result confirmed the experimental utility of peptidyl-MCE substrates as active-site titrants, but it also demonstrated that the pre-steady-state paradigm established with Kex2 does not hold for all processing proteases.

The possible role of the oxyanion hole Asp had also been addressed by mutagenesis of the Kex2 oxyanion hole (*18*). The resulting enzyme showed a striking defect in k_{cat} relative to wild-type Kex2 protease. However, the observed turnover numbers agree well with the observed acylation rates for PC2 (*108*). Thus, it seems likely that PC2 may be an intrinsically deficient enzyme, with a pronounced defect in acylation due to poor stabilization of negative charge on the scissile carbonyl at the transition state. The important observation that discrimination between Lys and Arg at P_1 is essentially absent with this enzyme (*116*) may be due to the compromised acylation step, which is the point at which this discrimination occurs (*79, 83, 84, 108*). Further work seems likely to shed more light on this interesting enzyme.

D.   PC1/3

Another enzyme of the regulated secretory pathway, PC1, has a normal oxyanion hole. Steady-state characterization of this enzyme has suggested that this enzyme will exhibit similar specificity to other members of this family, although reports vary as to specific activity and catalytic efficiency (*88, 114, 115, 117*). No pre-steady-state work has yet been reported with this enzyme. However, substrate inhibition has been reported with PC1 by Lazure, Leduc, and co-workers (*88*). In this case, the authors report that, despite substrate inhibition at high substrate concentrations, proper pseudo-first-order behavior could be obtained in the presence of detergent and carrier protein (BSA). This led them to suggest that aggregation could be the cause of nonclassical behavior in the processing proteases. Although BSA has not been examined in this regard with Kex2 or furin because of concerns that it will essentially act as a competing substrate, detergent is routinely present in assays with both these enzymes, and both enzymes are well behaved at the high concentrations used in pre-steady-state work. Thus, it seems likely that detergent will not prove to be a global panacea for this problem with processing proteases, and also likely that aggregation is not a significant concern in working with Kex2 and furin. It is also important to note that work with both furin and Kex2 has shown that substrate inhibition does not affect kinetics under pseudo-first-order conditions, as indeed should be the case with true substrate inhibition arising from interaction of a second substrate molecule with an enzyme–substrate complex (41, 79).

E. OTHER METAZOAN HOMOLOGS

There are a number of other homologs that have been isolated, including PC4, PC5/6, PC7, and PACE4 (22–29), but none of these enzymes has yet been characterized in similar detail. Thus, the Kex2 paradigm remains untested with these enzymes. However, work with Kex2 has already reaffirmed a number of important points for workers in the processing protease field to consider. First, characterization of specificity is best undertaken with extensive, internally consistent libraries of substrates, so that the contributions from individual subsites and from interplay among subsites may be examined separately. Second, substrate inhibition is likely to be a problem in many cases. Should it arise, it is critical to include this complication in experimental design and interpretation, even if it prevents the determination of k_{cat} and K_M as separate parameters. Third, it is important to examine a wide variety of conditions to ensure that the enzyme under study is stable during the course of the kinetic experiments. Finally, active-site titrations and other pre-steady state experiments are the ultimate means of distinguishing among possible models for kinetic behavior.

In addition to illuminating potential pitfalls for the processing protease researcher, work with Kex2 protease has also provided a body of reagents, techniques, and information that permits careful and rigorous study in spite of these difficulties. Methodologies developed during work with Kex2 protease permit the synthesis of small quantities of many peptidyl-MCA substrates from unprotected peptides (79), the active-site titration of multiple processing proteases using peptidyl-MCE substrates even when amide substrates have failed (108), and the direct examination of deacylation by pulse-chase (83). Moreover, work with Kex2 and furin has demonstrated that it is possible to measure k_{cat}/K_M values even in the presence of substrate inhibition through the use of pseudo-first-order conditions. In the future, the yeast system and Kex2 protease will permit further exploration of Kex2 structure, function, and specificity, making it likely that Kex2 will remain a vibrant research focus for years to come.

References

1. Fuller, R. S., Brake, A. J., and Thorner, J. (1989). *Science* **246,** 482.
2. Zhou, A., Webb, G., Zhu, X., and Steiner, D. F. (1999). *J. Biol. Chem.* **274,** 20745.
3. Wickner, R. B., and Leibowitz, M. J. (1976). *Genetics* **82,** 429.
4. Leibowitz, M. J., and Wickner, R. B. (1976). *Proc. Natl. Acad. Sci. U.S.A.* **73,** 2061.
5. Fuller, R. S., Sterne, R. E., and Thorner, J. (1988). *Ann. Rev. Physiol.* **50,** 345.
6. Julius, D., Brake, A., Blair, L., Kunisawa, R., and Thorner, J. (1984). *Cell* **37,** 1075.
7. Mizuno, K., Nakamura, T., Ohshima, T., Tanaka, S., and Matsuo, H. (1988). *Biochem. Biophys. Res. Commun.* **156,** 246.
8. Fuller, R. S., Brake, A., and Thorner, J. (1989). *Proc. Natl. Acad. Sci. U.S.A.* **86,** 1434.
9. Mizuno, K., Nakamura, T., Ohshima, T., Tanaka, S., and Matsuo, H. (1989). *Biochem. Biophys. Res. Comm.* **159,** 305.
10. Achstetter, T., and Wolf, D. H. (1985). *EMBO J.* **4,** 173.
11. Webb, E. C. (1992). "Enzyme Nomenclature, 1992." Academic Press, San Diego, p. 396.
12. Tanguy-Rougeau, C., Wesolowski-Louvel, M., and Fukuhara, H. (1988). *FEBS Lett.* **234,** 464.
13. Enderlin, C. S., and Ogrydziak, D. M. (1994). *Yeast* **10,** 67.
14. Davey, J., Davis, K., Imai, Y., Yamamoto, M., and Matthews, G. (1994). *EMBO J.* **13,** 5910.
15. Fuller, R. S., Brake, A. J., and Thorner, J. (1986) In: "Microbiology—1986" (L. Lieve, ed.), pp. 273–278. American Society for Microbiology, Washington, D.C.
16. Mizuno, K., Nakamura, T., Takada, K., Sakakibara, S., and Matsuo, H. (1987). *Biochem. Biophys. Res. Commun.* **144,** 807.
17. Germain, D., Dumas, F., Vernet, T., Bourbonnais, Y., Thomas, D. Y., and Boileau, G. (1992). *FEBS Lett.* **299,** 283.
18. Brenner, C., Bevan, A., and Fuller, R. S. (1993). *Curr. Biol.* **3,** 498.
19. Brenner, C., and Fuller, R. S. (1992). *Proc. Natl. Acad. Sci. U.S.A.* **89,** 922.
20. Gluschankof, P., and Fuller, R. S. (1994). *EMBO J.* **13,** 2280.
21. Brenner, C., Bevan, A., and Fuller, R. S. (1994). *Methods Enzymol.* **244,** 152.
22. Smeekens, S. P., and Steiner, D. F. (1990). *J. Biol. Chem.* **265,** 2997.
23. Smeekens, S. P., Avruch, A. S., LaMendola, J., Chan, S. J., and Steiner, D. F. (1991). *Proc. Natl. Acad. Sci. U.S.A.* **88,** 340.
24. Seidah, N. G., Gaspar, L., Mion, P., Marcinkiewicz, M., Mbikay, M., and Chretien, M. (1990). *DNA Cell Biol.* **9,** 415.
25. Kiefer, M. C., Tucker, J. E., Joh, R., Landsberg, K. E., Saltman, D., and Barr, P. J. (1991). *DNA Cell. Biol.* **10,** 757.
26. Rehemtulla, A., Barr, P. J., Rhodes, C. J., and Kaufman, R. J. (1993). *Biochem.* **32,** 11586.
27. Lusson, J., Vieau, D., Hamelin, J., Day, R., Chretien, M., and Seidah, N. G. (1993). *Proc. Natl. Acad. Sci. U.S.A.* **90,** 6691.
28. Meerabux, J., Yaspo, M. L., Roebroek, A. J., Van de Ven, W. J., Lister, T. A., and Young, B. D. (1996). *Cancer Res.* **56,** 448.
29. Nakayama, K., Kim, W. S., Torii, S., Hosaka, M., Nakagawa, T., Ikemizu, J., Baba, T., and Murakami, K. (1992). *J. Biol. Chem.* **267,** 5897.
30. Wilcox, C., and Fuller, R. S. (1991). *J. Cell Biol.* **115,** 297.
31. Leduc, R., Molloy, S. S., Thorne, B. A., and Thomas, G. (1992). *J. Biol. Chem.* **267,** 14304.
32. Creemers, J. W., Vey, M., Schafer, W., Ayoubi, T. A., Roebroek, A. J., Klenk, H. D., Garten, W., and Van de Ven, W. J. (1995). *J. Biol. Chem.* **270,** 2695.
33. Siezen, R. J., de Vos, W. M., Leunissen, J. A. M., and Dijkstra, B. W. (1991). *Protein Eng.* **4,** 719.
34. Siezen, R. J., and Leunissen, J. A. M. (1997). *Prot. Sci.* **6,** 501.

35. Shinde, U., and Inouye, M. (1996). *Adv. Exp. Med. Biol.* **349,** 147.
36. Lesage, G., Prat, A., Lacombe, J., Thomas, D. Y., Seidah, N. G., and Boileau, G. (2000). *Mol. Biol. Cell* **11,** 1947.
37. Germain, D., Thomas, D. Y., and Boileau, G. (1993). *FEBS Lett.* **323,** 129.
38. Leduc, R., Molloy, S. S., Thorne, B. A., and Thomas, G. (1992). *J. Biol. Chem.* **267,** 14304.
39. Anderson, E. D., VanSlyke, J. K., Thulin, C. D., Jean, F., and Thomas, G. (1997). *EMBO J.* **16,** 1508.
40. Siezen, R. J., Creemers, J. W. M., and Van de Ven, W. J. M. (1994). *Eur. J. Biochem.* **222,** 255.
41. Perona, J. J., and Craik, C. S. (1995). *Prot. Sci.* **4,** 337.
42. Fersht, A. R. (1985). "Enzyme Structure and Mechanism." W. H. Freeman and Co., New York.
43. Lipkind, G., Gong, Q., and Steiner, D. F. (1995). *J. Biol. Chem.* **270,** 13277.
44. Bajorath, J., Raghunathan, S., Hinrichs, W., and Saenger, W. (1989). *Nature* **337,** 481.
45. Strausberg, S. L., Alexander, P. A., Gallagher, D. T., Gilliland, G. L., Barnett, B. L., and Bryan, P. N. (1995). *Bio/Technol.* **13,** 669.
46. Robertus, J. D., Kraut, J., Alden, R. A., and Birktoft, J. J. (1972). *Biochem.* **11,** 4293.
47. Wells, J. A., Cunningham, B. C., Graycar, T. P., and Estell, D. A. (1986). *Phil. Trans. R. Soc. Lond.* **317,** 415.
48. Zhou, A., Martin, S., Lipking, G., LaMendola, J., and Steiner, D. F. (1998). *J. Biol. Chem.* **273,** 11107.
49. Lipkind, G. M., Zhou, A., and Steiner, D. F. (1998). *Proc. Natl. Acad. Sci. U.S.A.* **95,** 7310.
50. Wilcox, C. A., Redding, K., Wright, R., and Fuller, R. S. (1992). *Mol. Biol. Cell* **3,** 1353.
51. Brickner, J. H., and Fuller, R. S. (1997). *J. Cell Biol.* **139,** 23.
52. Zhu, Y.-S., Zhang, X.-Y., Zhang, M., Cartwright, C. P., and Tipper, D. J. (1991). *Mol. Microbiol.* **6,** 511.
53. Komano, H., and Fuller, R. S. (1995). *Proc. Natl. Acad. Sci. U.S.A.* **92,** 10752.
54. Komano, H., Rockwell, N., Wang, G., Krafft, G., and Fuller, R. S. (1999). *J. Biol. Chem.* **274,** 24431.
55. Conklin, D. S., Culbertson, M. R., and Kung, C. (1994). *FEMS Microbiol. Lett.* **119,** 221.
56. Martin, C., and Young, R. A. (1989). *Mol. Cell. Biol.* **9,** 2341.
57. Davie, J. K., and Kane, C. M. (2000). *Mol. Cell. Biol.* **20,** 5960.
58. Egel-Mitani, M., Flygenring, H. P., and Hansen, M. T. (1990). *Yeast* **6,** 127.
59. Thomas, G., Thorne, B. A., Thomas, L., Allen, R. G., Hruby, D. E., Fuller, R. S., and Thorner, J. (1988). *Science* **241,** 226.
60. Bevan, A., Brenner, C., and Fuller, R. S. (1998). *Proc. Natl. Acad. Sci. U.S.A.* **95,** 10384.
61. Redding, K., Holcomb, C., and Fuller, R. S. (1991). *J. Cell Biol.* **113,** 527.
62. Bankaitis, V. A., Aitken, J. R., Cleves, A. E., and Dowhan, W. (1990). *Nature* **347,** 561.
63. Graham, T. R., and Emr, S. D. (1991). *J. Cell Biol.* **114,** 207.
64. Wilcox, C. A. (1991). Ph.D. Thesis, Stanford University.
65. Redding, K., Brickner, J. H., Marschall, L. G., Nichols, J. W., and Fuller, R. S. (1996). *Mol. Cell. Biol.* **16,** 6208.
66. Redding, K., Seeger, M., Payne, G. S., and Fuller, R. S. (1996). *Mol. Biol. Cell* **7,** 1667.
67. Payne, G. S., and Schekman, R. (1989). *Science* **245,** 1358.
68. Seeger, M., and Payne, G. S. (1992). *J. Cell Biol.* **118,** 531.
69. Cooper, A., and Bussey, H. (1992). *J. Cell Biol.* **119,** 1459.
70. Nothwehr, S. F., Roberts, C. J., and Stevens, T. H. (1993). *J. Cell Biol.* **121,** 1197.
71. Roberts, C. J., Nothwehr, S. F., and Stevens, T. H. (1992). *J. Cell Biol.* **119,** 69.
72. Bryant, N. J., and Stevens, T. H. (1997). *J. Cell Biol.* **136,** 287.
73. Cereghino, J. L., Marcusson, E. G., and Emr, S. D. (1995). *Mol. Biol. Cell.* **6,** 1089.

74. Cooper, A. A., and Stevens, T. H. (1996). *J. Cell Biol.* **133,** 529.
75. Seaman, M. N., Marcusson, E. G., Cereghino, J. L., and Emr, S. D. (1997). *J Cell Biol.* **137,** 79.
76. Nothwehr, S. F., Bryant, N. J., and Stevens, T. H. (1996). *Mol. Cell. Biol.* **16,** 2700.
77. Nothwehr, S. F., Bruinsma, P., and Strawn, L. A. (1999). *Mol. Biol. Cell* **10,** 875.
78. Nothwehr, S. F., Seon-Ah Ha, S.-A., and Bruinsma, P. (2000). *J. Cell Biol.* **151,** 297.
79. Rockwell, N. C., Wang, G., Krafft, G., and Fuller, R. S. (1997). *Biochemistry* **36,** 1912.
80. Bravo, D. A., Gleason, J. B., Sanchez, R. I., Roth, R. A., and Fuller, R. S. (1994). *J. Biol. Chem.* **269,** 25830.
81. Jones, E. W. (1991). *Methods Enzymol.* **194,** 428.
82. Komiyama, T., and Fuller, R. S. (2000). *Biochem.* **39,** 15156.
83. Rockwell, N. C., and Fuller, R. S. (2001). *Biochemistry* **40,** 3657.
84. Rockwell, N. C., and Fuller, R. S. (1998). *Biochemistry* **37,** 3386.
85. Zimmerman, M., Ashe, B., Yurewicz, E. C., and Patel, G. (1977). *Anal. Biochem.* **78,** 47.
86. Krysan, D. J., Rockwell, N. C., and Fuller, R. S. (1999). *J. Biol. Chem.* **274,** 23229.
87. Matayoshi, E. D., Wang, G. T., Krafft, G. A., and Erickson, J. (1990). *Science* **247,** 954.
88. Denault, J. B., Lazure, C., Day, R., and Leduc, R. (2000). *Protein. Expres. Purif.* **19,** 113.
89. Grøn, H., Meldal, M., and Breddam, K. (1992). *Biochemistry* **31,** 6011.
90. Bathurst, I. C., Brennan, S. O., Carrell, R. W., Cousens, L. S., Brake, A. J., and Barr, P. J. (1987). *Science* **235,** 348.
91. Angliker, H., Wikstrom, P., Shaw, E., Brenner, C., and Fuller, R. S. (1993). *Biochem. J.* **293**(pt. 1), 75.
92. Rheinnecker, M., Baker, G., Eder, J., and Fersht, A. R. (1993). *Biochem.* **32,** 1199.
93. Rheinnecker, M., Eder, J., Pandey, P. S., and Fersht, A. R. (1994). *Biochem.* **33,** 221.
94. Ballinger, M. D., Tom, J., and Wells, J. A. (1996). *Biochem.* **35,** 13579.
95. Molloy, S. S., Bresnahan, P. A., Leppla, S. H., Klimpel, K. R., and Thomas, G. (1992). *J. Biol. Chem.* **267,** 16396.
96. Grøn, H., and Breddam, K. (1992). *Biochemistry* **31,** 8967.
97. Creemers, J. W. M., Siezen, R. J., Roebroek, A. J. M., Ayoubi, T. A. Y., Huylebroeck, D., and Van de Ven, W. J. M. (1993). *J. Biol. Chem.* **268,** 21826.
98. Wells, J. A., Cunningham, B. C., Graycar, T. P., and Estell, D. A. (1987). *Proc. Natl. Acad. Sci. U.S.A.* **84,** 5167.
99. Wells, J. A., Powers, D. B., Bott, R. R., Graycar, T. P., and Estell, D. A. (1987). *Proc. Natl. Acad. Sci. U.S.A.* **84,** 1219.
100. Bech, L. M., Sørensen, S. B., and Breddam, K. (1992). *Eur. J. Biochem.* **209,** 869.
101. Sørensen, S. B., Bech, L. M., Meldal, M., and Breddam, K. (1993). *Biochemistry* **32,** 8994.
102. Ballinger, M. D., Tom, J., and Wells, J. A. (1995). *Biochem.* **34,** 13312.
103. Philipp, M., and Bender, M. L. (1983). *Mol. Cell. Biochem.* **51,** 5.
104. Hartley, B. S., and Kilby, B. A. (1954). *Biochem J.* **56,** 288.
105. Hedstrom, L., Szilagyi, L., and Rutter, W. J. (1992). *Science* **255,** 1249.
106. Jencks, W. P. (1987). "Catalysis in Chemistry and Enzymology." Dover Publications, New York.
107. Fastrez, J., and Fersht, A. R. (1973). *Biochemistry* **12,** 2025.
108. Rockwell, N. C., Krysan, D. J., and Fuller, R. S. (2000). *Anal. Biochem.* **280,** 201.
109. Newport, G., and Agabian, N. (1997). *J. Biol. Chem.* **272,** 28954.
110. Enderlin, C. S., and Ogrydziak, D. M. (1994). *Yeast* **10,** 67.
111. Roebroek, A. J., Umans, L., Pauli, I. G., Robertson, E. J., van Leuven, F., Van de Ven, W. J., and Constam, D. B. (1998). *Development* **125,** 4863.
112. Robertson, B. J., Moehring, J. M., and Moehring, T. J. (1993). *J. Biol. Chem.* **268,** 24274.

113. Molloy, S. S., Anderson, E. D., Jean, F., and Thomas, G. (1999). *Trends Cell Biol.* **9,** 28.
114. Jean, F., Boudreault, A., Basak, A., Seidah, N. G., and Lazure, C. (1995). *J. Biol. Chem.* **270,** 19225.
115. Lazure, C., Gauthier, D., Jean, F., Boudreault, A., Seidah, N. G., Bennett, H. P. J., and Hendy, G. N. (1998). *J. Biol. Chem.* **273,** 8572.
116. Johanning, K., Juliano, M. A., Juliano, L., Lazure, C., Lamango, N. S., Steiner, D. F., and Lindberg, I. (1998). *J. Biol. Chem.* **273,** 22672.
117. Zhou, Y., and Lindberg, I. (1993). *J. Biol. Chem.* **268,** 5615.

11

The Enzymology of PC1 and PC2

A. CAMERON • E. V. APLETALINA • I. LINDBERG

Department of Biochemistry and Molecular Biology
Louisiana State University Health Sciences Center
New Orleans, Louisiana 70112

I. Introduction

The family of eukaryotic subtilisins contains two members expressed predominantly in neuroendocrine tissues, prohormone convertase 1 or PC1

THE ENZYMES, Vol. XXII

(also known as sPC3) and prohormone convertase 2 (PC2). These enzymes are thought to be largely responsible for the endoproteolytic conversion of neuropeptide precursors and prohormones to smaller fragments, which are then exoproteolytically processed to their final bioactive forms. The purpose of this review is to provide an overview of the biochemistry of these two convertases; the reader is referred to a recent review of the cell biology of PC1 and PC2 (*1*) for a discussion of the cellular trafficking of these enzymes as well as a complete discussion of the interaction of PC2 with its binding protein 7B2.

We first present a brief discussion of convertase enzyme structure and enzymatic characterization of PC1 and PC2. We next review the known substrate cleavage sites—of both naturally occurring and synthetic substrates—with particular emphasis on the similarities and differences in specificity between the two convertases. We then discuss the work that has been accomplished to date with regard to synthetic and natural inhibitors of PC1 and PC2, and describe recent evidence that both enzymes possess specific, naturally occurring inhibitors, which are structurally dissimilar in many ways to known substrates. We conclude with a theoretical comparison of the binding pockets of the two enzymes and highlight potential areas of interest in the coming years.

A. Comparison of the Various Domains of PC1 and PC2

PC1 and PC2, in common with other members of the eukaryotic subtilisin family, each contain four domains (Fig. 1): the propeptide, the catalytic domain, the P or homo B domain, and the carboxyl-terminal domain. These domains are discussed individually below.

1. *Propeptide Domains*

The subtilisin propeptide, like the propeptide of other enzymes, has been shown to actively participate in folding of the catalytic domain [reviewed in (*2*)]. This finding also appears to hold for the eukaryotic subtilisins, and several different propeptide mutations result in enzyme species that cannot exit the endoplasmic reticulum (*2a, 3, 4*). Whereas the propeptide of PC1 can be interchanged with that of furin (*5*), proPC2 maturation requires its own propeptide (*3*), not surprising in light of the fact that the propeptides of PC1 and PC2 exhibit only 27% sequence identity. Interestingly, we have found that mutation of a lysine within the primary cleavage site (see Fig. 1) results in the generation of an enzyme species that can exit the endoplasmic reticulum (ER) and that undergoes propeptide removal, but is catalytically inert— despite the lack of any mutated residue in the mature enzyme (*2a*). These

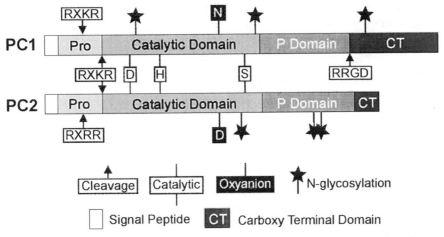

Fig. 1. Comparative structures of PC1 and PC2. The various domains are indicated.

data uphold the concept of "protein memory" proposed for subtilisin (6) and suggest that the primary cleavage site is involved in local folding of the catalytic domain.

2. Catalytic Domains

The catalytic domains of the two neuroendocrine convertases exhibit considerable homology with each other (51% identity for the human enzymes) but show only 26% identity with bacterial subtilisin itself. This relatively low conservation most likely contributes to the highly restricted specificity of the convertases compared to the subtilisins. PC1 and PC2 share many of the residues surrounding the catalytic triad residues, but also differ in certain respects. The most obvious of these differences is the presence of Asp *vs* the usual Asn in the oxyanion hole of PC2, the only enzyme in the family to contain this substitution. Studies with Kex2 have shown that this substitution does not affect intramolecular activation, but reduces the k_{cat} and the K_m of the enzyme for a peptide substrate (7). Substitution of Asp into PC2 resulted in an active enzyme species as judged by POMC cleavage (3, 8) and by cleavage of a fluorogenic peptide substrate (9). This mutation also generated an activatable zymogen when overexpressed in CHO/7B2 cells (Lamango and Lindberg, unpublished results). Other differences between PC1 and PC2 include the presence of an aromatic residue-rich loop in PC2 that contributes heavily to binding of the 21 kDa domain of the 7B2 protein (9, 8). Recent studies have identified a six-residue peptide stretch that is not conserved between PC2 and PC1 and is required for binding of the 7B2 CT peptide inhibitor (see below) (*10*).

3. *P or Homo B Domains*

The function of the P domain, also known as the Homo B domain (*11*), is relatively unexplored. This is the next most conserved region between the two neuroendocrine convertases (36% identity between human PC1 and PC2), suggesting similarities in structure. A study with the related enzyme Kex2 has shown that the presence of this domain is required for proper folding of the enzyme, i.e., ER exit (*12*). For PC1 and PC2, it is clear that the domain is also absolutely required for folding, as mutants truncated even two residues into this domain are retained in the ER [(*5, 13*); X. Zhu, unpublished results]. Few studies have addressed the potential function of this domain for the neuroendocrine enzymes; in the first, the P domain within PC1 was shown to be involved both in protein folding as well as in defining specificity and other enzymatic properties such as pH dependence and calcium sensitivity (*13*). This study also demonstrated that PC1 can function with P domains from other enzymes, but that PC2 requires its own P domain. In the second study, mutational blockade of the RRGD cleavage site that borders the carboxyl-terminal tail domain (see Fig. 1) was shown to hinder trafficking of the enzyme through the regulated pathway (*3, 14*).

4. *Carboxyl Terminal Domains*

The carboxyl-terminal domains of the two convertases differ considerably, both in length (see Fig. 1) and in overall sequence homology (25% between human PC1 and PC2). Late in the secretory pathway, most neuroendocrine-expressed PC1 undergoes an important carboxy-terminal cleavage event at an RRG motif separating the tail domain from the P domain. When this cleavage is accomplished *in vitro,* it results in a considerable increase in specific activity, but also produces a relatively unstable enzyme species (see below). One group has proposed that the carboxyl-terminal tail of PC1 represents an inhibitor of this enzyme, since prorenin processing in CHO cells by intact PC1 was inhibited by *in trans* coexpression of the tail region (*15*). However, we have generated GST and hexahistidine fusion proteins containing the PC1 tail region and found that they exhibit absolutely no inhibitory activity against any form of PC1 *in vitro* (M. Carroll, L. Muller, and I. Lindberg, unpublished results). There is evidence that the carboxyl-terminal tail is, however, involved in enzyme routing (*3, 14*). Although the RRG motif is conserved between PC1 and PC2, cleavage of PC2 at this site apparently does not occur (reviewed in Ref. *1*).

B. STRUCTURAL ANALYSIS

Although extensive data are available on the three-dimensional structure of subtilisin, the lack of a crystal structure for any of the eukaryotic subtilisins has hampered our understanding of the active site as well as the function of the additional domains contained within the eukaryotic family members. The catalytic domain of furin has been modeled based upon subtilisin (16) and predicted to contain a basic core containing similar structural elements to the bacterial enzymes, the presence of two calcium binding domains, and two disulfide bonds. Modeling of the neuroendocrine enzymes resulted in the tentative identification of the negatively charged active site residues that bind to the S1, S2, and S4 subsites; these predictions remain to be tested through experimentation (17). Lipkind et al. have proposed an eight-stranded beta barrel structure for the P domain of PC2, based upon predicted secondary structures (18). Although these models offer interesting starting points for a discussion of the major differences in specificity between the eukaryotic subtilisins and their bacterial counterparts, obtaining a crystal structure for even one family member would allow us to assign a molecular basis to the apparently much more extensive binding pockets of the eukaryotic enzymes. Unfortunately, despite the fact that milligram quantities of prohormone convertase 2 as well as of furin and Kex2 are currently available, as yet crystallization efforts have not proved successful. One difficulty may be a requirement of sugars for proper folding of the eukaryotic enzymes since deletion of glycosylation sites within the P domain results in a proPC2 species that is incapable of ER exit (Hwang and Lindberg, unpublished results). However, Fuller et al. have shown that active Kex2 can be synthesized in the presence of tunicamycin, implying that at least for the yeast enzyme, N-glycosylation is not critical (19). It is likely that one of the seven members of this family will be crystallized within the next few years, perhaps in complex with one of the two known naturally occurring peptide inhibitors.

C. ENZYMATIC CHARACTERIZATION

1. *Expression*

The prohormone convertases are relatively inabundant, and despite years of effort by many investigators, it was not until after the sequences of PC1 and PC2 had been established (20, 21) that significant quantities of enzymatically active protein became available using overexpression methods. Recombinant PC1 is now available from several sources. These include methotrexate-amplified transfected CHO cells (*22*), transfected L-cells (*23*), vaccinia virus-mediated overexpression (*24*), and baculovirally-mediated insect cell expression (*25*).

Active recombinant PC2 has been much more difficult to obtain because successful expression of enzyme activity depends upon the coexpression of a helper protein, 7B2 (see below). With or without 7B2, PC2 is apparently unable to be expressed in active form via baculoviral systems (Zhu and Lindberg, unpublished results; Lazure, personal communication); in a report to the contrary (*26*), the properties of the expressed enzymatic activity were not consistent with those of PC2. We have used the DES expression system of Invitrogen to generate active PC2 in Schneider S2 cells (*27*), indicating that the problem with baculoviral expression is not due to the use of insect cells per se. PC1 and PC2 have also been purified in active form from natural sources; Azaryan *et al.* (*28*) have reported the purification of both enzymes from chromaffin granules using antibody affinity columns, although some of the reported properties of PC2 (i.e., a neutral pH optimum) do not resemble those reported by other studies. Immunopurification using carboxyl-tail specific antiserum also represents a method for effective enzymatic isolation of PC2, but not of PC1 (*29, 30*).

2.   *Enzymatic Characterization: pH Optimum and Specific Activity*

a.   *PC1.* Recombinant PC1 has been isolated in several enzymatic forms that differ in length at the carboxyl terminus. The 87 kDa protein containing the carboxyl-terminal domain has been purified by several groups (*22–24*). This protein exhibits a very low specific activity against fluorogenic substrates compared to Kex2, which cleaves a fluorogenic substrate at a maximal rate of 2 mmol/mg/hr (*7*). In our laboratory, current preparations of this protein exhibit a specific activity of 200–400 nmol AMC/mg/hr protein, consistent with reported specific activity ranges from 134 nmol AMC/mg/hr protein (*23*) to 480 nmol AMC/mg/hr (*25*). By comparison, the activity of purified recombinant furin ranges between 11 μmol AMC/mg/hr (*31*) and 21 μmol AMC/mg/hr (*31a*) to 30 umol AMC/mg/hr (*32*), and purified recombinant PC2 exhibits specific activities comparable to those of furin (see below). The specific activity of PC1 can be improved by removal of the carboxyl terminal tail by autodigestion at the RRGD site (see Fig. 1) separating this domain from the P domain (*33, 34*), a process that requires an acid milieu and calcium, and is competed by substrate (*22, 35*). Carboxy-terminal truncation of PC1, which occurs spontaneously at low pH in the presence of calcium, results in the generation of a 66 kDa unstable enzyme with 4- to 40-fold increase in specific activity (*25, 33, 35*). Increased activity of the 66 kDa form of PC1 is a property that is also reflected *in vivo* (*3, 34*). Truncated PC1 exhibits altered biochemical properties, including an increased calcium requirement for maximal activity, a lowered pH optimum, from 6.0 to 5.5 (*22, 35, 25*), and possibly altered specificity (*36*). Boudreault *et al.* (*25*) have shown that the 66 kDa baculovirally expressed form (which in their hands exhibits a molecular mass of 71 kDa) is much less stable at pH 8 than at pH 6. Exactly how the carboxyl-terminal domain contributes to increased enzymatic activity and stability of PC1 is unclear.

   b. *PC2.* Purification of enzymatically active PC2 represented an elusive goal for many years; using the same method as for PC1, we were able to overexpress the zymogen, which was apparently correctly folded—as it could be efficiently secreted (*37*)—but this protein was unable to become activated, as defined by the assumption of an enzymatically active form. Using an immunopurification protocol we were able to characterize recombinant PC2 activity and show that it exhibited properties similar to, but also slightly different from, those of recombinant PC1 (*38*). These properties—a lower pH optimum (5.0) and calcium requirement (IC_{50} of 75 μM)—were consistent with those previously reported in work using insulinoma granule extracts and proinsulin as a substrate, the first enzymatic demonstration of PC2 activity [reviewed in (*39*)]. Trace quantities of PC2 activity with similar properties were also demonstrated via expression in oocytes (*40*). However, recovery of significant quantities of enzymatic activity from CHO/PC2 cells was not achieved until it was realized that expression of PC2 enzyme activity requires coexpression of 7B2, first identified as a PC2 binding protein by Braks and Martens (*41*). The finding that cotransfection of 7B2 resulted in the acquisition of enzyme activity (*42*) enabled the purification of milligram quantities of mouse proPC2 from the conditioned medium of DHFR-amplified CHO/PC2 cells stably transfected with 7B2 (*43*).
   Purified recombinant proPC2 can easily be activated *in vitro* (see below); recent work using bioreactor-generated proPC2 and omission of the iminodiacetic column step has resulted in preparations with a much higher specific activity than that originally reported, 17–29 μmol/h/mg protein (*44*), comparable to the activity of recombinant furin. Like PC1, recombinant PC2 is sensitive to thiol-directed reagents such as PCMB; the enzyme is resistant to even high concentrations of classical serine protease inhibitors such as PMSF (*43*).
   Like other enzymes of the mammalian convertase family (*13, 24, 33, 45–50*), but unlike the newer S1P1 enzyme (*51*), the enzymatic activity of both PC1 and PC2 is highly dependent on the presence of calcium and is inhibited by the action of chelators such as EDTA.

3. *The Activation of proPC1 and proPC2*

In common with the activation of subtilisins, the neuroendocrine convertases are thought to undergo autocatalytic processing of propeptide that
then results in activation, as defined by the production of an enzymatically
competent species. For PC1, this process occurs rapidly and within minutes
after synthesis, and zymogen forms cannot be isolated via overexpression
in eukaryotic cells [reviewed in (*1*)]. Although active PC1 can be generated in CHO cells without coexpression of other neuroendocrine proteins,
as discussed above, the specific activity of this convertase is much lower
than that of other family members produced using the same method, thus
raising the question of whether other neuroendocrine proteins are normally
required to generate comparably efficient forms of PC1. It is not yet clear
whether internal cleavage of the PC1 propeptide is required for propeptide dissociation and thus actual activation of the enzyme, as is the case for
furin (*52*).

An autocatalytic mechanism for PC1 and PC2 was initially indicated
by studies showing that active-site mutants cannot generate mature forms
(*53, 54, 3*); however, since these mutants also could not be secreted, correct folding of these mutants was not demonstrated. For proPC2, although
maturation to an inactive smaller form ("unproductive maturation") can
occur in the absence of intracellular encounter with 7B2, the resulting mature enzyme is inactive (*55*). Activation of recombinant proPC2 secreted
by overexpressing CHO/7B2 cells requires only acidification of the reaction
medium (*56*). The inability of any PC2 inhibitors to block this process—and
its lack of concentration dependence—support an intramolecular mechanism in which the propeptide primary cleavage site is buried within the cleft
of the zymogen. Cleavage of propeptide at an internal site is apparently not
required for effective dissociation of propeptide from the active site of PC2
(*2a*), as it is for furin (*52*) and for *Schizosaccharomyces pombe* Kex2 (*57*)
possibly because of the comparatively low affinity of PC2 propeptide for the
active enzyme (as assessed through inhibition studies) (*2a*).

   a. *proPC2 Activation Is Not Calcium Dependent: Implications for
Structure.* Calcium is not required for propeptide removal, as measured by
production of the 66 kDa PC2 form (*56*). We have carried out enzymatic
studies that differentiate the calcium independence of the activation process from the calcium dependence of substrate cleavage. In Fig. 2, data are
presented showing (i) the absence of activity against fluorogenic substrates in
the absence of calcium; (ii) the normal assumption of activity as a function of
time as the proform converts to the mature form in the presence of calcium;
and (iii) the virtually instantaneous assumption of full activity when calcium

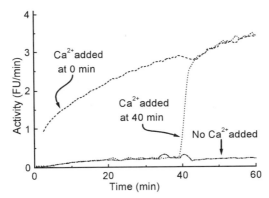

FIG. 2. The activation of proPC2. A stock preparation of proPC2 was diluted to 5 μM Ca^{2+} in 100 μM sodium acetate, 0.2% octyl glucoside, pH 5.0. Substrate (pERTKR-MCA) was added at a final concentration of 200 μM. Simultaneously, CaCl$_2$ (final concentration 11 mM in buffer) or buffer was added as indicated and fluorescence readings taken every 15 s in a 96-well fluorometer (Labsystems) with an excitation wavelength of 380 nm and an emission wavelength of 460 nm. At 40 min, CaCl$_2$ or buffer was added as indicated.

is added to proPC2 that has been incubated at low pH in the absence of calcium. The lack of lag time when calcium is added indicates that the propeptide is effectively cleaved in the absence of calcium. What these experiments do not reveal, however, is whether the propeptide is still associated with the enzyme after cleavage, only dissociating when calcium is added, or whether the propeptide has been cleaved and has dissociated prior to the addition of calcium. Homology modeling using thermitase as a template has suggested that both the strong (Ca1) and medium (Ca2) strength calcium-binding sites are present in furin (16). The primary structure corresponding to these binding sites is fairly well conserved between furin and PC2, implying that the subtilisin model could also be applicable to PC2. In this model, calcium presumably plays a role in forming the substrate binding pocket(s) of PC2, rather than affecting the catalytic cleft per se. This would permit cleavage of substrates that are already in place—such as the propeptide—but would not permit binding of other substrates. This model offers a satisfactory explanation for the lack of calcium dependence of proPC2 activation, but will require substantiation by solving the crystal structure of the zymogen.

II. Substrate Specificity of PC1 and PC2

Prohormone convertase-mediated cleavage of prohormones and neuropeptide precursors is known to occur most often at pairs of basic residues, as discussed below. Whereas several reviews have compiled protein cleavage sites known to result in bioactive peptides, in this review we have focused only on those precursors that were directly shown to be cleaved by PC1 and/or PC2. This information comes from two types of studies: *in vivo* and *in vitro* approaches. *In vivo* approaches include coexpression of prohormone convertase with a target precursor, antisense studies, and/or gene knockout studies. *In vitro* experiments involve incubation of purified PC1 or PC2 with a recombinant precursor or a synthetic peptide encompassing part of the precursor's sequence.

It should be mentioned that both *in vivo* and *in vitro* studies possess certain advantages and disadvantages. Under *in vivo* conditions, it is possible that the cleavage ascribed to the transfected protease is performed by a different cellular proprotein convertase. Second, the use of vaccinia virus vectors in coinfection experiments (an approach that has been widely employed for studying the cleavage preference of prohormone convertases) grossly alters the normal environment of the infected cells, because of extremely high expression of enzyme; high expression of recombinant viral proteins can also result in inhibition of endogenous protein synthesis and altered cell biology. Potentially negative effects may include a highly reduced possibility for overexpressed enzymes to interact with endogenous proteins (such as 7B2 and PC2) required for full enzymatic activity of the convertase. As a result, cleavage patterns observed in vaccinia virus–infected cells may differ from those using expression methods that offer lower levels of expression, such as transfection with plasmids (generally expressed at levels no higher than those of substrate). A distinct advantage to the study of precursor cleavage using transfected cell lines (as opposed to *in vitro* study) is that cleavages can be analyzed within their normal physiological setting, a subcellular compartment containing a set of proteins that may function to affect cleavage site usage or rate. For example, chromogranin B and secretogranin II, proteins residing in the secretory granules, have been shown to affect the cleavage of the nerve growth factor precursor by furin (*58*).

Certain caveats also apply with respect to the ability of prohormone convertase to cleave purified substrates *in vitro*. The results may be affected by impurities in the enzyme or the substrate preparations; other granular proteins that normally influence cleavage site preference might be lacking in the test tube. The advantage of *in vitro* experiments, of course, is the great degree of experimental control; both substrate and inhibitor concentrations can be carefully controlled, and thus kinetic considerations in substrate preference are much more easily established. Since specificity should ideally be established using kinetic measurements rather than single point assays, where possible, we have here focused on *in vitro* studies. To date, however, many more *in vivo* studies on precursor cleavage have been carried out, and few recombinant prohormones are available in purified form.

With these caveats in mind, we have assembled a substantial body of information relating to *in vivo* and *in vitro* studies of precursor and peptide cleavage by PC1 and PC2. We attempt below to present an analysis of the substrate binding preferences of these enzymes through a comprehensive study of the *in vivo* and *in vitro* cleavage data available to date. We follow this section with a review of the more systematic *in vitro* peptide cleavage studies that offer more information on residue preference through replacement analysis. Our final intent is to provide predictions that can be used for the cleavage site analysis of proteins containing putative cleavage sites encoded by novel cDNAs, as well as to provide specific information on the differences between residue subsites in the binding pockets of PC1 and PC2.

A. TRENDS FOR SUBSITE PREFERENCE: ANALYSIS OF *in Vivo* AND *in Vitro* DATA

Table I depicts the precursors known to be cleaved by PC1 and/or PC2, as well as the sequences around the cleavage sites. The ability of PC1 and PC2 to cleave these precursors was shown using *in vivo* and/or *in vitro* approaches, as indicated. We have not included data for substrate cleavages when the evidence for the involvement of PC1 or PC2 in the particular cleavage event was not strong.

PC1 and PC2 process hormone and peptide precursors C-terminally to a pair of basic residues or, much more rarely, C-terminally to a single Arg residue. Among the four dibasic sites, a Lys-Arg site is the most preferred for cleavage by PC1 as well as by PC2, followed by an Arg-Arg site. Cleavages at Lys-Lys and Arg-Lys sites occur relatively rarely. These data show that the S1 subsite of PC1 as well as PC2 exhibits a clear preference for Arg.

As is evident from Table I, the substrate specificity of PC1 is similar, but is not identical to that of PC2. In many cases, PC2 cleaves precursors with multiple cleavage sites at a greater number of sites as compared to PC1. Whereas processing of substrates such as proenkephalin, prodynorphin, and POMC by PC1 results in intermediate-sized products, processing by PC2 is much more extensive and thus gives rise to smaller peptides (*59–67*). More extensive processing by PC2 was also demonstrated for promelanin concentrating hormone (*68*). On the other hand, proglucagon is processed more extensively by PC1 than by PC2 (*69–73*). PC1 has been also shown to be more efficient in the cleavage of proneuropeptide Y (*74*).

Based upon the comparison of the sequences listed in Table I, certain conclusions that govern the cleavages (both di- and monobasic) by PC1 or PC2 can be drawn.

TABLE I

CLEAVAGE SITES FOR PC1 AND PC2 WITHIN PROTEIN PRECURSORS[a]

Protein precursor	Cleavage site $P_1 \downarrow P_1'$	Comments	References
Prohormones			
Proglucagon		Studies mostly *in vivo*	69–73
(rat or hamster)			
Lys31-Arg32	INEDKR↓HSQ	Both enzymes	
Lys62-Arg63	LMNTKR↓NRN	Cleaved by PC2 only[b]	
Lys70-Arg71	NNIAKR↓HDE	Both enzymes	
Arg77	HDEFER↓HAE	Cleaved by PC1 only	
Arg109-Arg110	KGRGRR↓DFP	Favored by PC1	
Arg124-Arg125	EELGRR↓HAD	Cleaved by PC1 only	
mProglucagon[c]		Studies *in vitro*	127
Lys31-Arg32	MNEDKR↓HSQ	Both enzymes	
Proinsulin I		Studies *in vivo*	128, 129
(mouse or rat)			
B-C junction	TPKSRR↓EVE	Favored by PC1	
C-A junction	VARQKR↓GIV	Favored by PC2	
mProinsulin II		Studies *in vivo*	128
B-C junction	TPMSRR↓EVE	Favored by PC1	
C-A junction	VAQQKR↓GIV	Favored by PC2	
rProinsulin II[d]		Studies *in vivo*	129
C-A junction	VARQKR↓GIV	Favored by PC2	
hProinsulin		Studies *in vivo* and *in vitro*[e]	130, 131, 46
B-C junction	TPKTRR↓EAE	Favored by PC1	
C-A junction	GSLQKR↓GIV	Favored by PC2	
rProTRH		Studies *in vitro* and *in vivo*	132–135
Arg51-Arg52	VRPERR↓FLW	Both enzymes	
Arg185	DPELQR↓SWE	Cleaved by PC2 only	
rProMCH		Studies *in vivo*	68
Lys129-Arg130	STQEKR↓EIG	Cleaved by PC2 only	
Arg145-Arg146	FPIGRR↓DFD	Both enzymes	
mPOMC		Studies *in vivo*	64–67
POMC(1-49)/γ-MSH	TENPRK↓YVM	Cleaved by PC2 only	
γ-MSH/JP	SAAQRR↓AEE	Both enzymes	
JP/α-MSH	PREGKR↓SYS	Both enzymes	
α-MSH/CLIP	PVGKKR↓RPV	Favored by PC2	
CLIP/γ-LPH	PLEFKR↓ELE	Favored by PC1	
γ-LPH/β-endo	PPKDKR↓YGG	Favored by PC2	
C-terminal to β-endo	KNAHKK↓GQ	Cleaved by PC2 only	
hProgastrin		Studies *in vivo*	136
Lys75-Lys76	ADPSKK↓QGP	Cleaved by PC2 only	
Proneuropeptides			
rProNT/NN		Studies *in vivo*	137
Lys140-Arg141	EEVIKR↓KIP	Both enzymes	

TABLE I (*continued*)

Protein precursor	Cleavage site $P_1\downarrow P_1'$	Comments	References
Lys148-Arg149	PYILKR↓QLY	Both enzymes	
Lys163-Arg164	PYILKR↓ASY	Both enzymes	
rProenkephalin		Studies mostly *in vitro;* some sites also *in vivo*	Johanning *et al.,* unpublished
Lys98-Lys99	HLLAKK↓YGG	Both enzymes	results; *60*
Lys105-Arg106	GGFMKR↓YGG	Both enzymes	
Lys112-Lys113	GGFMKK↓MDE	Cleaved by PC2 only	
Lys134-Arg135	EILAKR↓YGG	Favored by PC2	
Lys141-Lys142	GGFMKK↓DAD	Both enzymes	
Lys186-Arg187	DSTSKR↓YGG	Favored by PC2	
Lys196-Arg197	MRGLKR↓SPQ	Both enzymes	
Lys210-Arg211	KELQKR↓YGG	Both enzymes	
Arg217-Arg218	GGFMRR↓VGR	Cleaved by PC2 only	
Lys230-Arg231	MDYQKR↓YGG	Cleaved by PC2 only	
Lys237-Arg238	GGFLKR↓FAE	Both enzymes	
Lys261-Arg262	PEMEKR↓YGG	Favored by PC2	
hProenkephalin[f]		Studies *in vivo*	*59*
Lys184-Arg185	EEVSKR↓YGG	Favored by PC2	
hProdynorphin		Studies *in vivo* and *in vitro* with PC1; *in vitro* with PC2	*62, 63*
Lys164-Arg165	RKQAKR↓YGG	Both enzymes	
Lys175-Arg176	RKYPKR↓SSE	Favored by PC2	
Lys200-Arg201	EDLYKR↓YGG	Cleaved by PC2 only	
Arg210	FLRRIR↓PKL	Cleaved by PC2 only	
Lys219-Arg220	WDNQKR↓YGG	Cleaved by PC2 only	
Arg234	FKVVTR↓SQE	Both enzymes	
rProCCK		Studies *in vivo* with PC1, *in vitro* with PC2	*90, 138*
N-terminal to CCK8	HRISDR↓DYM	Both enzymes	
N-terminal to CCK58	PRRQLR↓AVL	Cleaved by PC1; no data on PC2	
rProneuropeptide Y		Studies *in vitro* and *in vivo*	*139, 140, 74*
Lys38-Arg39	QRYGKR↓SSP	Cleaved more efficiently by PC1 than by PC2	
rProvasopressin		Studies *in vitro* with PC1, *in vivo* with PC2	*36, 141*
VP-NP junction	PRGGKR↓ATS	Cleaved by PC1; indirect evidence of PC2 involvement	
NP-GP junction	FFRLTR↓ARE	Cleaved by PC1; no data on PC2	

(*continued*)

TABLE I (*continued*)

Protein precursors	Cleavage site $P_1 \downarrow P_1'$	Comments	References
rProoxytocin		Studies *in vitro*	*36*
OT-NP junction	PLGGKR↓AAL	Cleaved by PC1; no data on PC2	
rProsomatostatin		Studies *in vivo*	*142, 143, 78*
N-terminal to S-14	APRERK↓AGC	Both enzymes	
hProrenin		Studies *in vivo*	*144, 15*
Lys65-Arg66	SQPMKR↓LTL	Cleaved by PC1 only	
Proproteins			
mChromogranin B		Studies *in vivo*	*145*
Arg573-Arg574	DWWERR↓PFS	Both enzymes	
Lys586-Arg587	WGYEKR↓SFA	Both enzymes	
Anthrax toxin PA		Studies *in vitro*	
Lys167-Arg168	NSRKKR↓STA	Cleaved by PC1, no data on PC2	
hProalbumin		Studies *in vitro*	*23*
Arg23-Arg24	RGVFRR↓DAH	Cleaved by PC1, no data on PC2	
Chicken proalbumin		Studies *in vitro*	*23*
Arg23	LQRFAR↓DAE	Cleaved by PC1, no data on PC2	
Proenzymes			
mproPC1	KERSKR↓SVQ	Autocatalytically cleaved by PC1	*25*
mproPC2	FDRKKR↓GYR	Autocatalytically cleaved by PC2	*56*

[a] These precursors have been shown to be cleaved by both PC1 and/or PC2 *in vivo* and/or *in vitro*, unless otherwise noted.

[b] Ability of PC2 to cleave at this site was not confirmed in all studies.

[c] Cleavage was shown also at Lys70-Arg71 (by both PC1 and PC2) and Arg109-Arg110 (by PC1 only). The sequences at these cleavage sites are identical to those in rat proglucagon.

[d] Cleavage was shown also at the B-C junction (the sequence identical to that in mouse proinsulin II).

[e] *In Vitro* PC1 cleaved only at the B-C junction and PC2 only at the C-A junction.

[f] Cleavage by both PC1 and PC2 was also shown at Lys141-Lys142, Lys184-Arg185, Lys228-Arg229, and Lys235-Arg236 (the sequences at these sites are identical to those at the sites Lys141-Lys142, Lys186-Arg187, Lys230-Arg231, and Lys237-Arg238 in rat proenkephalin). Cleavage at the Lys259-Arg260 site (the sequence is identical to that around the Lys261-Arg262 site in rat proenkephalin) was shown to be performed mostly by PC2.

Conclusion 1: An aliphatic residue (Leu, Ile, Val, Met) and Thr almost never occurs at position P1'. Among the 63 sequences listed in Table I, only three sequences contain an aliphatic residue at this position. These consist of a site within human prorenin and the Lys112-Lys113 and Arg217-Arg218 cleavage sites in rat proenkephalin. The latter two sequences are cleaved only by PC2. In support of this conclusion, *in vitro* kinetic studies on internally quenched substrates containing proenkephalin cleavage site sequences have shown that the Lys112-Lys113 and Arg217-Arg218 sites—which possess this unfavorable feature—are the most poorly cleaved among the 12 sites present in this precursor and are cleaved only by PC2 [(*44*); Johanning, Hong, and Lindberg, unpublished data].

Conclusion 2: A Pro is very rare in close vicinity to the scissile bond. Proline rarely occurs at positions P1' and P2' and is also never present at position P2 within monobasic cleavage sites. This conclusion is especially true for PC1, since two out of four cleavage sites that contain Pro at either the P1' or P2' position are cleaved either exclusively or preferably by PC2.

Conclusion 3: Positively charged residues (Lys, Arg) are rarely present at position P1' and P3'. Only two sequences in Table I contain positively charged residues at the P1' position, one being cleaved preferably, if not exclusively, by PC2 (the α-MSH/CLIP junction in mouse POMC). The only two sequences with a positively charged residue (Arg) at position P3' flank a PC2 propeptide cleavage site and an Arg217-Arg218 cleavage site in rat proenkephalin that is cleaved exclusively by PC2.

Conclusion 4: Negatively charged residues (Asp, Glu) rarely occupy both positions P3 and P1' (or positions P2 and P1' in the case of a monobasic cleavage site). There are only two sequences in Table I in which a dibasic site or monobasic Arg is surrounded on both sides by negatively charged residues, and one of these sites (the Lys129-Arg130 site in rat proMCH) is cleaved only by PC2.

Conclusion 5: A Cys never occurs in the vicinity of the cleavage site, i.e., at positions P6 through P2'. An inspection of all of the sites listed in Table I reveals this to be the case.

Conclusion 6: In addition to an obligatory Arg or Lys residue at position P1, a basic residue must be present at position P2, P4, P6, or P8. Other authors have arrived at a similar conclusion (*75*).

Conclusion 7: Substrates with Phe or Tyr at the P1' position are processed preferentially by PC2. This is particularly obvious from an analysis of the cleavage of proenkephalin, in which the P1' residue is often Tyr.

1. Comparison with Previous Studies

Devi (*76*) has compared the primary sequences around monobasic cleavage sites in a variety of protein precursors, independently of the protease performing the cleavage, and has derived four rules and five tendencies that govern monobasic cleavages. These cleavages most likely include not only cleavage by PC1 and PC2 but also those of furin (*11*). Two of our conclusions, 1 and 6, are very similar to rules 2 and 3 cited in this study, and thus pertain to both mono- and dibasic cleavages. However, two other rules cited in the Devi study, although holding true for monobasic cleavages by PC1 and PC2, cannot be extended for the dibasic cleavages performed by these enzymes. One of these rules is the obligatory presence of a basic residue at the P4, P6, or P8 position. However, as seen in Table I, in many sequences containing dibasic sites processed by PC1 or PC2, a basic residue is *not* present at the P4, P6, or P8 position. This suggests that when position P2, in addition to P1, is occupied by a basic residue, the presence of a basic residue at positions P4, P6, or P8 is not necessary for PC1 or PC2 cleavage. It is, however, required when a basic residue is *not* present at the P2 position (monobasic processing): six out of seven sequences in Table I containing a monobasic cleavage site fit Devi's rule in this case. The only exception is the rat proTRH sequence flanking the Arg185 processing site.

The other rule that is applicable to monobasic cleavages performed by PC1 or PC2—but cannot be extended to dibasic cleavages—is that aromatic residues are never present at the P1' position. An overview of the sequences listed in Table I shows that Tyr or Phe (but not Trp) is not rare at the P1' position. However, most of these sites are processed either exclusively or preferentially by PC2 (see conclusion 7). Since an aromatic residue is never present at the P1' position for a monobasic cleavage by PC1 or PC2, we suggest that the presence of a basic residue at position P2 compensates for an unfavorable aromatic residue at P1' and thus renders cleavage by PC2 (and, rarely, by PC1) possible.

Rholam *et al.* (*77*) have analyzed the occurrence of given residues at positions P6 to P4' using a database of the primary sequences at 352 dibasic sites (both cleaved and uncleaved) in 83 prohormones and proproteins. Similarly to the study by Devi, the sequences were chosen independently of the protease performing the cleavage, and thus most likely also include furin sites. This group found that the sequences flanking cleaved dibasic sites, as compared to the sequences flanking uncleaved dibasic sites, have a higher proportion of hydrophilic residues at positions P5, P3, and P3', β-turn-forming amino acids at the majority of positions, and polar amino acids in positions P5, P3, and P1'. In addition, in the sequences flanking the cleaved basic sites, small amino acids are more frequent at position P2' and more abundant at position P4, and branched amino acids are more abundant at the majority of the positions, as compared to the uncleaved dibasic site sequences. These authors also observed that β-carbon branched side chain residues (Thr, Val, Leu, Ile), and Pro, Cys, Met, and Trp are either totally excluded or poorly represented at the P1′ position. The last conclusion is very similar to our conclusion 1 regarding PC1 and PC2 cleavages, and to Devi's rule for monobasic cleavages. This conclusion may therefore represent a general rule governing both the dibasic and monobasic cleavages (independently of the PC protease involved) and highlights the key role of the P1′ residue in precursor proteolytic processing.

2. *Predictive Value of Conclusions 1–7*

In the view of the preceding conclusions, it is instructive to consider those sites within naturally occurring precursors that either are not cleaved or are cleaved very poorly by PC1 or PC2. Rat prodynorphin, for example, contains three dibasic sites of that kind. The sequences flanking these sites are as follows: (1) GGFLRK↓YPK; (2) GGFLRR↓QFK; and (3) GGFLRR↓IRP. PC1 has been shown to process none of these sites, both *in vitro* and in coexpression vaccinia virus experiments (*62*). PC2, under *in vitro* conditions, was not able to cleave at site 1 and cleaved ineffectively at site(s) 2 and/or 3 (*63*). We suggest the following explanations for these sites being poorly, if at all, cleaved by PC1 and PC2. In sequence 3, position P1′ is occupied by an aliphatic residue (Ile), which according to conclusion 1 is very unfavorable for cleavage. In fact, the cleavage site in sequence 1 (Arg-Lys) is not very common for PC1 and PC2, in general. In this particular case, there is also an unfavorable Pro at the P2′ position (see conclusion 2) and an unfavorable Lys at position P3′ (conclusion 3). The combination of all these features could thus make cleavage at this site difficult. In sequence 2, similar to sequence 1, position P3′ is occupied by an unfavorable Lys residue.

Another example of a site that is not cleaved by PC1 and PC2 is a monobasic site in rat prosomatostatin (RLELQR↓SAN). Neither PC1 nor PC2 were able to cleave at this site in coexpression experiments using vaccinia virus, although this site could be cleaved by furin (*78*). The reason why PC1 and PC2 fail to process prosomatostatin at this site cannot be explained based on the conclusions listed earlier. This finding implies that although our analysis can be helpful in predicting cleavage sites for PC1 and PC2, it is not adequate for complete discrimination between cleaved and uncleaved sites. Other parameters, such as the secondary structure of polypeptide chain around the cleavage site (*79–83*), which may result in general surface exposure of the cleavage site, are also likely to contribute to recognition of cleavage sites by PC1 or PC2.

Last, it is interesting to consider the predictive value of the conclusions made earlier in determining a probable cleavage site pattern for a new precursor. Application of these conclusions for the novel precursor pre-pro-orexin (hypocretin; rat) (*84, 85*) results in the prediction of cleavage at two of three potential sites: TLGKRR↓PGP, which, because of the abundance of proline, is most likely better cleaved by PC2 than by PC1; LTMGRR↓AGA, a site predicted to be efficiently cleaved by both enzymes; and PCPGRR↓CPT, which is predicted to remain uncleaved because of the presence of Cys at the P1′ position. In fact, brain extracts have been shown to contain the first two peptides, indicating that the first two cleavages do indeed take place. Potential lack of cleavage of the last peptide has not been experimentally determined as yet.

3. *A Comparison of the Specificity of PC1 vs PC2*

At this point, although it is not always possible to conclusively distinguish PC1 from PC2 cleavage sites by the primary sequence, certain observations can be made. Sequences that contain Pro at position P1′ or P2′, or charged residues at P2′, or positively charged residues at P1′ or P3′, or negatively charged residues surrounding the cleavage site are cleaved exclusively or preferably by PC2. Sequences with an aromatic residue at the P1′ position are also generally processed exclusively by PC2. Taken together, these data suggest that the binding pocket of PC2 might be less sterically or structurally restricted as compared to PC1, and therefore can adapt to a greater variety of substrates.

B. *In Vitro* STUDIES ON SUBSTRATE SPECIFICITY OF PC1 AND PC2: SUBSTITUTIONAL ANALYSIS

1. *Synthetic Fluorogenic and Peptide Substrates Show Similar K_m Values*

The specificity of PC1 and PC2 has been studied *in vitro,* initially with partially purified enzymes and later with homogenous recombinant enzymes purified from various expression systems. An early use of fluorogenic substrates demonstrated extremely low cleavage rates of tripeptidyl-MCA substrates requiring overnight incubation (*40*) and led to an initial idea that large amounts of sequence information might be required for detection of convertase enzyme activity. In 1992, we showed that a tetrapeptidyl-MCA substrate, Cbz-RSKR-MCA, could be used to assay the paired-basic, calcium-sensitive processing endopeptidase present in insulinoma granules (*86*), which later was shown to correspond to PC2 (*46*). This rapid, convenient, and sensitive assay demonstrated that peptidyl-MCA substrates containing a minimum of four residues could be efficiently used for the characterization of proprotein convertases. Later studies performed with purified recombinant PC1 and PC2 confirmed that tripeptide— but not tetra- or pentapeptide—MCA-containing substrates are cleaved extremely ineffectively by both enzymes (*24, 23, 87, 44*). The K_m values for all studies employing tetra- to pentapeptidyl fluorogenic substrates for PC1 and PC2 are in the micromolar range. For example, the values obtained with the most commonly used substrate pERTKR-MCA vary from 8 to 23 μM for PC1 (24, 87, 88, 25) and from 18 to 131 μM for PC2 (43, 44, 88, 89). Interestingly, the K_m values obtained with the larger peptide substrates are also in the micromolar range. With proneuropeptide Y, the reported values are 96 and 69 μM for PC1 and PC2, respectively (74); with cholecystokinin 33, the value is 105 μM for PC2 (90); and with proenkephalin peptide B, the value is 89 μM for PC2 (44). These data suggest that the minimal substrate binding site of both these enzymes is not very extensive.

2. Substrate Binding Pocket Size

Limited kinetic studies of *in vitro* cleavage of IQ peptide substrates by PC1 have shown that this enzyme may have a slightly higher K_m for shorter peptides as compared to longer ones (74, 83, 91). On the other hand, a decrease in the affinity for the shorter peptides has often been accompanied by elevated V_{max} values, which result in unaltered or even elevated efficiency of cleavage. In studies on cleavage of proneuropeptide Y and two shorter peptides (74), or on the three peptides encompassing residues 505–514, 512–523, and 505–523 of HIV-1 gp160 (91), the K_m values for the shorter peptides were from 1.4 to 4 times higher than for the longer peptide. Efficiency of cleavage of shorter peptides was also increased up to 3-fold. In a study employing a series of 5 peptides reproducing the prooxytocin-neurophysin processing domain, truncation of the sequence on the C-terminal side of the scissile bond from 3 to 6 residues resulted in a 2- or 4-fold increase in the K_m value and, in this case, a decrease in cleavage efficiency (by 2 to 2.6 times) (83). Truncation of the peptidyl sequence at the N-terminal side from 12 to 6 residues led to an about 3-fold increase in the K_m value without affecting cleavage efficiency. Since the increase in the K_m values observed for the shorter peptides in these *in vitro* studies is not profound and is often compensated by increased cleavage efficiency, we suggest that precursor processing by PC1 is little affected by substrate length.

Little information is available on whether PC2 prefers longer or shorter peptides. Elongating the IQ substrate containing the best PC2 cleavage site in proenkephalin from 12 to 25 residues did not significantly affect its k_{cat}/K_m value for PC2 (44). On the other hand, truncation of proneuropeptide Y from 69 to 15 residues resulted in a peptide that, contrary to the parent molecule, was not cleaved by PC2 (74). Another proneuropeptide Y-derived peptide (29 residues) had a K_m value 3 times that of the parent molecule, but was cleaved with a 5-fold higher efficiency. Our own data indicate rapid and efficient cleavage of recombinant proenkephalin by purified recombinant PC2 (Johanning, K. *et al.*, unpublished data); however, it is not possible to derive kinetic constants for this complex precursor containing 12 cleavage sites (as compared to peptides containing a single site). Although it is clear that further studies are required with other prohormone and peptide substrates to conclusively determine whether PC2 cleavage is influenced by peptide length, the available data to date suggest that this effect will not be overwhelming in terms of either specificity or efficiency.

3. *Residue Preferences in Substrate Subsites*

In recent years, *in vitro* studies using purified recombinant enzymes have helped to obtain a considerable amount of information on the specificity of PC1 and PC2 at different subsites of the binding pocket. The studies reviewed next were performed using MCA-containing peptidyl substrates, internally quenched (IQ) synthetic peptides based upon the sequences of the natural precursors, or mutant variants of natural precursors. A systematic analysis of the available information leads to the following conclusions for each substrate or P position.

a. *P1′ Position.* Several studies performed with purified recombinant PC1 have shown high sensitivity of this enzyme to certain substitutions at the P1′ position. The natural human proalbumin variant with an Asp-to-Val mutation at this position (proalbumin Bleinheim) was cleaved very poorly by recombinant rat PC1 as compared to normal proalbumin (*23*). Substitution of the P1′ Asp by either Arg in human proalbumin or Lys in human proalbumin synthetic peptide prevented cleavage (*92*). Interestingly, the proalbumin peptide with the P1′ His (this residue would be basic at pH 5.5) was cleaved, suggesting that the constraints for Lys and Arg at the P1′ position were not solely due to the direct effect of charge (*92*). Incorporation of unnatural residues at the P1′ position of the deca- or dodecapeptide substrates (containing the sequence around the primary cleavage site for murine PC1 propeptide) generated inhibitors of PC1 (*93, 94*). The importance of a P1′ residue for PC1 cleavage was also shown in a study by Lazure *et al.* (*95*), using a series of internally quenched synthetic substrates based upon the sequences of proparathyroid hormone and proparathyroid hormone-related peptide. Replacement of the P1′ Ser with Val resulted in a compound that was poorly cleaved by murine PC1. The poor cleavage was due to an increased K_m as well as to a decreased V_{max}, suggesting that the presence of an aliphatic residue at position P1′ impairs both the deacylation and the recognition step (*95*).

The P1′ preference of PC2 has been examined *in vitro* less extensively as compared to PC1. Substitution of Ala for Pro at the P1′ position at the monobasic cleavage site in the dynorphin A 1–17 peptide enhanced the percentage of cleavage by recombinant murine PC2 from 75 to 100% (*63*). These data suggest that the P1′ Pro interferes with PC2 cleavage to a certain degree.

Other P1' substitutions in dynorphin A 1–17, with hydroxy-Pro or D-Pro, lowered the percentage of cleavage from 75% to 15 and 20%, respectively (63), also highlighting a potential involvement of the P' residue. It is worth noting that PC1 was unable to cleave dynorphin A 1–17 at its monobasic site (63). This finding may suggest that a Pro residue at the P1' position is more unfavorable for PC1 cleavage than it is for PC2 cleavage; however, the unfavorable presence of Lys at P2' could also contribute to the inability of PC1 to cleave this peptide (see below).

b. *P2' Position.* There are certain indications from *in vitro* studies that the presence of a charged residue at the P2' position might somewhat interfere with cleavage by PC1 and PC2. Replacement of Val with Lys at the P2' position in the proPTH-based internally quenched substrate resulted in a compound that was poorly cleaved by murine PC1, mostly because of an increased K_m (95). Substitution of the P2' Ala by Arg in the human proalbumin synthetic peptide completely prevented cleavage by PC1, and substitution by Glu resulted in a 4-fold decrease in the cleavage rate (92). Replacement of the P2' Lys with Ala in the dynorphin A 1–17 peptide enhanced the percentage of the peptide cleavage from 75 to 100% (63). Importantly, an overview of the naturally occurring substrates listed in the Table I shows that charged residues are not common at the P2' position. This is consistent with the predicted hydrophobic S2' pocket in subtilases (96).

c. *P2 Position.* Substitution of His for the P2 Arg in the human proalbumin synthetic peptide completely abolished its cleavage by PC1 (92), suggesting that the presence of a His residue at the P2 position is highly unfavorable for PC1 cleavage. This conclusion is consistent with the observation that His is never present at the P2 site in naturally occurring substrates (Table 1). The natural human proalbumin variant with an Arg-to-Cys mutation at this position (proalbumin Kaikoura) was also not cleaved by PC1 (23), supporting the idea that a Cys might be unfavorable in the vicinity of a PC1 cleavage site (see above). No data are available for PC2.

d. *P4 Position. In vitro* studies have also revealed a preference of PC1 for a positively charged residue, especially an Arg, at position P4. The P4 Arg preference is not as stringent as in the case of furin [reviewed in (11)]. Replacement of the P4 Arg with Orn or Lys in the synthetic substrate Ac-RSKR-MCA resulted in an 18-fold and 14-fold decrease in the k_{cat}/K_m rate constant for PC1, mostly due to an increased K_m (87). For comparison, a 280-fold and 538-fold decrease was observed for furin in this case (however, this was due to alterations in both k_{cat} and K_m) (87). Replacement of the

P4 Val with Arg in a human proalbumin peptide bearing a Phe-to-Arg substitution at P3 led to a 5-fold increase in the cleavage rate by PC1 (*92*). On the other hand, substitution of Arg for the P4 Val in the human proPTH substrates either did not affect the V_{max}/K_m value for PC1, or enhanced it by only 1.5 times (*95*). These data suggest that whether the presence of Arg at the P4 position will or will not enhance PC1 cleavage depends on the particular substrate.

The preference of PC2 at the P4 subsite has been addressed in far fewer *in vitro* studies as compared to PC1. The data that are available show no serious preference for a basic residue at P4. Substitution of Lys by Orn in Ac-RSKR-MCA reduced the k_{cat}/K_m value by 2.5 times by increasing the K_m (*44*), and replacement of the P4 Met in a human proenkephalin-based IQ substrate with Arg resulted in only a 1.5-fold decrease in the k_{cat}/K_m (*44*). An overview of the cleavage sites within naturally occurring substrates (Table I) shows that only a small number have a basic residue at P4. We can conclude that although a basic residue at the P4 position can be advantageous for PC1 or PC2 cleavage, in some cases, cleavage by both PC1 as well as PC2 does not depend on it to a considerable extent.

Interestingly, while replacement of the P4 Met with Arg in the human proenkephalin-based internally quenched substrate resulted in a 1.5-fold increase in the k_{cat}/K_m rate constant for PC2, replacement with Ala produced a significant 12-fold decrease (*44*). Inspection of the sequences of the naturally occurring substrates listed in Table I shows that small side-chain residues are, indeed, not very common at the P4 position. Further studies are required to determine whether the presence of a small side-chain residue at the P4 position is not very favorable for PC1 or PC2 cleavage in general, or only for particular substrates. It is interesting that a P4 Ala residue was not favored in tetrapeptidyl MCA-containing substrates by the yeast enzyme Kex2 (*97*).

e. *P3, P5, and P6 Positions.* No obvious preferences or restrictions have been shown so far for the positions P3, P5, and P6. It should be noted, however, that the specificity of these subsites for PC1 and, especially, for PC2, has not been well studied. Substitution of the P3 Ser with Glu, Lys, or Pro in the tetrapeptide MCA-containing substrate modifies the k_{cat}/K_m value for PC1 by no more than 2.5 times (*87*). With proPTH-related IQ substrates, replacement of Lys at position P3 with either Met or Glu resulted in only a 1.5-fold decrease in the V_{max}/K_m value, and replacement of Lys at position P6 with either Arg or Gly had almost no effect on the rate constant for PC1 (*95*). Substitution of the P5 Glu by Ala in a human proenkephalin-based IQ substrate did not alter its k_{cat}/K_m value toward PC2, and substitution by Arg resulted in a 2-fold increase (*44*). Although Arg may indeed have a positive influence at distal subsites, more studies are required to determine the generality of this conclusion.

C. CONCLUSIONS ON SUBSTRATE SPECIFICITY

In summary, the *in vitro* data on the specificity of PC1 and PC2 at different subsites of the substrate binding pocket generally support the conclusions drawn by reviewing the cleavage sites used in naturally occurring protein precursors. It should be noted, however, that most data were obtained using recombinant PC1; far fewer studies have been performed with recombinant PC2. The *in vitro* studies carried out to date support the idea that the presence of an aliphatic residue, proline, or a positively charged residue at position P1' is not favorable for cleavage by either PC1 or PC2 (conclusions 1–3). On the other hand, no *in vitro* data are available thus far to prove or disprove conclusion 4, that the presence of negatively charged residues at both positions P1' and P3 (or P2 for a monobasic cleavage site) would be unfavorable for cleavage by PC1 and PC2. Similarly, no *in vitro* data are available to test the idea that a charged residue is unfavorable at position P3' (conclusion 3). Further *in vitro* studies can also provide answers as to whether both enzymes indeed have no preference at subsites S6 and S5. Although the specificity displayed by PC1 and PC2 at these subsites for effective catalysis has been poorly studied, it has been shown that PC1 possesses a clear specificity at the S6 subsite for inhibition (see Section III).

Most importantly, further studies are required to elucidate whether interplay exists between different subsites within PC1 and PC2, i.e., whether the presence of a favorable residue at a particular position (for example, a basic residue at P4) can compensate for the presence of an unfavorable residue at other positions. Interplay between S1 and S4 subsites has been shown for Kex2 (97), and it is likely that similar side-chain interactions take place with the neuroendocrine convertases. Lastly, systematic studies using both PC1 and PC2 with precursors of different length are required to firmly establish the minimal and maximal substrate recognition sites of these two enzymes.

III. Inhibitory Specificity of PC1 and PC2

Much can be learned about the differences in the catalytic pockets of PC1 and PC2 by a comparison of inhibitory specificity. We here review the various studies of PC1 and PC2 inhibitors accomplished to date, focusing on *in vitro* data obtained with recombinant enzymes where possible. As the most work on convertase inhibitors has been performed using furin, relevant studies are included here for comparison.

A. PEPTIDE-BASED INHIBITORS

1. *Propeptide-Based Inhibitors*

In addition to their role as intramolecular chaperones [reviewed in (*98*)], propeptides of convertases are known to represent potent inhibitors of their cognate enzymes (*2a, 52, 99, 100*). Lazure's group showed (*99*) that a PC1 fragment containing the entire propeptide is a low-nanomolar inhibitor ($K_i = 6$ nM) of recombinant PC1 as well as of furin ($K_i = 10$ nM), exhibiting slow, tight-binding kinetics. Interestingly, inhibition of PC2 occurs only at high micromolar concentrations and is strictly competitive (*99*). Inhibition of all convertases tested thus far requires residues 72–84, the portion containing the primary cleavage site that flanks the catalytic domain (see Table II). Good inhibition of PC1 could be accomplished with residues 50–84 of the propeptide, suggesting that amino terminal extension of a good inhibitor does not increase potency (*101*). Cleavage of the propeptide occurs at the internal cleavage site only under conditions of enzyme excess and long incubation conditions (*101*).

We have used synthetic proPC2 propeptide fragments to show that a 28-residue fragment of the propeptide—residues 58–84 (see Table II), located directly adjacent to the catalytic domain—represents a micromolar inhibitor of PC2 (*2a*). Similarly to propeptide inhibition of PC7 (*100*), PC1

TABLE II

ALIGNMENT OF PROPEPTIDES FROM VARIOUS SPECIES

PC1		
76	PR**RSRR**SAFHIT**KRL**SDDDDRVIWAEQQYEKERSKRSALRD	HUMAN
76	PR**RSRR**SALHIT**KRL**SDDDDRVIWAEQQYEKERSKRSLLRD	PIG
76	PR**RSRR**SALHIT**KRL**SDDDDRVIWAEQQYEKERRKRSVPRD	RAT
76	PR**RSRR**SALHIT**KRL**SDDDDRVTWAEQQYEKERSKRSVQKD	MOUSE
79	PS**RMKR**SADHIT**KRL**SEDDRVLWAEQQYEKRRNKRASLGK	LOPHIUS
75	PR**RSRR**SAPAIT**KRL**YDDNRVSWAEQQYIKQRTKRGYVMN	RANA
75	PH**RSRR**GAHQHT**KRL**GDDERIQWVAQQVGRARSKRGPMGQ	AMPHIOXUS
76	PH**RSRR**SAHHHT**RKL**SEDERVAFVEQQQQKRRVKRGLVED	APLYSIA
PC2		
75	AKA**KRRR**SLHHKQQLERDPRVKMALQQEGFDRKKRGYRDI	HUMAN
75	AKA**KRRR**SLHHKQRLERDPRVKRALQQEGFDRKKRGYRDI	PIG
74	AKA**KRRR**SLHH**KR**QLERDPRIKMALQQEGFDRKKRGYRDI	RAT
74	AKA**KRRR**SLHH**KR**QLERDPRIKMALQQEGFDRKKRGYRDI	MOUSE
74	TTS**RSRR**SVNKKKHLAMDPKVNKVEQQEGFHRKKRGYRDI	XENOPUS
73	VHA**RTRR**SAGHHAKLHNDDEVLHVEQLKGYTRTKRGYRPL	C. ELEGANS
82	AHA**RSRR**SVPHTRQLRVHPHVVSAFQQNGYSRVKRGYKQT	APLYSIA
82	PHA**RSKR**SIPHTRQLRVHPQVRTAYQQSGYMRVKRGYKDA	LYMNAEA
108	PHA**RSRR**SIPHTRLLKSHPMIHTAVQQPGFKRVKRGLRPA	LUCILIA
88	PHA**RSRR**SLTHTRALKSHPAVHTAVQQPGFKRVKRGLRPA	DROSOPHILA
Furin		
67	FWH**RGVTKR**SLSPHRPRHSRLQREPQVQWLEQQVAKRRTKRDVYQE	HUMAN
269	FAH**HKVSKR**SLSPATHHQTRLDDDDRVHWAKQQRAKSRSKRDFIRM	DROSOPHILA

(*101*), and furin (*100*), the presence of carboxy-terminal basic residues is absolutely required for effective inhibition (*2a*). We also observed cleavage at the internal site after long incubation times by PC2; unlike results obtained for furin (*52*), this internal cleavage event did not correlate with the production of active enzyme (*2a*). It is interesting that only micromolar inhibitors of PC2 can be obtained by using propeptide fragments, whereas this approach results in the generation of nanomolar inhibitors of PC1 and furin. These data imply a fundamental difference in the activation mechanism of the latter two enzymes as opposed to PC2.

In Table II an alignment of the C-terminal sections of the propeptides of PC1, PC2, and, for comparison, furin is presented. The hexapeptide immediately N-terminal to the junction between the propeptide and catalytic domain scissile bond—the primary site—is underlined. The secondary and tertiary (where applicable) cleavage sites are in bold text. As the RXKR cleavage motif of the primary site is absolutely conserved between enzymes and between species, any inhibitory specificity must reside in the P3 or other residues N-terminal to this motif. By studying the conservation of these residues in different species it is possible to draw conclusions regarding which residues are preferred. At the P3 position both PC1 and furin contain small uncharged residues (rat PC1 is an anomaly with an Arg at P3), whereas in PC2 we see a split between higher species, containing a Lys at P3, and the lower species, which more closely resemble PC1 and furin. At the P5 position of PC1 and PC2 there is greater variability, but overall, acidic residues are preferred in higher species, whereas lower species more closely resemble furin (where the consensus among all species is Arg at the P5 position). At the P6 position there is a clear difference between PC1 and PC2. PC1 most resembles furin, with basic residues at the P6 position for all species, yet PC2 contains aromatic residues. Thus it would appear that inhibitory specificity lies in the residues N-terminal to P4, with the P6 position being important for all three enzymes, and that the P5 position is more important for PC1 and furin than for PC2.

2. Peptide Analogs

Reports that substrates with extended P′ residues are cleaved by furin with far higher k_{cat}/K_m values than substrates such as Boc-Arg-Val-Arg-Arg-MCA (*102*) led to the synthesis of compounds containing modified scissile bonds and extended P′ sequences. A series of deca- and dodecapeptides based on a partial sequence of the junction between the propeptide and catalytic domains of mouse PC1, containing a variety of unnatural amino acids in the P1 or P1′ positions, together with modifications of the N terminus, were tested for inhibition of PC1 and furin (*93, 94*). Kinetic studies demonstrated that these compounds were all competitive, reversible inhibitors; although all were only moderate inhibitors, with K_i's ranging from 0.8 to 10 μM for furin, and from 1.0 to 170 μM for PC1 (*94*), useful data regarding the binding pockets of the two convertases were obtained.

Replacements of the P1 Arg with ornithine (*93*) or citrulline (*94*) resulted in inhibition of PC1 and furin in the latter but not the former case. This was interpreted as implying that not only the chemical group but also the length of the P1 substituent was important to binding. Unexpectedly, considering that the sequence was based on the PC1 propeptide, substituting the P1' residue with a variety of unnatural amino acids resulted in inhibitors with greater potency for furin than PC1. The greatest difference in potency between PC1 and furin was seen when sarcosine, an N-alkylated amino acid with no side-chain, replaced Ser in the P1' position; in this instance the K_i's against PC1 and furin were 166 and 4.2 μM, respectively. However, when α-aminoisobutyric acid (where the alpha carbon bonds not involved in a peptide bond are linked to methyl groups) was in the P1' position, the K_i against PC1 was greatly decreased (to 1.2 μM), while against furin the K_i increased to 8.7 μM. These data strongly imply that, compared to furin, PC1 heavily relies upon hydrophobic contacts in the P1' position.

Various replacements of the P1' with L-Tic (as shown in the table below) yielded some of the most potent inhibitors. Omitting the glutamyl residue in P5 resulted in a sevenfold reduction in K_i against PC1, but less than a threefold reduction against furin. It thus appears that the substitution of a basic residue for an acidic residue in P5 is favorable to inhibition of both PC1 and furin. Omitting the glutamyl residue in P3' resulted in a reduced K_i against PC1 but little alteration in K_i against furin. However, interpretation of these findings is complicated by the concomitant movement of the P5 Asp into P4. The influence of an N-terminal hydrophobic Fmoc blocking group appeared to have contradictory effects, with the change in K_i against PC1 and furin ranging from 0 to 46-fold lower, depending on the P1' substitutions used. Overall, however, PC1 was more influenced by the hydrophobic N-terminal blocking group than was furin.

Example sequences used by Basak *et al.* (*94*) were as follows:

	7	6	5	4	3	2	1	−1	−2	−3	−4	−5	hFurin K_i (μM)	hPC1 K_i (μM)
	Y	K	E	R	S	K	R	S	V	Q	K	D	[Substrate]	
a]	Y	K	E	R	S	K	R	L-Tic[a]	V	Q	K	D	2.2	10.1
b]		Y	K	R	S	K	R	L-Tic	V	Q	K	D	0.79	1.4
c]	Y	K	E	R	S	K	R	L-Tic	V	K	D		2.3	4

[a] L-Tic = (3S)-1,2,3,4-tetrahydroisoquinoline-3-carboxylic acid.

Using a related peptide, also based on the PC1 propeptide sequence, but introducing a ketomethylene arginyl pseudopeptide bond, Jean *et al.* (*103*) identified a second series of inhibitors of PC1 and furin. Competitive kinetics with K_i's of 7.2 and 2.4 μM against PC1 and furin were obtained; thus, in agreement with the study cited above, this PC1-derived peptide represents a better inhibitor of furin than of PC1. These results are understandable in light of the work demonstrating that the furin and PC1 propeptides are interchangeable *in vivo* (*3*) as well as the foregoing analysis of propeptide sequences showing strong similarity between PC1 and furin in the C terminus of the propeptide.

The approach of altering a known cleavage sequence at the P-P' junction was further used to develop relatively specific PC1 inhibitors remarkably enough by using a known furin cleavage sequence. Decroly *et al.* used sequences from the furin-mediated cleavage site of the HIV-1 gp160 and human syncytial virus (*104*) as base peptides for substitution (see Table III). When gp160 was used as an *in vitro* substrate, the octapeptide analogs inhibited PC1 more potently than furin. In contrast, when the fluorogenic substrate pERTKR-MCA was used, furin was inhibited more potently than PC1, emphasizing the differences the choice of substrate can make to inhibitor studies (Table III). The substitution of an Ala for a Lys in P3 increased the potency against PC1 and furin by approximately the same amount against both substrates. However, the K_i using pERTKR-MCA of the most potent inhibitor, Dec-RKRR-ψ[CH$_2$NH]-FLGF-NH$_2$, against furin was almost 30-fold less than against PC1 (0.6 and 15 μM, respectively). *In vivo* studies

TABLE III

COMPARISON OF INHIBITION USING FLUOROGENIC VS PROTEIN SUBSTRATES [a,b]

	Inhibition of pERTKR-MCA cleavage			Inhibition of gp160 cleavage		
		Cleavage (%)			Cleavage (%)	
Peptide	Peptide (μM)	Furin	PC1	Peptide (μM)	Furin	PC1
1 Dec-RKRR-ψ[CH$_2$NH]-FLGF-NH$_2$	58	91	64	50	27	71
2 Dec-RAKR-ψ[CH$_2$NH]-FLGF-NH$_2$	62	35	30	50	3	48
3 Dec-REKR-ψ[CH$_2$NH]-AVGI-NH$_2$	61	36	44	50	5	72
4 yRSKR[β-Ala]VQKD	—	—		50	11	23
5 yRSKR[β-Cha]VQKD	—	—		50	2	23
6 yRSKR[γ-Abu]VQKD	—	—		50	5	8

[a] Adapted from (104).

[b] Not reported; Dec, decanoyl; y, D-tyrosine; β-Ala, β-alanine; β-Cha, β-cyclohexyl alanine; γ-Abu, γ-amino-butyric acid.

showed no apparent change in the ratio of cleaved to intact gp160, implying no effect against endogenous processing enzymes. Protein expression was strongly decreased, which was interpreted by the authors as indicating that these substances were toxic to the cells.

Despite the relatively low potency of PC1 inhibitors described in the preceding studies, the same approach was successfully used by Angliker to identify nanomolar inhibitors of furin (105). In the first case, the −NH− of the P−P′ bond was replaced with a methylene group using a ketomethylene pseudopeptide, ψ[COCH$_2$], whereas in the second instance a aminomethyl ketone ψ[COCH$_2$NH] group was used to insert a methylene group between the −CO− and −NH− of the scissile bond. Inhibitors of this type function by forming a tetrahedral semiketal with the active site serine. In both cases the base sequence was Dec-Arg-Val-Lys-Arg-Ala/Gly-Val-Gly-Ile, with the C terminus either amidated or methylated. The decanoyl group, introduced at the N terminus in an attempt to improve cell permeability, had little effect on the K_i against furin in vitro compared to an N-terminal acetyl group. In all instances the aminomethyl ketone series, containing an alanyl rather than a glycinyl residue at the P1′ position, were 100-fold more potent inhibitors than the ketomethylene series, with K_i values ranging from 3.4 to 7.5 nM. This was interpreted as increased binding resulting from the replacement of the P1′ amino group. Interestingly, a variation containing an P1 D-arginine had only a slightly reduced K_i compared with the L isomer, implying that the D conformation could be efficiently bound. In addition, the length of the P′ portion of the molecule could be varied from 1 to 4 residues with little effect on the K_i. When tested against subtilisin Carlberg and trypsin, no inhibition was seen against the former, whereas the latter was inhibited between 100- and 1000-fold less than furin. We speculate that combining the insertion of a methylene group into the scissile peptide bond [as used by Angliker (105)], with sequences specifically targeted to PC1 [as used by Decroly et al. (104)], would result in the generation of a potent PC1 inhibitor.

In summary, the use of uncleavable peptide derivatives has resulted in the identification of potent furin inhibitors, but has been less successful for PC1, and no data are available on the cross-inhibition of PC2 with any of these compounds. This is unfortunate, as the series of peptides available from this work could contribute greatly to the analysis of the subsite preferences for inhibitors of PC1 and PC2.

3. Combinatorial Library-Derived Inhibitors

Hexapeptide combinatorial libraries have been used in an attempt to identify inhibitors of PC1, PC2, and furin (31a, 88). Whereas the technique was remarkably successful for PC1—identifying a nanomolar inhibitor that was later shown to be the exact sequence seen in the endogenous PC1 inhibitor proSAAS (see below)—against PC2 no clear consensus was apparent. Nonetheless, valuable information in the subsite preferences of PC1 and PC2 for inhibition was gained from the analysis of a series of related acetylated and amidated hexapeptides, identified by compiling the most inhibitory amino acids for each enzyme at each of the six positions.

All of the peptides tested in the combinatorial peptide studies contained LysArg in the P2 and P1 positions; the other four positions were varied. In P3, inhibitory potency decreased in the order Val, Thr, His for both PC2 and PC1, with His substitutions resulting in a much weaker inhibitor of PC2 than PC1. Against both PC1 and PC2, Arg was much preferred to Met in P4. Although the P1–P4 sequence preferences for PC1 and PC2 are generally similar, there is a possible difference between these two convertases and furin at the P3 position, with furin having a greater preference for basic residues compared to the aliphatic valine preferred by PCs 1 and 2. In positions P5 and P6, the potency of inhibition against PC2 decreased in the order LeuLeu>LeuMet = IleIle>>LeuLys>LeuTyr>IleLys>IlePro>GluIle; for peptides tested against PC1, the order was LeuLeu>LeuMet>LeuLys> LeuTyr>>IleIle>IleLys; and against furin, LeuLys>>LeuLeu>LeuMet> LeuTyr. The last data are in line with work in our laboratory showing that basic residues in all positions are much preferred for inhibition of furin (*31a*). Whereas the P5–P6 Leu-Leu combination resulted in the greatest inhibition of both enzymes, the combination of P5–P6 IleIle produced a relatively potent PC2 inhibitor but a much weaker PC1 inhibitor. The basic residue Lys in the P5 position severely reduced the inhibitory potency against PC2, but maintained much of its potency against PC1, more so than a P5 Tyr. Against PC2, Pro in P5 and Glu in P6 were both very unfavorable. Thus, in agreement with the propeptide analysis presented earlier, a major difference between PC1 and PC2 is located at the S5 and S6 subsites.

The peptides just described all contain amidated and acetylated termini. We have found that these terminating groups contribute to inhibition of PC2, with the K_i of the hexapeptide Ac-LeuLeuArgValLysArg-NH$_2$ increasing fourfold when the terminating groups were removed (*31a*). The effect could largely be attributed to the presence of the N-terminal acetyl group: when this group was removed, the K_i of the resulting inhibitor more than doubled. Presumably, the greater hydrophobicity of the N-terminally amidated peptide derivative results in higher affinity binding to a PC2 subsite, while the increased positive charge of the unmodified peptide assists binding to furin. Against PC1, amidation and acetylation had little effect on the inhibitory potency (Cameron, unpublished observation).

We have also shown that while L-polyarginines with a chain length of six or more represent potent furin inhibitors, with K_i values ranging from 40 to 100 nM, the same compounds actually stimulate PC2, with increases in activity up to 140% relative to controls dependent on concentration (*31a*). On the other hand, polyarginines constitute relatively poor PC1 inhibitors, with a minimum K_i of 4 μM seen for all the peptides tested. Polyarginines thus appear to be promising lead compounds for the development of specific and potent furin inhibitors.

4. *Chloromethanes and Other Irreversible Small Inhibitors*

In vivo inhibition of prohormone convertases with peptidyl chloromethane derivatives was first described in the early 1990s (*106, 107*). Inhibitors of this type form irreversible complexes with the enzyme in a competitive fashion, covalently labeling the active-site His and Ser residues. The typical K_i values of chloromethanes are in the low nanomolar range, but their use *in vivo* is often compromised by their toxicity. Initially targeted toward Kex2 and furin, the chloromethane derivatives reported to date contain variations of RXK/RR, the recognition sequence of these target enzymes. Despite their toxicity, some of these compounds have proved to be valuable tools for the elucidation of mechanism and specificity, both *in vivo* and *in vitro*. Few studies have compared the specificity against different convertases, but inhibitors of this type were shown to preferentially inhibit Kex2 over serine proteases of the chymotrypsin family, or nonconvertase members of the subtilisin family. An early example of the utility of peptidyl chloromethane derivatives is provided by Hallenberger *et al.*, who demonstrated the involvement of furin in the cleavage and activation of the HIV-1 glycoprotein gp160 (*108*). Of those inhibitors tested, decanoyl (dec)-REKR-chloromethylketone (cmk) and dec-REKR-cmk potently inhibited cleavage of the HIV glycoprotein, while dec-RAIR-cmk and dec-FAKR-cmk showed insignificant inhibition.

An octapeptidyl chloromethane derivative, Ac-YEKERSKR-CH$_2$Cl, that in its corresponding MCA form was shown to be a relatively efficient substrate, when derivatized with chloromethyl ketone was transformed into a potent irreversible inhibitor of PC1 and furin (*87*). The observation that the compound bound to all three forms of PC1 (87, 71, and 66 kDa), in combination with the $K_{iapp}[I]$ of 1.6×10^6 s^{-1} M^{-1}, allowed the compound to be used as an active site titrant to estimate the concentration of active PC1. Further analysis of the kinetics of inhibition demonstrated that deacylation of the acylenzyme is the rate-limiting step for both furin and PC1.

B.   OTHER SMALL-MOLECULE INHIBITORS (ANDROGRAPHOLIDE)

One of the diterpenes of *Andrographis paniculata* (andrographolide) was shown to inhibit furin with a K_i of 200 μM, a value that could be reduced to <30 μM on succinylation of the compound. However, the other PCs tested, PC1 and PC7, were inhibited with comparable efficiency, implying a relatively nonspecific interaction (*109*).

C.   PROTEIN-BASED INHIBITORS

1. *Endogenous Proteins*

a. *7B2.* The most potent protein inhibitor of a neuroendocrine convertase is the PC2 binding protein 7B2, which contains two domains: a 21 kDa amino-terminal domain which is required for successful synthesis of active PC2, and a carboxyl-terminal domain (CT peptide), which represents a nanomolar inhibitor of PC2 (*110, 38*). Inhibition by this protein—as well as the 31-residue CT peptide (see Table IV)—is absolutely specific for PC2; PC1 is not inhibited even at micromolar concentrations (*110*). Initial mutagenesis and peptide studies showed that the peptide can be C-terminally truncated until the 18-residue peptide terminating in the KK pair is reached; however, this dibasic pair is absolutely required for inhibition (*111, 112*). Based on a comparison of this peptide sequence within vertebrate and invertebrate species, and the fact that substitution of this Val with Arg abolished inhibitory potency (*111*), we have speculated that there is a requirement for hydrophobic aliphatic residues at P4 (*113*). Inhibitory potency is distributed within a surprisingly large region of the 18-residue CT amino-terminal peptide; only two residues can be substituted by Ala without great loss in inhibitory potency (*10*). These findings help to explain why a relatively distally located sequence within this peptide, VNPYLQG, 17 residues away from the scissile bond, represents one of the most conserved regions within the 7B2 family (*114, 113*) (see Table IV). Only two residues may be removed from the amino terminus of the CT peptide without a dramatic loss of inhibitory potency (*113, 10*). The length requirement for inhibition is difficult to explain; dual binding to the VNPYLQG-recognizing exosite

TABLE IV

CT PEPTIDES IN VARIOUS SPECIES

Rat	SVNPYLQGKRLDNVVAKKSVPHFSEEEKEPE
C. elegans	SIPSSAHKVNPYLQGEPLRSMQ-KKNGKIIS
Lymnaea	SEMSDHGNPFLQGEQMD-IAAKKDPLLAKNAMHNWRLNHH
Drosophila	AGYPVMPDPRLDDAVINPFLQGDRL-PIAAKKGNLLFH

and to the substrate binding pocket might be required for tight binding to the enzyme. Alternatively, perhaps effective inhibition requires interaction of side-chain residues of the CT peptide with each other. Crystallization of PC2 together with the CT peptide will be necessary to conclusively establish why inhibition requires such a lengthy peptide.

The 7B2 CT peptide exhibits the typical kinetics of a tight-binding competitive inhibitor, resembling noncompetitive kinetics when plotted by Lineweaver–Burk analysis [Lindberg, unpublished results (*88*)]. Binding of the CT peptide is relatively rapid and long preincubation times are not required, in contrast to inhibition by serpin derivatives that exhibit much slower binding kinetics following exposure to convertase [(*115, 116*); Lindberg and Cameron, unpublished results].

Our laboratory has succeeded in generating a PC2 mutant that is no longer inhibited by the CT peptide; this was accomplished by mutating residues 242–248 to the corresponding residues in PC1 (*117*). The resulting enzyme exhibited no major differences in reactivity toward a fluorogenic substrate or toward recombinant proenkephalin, indicating that this particular mutation did not affect the residues directly bordering the scissile bond. We speculate that the mutation most likely affects the exosite binding site, i.e., the binding of VNPYLQG, rather than the binding of the residues neighboring the LysLys pair directly to the S1–S4 sites within the enzyme pocket. The converse exchange, placing PC2-specific residues within PC1, did not result in acquired susceptibility to inhibition by the CT peptide (Apletalina, unpublished results); this result might be expected if two-site binding is required for inhibition and only one site was created by the mutation.

b. *proSAAS.* The low specific activity of purified recombinant PC1 led us to suspect that a "helper" (a nonclassical protein chaperone that assists in enzyme maturation) protein equivalent to 7B2 existed for PC1, yet coimmunoprecipitation experiments using ^{35}S-labeled proteins synthesized by AtT-20 cells—a neuroendocrine cell line rich in PC1—consistently failed to reveal such a protein (Lindberg, unpublished results). It was only when Fricker and co-workers undertook a systematic investigation of novel peptides in the brains of mutant mice lacking carboxypeptidase E that an endogenous PC1-interacting protein was identified (*118*). This 24 kDa protein, proSAAS, located primarily in neuroendocrine tissues, is secreted via the regulated pathway in AtT-20 cells and is processed at dibasic cleavage sites (*118*). Further work in our laboratory has revealed that the inhibitory portion of intact proSAAS can be almost entirely reduced to a hexapeptide, LeuLeuArgValLysArg, located 16 residues from the C terminus (*89*). Remarkably, this sequence is identical to a peptide previously identified by combinatorial library screening to be a low-nanomolar inhibitor of PC1 (*88*).

The SAAS protein shares no sequence homology with 7B2, but exhibits some structural similarity. In both instances the inhibitory portion is located in the carboxy-terminal 40 residues of the protein, and cleavage of either protein at furin recognition sites creates peptides that, as inhibitors against their target enzymes, are as potent as the original protein. Peptides derived from 7B2 and proSAAS are tight binding, competitive inhibitors (the K_i values of SAAS peptides range from 1.5 to 7.5 nM, depending on chain length) and inhibition is abolished by the action of carboxypeptidase B [(89); Cameron, unpublished results]. The inhibitory potency of proSAAS fragments toward PC1 is 600–1500 times greater than toward PC2. As well as sharing a similarly situated furin cleavage site, both 7B2 and proSAAS contain proline-rich sequences located in the middle of the proteins. It is possible that proSAAS may offer biosynthetic assistance to proPC1 maturation in a similar manner to the proPC2/7B2 interaction; however, the fact that enzymatically active PC1 can be obtained from cells not expressing proSAAS indicates that this interaction is not absolutely required. It is noteworthy that the P5 and P6 residues of the nanomolar PC1 hexapeptide inhibitor, LeuLeuArgValLysArg, do not resemble any sequence within the PC1 propeptide, implying that the mechanism of inhibition of the complete propeptide, as described by Lazure's group (99), must rely on an alternative binding mechanism than that of proSAAS-derived peptides.

2. Bioengineered Proteins

a. *Antitrypsin Portland.* Bioengineered protein-based inhibitors have been successfully developed against convertases, but to date all have been directed against furin. The best-characterized inhibitor of this category is an α1-antitrypsin variant, designated α1-antitrypsin Portland (the acronym, α1-PDX, is taken from the three-letter code for Portland International Airport). This serpin was engineered to contain a copy of the minimal furin recognition sequence (LEAIMPS^{359}→LERIMRS^{359}). Reported K_i's of α1-PDX are 1.4 nM against furin, 206 nM against PC1, and 1000 nM against PC2 (116). *In vivo* studies with AtT-20 cells that contain secretory granules demonstrated the ability of PC1 but not PC2 to partially (~10%) cleave α1-PDX into an inactive fragment, presumably at the engineered site (119). Studies with PC1 expressed in constitutively secreting BSC40 cells, and with PC2 in both BSC40 and AtT-20 cells showed that α1-PDX was not inhibitory to POMC cleavage most likely because of targeting issues (119). These data demonstrate distinct differences between α1-PDX binding to PC1 and PC2 and are in agreement with other data suggesting that PC1 is more closely related to furin.

b. *Others (Ovomucoid, Proteinase Inhibitor 8, α_2-Macrogloblulin).* The third domain of turkey ovomucoid has been engineered (KPACTLE[19] KP**RCKR**E[19]) in an attempt to increase its reactivity with furin (*120*); however the equilibrium constant of $1.1 \times 10^7 \ M^{-1}$ was representative of a moderate, rather than a potent, inhibitor. No data have been published regarding the specificity toward convertases other than furin, although limited collaborative studies with PC1 provided preliminary data of similarly modest inhibitory potency (Laskowski and Lindberg, unpublished results).

Two other proteins have been engineered to generate furin inhibitors: the ovalbumin-type serpin human proteinase inhibitor 8 (PI8) and the general protease inhibitor, α_2-macroglobulin (α_2M) (*115, 121*). Although mutation of the reactive loops of these proteins into furin recognition sequences effectively generated potent furin inhibitors (with picomolar inhibitory potency in the case of PI8), the activity of these proteins against related convertases was not tested. It would be interesting to probe the effectiveness of PI8 against PC1 and PC2 to provide additional data on the ability of serpins to cross-inhibit the various members of the convertase family. It is remarkable that serpins, which normally inhibit members of the chymotrypsin-like enzyme superfamily, can, with minimal modification, potently cross-inhibit eukaryotic subtilisin-like enzymes, which are thought to have evolved independently.

D. Conclusions on Inhibitory Specificity

The data presented in the studies cited above collectively indicate that for inhibition, the consensus sequence RXKR is preferred for PC1 and PC2 as well as for furin. At the P3 position, small uncharged and preferably aliphatic residues are preferred by PC1, whereas PC2 only exhibits this preference in lower species. The P5 residue is less well conserved for both PC1 and PC2, but again, aliphatic residues, especially Leu, seem to be the most strongly bound. At the P5 position, there is evidence that the shape and hydrophobicity of the side chain plays an important role, with Leu being preferred to Ile. Whereas the P6 residue of PC1's propeptide closely resembles that of furin, with basic resides strongly conserved, PC2s generally contain aromatic residues at this position. However, our combinatorial peptide studies indicate that PC1 greatly prefers Leu to basic residues at the P6 position, implying that propeptides and small peptides exhibit distinctly different modes of binding. With the exception of *in vitro* studies of the endogenous 7B2 CT inhibitor, which appears to be a unique inhibitor because of its unprecedented length requirement, little systematic replacement analysis of inhibitory short peptides has been done with PC2 to date, and this is an area requiring further study.

IV. General Summary and Conclusions

The preceding review has attempted to focus on the catalytic pockets of PC1 and PC2. Our intent has been not just to define the differences in specificity that distinguish these two interesting neuroendocrine enzymes, but also to provide an initial algorithm for use in the prediction of cleavage of new neuroendocrine precursor proteins, which are sure to emerge through the exponential growth of sequence information via human genome sequencing efforts. Although our understanding of cleavage site preferences by PC1 and PC2 is clearly incomplete, largely because most studies to date have focused on furin and PC1, we can nonetheless draw a limited number of conclusions for cleavage by both enzymes. Table V presents an overview of residue preferences for substrate cleavage and for inhibition; it is obvious that both PC1 and PC2 share a number of residue preferences in the vicinity of the cleavage site. These include a requirement for a P1 Arg or Lys with a strong preference for Arg, a preference for other basic residues at P4 or P6, a strong preference for Lys at P2, and a dislike for bulky aliphatic residues at P1'. Preferences at P5 and P6 cannot be distinguished from inspections of precursors, but have not actually been systematically studied *in vitro*.

Most sites that are cleaved by PC1 will also be cleaved by PC2; few examples of sites only cleaved by PC1 have been described to date. This redundant function possibly provides a mechanism for ensuring that critical cleavages will indeed occur. In addition, PC2 will also accept residues that are distinctly disfavored by PC1, such as aromatics in the P1' position. These data suggest that the binding pocket of PC2 may be more accommodating to many types of residues and fits with the idea that the less specialized PC2 is evolutionarily the older enzyme (*122*). Indeed, it has not been possible thus far to identify a PC1 equivalent in *Drosophila, C. elegans,* or *Lymnaea* [although *Aplysia* (*123*) and hydra (*124*) appear to express a PC1-like enzyme]. In the former invertebrates perhaps invertebrate furins and PC2s act to fill the role of PC1. It is interesting to note that despite the fact that PC2 appears to be more tolerant of variation in substrates, all studies carried out to date indicate extensive discrimination at various subsites within inhibitors. Indeed, the only nanomolar PC2 inhibitor—the CT peptide—may require complex interaction involving multiple sites within the binding pocket, since alanine scanning has shown that 14 residues are necessary for inhibition. This paradoxical requirement for extensive amino-terminal residue interaction in inhibitors, but not in substrates, may hold for all peptide inhibitors of PC2, as no potent hexapeptides could be identified during combinatorial peptide library screens with PC2. By contrast, PC1 may discriminate less extensively for inhibitors in terms of length; for example, length beyond six

TABLE V

<small>SUBSTRATE AND INHIBITOR RESIDUE PREFERENCES FOR PC1 AND PC2</small>

	PC1	PC2
A. Substrate preferences		
P6	Arg preferred? (few studies)	Same as PC1
P5	No preference noted (few studies)	Same as PC1
P4	Arg preferred depending on substrate. Small side-chains unlikely	Same as PC1
P3	No preference except no Cys	Same as PC1
P2	Most often Arg or Lys	Same as PC1, but PC2 shows less preference for Arg
P1	Arg or Lys required, Arg preferred	Same as PC1, but PC2 shows less preference for Arg
P1′	Small side chain preferred. Not aliphatic. Rarely aromatic. Not Pro	Same as PC1, but less preference (Tyr acceptable, Pro tolerated)
P2′	Small side chain preferred. Cys excluded. Pro and charged residues unlikely	Same as PC1
B. Peptide inhibitor preferences		
P6	Aliphatic hydrophobic; Leu much better than Ile; not charged	Same as PC1, but PC2 tolerates Ile better than PC1
P5	As P6	Same as P5. Val in 7B2 CT
P4	Arg greatly preferred	Val required, Arg not tolerated in 7B2 CT Arg preferred in combinatorial library screens
P3	Aliphatic hydrophobic preferred	Same as PC1
P2	Basic, Arg preferred	Same as PC1, but PC2 shows less preference for Arg. Lys in 7B2 CT
P1	As P2	Same as P2
P1′	Aliphatic hydrophobic (Leu in SAAS CT)	Ser in rat 7B2 CT but generally unconserved with no apparent preference
P2′	Charged (Glu in SAAS CT)	Val in 7B2 CT

residues does not appear to play a significant role in the inhibition of this enzyme by proSAAS (*89*). Whether PC1 genuinely prefers longer substrates or not is not yet clear, and more work needs to be done to establish whether the binding pocket of PC1 actually contains extended binding determinants that contribute to effective cleavage of longer substrate. An additional area of interest is the effect of carboxyl terminal truncation of PC1 on substrate discrimination; the potentially broader spectrum of activity (*36*) should be further substantiated using a variety of different substrates.

Also noteworthy is the fact that for PC1, sequences that constitute poor substrates—such as peptides containing aliphatic hydrophobics at the P1′ position and a charged residue at P2′—are precisely those that make excellent inhibitors, implying that not binding, but catalytic events are adversely affected by the presence of these P′ residues. This conclusion cannot be generalized to PC2, in part because PC2 has so few residue preferences determined to date, and because so much amino-terminal length is apparently required to build a potent PC2 inhibitor. These intriguing findings suggesting profound differences in the architecture of the catalytic pockets of the two neuroendocrine enzymes will represent an interesting area for further mapping studies. The use of active-site photoaffinity reagents coupled to proSAAS or to the 7B2 CT peptide represents a promising direction in this regard.

Another interesting area for future study is the elucidation of the molecular mechanism of the interaction of 21 kDa 7B2 with proPC2. Because this phenomenon cannot be duplicated with purified reagents, it is quite difficult to study. One preliminary idea is that 21 kDa 7B2 might protect proPC2 from an unfavorable conformational rearrangement at the relatively low pH of the Golgi; the adoption of this unfavorable conformation might somehow result in the unproductive propeptide cleavage event associated with a lack of enzymatic activity. The yeast PBN1 gene product has been shown to be required for proper maturation of the zymogen to a vacuolar hydrolase, protease B (125); although this is a somewhat different mechanism, it illustrates the use of helper proteins in zymogen maturation. However, there are few parallels in the prokaryotic proteinase field for zymogen conformational rearrangement; one possible exception is the maturation protein PrtM, a peptidyl proline isomerase, which is required for SK11 serine proteinase maturation in *Lactococcus* (126). Until direct evidence for multiple conformations for proPC2 is obtained, the idea of conformational rearrangement must continue to be viewed as highly speculative. The very recent discovery of proSAAS, a binding protein for PC1 (118), highlights the importance of elucidating the biochemical mechanisms of protein–protein interaction in the secretory pathway. Further studies will determine whether proSAAS and PC1 interact in a similar or in a different manner to the 7B2/proPC2 interaction. Most likely there will be both similarities and differences in the interaction—similarities due to the potential structural homology of proSAAS and 7B2 (89), and differences because 7B2 interacts with a zymogen, whereas proSAAS most likely interacts with a mature form of PC1. The question of whether similar binding proteins exist for other convertases—especially for furin—and how these function to direct convertase activity in a temporally correct fashion is sure to represent an exciting area for further development in the coming years.

ACKNOWLEDGMENTS

This work was supported by DK49703 and DA05084; IL was supported by a Research Career Development Award from NIDA.

REFERENCES

1. Muller, L., and Lindberg, I. (1999). *In* "Progress in Nucleic Acids Research" (K. Moldave, ed.). Academic Press, San Diego, CA.
2. Inouye, M. (1991). *Enzyme* **45**, 314.
2a. Muller, L.*, Cameron, A.*, Apletalina, E., Fortenberry, Y., and Lindberg, I. (2000). *J. Biol. Chem.* **275**, 39213. (*These authors contributed equally to the work.)
3. Zhou, A., Paquet, L., and Mains, R. E. (1995). *J. Biol. Chem.* **270**, 21509.
4. Taylor, N. A., Shennan, K. I. J., Cutler, D. F., and Docherty, K. (1997). *Biochem. J.* **312**, 367.
5. Creemers, J. W., Usac, E. F., Bright, N. A., Van de Loo, J. W., Jansen, E., Van de Ven, W. J. M., and Hutton, J. C. (1996). *J. Biol. Chem.* **271**, 25284.
6. Shinde, U. P., Liu, J. J., and Inouye, M. (1997). *Nature* **389**, 520.
7. Brenner, C., and Fuller, R. S. (1992). *Proc. Natl. Acad. Sci. USA* **89**, 922.
8. Benjannet, S., Mamarbachi, A. M., Hamelin, J., Savaria, D., Munzer, J. S., Chretien, M., and Seidah, N. S. (1998). *FEBS Lett.* **428**, 37.
9. Zhu, X., Muller, L., Mains, R. E., and Lindberg, I. (1998). *J. Biol. Chem.* **273**, 1158.
10. Apletalina, E. V., Juliano, M. A., Juliano, L., and Lindberg, I. (2000). *Biochem. Biophys. Res. Commun.* **267**, 940.
11. Nakayama, K. (1997). *Biochem. J.* **327(Pt 3)**, 625.
12. Gluschankof, P., and Fuller, R. S. (1994). *EMBO J.* **13**, 2280.
13. Zhou, A., Martin, S., Lipkind, G., LaMendola, J., and Steiner, D. F. (1998). *J. Biol. Chem.* **273**, 11107.
14. Lusson, J., Benjannet, S., Hamelin, J., Savaria, D., and Chretien, M. (1997). *Biochem. J.* **326**, 737.
15. Jutras, I., Seidah, N. G., Reudelhuber, T. L., and Brechler, V. (1997). *J. Biol. Chem.* **272**, 15184.
16. Siezen, R. J., Creemers, J. W. M., and Van de Ven, W. J. M. (1994). *Eur. J. Biochem.* **222**, 255.
17. Lipkind, G., Gong, Q., and Steiner, D. F. (1995). *J. Biol. Chem.* **270**, 13277.
18. Lipkind, G. M., Zhou, A., and Steiner, D. F. (1998). *Proc. Natl. Acad. Sci. USA* **95**, 7310.
19. Wilcox, C. A., and Fuller, R. S. (1991). *J. Cell Biol.* **115**, 297.
20. Smeekens, S. T., and Steiner, D. F. (1990). *J. Biol. Chem.* **265**, 2997.
21. Seidah, N. G., Gaspar, L., Mion, P., Marcinkiewicz, M., Mbikay, M., and Chretien, M. (1990). *DNA Cell Biol.* **9**, 415.
22. Zhou, Y., and Lindberg, I. (1993). *J. Biol. Chem.* **268**, 5615.
23. Rufaut, N. W., Brennan, S. O., Hakes, D. J., Dixon, J. E., and Birch, N. P. (1993). *J. Biol. Chem.* **268**, 20291.
24. Jean, F., Basak, A., Rondeau, N., Benjannet, S., Hendy, G. N., Seidah, N. G., Chretien, M., and Lazure, C. (1993). *Biochem. J.* **292**, 891.
25. Boudreault, A., Gauthier, D., Rondeau, N., Savaria, D., Seidah, N. G., Chretien, M., and Lazure, C. (1998). *Protein. Expr. Purif.* **14**, 353.
26. Fahnestock, M., and Zhu, W. (1999). *DNA Cell Biol.* **18**, 409.

27. Hwang, J. R., Siekhaus, D. E., Fuller, R. S., Taghert, P. H., and Lindberg, I. (2000). *J. Biol. Chem.* **275,** 17886.
28. Azaryan, A. V., Krieger, T. J., and Hook, V. Y. (1995). *J. Biol. Chem.* **270,** 8201.
29. Bruzzaniti, A., Marx, R., and Mains, R. E. (1999). *J. Biol. Chem.* **274,** 24703.
30. Xu, H., and Shields, D. D. (1994). *J. Biol. Chem.* **269,** 22875.
31. Nakayama, K., Watanabe, T., Nakagawa, T., Kim, W. S., Nagahama, M., Hosaka, M., Hatsuzawa, K., Kondoh-Hashiba, K., and Murakami, K. (1992). *J. Biol. Chem.* **267,** 16335.
31a. Cameron, A., Appel, J., Houghten, R. A., and Lindberg, I. (2000). *J. Biol. Chem.* **275,** 36741.
32. Bravo, D. A., Gleason, J. B., Sanchez, R. I., Roth, R. A., and Fuller, R. S. (1994). *J. Biol. Chem.* **269,** 25830.
33. Zhou, Y., and Lindberg, I. (1994). *J. Biol. Chem.* **269,** 18408.
34. Zhou, Y., Rovere, C., Kitabgi, P., and Lindberg, I. (1995). *J. Biol. Chem.* **270,** 24702.
35. Coates, L. C., and Birch, N. P. (1997). *J. Neurochem.* **68,** 828.
36. Coates, L. C., and Birch, N. P. (1998). *J. Neurochem.* **70,** 1670.
37. Shen, F.-S., Lindberg, I., and Seidah, N. G. (1993). *J. Biol. Chem.* **268,** 24910.
38. Lindberg, I., Van den Hurk, W. H., Bui, C., and Batie, C. J. (1995). *Biochemistry* **34,** 5486.
39. Hutton, J. C. (1990). *Curr. Opin. Cell Biol.* **2,** 1131.
40. Shennan, K. I. J., Smeekens, S. P., Steiner, D. F., and Docherty, K. (1991). *FEBS Lett.* **284,** 277.
41. Braks, J. A. M., and Martens, G. J. M. (1994). *Cell* **78,** 263.
42. Zhu, X., and Lindberg, I. (1995). *J. Cell Biol.* **129,** 1641.
43. Lamango, N. S., Zhu, X., and Lindberg, I. (1996). *Arch. Biochem. Biophys.* **330,** 238.
44. Johanning, K., Juliano, M. A., Juliano, L., Lazure, C., Lamango, N. S., Steiner, D. F., and Lindberg, I. (1998). *J. Biol. Chem.* **273,** 22672.
45. Shennan, K. I. J., Taylor, N. A., Jermany, J. L., Matthews, G., and Docherty, K. (1995). *J. Biol. Chem.* **270,** 1402.
46. Bailyes, E. M., Shennan, K. I. J., Usac, E. F., Arden, S. D., Guest, P. C., Docherty, K., and Hutton, J. C. (1995). *Biochem. J.* **309,** 587.
47. Molloy, S. S., Bresnahan, P. A., Leppla, S. H., Klimpel, K. R., and Thomas, G. (1992). *J. Biol. Chem.* **267,** 16396.
48. Munzer, J. S., Basak, A., Zhong, M., Mamarbachi, A., Hamelin, J., Savaria, D., Lazure, C., Benjannet, S., Chretien, M., and Seidah, N. G. (1997). *J. Biol. Chem.* **272,** 19672.
49. Basak, A., Toure, B. B., Lazure, C., Mbikay, M., Chretien, M., and Seidah, N. G. (1999). *Biochem. J.* **343(Pt 1),** 29.
50. Sucic, J. F., Moehring, J. M., Inocencio, N. M., Luchini, J. W., and Moehring, T. J. (1999). *Biochem. J.* **339(Pt 3),** 639.
51. Cheng, D., Espenshade, P. J., Slaughter, C. A., Jaen, J. C., Brown, M. S., and Goldstein, J. L. (1999). *J. Biol. Chem.* **274,** 22805.
52. Anderson, E. D., VanSlyke, J. K., Thulin, C. D., Jean, F., and Thomas, G. (1997). *EMBO J.* **16,** 1508.
53. Goodman, L. J., and Gorman, C. M. (1994). *Biochem. Biophys. Res. Commun.* **201,** 795.
54. Matthews, G., Shennan, K. I. J., Seal, A. J., Taylor, N. A., Colman, A., and Docherty, K. (1994). *J. Biol. Chem.* **269,** 588.
55. Muller, L., Zhu, X., and Lindberg, I. (1997). *J. Cell Biol.* **139,** 625.
56. Lamango, N. S., Apletalina, E., Liu, J., and Lindberg, I. (1999). *Arch Biochem. Biophys.* **362,** 275.
57. Powner, D., and Davey, J. (1998). *Mol. Celll. Biol.* **18,** 400.

58. Seidah, N. G., Benjannet, S., Pareck, S., Savaria, D., Hamelin, J., Goulet, B., Laliberte, J., Lazure, C., Chretien, M., and Murphy, R. A. (1996). *Biochem. J.* **314,** 951.
59. Breslin, M. B., Lindberg, I., Benjannet, S., Mathis, J. P., Lazure, C., and Seidah, N. G. (1993). *J. Biol. Chem.* **268,** 27084.
60. Johanning, K., Mathis, J. P., and Lindberg, I. (1996). *J. Neurochem.* **66,** 898.
61. Johanning, K., Mathis, J. P., and Lindberg, I. (1996). *J. Biol. Chem.* **271,** 27871.
62. Dupuy, A., Lindberg, I., Zhou, Y., Akil, H., Lazure, C., Chretien, M., Seidah, N. G., and Day, R. (1994). *FEBS Lett.* **337,** 60.
63. Day, R., Lazure, C., Basak, A., Boudreault, A., Limperis, P., Dong, W., and Lindberg, I. (1998). *J. Biol. Chem.* **273,** 829.
64. Benjannet, S., Rondeau, N., Day, R., Chretien, M., and Seidah, N. G. (1991). *Proc. Natl. Acad. Sci. USA* **88,** 3564.
65. Thomas, L., Leduc, R., Thorne, B. A., Smeekens, S. P., Steiner, D., and Thomas, G. (1991). *Proc. Natl. Acad. Sci. USA* **88,** 5297.
66. Zhou, A., Bloomquist, B. T., and Mains, R. E. (1993). *J. Biol. Chem.* **268,** 1763.
67. Zhou, A., and Mains, R. E. (1994). *J. Biol. Chem.* **269,** 17440.
68. Viale, A., Ortola, C., Hervieu, G., Furuta, M., Barbero, P., Steiner, D. F., Seidah, N. G., and Nahon, J. L. (1999). *J. Biol. Chem.* **274,** 6536.
69. Rouille, Y., Martin, S., and Steiner, D. F. (1995). *J. Biol. Chem.* **270,** 26488.
70. Dhanvantari, S., Seidah, N. G., and Brubaker, P. L. (1996). *Mol. Endocrinol.* **10,** 342.
71. Rouille, Y., Bianchi, M., Irminger, J. C., and Halban, P. A. (1997). *FEBS Lett.* **413,** 119.
72. Rouille, Y., Kantengwa, S., Irminger, J. C., and Halban, P. A. (1997). *J. Biol. Chem.* **72,** 32810.
73. Dhanvantari, S., and Brubaker, P. L. (1998). *Endocrinology* **139,** 1630.
74. Brakch, N., Rist, B., Beck-Sickinger, A. G., Goenaga, J., Wittek, R., Burger, E., Brunner, H. R., and Grouzmann, E. (1997). *Biochemistry* **36,** 16309.
75. Seidah, N. G., and Chretien, M. (1999). *Brain Res.* **848,** 45.
76. Devi, L. (1991). *FEBS Lett.* **280,** 189.
77. Rholam, M., Brakch, N., Germain, D., Thomas, D. Y., Fahy, C., Boussetta, H., Boileau, G., and Cohen, P. (1995). *Eur. J. Biochem.* **227,** 707.
78. Brakch, N., Galanopoulou, A. S., Patel, Y. C., Boileau, G., and Seidah, N. G. (1995). *FEBS Lett.* **362,** 143.
79. Rholam, M., Nicolas, P., and Cohen, P. (1986). *FEBS Lett.* **207,** 1.
80. Bek, E., and Berry, R. (1990). *Biochemistry* **29,** 178.
81. Rholam, M., Cohen, P., Brakch, N., Paolillo, L., Scatturin, A., and Di Bello, C. (1990). *Biochem. Biophys. Res. Commun.* **168,** 1066.
82. Di Bello, C., Simonetti, M., Dettin, M., Paolillo, L., D'Aurla, G., Falcigno, L., Saviano, M., Scatturin, A., Vertuani, G., and Cohen, P. (1995). *J. Pept. Sci.* **1,** 251.
83. Brakch, N., Rholam, M., Simonetti, M., and Cohen, P. (2000). *Eur. J. Biochem.* **267,** 1626.
84. de Lecea, L., Kilduff, T. S., Peyron, C., Gao, X., Foye, P. E., Danielson, P. E., Fukuhara, C., Battenberg, E. L., Gautvik, V. T., Bartlett II, F. S., Frankel, W. N., van den Pol, A. N., Bloom, F. E., Gautvik, K. M., Sutcliffe, J. G. (1998). *Proc. Natl. Acad. Sci. USA* **95,** 322.
85. Sakurai, T., Amemiya, A., Ishii, M., Matsuzaki, I., Chemelli, R. M., Tanaka, H., Williams, S. C., Richardson, J. A., Kozlowski, G. P., Wilson, S., Arch, J. R., Buckingham, R. E., Haynes, A. C., Carr, S. A., Annan, R. S., McNulty, D. E., Liu, W. S., Terrett, J. A., Elshourbagy, N. A., Bergsma, D. J., Yanagisawa, M. (1998). *Cell* **92,** 573.
86. Lindberg, I., Lincoln, B., and Rhodes, C. J. (1992). *Biochem. Biophys. Res. Commun.* **18,** 1.
87. Jean, F., Boudreault, A., Basak, A., Seidah, N. G., and Lazure, C. (1995). *J. Biol. Chem.* **270,** 19225.

88. Apletalina, E., Appel, J., Lamango, N. S., Houghten, R. A., and Lindberg, I. (1998). *J. Biol. Chem.* **273,** 26589.
89. Cameron, A., Fortenberry, Y., and Lindberg, I. (2000). *FEBS Lett.* **473,** 135.
90. Wang, W., and Beinfeld, M. C. (1997). *Biochem. Biophys. Res. Commun.* **231,** 149.
91. Brakch, N., Dettin, M., Scarinci, C., Seidah, N. G., and Di Bello, C. (1995). *Biochem. Biophys. Res. Commun.* **213,** 356.
92. Ledgerwood, E. C., Brennan, S. O., Birch, N. P., and George, P. M. (1996). *Biochem. Mol. Biol. Int.* **39,** 1167.
93. Basak, A., Jean, F., Seidah, N. G., and Lazure, C. (1994). *Int. J. Pept. Protein Res.* **44,** 253.
94. Basak, A., Schmidt, C., Ismail, A. A., Seidah, N. G., Chretien, M., and Lazure, C. (1995). *Int. J. Pept. Protein Res.* **46,** 228.
95. Lazure, C., Gauthier, D., Jean, F., Boudreault, A., Seidah, N. G., Bennett, H. P., and Hendy, G. N. (1998). *J. Biol. Chem.* **273,** 8572-80.
96. Siezen, R., and Leunissen, J. A. (1997). *Protein Sci.* **6,** 501.
97. Rockwell, N. C., and Fuller, R. S. (1998). *Biochem. J.* **37,** 3386.
98. Shinde, U., and Inouye, M. (2000). *Semin. Cell Dev. Biol.* **11,** 35.
99. Boudreault, A., Gauthier, D., and Lazure, C. (1998). *J. Biol. Chem.* **273,** 31574.
100. Zhong, M., Munzer, J. S., Basak, A., Benjannet, S., Mowla, S. J., Decroly, E., Chretien, M., and Seidah, N. G. (1999). *J. Biol. Chem.* **274,** 33913.
101. Basak, A., Gauthier, D., Seidah, N. G., and Lazure, C. (1997). *In* "Peptides: Frontiers of Peptide Science" (J. P. Tam and P. T. P. Kauyama, eds.), *Proceedings of the XVth American Peptide Symposium.* Kluwer Academic Publishers, Dordrecht, The Netherlands.
102. Angliker, H., Neumann, U., Molloy, S. S., and Thomas, G. (1995). *Anal Biochem.* **224,** 409.
103. Jean, F., Basak, A., DiMaio, J., Seidah, N. G., and Lazure, C. (1995). *Biochem. J.* **307(Pt 3),** 689.
104. Decroly, E., Vandenbranden, M., Ruysschaert, J. M., Cogniaux, J., Jacob, G. S., Howard, S. C., Marshall, G., Kompelli, A., Basak, A., Jean, F., and others (1994). *J. Biol. Chem.* **269,** 12240.
105. Angliker, H. (1995). *J. Med. Chem.* **38,** 4014.
106. Stieneke-Grober, A., Vey, M., Angliker, H., Shaw, E., Thomas, G., Roberts, C., Klenk, H. D., and Garten, W. (1992). *EMBO J.* **11,** 2407.
107. Angliker, H., Wikstrom, P., Shaw, E., Brenner, C., and Fuller, R. S. (1993). *Biochem. J.* **293(Pt 1),** 75.
108. Hallenberger, S., Bosch, V., Angliker, H., Shaw, E., Klenk, H. D., and Garten, W. (1992). *Nature* **360,** 358.
109. Basak, A., Cooper, S., Roberge, A. G., Banik, U. K., Chretien, M., and Seidah, N. G. (1999). *Biochem. J.* **338(Pt 1),** 107.
110. Martens, G. J. M., Braks, J. A. M., Eib, D. W., Zhou, Y., and Lindberg, I. (1994). *Proc. Natl. Acad. Sci. USA* **91,** 5784.
111. Van Horssen, A. M., Van den Hurk, W. H., Bailyes, E. M., Hutton, J. C., Martens, G. J. M., and Lindberg, I. (1995). *J. Biol. Chem.* **270,** 14292.
112. Zhu, X., Rouille, Y., Lamango, N. S., Steiner, D. F., and Lindberg, I. (1996). *Proc. Natl. Acad. Sci. USA* **93,** 4919.
113. Lindberg, I., Tu, B., Muller, L., and Dickerson, I. (1998). *DNA Cell Biol.* **17,** 727.
114. Spijker, S., Smit, A. B., Martens, G. J., and Geraerts, W. P. (1997). *J. Biol. Chem.* **272,** 4116.
115. Dahlen, J. R., Jean, F., Thomas, G., Foster, D. C., and Kisiel, W. (1998). *J. Biol. Chem.* **273,** 1851.
116. Jean, F., Stella, K., Thomas, L., Liu, G., Xiang, Y., Reason, A. J., and Thomas, G. (1998). *Proc. Natl. Acad. Sci. USA* **95,** 7293.
117. Apletalina, E. V., Muller, L., and Lindberg, I. (2000). *J. Biol. Chem.* **275,** 14667.

118. Fricker, L. D., McKinzie, A. A., Sun, J., Curran, E., Qian, Y., Yan, L., Patterson, S. D., Courchesne, P. L., Richards, B., Levin, L., Mzhavia, N., Devi, L. A., and Douglass, J. (2000). *J. Neurosci.* **20,** 639.
119. Benjannet, S., Savaria, D., Laslop, A., Munzer, J. S., Chretien, M., Marcinkiewicz, M., and Seidah, N. G. (1997). *J. Biol. Chem.* **272,** 26210.
120. Lu, W., Zhang, W., Molloy, S. S., Thomas, G., Ryan, K., Chiang, Y., Anderson, S., and Laskowski, Jr., M. (1993). *J. Biol. Chem.* **268,** 14583.
121. Van Rompaey, L., Ayoubi, T., Van De Ven, W., and Marynen, P. (1997). *Biochem. J.* **326(Pt 2),** 507.
122. Rouille, Y., Duguay, S. J., Lund, K., Furuta, M., Gong, Q., Lipkind, G., Oliva, A. A. J., Chan, S. J., and Steiner, D. F. (1995). *Front. Neuroendocrinol.* **16,** 322.
123. Gorham, E. L., Nagle, G. T., Smith, J. S., Shen, H., and Kurosky, A. (1996). *DNA Cell Biol.* **15,** 339.
124. Chan, S. J., Oliva, A. A. Jr., LaMendola, J., Grens, A., Bode, H., and Steiner, D. F. (1992). *Proc. Natl. Acad. Sci. USA* **89,** 6678.
125. Naik, R. R., and Jones, E. W. (1998). *Genetics* **149,** 1277.
126. Vos, P., van Asseldonk, M., van Jeveren, F., Siezen, R., Simons, G., and de Vos, W. M. (1989). *J. Bacteriol.* **171,** 2795.
127. Rothenberg, M. E., Eilertson, C. D., Klein, K., Zhou, Y., Lindberg, I., McDonald, J., K. Mackin, R. B., and Noe, R. B. (1995). *J. Biol. Chem.* **270,** 10136.
128. Furuta, M., Carroll, R., Martin, S., Swift, H. H., Ravazzola, M., Orci, L., and Steiner, D. F. (1998). *J. Biol. Chem.* **273,** 1.
129. Irminger, J. C., Meyer, K., and Halban, P. (1996). *Biochem. J.* **320(Pt 1),** 11.
130. Kaufmann, J. E., Irminger, J. C., Mungall, J., and Halban, P. A. (1997). *Diabetes* **46,** 978.
131. Bailyes, E., Shennan, K. I. J., Seal, A. J., Smeeklens, S. P., Steiner, D. F., Hutton, J. C., and Docherty, K. (1992). *Biochem. J.* **285,** 391.
132. Friedman, T. C., Loh, Y. P., Cawley, N. X., Birch, N. P., Huang, S. S., Jackson, I. M., and Nillni, E. A. (1995). *Endocrinology* **136,** 4462.
133. Nillni, E. A., Friedman, T. C., Todd, R. B., Birch, N. P., Loh, Y. P., and Jackson, I. M. (1995). *J. Neurochem.* **65,** 2462.
134. Schaner, P., Todd, R. B., Seidah, N. G., and Nillni, E. A. (1997). *J. Biol. Chem.* **272,** 19958.
135. Nillni, E. A. (1999). *Endocrine* **10,** 185.
136. Dickinson, C. J., Sawada, M., Guo, Y. J., Finniss, S., and Yamada, T. (1995). *J. Clin. Invest* **96,** 1425.
137. Rovere, C., Barbero, P., and Kitabgi, P. (1996). *J. Biol. Chem.* **271,** 11368.
138. Wang, W., Birch, N. P., and Beinfeld, M. C. (1998). *Biochem. Biophys. Res. Commun.* **248,** 538.
139. Paquet, L., Zhou, A., Chang, E. Y., and Mains, R. E. (1996). *Mol. Cell Endocrinol.* **120,** 161.
140. Paquet, L., Massie, B., and Mains, R. E. (1996). *J. Neurosci.* **16,** 964.
141. Gabreels, B. A., Swaab, D. F., de Kleijn, D. P., Dean, A., Seidah, N. G., Van de Loo, J. W., Van de Ven, W. J., Martens, G. J., and Van Leeuwen, F. W. (1998). *J. Clin. Endocrinol. Metab.* **83,** 4026.
142. Galanopoulou, A. S., Kent, G., Rabbani, S. N., Seidah, N. G., and Patel, Y. C. (1993). *J. Biol. Chem.* **268,** 6041.
143. Galanopoulou, A. S., Seidah, N. G., and Patel, Y. C. (1995). *Biochem. J.* **311,** 111.
144. Benjannet, S., Reudelhuber, T., Mercure, C., Rondeau, N., Chretien, M., and Seidah, N. G. (1992). *J. Biol. Chem.* **267,** 11417.
145. Laslop, A., Weiss, C., Savaria, D., Eiter, C., Tooze, S. A., Seidah, N. G., and Winkler, H. (1998). *J. Neurochem.* **70,** 374.

Section III

Other Proteases That Cleave Proteins

12

Self-Processing of Subunits of the Proteasome

ERIKA SEEMÜLLER • PETER ZWICKL •
WOLFGANG BAUMEISTER

Max-Planck-Institut für Biochemie
D-82152 Martinsried, Germany

I. Introduction

In living cells, balance must be maintained between the rates of protein synthesis and protein breakdown. Beyond this "housekeeping" function,

335

THE ENZYMES, Vol. XXII
Copyright © 2001 by Academic Press
All rights of reproduction in any form reserved.

protein degradation serves other roles of vital importance. Misfolded proteins ensuing from mutations or environmental stress must be removed in a selective manner because they are prone to aggregation. Many regulatory proteins, such as transcription factors or elements of signal transduction pathways, appear on the "scene," and when their act is over they have to disappear. Obviously, their irreversible removal must be subject to a tight spatial and temporal control.

In eukaryotic cells the majority of proteins are degraded via the ATP-dependent ubiquitin–proteasome pathway (for reviews, see Refs. *1–3*). This pathway combines two basic elements: The ubiquitin system selects proteins destined for degradation, and, in a sequence of activating and ligating steps involving arrays of different enzymes (E1, E2, E3), marks them by the covalent attachment of ubiquitin, a small, extraordinarily conserved protein (for reviews, see Refs. *3–5*). The second element, the 26S proteasome, recognizes multiubiquitylated proteins and degrades them in an ATP-dependent manner while the ubiquitin moieties are recycled.

The 26S proteasome is a huge (2.5 MDa) molecular machine built from more than 30 different subunits (for reviews, see Refs. *6–8*). It comprises two subcomplexes, the barrel-shaped proteolytic core complex, also referred to as 20S proteasome, and one or two regulatory complexes, which, in the presence of ATP, associate with the polar ends of the (C2) symmetric core complex (Fig. 1). The (19S) regulatory or "cap" complexes prepare substrates for degradation by the 20S proteasome. The preparation process is little understood at present; it is certain to include the binding of multiubiquitylated proteins, the removal of ubiquitin moieties by deubiquitylating enzymes, the local unfolding of the substrate's polypeptide chain, and its translocation into the interior of the 20S proteasome. A key role in the unfolding and translocation function is assigned to an array of six distinct but homologous ATPases that are assumed to form a heterohexameric ring at the "base" of the asymmetric regulatory complex, i.e, at the interface between core and cap. 20S proteasomes are found in all domains of existing organisms, including archaea, bacteria, and eukaryotes. Their overall architecture is highly conserved, although their subunit complexity is surprisingly different. 20S proteasomes are built from two types of subunits, α and β, which segregate into seven-membered rings and collectively form a barrel-shaped complex with a hollow interior. The internal cavities or "nanocompartments" harbor the active sites and segregate the proteolytic action from the surrounding cytoplasm (Fig. 1). This regulatory principle, referred to as "self-compartmentalization" (*9, 10*), is for prokaryotes, which lack organelles, the only means to confine the proteolytic action spatially. Proteasomes are probably the best-studied, but not the only, representatives of self-compartmentalizing proteases. Members of different hydrolase

26S proteasome

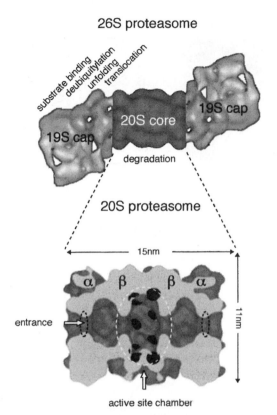

Fig. 1. Composite model of the three-dimensional (3-D) structure of the 26S proteasome combining a 3-D reconstitution from electron micrographs of *Drosophila* 26S complexes and the crystal structure of the *Thermoplasma* 20S proteasome (*126*). Below, the 20S structure is cut along the sevenfold axis for display of the three inner compartements; in the central compartment 8 of the 14 active sites are marked in black.

families, e.g., hs1UV, clpAP, gal6/bleomycin hydrolase, or tricorn protease, have converged toward the same barrel-shaped architecture (*10*).

In addition to self-compartmentalization, proteasomes rely on another mechanism to prevent the uncontrolled breakdown of proteins which is not uncommon among proteases. The β-type subunits of 20S proteasomes, which bear the active site residues, are synthesized in an inactive precursor form; for the formation of the active sites, the propeptides must be removed posttranslationally. This process is tied in with the assembly of the 20S proteasome in such a manner that activation is delayed until assembly is complete and the active sites are sequestered from the cellular environment. In mature 20S proteasomes access to the inner proteolytic compartment is through

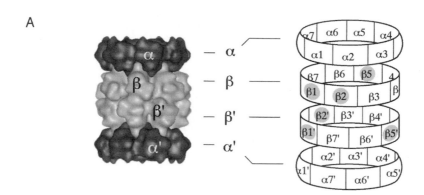

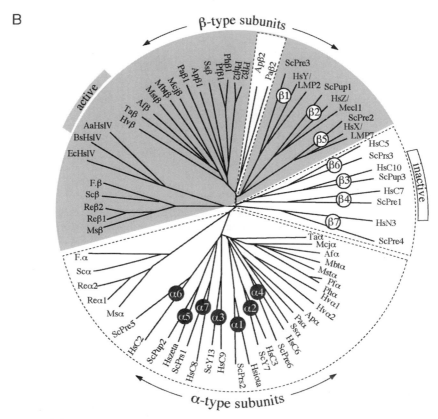

FIG. 2. (A) Surface model of yeast 20S proteasomes as obtained by low-pass filtering (1.2-nm cutoff) of the atomic model and a schematic representation of the subunit arrangement (35). In the scheme, proteolytically active subunits are marked by shadowing. (B) Dendrogram of the superfamily of proteasome subunits showing their classification into α- and β-type subunits.

narrow gates at the polar ends of the complex and thus restricted to small peptides and unfolded proteins (*11*). Therefore the proteasome behaves, with repect to folded proteins, as a "latent" protease that is dependent on regulatory complexes to assume its function. At the heart of the regulatory complexes, or "activators," are members of the AAA family of ATPases (*12, 13*). In prokaryotes, members of this family form homohexameric complexes (*14–16*). In eukaryotes, six paralogs form heterohexameric complexes, which are integrated into the larger and multifunctinal 19S regulatory complexes (*17*).

Significant progress has been made in recent years in our understanding of the cellular roles of the proteasome, its structure and mechanism of action. Our understanding of proteasome biogenesis, however, in particular the roles of the β propeptides in the maturation process, is still substantially incomplete. These topics will be the focus of this review.

II. The 20S Proteasome

A. Subunit Composition and Relatedness

20S proteasomes from all three domains of life have a common overall architecture. They are built of 28 subunits arranged in a stack of four seven-membered rings. All subunits are members of the same superfamily of proteins, which group into two families, designated α and β (Fig. 2). The two polar rings of the proteasome are formed by α-type subunits; the two equatorial rings are formed by β-type subunits. Both families comprise a number of subfamilies. The sequence similarity between members of a subfamily, even from evolutionary distant species, is considerably higher than the similarity between the members of different subfamilies from one species. Whereas human and yeast proteins of the same subfamily have at least 50% similarity, there is only around 30% similarity between the seven subfamilies of humans, and even less among the two families α and β.

The main difference between eukaryotic and prokaryotic 20S proteasomes is one of complexity. Prokaryotic proteasomes are mostly composed of

---

Proteolytically active subunits are highlighted by shadowing. The 14 eukaryotic subfamilies ($\alpha1$ to $\alpha7$, $\beta1$ to $\beta7$) are represented by human (Hs) and yeast (Sc) subunits. Archaeal subunits are from *Thermoplasma acidophilum* (Ta), *Methanococcus jannaschii* (Mcj), *Archeoglobus fulgidus* (Af), *Methanobacterium thermoautotrophicum* (Mbt), *Methanosarcina thermophila* (Mst), *Pyrococcus furiosus* (Pf), *Pyrococcus horikoshii* (Ph), *Haloferax volcanii* (Hv), *Aeropyrum pernix* (Ap), *Pyrobaculum aerophilum* (Pa), and *Sulfolobus solfataricus* (Ss). Bacterial subunits are represented by *Frankia* strain ACN14a (F.), *Streptomyces coelicolor* (Sc), *Rhodococcus erythropolis* (Re), and *Mycobacterium smegmatis* (Ms). HslV proteins are from *Aquifex aeolicus* (Aa), *Bacillus subtilis* (Bs), and *Escherichia coli* (Ec).

identical copies of 14 α subunits and 14 β subunits; eukaryotic proteasomes recruit α and β subunits out of 14 different subfamilies. Thus, the multiple axes of symmetry of the *Thermoplasma* proteasome (*18*) are reduced to C2 symmetry in the eukaryotic proteasome (*19, 20*); each of the 14 different subunits is found twice within the complex and occupies well-defined positions (*20*). According to their location within the α and β rings, yeast subunits are sequentially numbered $\alpha1/\beta1$ through $\alpha7/\beta7$, and those related by C2 symmetry are distinguished by the prime symbol ($\alpha1'/\beta1'$ through $\alpha7'/\beta7'$) (Fig. 2A). The subunit topology in higher eukaryotes corresponds to that in yeast, i.e., members of the same subfamily occupy the same positions (*21, 22*). Hence, the systematic nomenclature is generally applicable.

Whereas lower eukaryotes assume a stoichiometry of $(\alpha1-\alpha7)(\beta1-\beta7)$ $(\beta1-\beta7)(\alpha1-\alpha7)$, vertebrates have achieved an even higher degree of complexity. In addition to the 14 constitutive subunits, they contain three γ-interferon-inducible β-type subunits, which can replace their constitutive counterparts ($\beta1$, $\beta2$, and $\beta5$) to form immunoproteasomes (for reviews, see Refs. *23–25*). Sequence similarity within such pairs is near 60%. According to the systematic nomenclature the inducible subunits are designated $\beta1i$, $\beta2i$, and $\beta5i$.

Among bacteria, proteasomes have hitherto only been found in actinomycetes. Typically, bacterial proteasomes are built of one type of α and one type of β subunits; *Rhodococcus erythropolis* appears to be an exception, containing two α-type and two β-type subunits (*26*). The exact subunit topology of *Rhodococcus* proteasomes is not known, but it is likely that all four subunits coexist in one particle (*27*). Initially, archaeal proteasomes were thought to have the $\alpha7\beta7\beta7\alpha7$ stoichiometry, too (e.g., *Thermoplasma acidophilum, Methanococcus jannaschii*), but genomic sequencing has revealed more and more examples for two types of α (e.g., *Haloferax volcanii*) and two types of β genes (e.g., *Pyrococcus horikoshii* and *P. furiosus, Aeropyrum pernix*). Whether multiple subunits are incorporated into a single proteasome, or whether two population of proteasomes coexist, remains to be investigated. Proteasomes isolated from *P. furiosus* were found to contain only one type of β subunit (*28*), whereas both α subunits were found in proteasomes purified from *Haloferax volcanii* (*29*).

The eubacterial HslV proteins, which are members of the β-type family, assemble in form of a homododecameric complex built of two hexameric rings. They lack the equivalent of α-subunit rings, but associate directly with an ATPase, HslU, a member of the Hsp100(Clp) family (*30–32*).

B. Architecture of the Complex and Subunit Fold

On electron micrographs, 20S proteasomes from prokaryotes and eukaryotes are virtually indistinguishable (*26, 33, 34*), and the crystal structures of

Thermoplasma and yeast proteasomes are also very similar (*18, 35*). Proteasomes are barrel-shaped particles, with overall dimensions of 15 nm in length and 11 nm in diameter. A channel traverses the particle from end to end and widens into three internal cavities, each approx. 5 nm in diameter (Fig. 1). The two outer cavities are formed jointly by one α and one β ring; the central cavity is formed by the two adjacent β rings. The most conspicious difference between the crystal structures of the *Thermoplasma* and the yeast proteasome is at the center of the α rings. In the archaeal proteasome 1.3 nm wide openings are visible, whereas in the yeast proteasome this channel appears to be closed. Here, one has to keep in mind, that in the crystal structure of the *Thermoplasma* proteasome the 12 N-terminal residues of the α subunits remained invisible because of disordering; the corresponding residues in yeast plug the channel and may act as a gate.

As expected from their sequence similarity, proteasome α and β subunits have the same fold (Fig. 3): a four-layer structure with two five-stranded antiparallel β sheets (S1 to S10) flanked on either side by α helices, two on the one (H1 and H2) and three on the other side (H3, H4, and H5). The β-sheet package is unusual in that one sheet is rotated relative to the other through a positive dihedral angle of +30°, in contrast to the value of the typical β-sheet twist, which is −30°. In the β-type subunits, the β-sheet

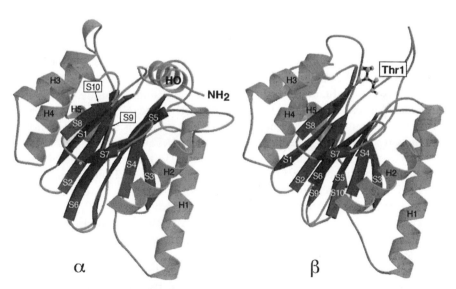

FIG. 3. Folds of the α and β subunits from *Thermoplasma* 20S proteasomes (*18*). Helices are numbered H0 to H5 in the α subunit, and H1 to H5 in the β subunit; strands are numbered S1 to S10. H0 occupies the cleft on top of the β-sheet sandwich in the α subunit, which forms the active site cleft in the β subunit. The active site threonine (Thrl) is shown in ball-and-stick representation.

sandwich is open at one end to form the active site cleft and closed at the other end by four hairpin loops. In α-type subunits, however, an additional helix (H0) crosses the top of the β-sheet sandwich and fills the cleft. The fold of proteasome subunits was initially considered to be unique, but as it turned out it is prototypical of a new family of proteins referred to as Ntn (N-terminal nucleophile) hydrolases (*36*). Currently known members of this family, whose primary structures have diverged beyond recognizable similarity, are aspartylglucosaminidase (AGA) (*37*), glutamine phosphoribosylpyrophosphate amidotransferase (GAT) (*38*), penicillin acylase (PA) (*39*), and L-aminopeptidase D-Ala-esterase/amidase (DmpA) (*40*). Beyond the common fold, these enzymes share the mechanism of the nucleophilic attack and autocatalytic processing (see below), but none has a quaternary structure similar to the proteasome.

C. The Active Site

A characteristic feature of Ntn hydrolases is a "single-residue" active site: both the nucleophile and the primary proton acceptor are provided by the same N-terminal residue of the proteins. The nucleophilic attack is initiated when the free N terminus, the primary proton acceptor, strips the proton off the catalytic side chain, the nucleophile; for steric reasons a water molecule is supposed to mediate the proton shuttle, at least in the proteasome and in penicillin acylase. Details of the mechanism are depicted in Fig. 7B and discussed in context of autolysis in Section III,C,2. Different N-terminal residues are used as nucleophiles in Ntn hydrolases: serine in PA and DmpA, cysteine in GAT, and threonine in AGA and in the proteasome. In agreement with a catalytic mechanism that relies on a single residue, there is no consensus of residues in the vicinity of the active site. Nevertheless, several other residues are critical for activity of proteasomes, as consistently shown by site-directed mutagenesis with archaeal (*41–43*), bacterial (*44*), yeast (*20, 45*), and mammalian proteasomes (*46, 47*). The exact roles of these residues, all highly conserved and in close proximity to Thr1, remain to be clarified. Lys33, which forms a salt bridge to Glu17, may lower the pK_a of the N terminus by electrostatic effects or may be part of the charge relay system in delocalization of the proton from the Thr1 hydroxyl group. Asp166 may be required for the structural integrity of the active site or may be directly involved in catalysis (*18, 42*).

Although all proteasome subunits have the Ntn-hydrolase fold, not all are active. As mentioned above, in α subunits, the active site cleft is occupied by an additional N-terminal α helix, H0 (Fig. 3). Of the different eukaryotic β-type subunits, only three (β1, β2, and β5 and their interferon-inducible counterparts) display proteolytic activity, whereas subunits β3, β4, β6, and β7 lack one or more of the critical residues and are inactive. This assignment, initially derived from the conservation pattern of the active site residues (41), has been confirmed by the crystal structure of yeast 20S proteasomes soaked in N-acetyl-Leu-Leu-norleucinal; only three subunits, β1/Pre3, β2/Pup1, and β5/Doa3, contain the inhibitor bound to Thr1 (35). Since the backbone geometry is well conserved between all subunits, active or inactive, one would expect that introducing all residues regarded as critical would render inactive subunits active. However, this is not the case; the yeast subunits β3/Pup3, β6/Pre7, and β7/Pre4 remained unprocessed and consequently inactive, when the respective mutations were made (20).

Most of the prokaryotic β subunits are active, but genomic sequencing has revealed the existence of two β-type genes (in the genomes of *Aeropyrum pernix* and *Pyrobaculum aerophilum*) lacking the catalytic threonine at the N terminus (Figs. 2B and 4). It remains to be investigated whether inactive subunits are in fact incorporated into prokaryotic proteasomes.

D. SUBSTRATE SPECIFICITY

Before the name proteasome was coined (48), this large enzyme complex was known as the "multicatalytic protease" (49, 50), a name reflecting its multiple catalytic activities. Three major peptidase activities had been defined using fluorogenic peptides: a chymotryptic activity, which cleaves after hydrophobic residues; a tryptic activity, which cleaves after basic residues; and a peptidylglutamyl-hydrolyzing (PGPH) activity, which cleaves after acidic residues (51). Two additional specificities were found with mammalian proteasomes: one cleaving after branched-chain amino acids (BrAAP activity), and one cleaving between small neutral amino acids (SNAAP activity) (52). Although this spectrum of activity was initially attributed to the combined action of different proteolytic enzymes within the complex (e.g., trypsin-like and chymotrypsin-like proteases), it is now clear that all active proteasome subunits are N-terminal threonine proteases utilizing the same catalytic mechanism. The physical basis for the selection of different cleavage sites by different β-type subunits is revealed by the crystal structure of the yeast 20S proteasome. Several residues varying between the three active β-type subunits determine the properties of the "specificity pocket" and confer selectivity; among them, residue 45 at the end of β strand S4 appears to have a dominant role (35).

Meanwhile, all activities of eukaryotic proteasomes have been assigned to distinct β-type subunits through a number of mutagenesis, inhibitory, and X-ray diffraction studies (20, 45, 47, 53–60): chymotryptic activity to β5, PGPH activity to β1, and tryptic activity to β2. BrAAP activity was found at two sites, β5 and β1; SNAAP activity correlates with β2, but is probably not restricted to one subunit. When mammalian cells are exposed to the cytokine γ-interferon, the subunits β1/Y, β2/Z, and β5/X are replaced

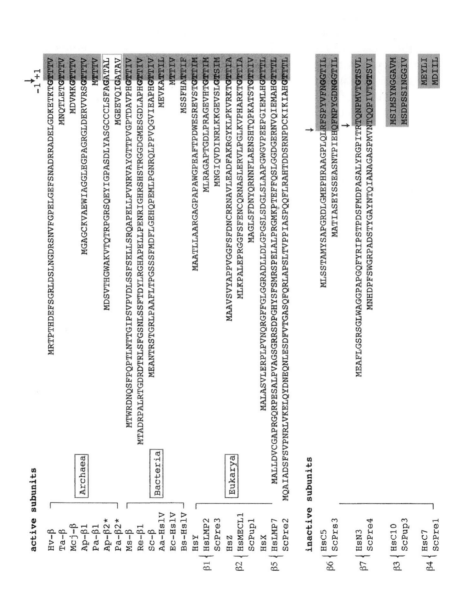

by subunits LMP2, MECL1, and LMP7, and substrate specificity of these immunoproteasomes is modulated. PGPH activity seems to be reduced, whereas chymotryptic, tryptic and BrAAP activities are enhanced (for a review, see Ref. *25*).

Since prokaryotic proteasomes are typically built of only one type of β subunits, they have only one type of specificity. Mostly, "chymotryptic" substrates are preferred (*26, 28, 29, 61–63*), but additional cleavage after acidic residues (PGPH activity) has also been reported (*16, 43*).

It must be noted, however, that specificities as assigned by means of small synthetic peptides do not correlate in an obvious manner with the cleavage patterns observed when longer peptides or entire proteins are degraded by prokaryotic or eukaryotic proteasomes. An enormous variety of overlapping fragments is produced from natural polypeptide substrates, indicating that both C- and N-terminal residues, and also those beyond the P1 and P1′ positions (the respective residues at the amino and carboxyl side, respectively, of the scissile bond), contribute to the selection of cleavage sites (*64–72*). Taken together with the characteristic size range of the generated products, this led to the proposal that a molecular ruler might dictate the site of cleavage rather than the substrate's amino acid sequence (*66*). Since the average length of products, octa- to nonapeptides, corresponds well to the distance between neighboring active sites in the *Thermoplasma* proteasome (2.8 nm), this was at first believed to provide the distance criterion (*18*). Later, more comprehensive studies revealed a size distribution of the released products broader than initially assumed, nevertheless averaging at 7–9 residues (*71, 73*). Moreover, the size distribution turned out to be the same with *Thermoplasma* proteasomes, containing 14 active sites, and with wild-type or mutant yeast proteasomes, containing six or fewer active sites (*71*). This is obviously difficult to reconcile with the aforementioned molecular ruler model invoking a concerted action of neighboring active subunits. The fact remains, however, that the degradation products fall into a certain size range, and a diffusion-controlled mechanism may account for this (*66*). It has been shown experimentally that longer peptides are degraded faster than shorter ones (*70*). A higher affinity of longer polypeptides correlates with longer association times and increases the probability of being degraded. Below a certain threshold in length the peptides have a higher probability to exit the central cavity of the proteasome and thus escape degradation.

FIG. 4. Variability among prosequences of proteasome β subunits. The cleavage sites in active and inactive subunits are marked by arrows; shading indicates the N-terminal sequences of mature proteins. Although listed among active subunits, the second β-type subunits of *Aeropyrum pernix* (Ap-β2*) and *Pyrobaculum aerophilum* (Pa-β2*) are likely to be inactive; the exact N termini of the mature proteins are not known. For species abbreviations see Fig. 2.

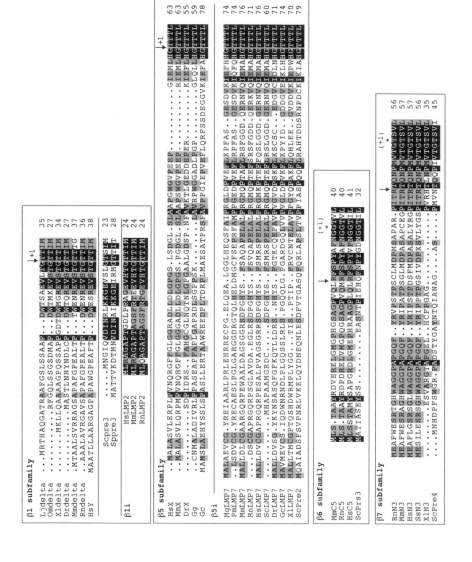

III. Maturation of β-Type Subunits

A. THE β PROREGIONS

In all of the Ntn hydrolases, posttranslational processing of inactive precursors exposes the α-amino group of the residue with the nucleophilic side chain, which then participates in the catalytic process (36). Almost all proteolytically active proteasome β-type subunits are synthesized in a precursor form, with the exceptions of HslV proteins of *E. coli* and closely related bacteria (Fig. 4). Of the inactive β-type subfamilies two are synthesized in their mature form, $\beta3$ and $\beta4$, while the two others, $\beta6$ and $\beta7$, are posttranslationally processed; however, cleavage occurs 8 to 9 residues upstream as compared to the cleavage sites in the active subunits. The β proregions exhibit strong variations in length and sequence: some are only a few residues long, while others comprise more than 60 residues. In spite of the high degree of conservation among mature proteasome subunits, there is no recognizable similarity between most prosequences. Only the site of cleavage, Gly-1/Thr$+1$, is conserved throughout all active β subunits, and mutation of either residue has drastic effects on processing. Mutation of Thr1 abolishes proteolytic as well as autolytic activity, whereas mutation of Gly-1 severely impairs precursor processing, which in turn affects catalytic activity (42, 45, 46).

The divergence of the proregions of proteasome β subunits precludes a meaningful sequence comparison across subfamily boundaries. In the following, a few structural features of β propeptides will be highlighted:

1. The propeptides of archaea β-type subunits are quite variable in length; most are short, typically 8 to 10 residues, but a 49-residue propeptide precedes the β subunits of *Haloferax volcanii,* a halophilic euryarcheaon (29), and both β-type genes of *Aeropyrum pernix,* a hyperthermophilic crenarcheon, encode propeptides of more than 30 residues (Fig. 4).

2. In eukaryotic propeptides, regions of significant sequence similarity appear to be restricted to the C-terminal half of the propeptides; an accumulation of highly conserved residues, proline in particular, is often found immediately upstream to the site of cleavage (Figs. 5 and 6).

FIG. 5. Alignments of the prosequences available for subfamilies $\beta1$, $\beta5$, $\beta6$, and $\beta7$. Alignments were performed using ClustalX (127). Residues in reverse type (white or black) are 100% conserved, shaded residues are $>60\%$ conserved. Arrows indicate the posttranslational cleavage sites in active subunits and the corresponding positions in the inactive subunits. Sequences are from *Lampetra japonica* (Lj), *Oncorhynchus mykiss* (Om), *Xenopus laevis* (Xl), *Danio rerio* (Dr), *Mus musculus* (Mm), *Rattus norwegicus* (Rn), *Homo sapiens* (Hs), *Saccharomyces cerevisiae* (Sc), *Schizosaccharomyces pombe* (Sp), *Gallus gallus* (Gg), *Ginglymostoma cirratum* (Gc), *Myxine glutionosa* (Mg), and *Petromyzon marinus* (Pm)

Actinomycetes β subunits

```
                                                                              ↓+1
Mt     MTW....PLPDRLSIN.SLSGTPAV.DLSSFTDFLRRQAPELLPAS.ISG.....GAPLAGGDAQLPHGTTIV  61
Ml     MTR...SFPDRLPTNLAFPGISVI.NQSSFVDLIRRQAPELLPVS.LGG.....GQ..SGGGQQLSHGTTIV  60
Ms     MTWRDNQSFPQPTLNTTGIPSVP.V.DLSSFSELLSQAPELLPVNRVAY....GTTPVGPTDAVPHGTTIV  66
Re1    MTADRP..ALRTGDRDTRLSFGS...NLSSFTDYLRGHAPELLPENRIGHRSHSTRGDGMESGDLAPHGTTIV  69
Re2    MTVDR...APRITGDDRLSFGS..NLSSFSEYLRVHAPEHLPQNRFAD....T.GVVMGGGDVAPHGTTIV  63
Sc     .......MEANTRSTGRLPAAFLTPGSSFMDFLGEHQPEMLPGN.......RQLPPVQGVIEAPHGTTIV  57
F.     .......MADPMGGAGRLPAVFMTPGTSSFTDFLSQSAPHLLPGA.......RGGLPGP.VTEVAHGTTIV  56

dsc    :----------------------------------------HHHHHHHHHHHHHHHHHHH--
jalign :---------------------------------------------------HHHHH--
Jhmn   :------------------------------HHHHH--
jnet   :------------------------------HHHHH--
mul    :----------------------------------HHH--
phd    :-----------------------E-----HHHHHHHH--
pred   :-----------------EE-------HHHHHHHHH--
zpred  :HH------------------EEEEEH----EEE--EEEEEEEEE--
Jpred  :--------------------------------HHHHHH--E--
```

β2 (β2i) subfamily

```
                                                                         ↓+1
HsZ      .MAAVSVYAPPVGGFSPDNCRRNAVEADFAKRGYKLPKVRKTGTTIA  47
MmZ      .MAAVSVFQPPVGGFSFDNCRRNAVLEADFAKKGFKLPARKTGTTIA  47
HsMECL1  ...MLKPALEPRGGFSFENCQRNASLERVLP.GLKVPHARKTGTTIA  43
MmMECL1  ...MLKQAVEPRGGFSFENCQRNASLEHVLP.GLRVPHARKTGTTIA  43
Tb       .........MTGFSFENVQRNLNEQQGL...HPPRTLKTGTTIV  33
At       .MSQSSVDIPFKGGFSFDCKRNDMTQKGL...KASFLKTGTTIV  43
Dm       .MDLDNARELPRAGFNFDNCKRNATLLNRGF..KPPITTKTGTTIV  43
Ci       MATTLTCVEIPNGGFTFDNCTRNAHLEKKGM..VAPKTRKTGTTIV  44
Dr       ...LTSHVLEPSLCGFNFRATRNIVLENGAEEGKIKPDKALKTGTTIV  46
ScPupl   .........MAGLSFDNYQRNFLAENSH...TQPKATSTGTTIV  33
SpPupl   ......MMGINERKGFDFYYQRNLLQEKGF...PTPKATSTGTTIV  39

dsc      :--------HHHHHHHHHHHHHHHHHHH--
jalign   :--------HHHHHHHHHHHHH--
jhnm     :--------HHHHHHHHHHHH--E--
jnet     :--------HHHHHHHHHHHH--E--
mul      :--------HHHH-----H---EEE--
phd      :--------------H-----EE--
pred     :--------HHHHHHHHHHH--
zpred    :--------HHHHHHHHHH--
Jpred    :HH------EE---HHHH------EEEEE
```

3. A high frequency of charged residues was noted in the prosequences of active subunits, especially toward their C termini (*74*). Such a motif is absent in inactive subunits, but its functional importance remains questionable, since replacement of charged residues by neutral ones in mouse β1i had no effect on correct processing and subunit incorporation (Figs. 5 and 6).

4. Propeptides of the constitutive mammalian β2 subunits are closely related to their γ-interferon-inducible counterparts (β2i), in this particular case, matching the similarity between the mature proteins. In contrast, there is no similarity between the propeptides of β1 and β1i, and also those of β5 and β5i form two separate groups. Interestingly, the yeast β5/Pre2 propeptide branches with the mammalian immune response subunits, rather than the constitutive subunits (Figs. 5 and 6).

5. Regions of significant similarity emerge from alignments of actinomycetes β propeptides and of those of the eukaryotic β2 subfamily (Fig. 6). In both groups, a region comprising about 15 amino acids near the center of the prosequence shows a remarkable degree of conservation. A functional role of these regions is likely, and its deletion from the *Rhodococcus* β1 prosequence indeed prevents subunit incorporation, while deletion of the preceding N-terminal segment is tolerated (Mayr *et al.*, submitted).

Data on the three-dimensional structure of the β propeptides are scarce. Only alignments of the prosequences of actinomycetes and the β2 group provide the basis for a reliable secondary structure prediction; about 10 sequences are available for each group covering a sufficiently broad range of identities (40% to 90%). Jpred, an Internet Web server (http://jura.ebi.ac.uk:8888) that combines several algorithms for secondary structure prediction, returns very similar results for both groups: the highly conserved core regions are predicted to form α helices (Fig. 6).

Hitherto, only the propeptide of the *Rhodococcus* subunit β1 has been separately expressed in *E. coli* and studied by various spectroscopic methods, CD, NMR, and IR spectroscopy (Mayr *et al.*, submitted). In solution, the propeptide appears to be largely unfolded, but association with the protease domain of β subunits it might assume a defined fold. Processing-incompetent mutants of the *Rhodococcus* proteasome harboring the complete 64-residue propeptides of all 14 β subunits are able to accomodate the extra mass of

---

Fig. 6. Prosequences of β subunits of actinomycetes and of the eukaryotic β2 subfamily. Alignments and secondary structure predictions were obtained from the Internet Web server Jpred (http://jura.ebi.ac.uk:8888) (*128*). H $= \alpha$ helix, E $= \beta$ strand. For alignment features see Fig. 5. Bacterial sequences are from *Mycobacterium tuberculosis* (Mt), *M. leprae* (Ml), *M. smegmatis* (Ms), *Rhodococcus erythropolis* (Re), *Streptomyces coelicolor* (Sc), and *Frankia* strain ACN14a (F.). Eukaryotic sequences are from *Trypanosoma brucei* (Tb), *Arabidopsis thaliana* (At), *Drosophila melanogaster* (Dm), and *Ciona intestinalis* (Ci); for others, see Fig. 5.

approximately 100 kDa in the internal cavities (44); this would hardly be possible if the propeptides would exist in an unfolded state. Analogous to *Rhodoccocus*, yeast proteasome subunits have been mutated and crystallized in order to catch a glimpse of the propeptide structure *in situ* (20, 75). However, information on the propeptide's fold remained very limited, since full-length precursors cannot be found in eukaryotic proteasomes. Because of the action of neighboring subunits, precursors of mutant subunits are cleaved within their prosequences, leaving short fragments of the propeptides only, and the same holds for the naturally inactive subunits $\beta6$ and $\beta7$ (intersubunit cleavage is discussed in Section III,D).

B.   PROCESSING AND ASSEMBLY

1.   *Assembly Pathways*

Processing of the β-precursor subunits is intimately coupled to the formation of 20S complexes, and must therefore be discussed in the context of proteasome assembly.

It appears that a single pathway does not exist. The assembly pathways of the archeabacterial *Thermoplasma* proteasome and the bacterial *Rhodococcus* proteasome, for instance, show some basic differences: recombinant *Thermoplasma* α subunits assemble into seven-membered rings in the absence of β subunits (76). β subunits, in turn, are unable to assemble, remain unprocessed and inactive, and do not even fold properly in the absence of α subunits. This suggests that preassembled α rings provide a platform or a template onto which the β subunits are mounted. Self-assembly of α rings has also been reported for other archaeal proteasomes and appears to be a common feature (16, 43). In contrast, neither α nor β subunits of the *Rhodococcus* proteasome assemble by themselves (77). Only when both types of subunits are present are assembly products observed. Thus, it appears that the bacterial pathway starts with the formation of α/β heterodimers that assemble further into half proteasomes, containing one ring of α subunits and one ring of β precursors. The association of two half proteasomes yields preholoproteasomes, which convert into active proteasomes in a two-step process. *In vitro* experiments indicate that the docking of two half proteasomes is sufficient to trigger activation, a mechanism that is different from those of most other proteases, which require external signals, such as pH changes, to initiate the conversion of the zymogen into the active enzyme (for a review, see ref. 78). In the first step, the propeptide is cleaved off, exposing the nucleophile, and in a subsequent step, the active site assumes its final conformation. While the first step can be monitored directly, it is concluded from kinetic considerations that a conformational rearrangement follows (44, 77):

the rate-limiting step of the whole process was found to be unimolecular. Yet, the time needed for the initial folding of individual subunits is much shorter than for subsequent association steps, and also precursor processing itself, another unimolecular reaction along the pathway, proceeds rapidly as indicated by the short lifetime of preholoproteasomes. Therefore, the final conformational change is likely to determine the overall velocity of proteasome maturation. There are other lines of evidence indicating that the removal of the propeptides is necessary, but not sufficient, to acquire proteolytic activity. *Rhodococcus* half proteasomes built of propeptide deleted β subunits are inactive until they dimerize and form holoproteasomes (*77*). Yeast proteasomes lose chymotryptic activity, when the correct contact between $\beta5$/Doa3 and the dyad related subunit $\beta4'$/Pre1 is destroyed, even when the Doa3 propeptide has been genetically deleted (*45*). These results are in agreement with the idea that proteasomes undergo a conformational change upon β processing, a sort of interlocking, that is required to establish the critical β–β contacts.

It is not yet clear how exactly the assembly of eukaryotic proteasomes proceeds. Do α/β heterodimers precede ring formation, or are α rings initially formed, which then act as templates for the assembly of the β-type subunits? For some eukaryotic α-type subunits [human $\alpha7$/C8 (*79*) and $\alpha5$ of *Trypanosoma brucei* (*80*)] it has been shown that they are capable of assembling into seven-membered rings when expressed separately in *E. coli*, supporting the latter hypothesis. However, intermediates of the eukaryotic proteasome assembly pathway have mostly been studied *in vivo*, and their exact composition remained ill-defined. Complexes, often referred to as "half proteasomes," do not necessarily have $\alpha1$–7, $\beta1$–7 stoichiometry. For instance, 15S complexes have been described that contain the seven murine α subunits and at least four β-subunit precursors (*81*). Of the 12 proteins of the so-called 13S precursor complexes, only six are definitely α-type subunits, while the identity of the others remained unclear; they may represent β-subunit precursors, or transiently associated assembly factors (*82*). When 13S precursor complexes (with the approximate molecular mass of half proteasomes) and 16S precursor complexes (with the approximate molecular mass of holoproteasomes) purified from mammalian cells were tested for their ability to reach a mature state *in vitro*, both failed; neither of them yielded active proteasomes, although processing of β precursors took place in the 16S species (*83*). This suggests that the isolated intermediates lack one or several subunits, and/or that maturation of eukaryotic proteasomes depends critically on extrinsic factors (see below). By means of subunit-specific antibodies the subunit composition of mammalian proteasome precursor complexes has been examined in depth (*84*). This work lends support to a template role of α rings and it proposes a schedule for the incorporation of β subunits. Early assembly intermediates contain the whole set of α subunits, together with β1i and β2 (or β2i), both in their precursor form, as well as β3 and β4, which are synthesized in their mature form. Subsequently, the remaining β subunits, the precursor forms of β5 (or β5i), β6, β7, and β1, are added, yielding complete half proteasomes; hence, the chronological order of subunit incorporation reflects their spatial arrangement within the complex. Dimerization of half proteasomes and precursor processing are supposed to follow. The shorter the lifetime of the precursors was found to be, the later they were incorporated, suggesting that the conversion of the preholoproteasome to the mature proteasome is fast. According to the time table proposed by Nandi *et al.* (*84*), the immune response subunit β1i belongs to the group of early subunits, whereas its constitutive counterpart β1 is one of the late subunits. Given the high degree of similarity between the mature subunits, the different propeptides of β1 and β1i are likely to be responsible for the timing (such a role is discussed in Section IV,C). Two reports describing the maturation of eukaryotic proteasomes are not in agreement with the proposal of half proteasome intermediates, built of a ring of α subunits and a ring of β-subunit precursors (*85, 86*). By immunoprecipitation and pulse-chase labeling, these authors show that precursor forms of the mammalian subunits β6/C5 and β7/N3 exist neither in mature proteasomes nor in proteasome subcomplexes, not even transiently, but only as free monomers. Consequently, the processing of these two subunits must occur concomitantly with their incorporation into half proteasome intermediates, and not further down the pathway. It is noteworthy that these two exceptions, C5 and N3, are proteolytically inactive β subunits, processing of which is based on a mechanism different from that which applies to active β-type subunits (see Sections III,C,1 and III,D).

2. Assembly-Assisting Factors

The efficient *in vitro* assembly and maturation of archaeal and bacterial proteasomes proves its autonomous character. Yet, the assembly of eukaryotic proteasomes poses much greater problems since seven distinct α and seven distinct β subunits must be positioned correctly. The orchestration of this task may explain the need for extrinsic assembly factors. In early reports, the chaperone Hsc73 was found to associate with mammalian 16S proteasome precursor complexes, but not with mature proteasomes (*83*). Its precise role, however, remained unclear; it was proposed to prevent aggregation of precursor complexes. More recently, a proteasome maturation factor, named Ump1, was identified in yeast (*87*). The small protein (16.8 kDa) is contained within half proteasome precursor complexes and it becomes degraded (by the proteasome) in the course of proteasome maturation. Although

Ump1 is not essential, it was shown to be involved in the maturation of the three active β subunits of the yeast proteasome, interacting in a special manner with the propeptide of $\beta2$/Doa3. Interestingly, deletion of *UMP1* suppresses the lethal phenotype of a deletion of the $\beta2$/Doa3 propeptide (*45*) (see below). Possibly, the Doa3 propeptide acts on Ump1, preparing it for degradation; in the absence of the propeptide, Ump1 may remain in a position or conformation that is incompatible with the formation of active proteasomes.

C. The Mechanism of Processing

1. *Self-Processing*

The fact that *Thermoplasma* and *Rhodococcus* proteasomes expressed in *E. coli* were correctly processed at the Gly-1/Thr1 positions was a first indication of a self-processing mechanism (*27, 88*). Later, the efficient *in vitro* maturation of bacterial proteasomes from purified precursor subunits clearly demonstrated the autocatalytic character of the process (*77*). This does not rule out, of course, that in eukaryotes assembly factors such as Ump1 are involved. It was further shown that processing is inhibited by proteasome-specific inhibitors (*46*) and abolished in most active-site mutants, such as Thr1Ala or Lys33Ala mutants (*20, 42, 44–47*). Interestingly, the replacement of the active site threonine by cysteine, which yields proteasomes unable to cleave peptide substrates, does not affect precursor processing (*42*). Conversely, a serine residue at the N terminus of yeast or *Thermoplasma* β subunits does not allow precursor processing to proceed efficiently, whereas propeptide-deleted Thr1Ser mutants cleave standard peptide substrates at nearly wild-type rates (*42, 43, 45, 47*). These data suggest that there are subtle differences in the conformations of these mutants that interfere with the dual role of the active site nucleophile in propeptide and substrate cleavage. In this context, it seems noteworthy that Thr1Ser mutants are significantly slower in degrading proteins, such as lactalbumin or lysozyme, than wild-type proteasomes (*43, 89*).

In principle, there are two alternative routes by which self-processing can proceed, namely, intermolecularly (*in trans*), or intramolecularly (*in cis*). Initial results obtained with *Thermoplasma* proteasomes were indicative of an intermolecular mechanism. A small portion of β subunits that were rendered incompetent for self-processing by mutations were found to be correctly processed in the presence of (active) wild-type subunits, but all remained immature in the absence of wild-type subunits (*42*). Yet, in view of our current understanding that processing takes place in a fully assembled proteasome complex, an intersubunit action is difficult to envisage. Also, inactivated mammalian β1i/LMP2 is not correctly processed, despite incorporation into the complex along with active neighbors (46), which is not in line with the results obtained with archaeal proteasomes. Moreover, when yeast 20S proteasomes were modified such that they contained only two active β subunits, one copy per β ring, these two were correctly processed (20), and the same observation was made with *Rhodococcus* proteasomes containing a single wild-type β subunit among 13 inactive subunits (Mayr *et al.*, submitted). All these data are indicative of an intramolecular mechanism of processing. Then, the question arises as to what might account for the aforementioned results obtained with recombinant *Thermoplasma* proteasomes. First, archaeal proteasomes may indeed use a mechanism that is different from that of eukaryotic and bacterial proteasomes. Second, the observed maturation of inactive *Thermoplasma* precursors could result from some residual activity of the mutant subunits. Also in yeast proteasomes, a small fraction of the inactive mutant subunit β5/Doa3(Lys33Ala) was found to have a threonine at its N terminus, implying that it underwent autolysis (20).

2. The Autolytic Reaction

The active sites of proteasomes, like those of other Ntn hydrolases, are designed to serve a dual role, namely, propeptide cleavage and substrate cleavage. A view of an autolysis site of proteasomes was provided by the molecular structure of the processing-incompetent yeast β1/Pre3(Thr1Ala) mutant with the attached prosegment well defined by electron density from its first residue Leu-9 to the γ-turn conformation at Gly-1 (75). By a simple Ala-to-Thr exchange the self-cleavage competent wild-type structure was derived. Briefly, residues Lys33, Asp17, Gly128, Ser129, Gly130, and Asp166 are involved, surrounding the three-residue γ turn (Leu-2, Gly-1, and Thr1) and connected via various hydrogen bonds. These residues are conserved among all active proteasome subunits. The hydroxyl group of Thr1 is located above the γ-turn bulge close to the carbonyl carbon of Gly-1 such that a small displacement allows it to add to the electrophilic center. The carbonyl carbon of Gly-1, in turn, is strongly polarized by a hydrogen bond facilitating the nucleophilic addition.

Figures 7A and B compare in a schematic manner the proposed one-time mechanism of self-processing and the repetitive mechanism of substrate hydrolysis. In both schemes, the hydroxyl group of Thr1 acts as the nucleophile. In autolysis, Thr1Oγ adds to the carbonyl carbon of Gly-1, in substrate hydrolysis, to the carbonyl carbon of the scissile peptide bond. To initiate this attack, the amino group of Thr1 serves as a proton acceptor in the mature enzyme and abstracts a proton from the hydroxyl group, but there is no

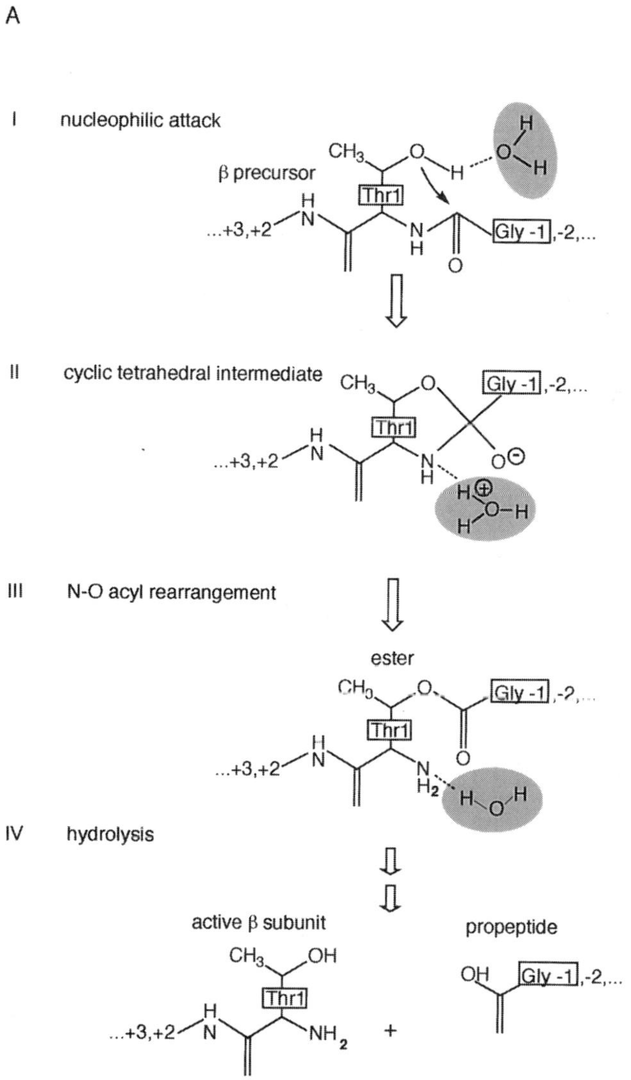
A
I nucleophilic attack
β precursor
CH3
Thr1
...+3,+2
Gly -1,-2,...
II cyclic tetrahedral intermediate
III N-O acyl rearrangement
ester
IV hydrolysis
active β subunit
OH
NH2
propeptide

FIG. 7. (A) Putative mechanism of self-processing of the proteasome β-type subunits. (B) Putative mechanism of substrate hydrolyis by 20S proteasomes.

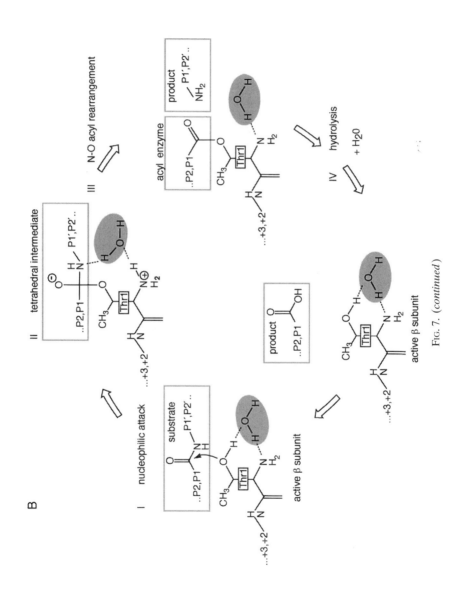

F IG . 7. (*continued*)

amino acid base close to Thr1Oγ in the immature protein; Lys33Nε, which was initially considered as a possible candidate, is fully engaged otherwise. Therefore, a water molecule, found in the crystal structures of both unprocessed mutant and mature wild-type subunits, is likely to serve as the proton acceptor in autolysis, and possibly as a proton shuttle between Thr1Oγ and Thr1N in substrate hydrolysis (Fig. 7A and B,I). Through nucleophilic addition tetrahedral intermediates are formed, a linear intermediate in the substrate hydrolysis pathway and a cyclic compound, hydroxyoxazolidine, in the self-cleavage pathway (Fig. 7A and B,II). Next, the C–N bonds are cleaved and ester bonds are formed. In substrate hydrolysis, this N–O acyl rearrangement yields an acyl enzyme and releases the C-terminal part of the substrate (Fig. 7A and B,III). Cleavage of the acyl enzyme proceeds in analogy to its formation. H_2O adds to the electrophilic center of the ester bond and a tetrahedral intermediate is formed, which decays into the N-terminal portion of the substrate and the enzyme, ready for another catalytic cycle. Hydrolysis of the ester compound in the autolysis pathway produces the free propeptide (corresponding to the N-terminal part of a polypeptide substrate) and the mature enzyme (corresponding to the C-terminal part of a polypeptide substrate), now exposing the amino group of Thr1; the water molecule seen in the crystal structure may be incorporated into the products (Fig. 7A and B,IV).

Initial N–O or N–S acyl rearrangements are a common feature of many self-catalyzed protein rearrangements serving different purposes (for a review, see ref. 90). For example, self-cleavage of aspartylglucosaminidase, another Ntn hydrolase, separates the two polypeptides, α and β, that form the mature enzyme and generates the N-terminal threonine required for activity (91). Autoprocessing of hedgehog proteins, animal signaling proteins, removes a nonfunctional C-terminal domain and allows the addition of a cholesterol moiety to the newly generated C terminus, which mediates contact to cell surfaces (92, 93). By far, the most complex protein self-processing mechanism is protein splicing, because it requires the cleavage of two peptide bonds, the excision of the splicing element, the intein, and the ligation of the two flanking domains, the exteins, to form the functional protein (for reviews, see Refs. 94, 95). In all systems, the initial formation of an ester or thioester requires a threonine, serine, or cysteine, but there is no consensus for the residue immediately upstream of the scissile peptide bond. In fact, it is not clear why Gly-1 is conserved in all active proteasome β subunits; its mutation to alanine causes severe processing defects as shown for yeast β5/Doa3, mammalian LMP2/β1i and the *Thermoplasma* β subunit (42, 45, 46). Among the other members of the Ntn-hydrolase family, glycine is only found in DmpA; a glutamate takes the position in GAT, an aspartate in AGA, and a threonine in PA.

D. INTERSUBUNIT PROCESSING AND DEGRADATION OF PROPEPTIDES

Mature eukaryotic proteasomes contain no full-length β-precursor subunits. Thus, not only are the three active β subunits processed, but also the precursor forms of the two inactive subunits, β6 and β7, have their N termini removed, and the same applies to inactive mutant subunits (*20, 46, 47, 75, 85*). However, cleavage of all inactive precursor subunits occurs at some distance upstream of the consensus cleavage site of the active subunits, Gly-1/Thr1, leaving propeptide remnants of typically 8 to 18 residues. Furthermore, cleavage of precursor forms of inactive subunits appears to occur at an earlier stage of proteasome assembly than processing of active β subunits (see above) (*85, 86*). Taken together, this suggests that the mechanism for propeptide removal from the inactive subunits is different from the self-cleavage mechanism of active subunits. Initially, it was thought that the location of the residual propeptide segments, that is, the position of their newly generated N termini, would directly lead to the catalytic sites responsible for the cleavage. In the crystal structure of yeast wild-type proteasomes, the prosegments of β6/Pre7 and β7/Pre4 are tightly attached to the protein walls inside the central cavity with their first residues close to the inner annulus of the β rings; a second hydrolytic site was therefore postulated to exist, far removed from the Thr1 catalytic site (*35*). Yet this proposal proved to be incorrect; it is in fact in conflict with several mutagenesis studies, which taken together provide conclusive evidence that processing of inactive precursor subunits is performed by the intact Thr1 active site of neighboring subunits (*20, 46, 47*). In summary, the following picture emerges: the propeptides, which have some freedom to move inside the central cavity, are cleaved by the nearest catalytic center to which they gain access. This distance determines the length of the residual segment, whereas, in accordance with the proteasome's cleavage behavior, the P1 amino acid is of minor importance for the cleavage. Hence, the final positions of the prosegments of β6/Pre7 and β7/Pre4, and those of inactive mutant subunits, seen in the crystal structures were misleading and are likely to result from a conformational rearrangement following intersubunit cleavage (*20, 35, 75*). In yeast wild-type proteasomes, the propeptide of β7/Pre4 is cleaved between Asn-9 and Thr-8 by its nearest active neighbor, β2'/Pup 1, located directly opposite to it. When the active site of β2'/Pup 1 is mutated, cleavage of β7/Pre4 occurs somewhat upstream of the former cleavage site, between Val-10 and Asn-9; β1'/Pre3, the subunit next to β2'/Pup 1, is now the obvious candidate to execute the cleavage. In a double mutant, when both β2/Pup 1 and β1/Pre3 are inactive, β7/Pre4 is still processed; in this case, cleavage occurs between Ile-19 and Ala-18, which is a rather long distance away from the only available active site, β5/Pre2. While cleavages after valine and isoleucine can

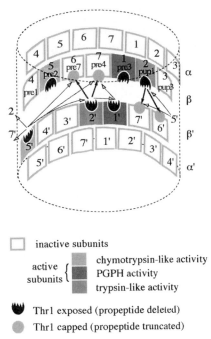

inactive subunits

active subunits { chymotrypsin-like activity
  PGPH activity
  trypsin-like activity

Thr1 exposed (propeptide deleted)

Thr1 capped (propeptide truncated)

FIG. 8. Current model for intersubunit processing in 20S proteasomes. In wild-type proteasomes, inactive precursor subunits $\beta6$ and $\beta7$ are processed by the active subunit $\beta2$ of the opposite ring (cleavages are indicated by bold arrows with filled heads). Mutations of $\beta2$ (and other active subunits) induce alternate cleavages exerted by the nearest active neighbor, respectively (cleavages are indicated by open arrow heads). For the sake of simplicity alternate cleavages are shown in one direction only, $\beta' \rightarrow \beta$ (20, 46, 47).

be assigned to the BrAAP activities of subunits $\beta1$/Pre3 and $\beta5$/Pre2, asparagine is hardly a typical "tryptic" P1 position, as it is preferred by $\beta2$/Pup1 (20, 47). A schematic summary of intersubunit processing in yeast proteasomes derived from genetic, biochemical, and structural studies is shown in Fig. 8. Although based on reliable data (from yeast), intersubunit cleavage across the two β rings is difficult to reconcile with the proposal that length reduction of the two inactive mammalian subunits, $\beta6$/C5 and $\beta7$/N3, occurs concomitantly with the formation of half proteasomes (see Section III,B,1), even if one assumes that an association of two half proteasomes is fully reversible at this stage.

The existence of processing intermediates has prompted the proposal of a two-step mechanism for β precursor maturation (46). However, it is not clear whether initial length reduction generally precedes self-processing of active β subunits, or whether it only serves to remove bulky protrusions from inactive subunits obstructing access to active sites. For a few

wild-type β subunits, β5i/LMP7, βli/LMP2, and β1/Y, processing intermediates have, in fact, been detected, supporting the two-step model (*82, 83*).
However, in yeast active site double mutants (e.g., β2/Pup1 β1/Pre3 mutant),
the remaining active subunit (e.g., β5/Pre2) would not be correctly processed,
if an initial length reduction were indeed necessary for final self-activation
(*20, 47*); it remains to be investigated whether such a step facilitates subsequent autolysis.

Although the coexistence of active and inactive β subunits is a special feature of eukaryotic proteasomes, the mechanism of intermolecular precursor
shortening is also seen at work in prokaryotes, as demonstrated by proteasomes assembled *in vitro* from wild-type and inactive mutant β subunits
of *Rhodococcus erythropolis* (Mayr *et al.*, submitted). However, processing
intermediates are not detected with wild-type subunits alone, and a single
wild-type subunit surrounded by inactive mutant subunits is activated, indicating that an initial length reduction is not essential.

Regardless of whether propeptides are liberated through self-cleavage of
active subunits or through intersubunit cleavage, they have to be removed
from the catalytic chamber so they will not interfere with efficient substrate
uptake. Proteasomes from *Rhodococcus erythropolis* are particularly well
suited to study the fate of cleaved propeptides, as subunit assembly and
maturation proceeds *in vitro,* and identical copies of β-precursor subunits
facilitate the identification of products. Both *Rhodococcus* β-type subunits
encode propeptides of more than 60 amino acids in length (*44*). Interestingly,
the degradation products derived from the propeptides are very similar to
degradation products from protein substrates. Cleavage occurs in a processive manner, i.e., the polypeptide chain is cleaved several times before being
discharged (*96*), and also the size distribution of the resulting products is
reminiscent of that of protein substrates.

IV. Functional Aspects: Why Proregions?

As they are removed from the precursor during maturation, propeptides
do not participate in the final function of their cognate proteins; nevertheless, in many cases, they are involved in the maturation process itself and
are often essential for attaining the "active form." Many protein precursors
are synthesized in a prepro form, with the presequence, usually the very
N-terminal segment, acting as a signal peptide for intracellular targeting or
export, and the prosequence, located between the signal peptide and the region that forms the mature protein, mediating the correct folding of the latter
(for reviews, see Refs. *97–99*). Particularly well studied examples of this concept are the serine proteases subtilisin and α-lytic protease (*100*), but also
several nonprotease proteins, such as peptide hormones and neuropeptides,

are synthesized in a prepro form and depend on their prosequence to reach the final conformation (see, e.g., Ref. *101*). Although propeptide-assisted folding is a widespread phenomenon, it is not obvious when and why it is essential; the serine proteases trypsin, chymotrypsin, and elastase, for example, do not require their propeptides for proper folding. Sometimes, the propeptide itself contains the signal for targeting proteins to specific subcellular locations (*102–104*), but often the effect of propeptides on cellular localization appears to be an indirect one, i.e., to depend on the folding state. Another function of propeptides is to maintain proteases or hormones in an inactive state (*78*). Steric hindrance is one mechanism to suppress activity, as structural data for enzymes in complex with their propeptides show (see, e.g., Refs. *105–107*). Parts of the propeptides occupy the substrate binding sites, thereby preventing access of substrate. The active sites are already completely formed in these proenzymes, i.e., they do not undergo conformational changes upon activation. Through a perfect fit of the respective parts of the propeptide and the enzyme's substrate binding pocket, many propeptides are very potent inhibitors with K_i values ranging from 10^{-7} to 10^{-11} M (see, e.g., Refs. *108–110*). *In vivo*, mechanisms must exist for relieving inhibition. In cathepsin B and related proteases, for instance, a structural rearrangement of the propeptide part is triggered by changes in pH loosening the interaction between the pro- and mature domains, such that the propeptide becomes susceptible to degradation (see, e.g., Refs. *105, 111*).

A deeper understanding of the functional role(s) of the propeptides of proteasome β-type subunits is just emerging, and some of the aforementioned "typical" propeptide functions do apply to them. In view of the enormous divergence of proteasome propeptides, it is unlikely that they serve a single function. Different propeptides appear to have different roles, and some are probably multifunctional. Table I summarizes currently available experimental facts.

A. PROREGIONS CONTROL ACTIVITY BY CAPPING THE N TERMINUS

More than with other enzymes synthesized in an inactive pro form, it is obvious how Ntn hydrolases are maintained in an inactive form: since the N-terminal amino groups of the mature proteins participate in the catalytic mechanism, acting as primary proton acceptors, proregions of any length and sequence will prevent the initial step of catalysis; once cleaved off, the inhibitory effect is irreversibly lost. It has been shown for mature *Thermoplasma acidophilum* proteasomes that they cannot be inactivated by adding an excess of propeptide, although the octapeptide is certainly small enough to enter the archaeal proteasome (*76*). To maintain β subunits in an inactive form, until they are properly assembled and activity is sequestered from the cellular environment, may be a fundamental function of all propeptides. It

TABLE I

EVIDENCE OF PROPEPTIDE FUNCTION

Subunit (species, length of propeptide)	Phenotype of propeptide deletion/replacement	→	Proposed function of propeptide
β1/Pre3 (yeast, 19aa)	No severe growth defect, but total loss of PGPH activity ($112, 60, 113$)		Not essential for folding/assembly
	N^{α}-acetylation of Thr1 ($112, 60, 113$)		Protects Thr1 against acetylation
βli/LMP2 (human, 20aa)	Incorporation efficiency of βli/LMP2 reduced (to 66%), which causes reduced incorporation of β2i/MECL1 ($124, 74$)		Supports folding/assembly
			Favors generation of immunoproteasomes
	Replacement by β5i/LMP7 propeptide: LMP2 positioning unchanged, processing impaired (46)		No subunit positioning signal
			Provides information for processing
β2/Pup1 (yeast, 29aa)	Poor growth, partly restored by expression of propeptide *in trans* (60)		May support folding/assembly
	N^{α}-acetylation of Thr1, trypsin-like activity reduced to 40% ($112, 113$)		Protects Thr1 against acetylation
β2i	No published data		
β5/Doa3 (yeast, 76aa)	Lethal, Doa3 not incorporated [Doa3 propeptide fully functional *in trans* ($45, 47$)]		Essential for folding/assembly "chaperone-like" function
	Effect suppressed by Δ*UMP1* (87)		Essential for Ump1-assisted proteasome maturation
	Not replaceable by the β1/pre3 propeptide (45)		Subunit-specific function
	N^{α}-acetylation of Thr1 ($112, 113$)		Protects Thr1 against acetylation

β5i/LMP7E2 (human, 71aa)	No incorporation of subunit (115)	Essential for folding/assembly "Chaperone-like" function
	Replacement by propeptide β5i/LMP7E1: inefficient incorporation, processing of βli/LMP2 and β2i/MECL1 impaired (125, 74)	Favors generation of a homogeneous population of immunoproteasome
β6	No published data	
β7/Pre4 (yeast, 41aa)	Minute effect on growth (60)	No essential function
β1 (Rhodococcus erythropolis, 65aa)	In vitro formation of proteasomes drastically retarded subunit folding impaired (77)	Supports folding and assembly "Chaperone-like" function
β (Thermoplasma acidophilum, 8aa; Methanococcus jannaschii, 6aa; Methanosarcina thermophila, 9aa)	In the presence of α subunits no effect on formation of recombinant proteasomes (76, 43, 16) In the absence of α subunits no ordered assembly but minimal proteolytic activity of βΔ subunits (76)	Not essential for folding/assembly[a] Prevents premature activation[a]
β (Haloferax volcanii, 49aa)	Only 70% reassembly from dissociated mature proteasomes (29)	May support folding/assembly

[a] To prevent premature activation is probably a fundamental function common to all proteasome propeptides.

may be the only role of short archaeal propeptides, which have been shown to be dispensable for the formation of functional proteasomes (*16, 43, 76*). On the other hand, such a function would seem to be unnecessary, given that complete particle assembly is required to achieve an active conformation. In fact, members of the bacterial HslV family are synthesized without propeptides, and this does not result in uncontrolled protein degradation by nonassembled subunits. One could imagine *in vivo* conditions, however, under which β subunits associate improperly, yielding partially active complexes. For instance, a weak hydrolytic activity has been reported in irregular protein aggregates formed by separately expressed *Thermoplasma* β subunits in *E. coli* (*76*).

More recent investigations revealed an additional function of β propeptides, related to its role as a protective device for the N-terminal threonine, but of an opposite character: Propeptides prevent inactivation. The three potentially active β subunits of yeast proteasomes were found to become targets of cellular N^{α}-acetyltransferases when the propeptides were genetically deleted, and all three subunits lost activity by this modification (*112, 60, 113*). N^{α}-acetylation, which immediately follows cotranslational processing of the initiator methionine by aminopeptidases, is a common modification of eukaryotic proteins but rarely used in prokaryotic cells (*114*). In fact, nine of the yeast proteasome subunits (seven α subunits, as well as subunits $\beta3$ and $\beta4$) are acetylated, whereas the remainig five subunits (the active β subunits, $\beta1$, $\beta2$, and $\beta5$, as well as the inactive subunits $\beta6$ and $\beta7$) are obviously protected by their propeptides (*113*).

B.   THE PROREGIONS AND THEIR ROLE IN FOLDING

By analogy with molecular chaperones, which assist the folding of other polypeptide chains but do not become a part of the final structures, prosequences involved in the folding of covalently linked protein domains are often referred to as intramolecular chaperones (*98–100*). However, the action of prosequences in protein folding is different from that of molecular chaperones in some respects. Most importantly, the latter assist folding by preventing off-pathway reactions, whereas prosequences directly accelerate the forward folding reaction; as they work in a 1 : 1 fashion and are degraded after mediating folding, they can be seen as single turnover catalysts.

The Ntn fold does not generally rely on propeptide assistance, as the proteasome α-type subunits or HslV proteins demonstrate, both lacking proregions. In addition, a number of propeptide deletions in β-type subunits have been reported to have no or only a modest effect on the mutants' phenotype (Table I). Yet, three proteasome propeptides have clearly been implicated in subunit folding: both the propeptides of yeast $\beta5$/Doa3 (75 amino acids) and those of human β5i/LMP7 (72 amino acids) are indispensable for incorporation of the respective subunits into proteasomes (45, 115), and the third, the propeptide of *Rhodococcus* β1 subunit (65 amino acids), is critical, though not essential, for proteasome assembly. Hitherto, only the bacterial propeptide has directly been shown to facilitate subunit folding (77). With their unusual lengths, propeptides of these β subfamilies (β5 and actinomycetes β subunits) may in fact be particularly competent to exert such a function (Figs. 5 and 6). For comparison, the propeptides of α-lytic protease, subtilisin, or carboxypeptidase Y, classical examples of intramolecular chaperones, are 166, 77, and 91 residues long (116–118).

Although it remains to be established at which step in the folding and assembly pathway the propeptide of yeast β5/Doa3 is required, its role, studied *in vivo,* appears to fit into the concept of intramolecular chaperones (45). β5/Doa3 propeptide cannot be replaced by the propeptide of another subunit, β1/Pre3, but functions equally well when added as a separate polypeptide (*in trans*). Both specific and *in trans* functions are characteristics of intramolecular chaperones. Interestingly, the lethal deletion of Doa3 propeptide is suppressed in the $\Delta UMP1$ mutant, indicating an interaction between the two polypeptides (87) (see, Section III,B,2). The role of the propeptide of *Rhodococcus* β1 has been studied *in vitro* (77). The propeptide deleted subunit turned out to be unable to reach its native fold, and association with α-type subunits, i.e., formation of proteasomes, is drastically affected. Stoichiometric addition of the propeptide as a separate entity, however, restores proper folding and allows efficient proteasome assembly. It is noteworthy that dimerization of half proteasomes is also retarded in the absence of the propeptide, indicating that the propeptide has another role at a later stage of proteasome assembly.

Work on α-lytic protease and subtilisin suggested that the propeptides accelerate folding of the protease domains by lowering the height of energy barriers along the pathway (119, 120). The *Rhodococcus* propeptide may act in a similar manner, binding to, and thus stabilizing, a transition state of folding. Like the propeptides of the two serine proteases, the *Rhodococcus* β1 propeptide is fully functional without covalent linkage, although it is probably not able to adopt a defined structure autonomously (Seemüller, unpublished data); this is also the case for the propeptide of subtilisin (120). All proregions, examined so far that promote folding are also potent inhibitors of their respective enzymes, since they interact strongly with the native state. For the propeptide of *Rhodococcus* β1 such an inhibitory effect has not been shown, because of experimental problems: β subunits are inactive as monomers, and any effect of the propeptide on assembled proteasomes would primarily reflect the competition between substrate and inhibitor for entry into the interior (through the narrow pore in the outer α

rings). When the propeptide is added to individual α- and β-type subunits it is instantly degraded on proteasome assembly, clearing the proteolytic chamber for substrate uptake (*44*). Interestingly, the same happens to the propeptide of subtilisin, the inhibitory effect of which is only temporary, and decreases as the propeptide is degraded by mature subtilisin (*121*).

C. THE PROREGIONS AND THEIR ROLE IN ASSEMBLY

In eukaryotes, each of the 14 different subunits takes a unique position within the proteasome complex. In view of the similarity among the subunits and their practically identical fold, it is tempting to speculate that the diversity of their propeptides might help them find the correct position within the rings. Yet, such a hypothesis is in conflict with the observation that α subunits, which lack these distinguishing features, assemble spontaneously and that their assembly precedes that of the β subunits. Moreover, at least for one β-subunit precursor, a positioning function of the propeptide can be excluded. Fusion of the propeptide of human $\beta5i$/LMP7 to the mature sequence of human $\beta1i$/LMP2 did not exchange the position of the subunit (*46*). Like the yeast subunits $\beta1$/Pre3, $\beta2$/Pup 1, and $\beta7$/Pre3 (Table I), mammalian $\beta1i$/LMP2 finds its correct neighbors within the β ring even in the absence of the propeptide, although the incorporation efficiency may be reduced. This latter effect is likely to reflect minor folding defects of the propeptide-deleted subunits, indicating a folding function of the respective propeptides rather than a positioning function. The fact that propeptides do not represent signals for the correct positioning of subunits within the proteasome does not exclude the possibility that they exert an influence on the chronology of subunit assembly; the need for such an orchestration function is most obvious for the generation of immunoproteasomes. It has been shown that the exchange of a constitutive subunit by its interferon inducible counterpart is not regulated on the protein expression level; induced expression of $\beta1i$/LMP2, for example, does not down regulate $\beta1$/Y expression (*122*). Rather, both proteins appear to compete for incorporation into the complex, and on overexpression $\beta1i$/LMP2 can completely displace $\beta1$/Y (*123*). In agreement with this scenario, $\beta1i$/LMP2 is found early in proteasome precursor complexes, while $\beta1$/Y appears to be incorporated late (*84*). Since these two proteins differ mainly in their prosequences (Fig. 5), it is tempting to hold them responsible for the different incorporation characteristics. The mode of action is most likely related to the chaperone-like activity of propeptides: accelerating the folding of subunit $\beta1i$ increases its chance of being incorporated. Consequently, the $\beta1i$/LMP2 propeptide would support the generation of immunoproteasomes, because the two other inducible subunits, $\beta2i$/MECL1 and $\beta5i$/LMP7, both require the presence of $\beta1i$/LMP2 in the complex for efficient incorporation (*123–125*). In addition, proteasome precursor complexes containing βli/LMP2 and β2i/MECL1 are long-lived in the absence of β5i/LMP7, whereas the final integration of β5i/LMP7 causes rapid processing of βli/LMP2 and β2i/MECL1. This effect is not dependent on the subunit's catalytic activity, but specifically requires the prosequence of β5i/LMP7E2. Neither β5i/LMP7E1, another splicing form of LMP7 that is identical to β5i/LMP7E2 except for the prosequence, nor the constitutive counterpart of LMP7, subunit β5/X, are efficient (74, 125). This complex interplay favors the generation of a homogeneous population of immuno-proteasomes containing all three inducible subunits.

REFERENCES

1. Coux, O., Tanaka, K., and Goldberg, A. L. (1996). *Annu. Rev. Biochem.* **65**, 801.
2. Hilt, W., and Wolf, D. (1996). *Trends Biochem. Sci.* **21**, 96.
3. Hershko, A., and Ciechanover, A. (1998). *Annu. Rev. Biochem.* **67**, 425.
4. Hochstrasser, M. (1996). *Annu. Rev. Gen.* **30**, 405.
5. Varshavsky, A. (1997). *Trends Biochem. Sci.* **22**, 383.
6. Baumeister, W., Walz, J., Zühl, F., and Seemüller, E. (1998). *Cell* **92**, 367.
7. Tanaka, K. (1998). *J. Biochem.* **123**, 195.
8. Voges, D., Zwickl, P., and Baumeister, W. (1999). *Annu. Rev. Biochem.* **68**, 1015.
9. Baumeister, W., Cejka, Z., Kania, M., and Seemüller, E. (1997). *Biol. Chem.* **378**, 121.
10. Lupas, A., Flanagan, J., Tamura, T., and Baumeister, W. (1997). *Trends Biochem. Sci.* **22**, 399.
11. Wenzel, T., and Baumeister, W. (1995). *Nat. Struct. Biol.* **2**, 199.
12. Confalonieri, F., and Duguet, M. (1995). *Bioessays* **17**, 639.
13. Beyer, A. (1997). *Protein Sci.* **6**, 2043.
14. Wolf, S., Nagy, I., Lupas, A., Pfeifer, G., Cejka, Z., Müller, S., Engel, A., DeMot, R., and Baumeister, W. (1998). *J. Mol. Biol.* **277**, 13.
15. Zwickl, P., Ng, D., Woo, K. M., Klenk, H. P., and Goldberg, A. L. (1999). *J. Biol. Chem.* **274**, 26008.
16. Wilson, H. L., Ou, M. S., Aldrich, H. C., and Maupin-Furlow, J. (2000). *J. Bacteriol.* **182**, 1680.
17. Dubiel, W., Ferrell, K., Pratt, G., and Rechsteiner, M. (1992). *J. Biol. Chem.* **267**, 22699.
18. Löwe, J., Stock, D., Jap, B., Zwickl, P., Baumeister, W., and Huber, R. (1995). *Science* **28**, 533.
19. Schauer, T. M., Nesper, M., Kehl, M., Lottspeich, F., Muller, T. A., Gerisch, G., and Baumeister, W. (1993). *J. Struct. Biol.* **111**, 135.
20. Groll, M., Heinemeyer, W., Jäger, S., Ullrich, T., Bochtler, M., Wolf, D. H., and Huber, R. (1999). *Proc. Nat. Acad. Sci. USA* **96**, 10976.
21. Kopp, F., Hendil, K. B., Dahlmann, B., Kristensen, P., Sobek, A., and Uerkvitz, W. (1997). *Proc. Nat. Acad. Sci. USA* **94**, 2939.
22. Dahlmann, B., Kopp, F., Kristensen, P., and Hendil, K. B. (1999). *Arch. Biochem. Biophys.* **363**, 296.
23. Goldberg, A. L., Gaczynska, M., Grant, E., Michalek, M., and Rock, K. L. (1995). *Cold Spring Harbor Sym. Quant. Biol.* **60**, 479.
24. Tanaka, K., Tanahashi, N., Tsurumi, C., Yokota, K. Y., and Shimbara, N. (1997). *Adv. Immunol.* **64**, 1.
25. Rock, K., and Goldberg, A. (1999). *Annu. Rev. Immunol.* **17**, 739.

26. Tamura, T., Nagy, I., Lupas, A., Lottspeich, F., Cejka, Z., Schoofs, G., Tanaka, K., Demot, R., and Baumeister, W. (1995). *Curr. Biol.* **5,** 766.

27. Zühl, F., Tamura, T., Dolenc, I., Cejka, Z., Nagy, I., Demot, R., and Baumeister, W. (1997). *FEBS Lett.* **400,** 83.

28. Bauer, M. W., Bauer, S. H., and Kelly, R. M. (1997). *Appl. Environ. Microbiol.* **63,** 1160.

29. Wilson, H. L., Aldrich, H. C., and Maupin-Furlow, J. (1999). *J. Bacteriol.* **181,** 5814.

30. Kessel, M., Wu, W. F., Gottesman, S., Kocsis, E., Steven, A. C., and Maurizi, M. R. (1996). *FEBS Lett.* **398,** 274.

31. Rohrwild, M., Coux, O., Huang, H. C., Moerschell, R. P., Yoo, S. J., Seol, J. H., Chung, C. H., and Goldberg, A. L. (1996). *Proc. Nat. Acad. Sci. USA* **93,** 5808.

32. Bochtler, M., Hartmann, C., Song, H. K., Bourenkov, G. P., Bartunik, H. D., and Huber, R. (2000). *Nature* **403,** 800.

33. Baumeister, W., Dahlmann, B., Hegerl, R., Kopp, F., Kuehn, L., and Pfeifer, G. (1988). *FEBS Lett.* **241,** 239.

34. Pühler, G., Weinkauf, S., Bachmann, L., Müller, S., Engel, A., Hegerl, R., and Baumeister, W. (1992). *EMBO J.* **11,** 1607.

35. Groll, M., Ditzel, L., Löwe, J., Stock, D., Bochtler, M., Bartunik, H. D., and Huber, R. (1997). *Nature* **386,** 463.

36. Brannigan, J. A., Dodson, G., Duggleby, H. J., Moody, P. C. E., Smith, J. L., Tomchick, D. R., and Murzin, A. G. (1995). *Nature* **378,** 416.

37. Oinonen, C., Tikkanen, R., Rouvinen, J., and Peltonen, L. (1995). *Nat. Struct. Biol.* **2,** 1102.

38. Smith, J. L., Zaluzec, E. J., Wery, J.-P., Niu, L., Switzer, R. L., Zalkin, H., and Satow, Y. (1994). *Science* **264,** 1427.

39. Duggleby, H. J., Tolley, S. P., Hill, C. P., Dodson, E. J., Dodson, G., and Moody, P. C. E. (1995). *Nature* **373,** 264.

40. Bompard-Gilles, C., Villeret, V., Davies, G. J., Fanuel, L., Joris, B., Frere, J. M., and Van Beeumen, J. (2000). *Structure* **8,** 153.

41. Seemüller, E., Lupas, A., Stock, D., Löwe, J., Huber, R., and Baumeister, W. (1995). *Science* **268,** 579.

42. Seemüller, E., Lupas, A., and Baumeister, W. (1996). *Nature* **382,** 468.

43. Maupin-Furlow, J. A., Aldrich, H. C., and Ferry, J. G. (1998). *J. Bacteriol.* **180,** 1480.

44. Mayr, J., Seemüller, E., Müller, S. A., Engel, A., and Baumeister, W. (1998). *J. Struct. Biol.* **124,** 179.

45. Chen, P., and Hochstrasser, M. (1996). *Cell* **86,** 961.

46. Schmidtke, G., Kraft, R., Kostka, S., Henklein, P., Frommel, C., Löwe, J., Huber, R., Kloetzel, P. M., and Schmidt, M. (1996). *EMBO J.* **15,** 6887.

47. Heinemeyer, W., Fischer, M., Krimmer, T., Stachon, U., and Wolf, D. H. (1997). *J. Biol. Chem.* **272,** 25200.

48. Arrigo, A. P., Tanaka, K., Goldberg, A. L., and Welch, W. J. (1988). *Nature* **331,** 192.

49. Dahlmann, B., Kuehn, L., Ishiura, S., Tsukahara, T., Sugita, H., Tanaka, K., Rivett, A. J., Hough, R. F., Rechsteiner, M., Mykles, D. L., Fagan, M., Waxman, L., Ishii, S., Sasaki, M., Kloetzel, P.-M., Harris, H., Ray, K., Behal, F. J., DeMartino, G. N., and McGuire, M. J. (1988). *Biochem. J.* **255,** 750.

50. Orlowski, M., and Wilk, S. (1988). *Biochem. J.* **255,** 751.

51. Cardozo, C. (1993). *Enzyme Protein* **47,** 296.

52. Orlowski, M., Cardozo, C., and Michaud, C. (1993). *Biochemistry* **32,** 1563.

53. Heinemeyer, W., Kleinschmidt, J. A., Saidowsky, J., Escher, C., and Wolf, D. H. (1991). *EMBO J.* **10,** 555.

54. Heinemeyer, W., Gruhler, A., Mohrle, V., Mahe, Y., and Wolf, D. H. (1993). *J. Biol. Chem.* **268,** 5115.

55. Hilt, W., Enenkel, C., Gruhler, A., Singer, T., and Wolf, D. H. (1993). *J. Biol. Chem.* **268,** 3479.
56. Enenkel, C., Lehmann, H., Kipper, J., Guckel, R., Hilt, W., and Wolf, D. H. (1994). *FEBS Lett.* **341,** 193.
57. Arendt, C. S., and Hochstrasser, M. (1997). *Proc. Nat. Acad. Sci. USA* **94,** 7156.
58. Dick, T. P., Nussbaum, A. K., Deeg, M., Heinemeyer, W., Groll, M., Schirle, M., Keilholz, W., Stevanovic, S., Wolf, D. H., Huber, R., Rammensee, H. G., and Schild, H. (1998). *J. Biol. Chem.* **273,** 25637.
59. McCormack, T. A., Cruikshank, A. A., Grenier, L., Melandri, F. D., Nunes, S. L., Plamondon, L., Stein, R. L., and Dick, L. R. (1998). *Biochemistry* **37,** 7792.
60. Jäger, S., Groll, M., Huber, R., Wolf, D. H., and Heinemeyer, W. (1999). *J. Mol. Biol.* **291,** 997.
61. Dahlmann, B., Kuehn, L., Grziwa, A., Zwickl, P., and Baumeister, W. (1992). *Eur. J. Biochem.* **208,** 789.
62. Nagy, I., Tamura, T., Vanderleyden, J., Baumeister, W., and Demot, R. (1998). *J. Bacteriol.* **180,** 5448.
63. Pouch, M. N., Cournoyer, B., and Baumeister, W. (2000). *Mol. Microbiol.* **35,** 368.
64. Cardozo, C., Vinitsky, A., Michaud, C., and Orlowski, M. (1994). *Biochemistry* **33,** 6483.
65. Dick, L. R., Aldrich, C., Jameson, S. C., Moomaw, C. R., Pramanik, B. C., Coyle, C. K., DeMartino, G. N., Bevan, M. J., Forman, J. M., and Slaughter, C. A. (1994). *J. Immunol.* **152,** 3884.
66. Wenzel, T., Eckerskorn, C., Lottspeich, F., and Baumeister, W. (1994). *FEBS Lett.* **349,** 205.
67. Leibovitz, D., Koch, Y., Fridkin, M., Pitzer, F., Zwickl, P., Dantes, A., Baumeister, W., and Amsterdam, A. (1995). *J. Biol. Chem.* **270,** 11029.
68. Niedermann, G., Butz, S., Ihlenfeldt, H., Grimm, R., Lucchiari, M., Hoschützky, H., Jung, G., Maier, B., and Eichmann, K. (1995). *Immunity* **2,** 289.
69. Ehring, B., Meyer, T. H., Eckerskorn, C., Lottspeich, F., and Tampe, R. (1996). *Eur. J. Biochem.* **235,** 404.
70. Dolenc, I., Seemüller, E., and Baumeister, W. (1998). *FEBS Lett.* **434,** 357.
71. Nussbaum, A. K., Dick, T. P., Keilholz, W., Schirle, M., Stevanovic, S., Dietz, K., Heinemeyer, W., Groll, M., Wolf, D. H., Huber, R., Rammensee, H. G., and Schild, H. (1998). *Proc. Nat. Acad. Sci. USA* **95,** 12504.
72. Altuvia, Y., and Margalit, H. (2000). *J. Mol. Biol.* **295,** 879.
73. Kisselev, A. F., Akopian, T. N., and Goldberg, A. L. (1998). *J. Biol. Chem.* **273,** 1982.
74. Schmidt, M., Zantopf, D., Kraft, R., Kostka, S., Preissner, R., and Kloetzel, P. M. (1999). *J. Mol. Biol.* **288,** 117.
75. Ditzel, L., Huber, R., Mann, K., Heinemeyer, W., Wolf, D. H., and Groll, M. (1998). *J. Mol. Biol.* **279,** 1187.
76. Zwickl, P., Kleinz, J., and Baumeister, W. (1994). *Nat. Struct. Biol.* **1,** 765.
77. Zühl, F., Seemüller, E., Golbik, R., and Baumeister, W. (1997). *FEBS Lett.* **418,** 189.
78. Khan, A. R., and James, M. N. G. (1998). *Protein Sci.* **7,** 815.
79. Gerards, W. L. H., Enzlin, J., Haner, M., Hendriks, I. L. A. M., Aebi, U., Bloemendal, H., and Boelens, W. (1997). *J. Biol. Chem.* **272,** 10080.
80. Yao, Y., Toth, C. R., Huang, L., Wong, M. L., Dias, P., Burlingame, A. L., Coffino, P., and Wang, C. C. (1999). *Biochem. J.* **344,** 349.
81. Yang, Y., Fruh, K., Ahn, K., and Peterson, P. A. (1995). *J. Biol. Chem.* **270,** 27687.
82. Frentzel, S., Pesold, H. B., Seelig, A., and Kloetzel, P. M. (1994). *J. Mol. Biol.* **236,** 975.
83. Schmidtke, G., Schmidt, M., and Kloetzel, P. M. (1997). *J. Mol. Biol.* **268,** 95.
84. Nandi, D., Woodward, E., Ginsburg, D. B., and Monaco, J. J. (1997). *EMBO J.* **16,** 5363.

85. Thomson, S., and Rivett, A. J. (1996). *Biochem. J.* **315,** 733.
86. Rodriguez-Vilarino, S., Arribas, J., Arizti, P., and Castano, J. G. (2000). *J. Biol. Chem.* **275,** 6592.
87. Ramos, P. C., Hockendorff, J., Johnson, E. S., Varshavsky, A., and Dohmen, R. J. (1998). *Cell* **92,** 489.
88. Zwickl, P., Lottspeich, F., and Baumeister, W. (1992). *FEBS Lett.* **312,** 157.
89. Kisselev, A. F., Songyang, Z., and Goldberg, A. L. (2000). *J. Biol. Chem.* **275,** 14831.
90. Perler, F. B., Xu, M.-Q., and Paulus, H. (1997). *Curr. Opin. Chem. Biol.* **1,** 292.
91. Guan, C., Cui, T., Rao, V., Liao, W., Benner, J., Lin, C.-L., and Comb, D. (1996). *J. Biol. Chem.* **271,** 1732.
92. Lee, J. J., Stephen, E. E., von Kessler, D. P., Porter, J. A., Sun, B. I., and Beachy, P. A. (1994). *Science* **266,** 1528.
93. Porter, J. A., Young, K. E., and Beachy, P. A. (1996). *Science* **274,** 255.
94. Shao, Y., and Kent, S. B. H. (1997). *Chem. Biol.* **4,** 187.
95. Perler, F. B. (1998). *Cell* **92,** 1.
96. Akopian, T. N., Kisselev, A. F., and Goldberg, A. L. (1997). *J. Biol. Chem.* **272,** 1791.
97. Neurath, H. (1989). *Trends Biochem. Sci.* **14,** 268.
98. Shinde, U., and Inouye, M. (1993). *Trends Biochem. Sci.* **18,** 442.
99. Eder, J., and Fersht, A. R. (1995). *Mol. Microbiol.* **16,** 609.
100. Baker, D., Shiau, K., and Agard, D. A. (1993). *Curr. Opin. Cell. Biol.* **5,** 966.
101. Dodson, G., and Steiner, G. (1998). *Curr. Opin. Struct. Biol.* **8,** 189.
102. Valls, L. A, Winter, J. R, and Stevens, T. H. (1990). *J. Cell Biol.* **111,** 361.
103. Kizer, J. S, and Tropsha, A. (1991). *Biochem. Biophys. Res. Commun.* **174,** 585.
104. Koelsch, G., Mares, M., Metcalf, P., and Fusek, M. (1994). *FEBS Lett.* **343,** 6.
105. Coulombe, R., Grochulski, P., Sivaraman, J., Menard, R., Mort, J. S., and Cygler, M. (1996). *EMBO J.* **15,** 5492.
106. Groves, M. R., Gaylor, M. A. J., Scott, M., Cummings, N. J., Pickersgill, R. W., and Jenkins, J. A. (1996). *Structure* **4,** 1193.
107. Sauter, N., Mau, T., Rader, S., and Agard, D. (1998). *Nat. Struct. Biol.* **5,** 945.
108. Ohta, Y., Hojo, H., Aimoto, S., Tobayashi, T., Zhu, X., Jordan, F., and Inouye, M. (1991). *Mol. Microbiol.* **5,** 1507.
109. Baker, D., Silen, J. L., and Agard, D. A. (1992). *Proteins* **12,** 339.
110. Maubach, G., Schilling, K., Rommerskirch, W., Wnez, I., Schultz, J. E., Weber, E., and Wiederanders, B. (1997). *Eur. J. Biochem.* **250,** 745.
111. Cygler, M., Sivaraman, J., Grochulski, P., Coulombe, R., Storer, A. C., and Mort, J. S. (1996). *Structure* **4.**
112. Arendt, C. S., and Hochstrasser, M. (1999). *EMBO J.* **18,** 3575.
113. Kimura, Y., Takaoka, M., Tanaka, S., Sassa, H., Tanaka, K., Polevoda, B., Sherman, F., and Hirano, H. (2000). *J. Biol. Chem.* **275,** 4635.
114. Bradshaw, R. A., Brickey, W. W., and Walker, K. W. (1998). *Trends Biochem. Sci.* **23,** 263.
115. Cerundolo, V., Kelly, A., Elliot, T., Trowsdale, J., and Townsed, A. (1995). *Eur. J. Immunol.* **25,** 554.
116. Silen, J. L., and Agard, D. A. (1989). *Nature* **341,** 462.
117. Zhu, X., Ohta, Y., Jordan, F., and Inouye, M. (1989). *Nature* **339,** 483.
118. Winther, J. R., and Sorensen, P. (1991). *Proc. Nat. Acad. Sci. USA* **88,** 9330.
119. Baker, D., Sohl, J. L., and Agard, D. A. (1992). *Nature* **356,** 263.
120. Eder, J., Rheinnecker, M., and Fersht, A. R. (1993). *J. Mol. Biol.* **233,** 293.
121. Kojima, S., Minagawa, T., and Miura, K. (1997). *FEBS Lett.* **411,** 128.
122. Früh, K., Gossen, M., Wang, K., Bujard, H., Peterson, P. A., and Yang, Y. (1994). *EMBO J.* **13,** 3246.

123. Kuckelkorn, U., Frentzel, S., Kraft, R., Kostka, R., Groettrup, M., and Kloetzel, P. M. (1995). *Eur. J. Immunol.* **25,** 2605.
124. Groettrup, M., Standera, S., Stohwasser, R., and Kloetzel, P. M. (1997). *Proc. Natl. Acad. Sci. USA* **94,** 8970.
125. Griffin, T. A., Nandi, D., Cruz, M., Fehling, H. J., Vankaer, L., Monaco, J. J., and Colbert, R. A. (1998). *J. Experiment. Med.* **187,** 97.
126. Walz, J., Erdmann, A., Kania, M., Typke, D., Koster, A. J., and Baumeister, W. (1998). *J. Struct. Biol.* **121,** 19.
127. Thompson, J. D., Gibson, T. J., Plewniak, F., Jeanmougin, F., and Higgins, D. G. (1997). *Nucl. Acids Res.* **25,** 4876.
128. Cuff, J. A., Clamp, M. E., Siddiqui, S. S., Finlay, M., and Barton, G. J. (1998). *Bioinformatics* **14,** 892.

13

Tsp and Related Tail-Specific Proteases

KENNETH C. KEILER • ROBERT T. SAUER

I. Introduction

Tsp is a bacterial periplasmic protease that processes C-terminal extensions from precursor proteins and also acts as part of a general surveillance pathway to degrade incomplete proteins that have been tagged by the SsrA system (*1, 2*). Both functions require selective recognition of proteins with specific sequences at the carboxyl terminus. Tsp efficiently degrades substrate proteins by making endoproteolytic cleavages at sites distant from the C-terminal recognition motif. The active site appears to involve a serine–lysine dyad mechanism. Most biochemical characterization has been performed on the enzyme from *Escherichia coli*, but Tsp homologs are broadly represented in bacteria and are also present in plant chloroplasts. Structural

THE ENZYMES, Vol. XXII

information has not been obtained for Tsp, but the structure of an algal homolog, the D1 C-terminal processing protease, has been determined (3).

II. Identification and General Structural Properties

The structural gene for Tsp from *E. coli* was identified as the site of a mutation that caused a defect in the C-terminal processing of penicillin-binding protein 3 and was named *prc* (processing involving C-terminal cleavage) (1). Independently, the Tsp (tail-specific protease) protein was purified from *E. coli* extracts using an assay for specific degradation of a variant of the N-terminal domain of λ repressor containing hydrophobic residues at its five C-terminal positions (4). Sequencing of the *tsp* gene revealed its identity with *prc*. The protein has been called both Tsp and Prc, and the gene both *tsp* and *prc*. Tsp is also likely to be identical to two other *E. coli* proteases described earlier in the literature, protease Re (5) and a protease that degrades oxidized glutamine synthetase (6). All of these enzymes have a similar molecular weights, subcellular localization patterns, and interactions with protease inhibitors, but sequence data are not available for Re or the glutamine synthetase degrading activity.

Tsp from *E. coli* is a 74-kDa protein containing 660 residues (4). Cellular fractionation experiments show that Tsp is located in the periplasmic compartment (4, 7), and the first residue, Val1, of the mature enzyme is encoded by a GTA codon roughly 60 bases upstream from the nearest in-frame ATG codon (4). In fact, there are two closely spaced potential initiator codons, each with a consensus Shine–Dalgarno sequence. Translational initiation at either ATG would result in synthesis of 20 and 22 residue leader sequences, respectively, with characteristics of a typical periplasmic-targeting signal peptide (4, 7).

Gel filtration chromatography and sedimentation-equilibrium centrifugation show that Tsp is monomeric, and circular dichroism studies indicate that the protein is approximately 35% helical (8). Structural information is not yet available for Tsp, but the crystal structure of a homologous protease from the algae *Scenedesmus obliquus* has been determined (3). This enzyme removes a C-terminal extension from the photosystem II D1 protein and is called D1P (gene *ctpA;* C-terminal processing). *Scenedesmus obliquus* D1P contains three structural domains (Fig. 1A, see color plate): a largely α-helical domain, a PDZ-like domain, and a C-terminal domain that contains the protease active site (see below). Figure 1B shows a schematic alignment of Tsp and D1P sequence homologs. The smallest members of this family are similar in size to *S. obliquus* D1P. *E. coli* Tsp is one of the larger enzymes and presumably contains one or more additional domains compared to

S. obliquus D1P. No members of this protease family share significant homology with any of the classical protease families. The PDZ-like domain of the Tsp/D1P family is homologous to mammalian PDZ domains (9), which mediate a variety of protein–protein interactions involving recognition of specific C-terminal sequences (10–17). Tsp also uses its PDZ domain to bind to the C-terminal sequences of substrate proteins [(18); see below]. In addition, a portion of the active-site domain is homologous to the tandem-repeat domains of interphotoreceptor retinol binding protein (IRBP) (4). IRBP binds hydrophobic ligands such as retinol, cholesterol, and fatty acids but has no known enzymatic activity (19).

III. Activity and Active-Site Residues

Assays for Tsp-mediated degradation *in vitro* typically follow the loss of full-length protein substrates, monitored by SDS–polyacrylamide gels, reverse-phase HPLC, or circular dichroism (4, 8, 20, 21). A continuous fluorimetric assay has also been developed using a fluorescence donor/quencher system to monitor accumulation of cleavage products (22). Assays *in vivo* have relied on degradation of cytochrome b_{562}, a periplasmic heme protein. When cytochrome b_{562} is cleaved by Tsp, the heme is released and its absorbance decreases, providing a convenient means to detect degradation either *in vivo* or *in vitro* (21).

Tsp has maximal activity against protein and peptide substrates between pH 5 and 9; there is little effect of ionic strength from 20 mM to 300 mM; and the enzyme is inhibited by carboxylic acid buffers (e.g., glycine and citrate) but is active in most other buffers tested (unpublished observations). Tsp is not inhibited by conventional protease inhibitors such as diisopropylfluorophosphate, phenylmethylsulfonyl fluoride, 3,4-dichloroisocoumarin, pepstatin, or *o*-phenanthroline (4), nor does it require ATP or other cofactors for activity (8). Tsp activity is stimulated by some multivalent cations (Mn^{2+}, Co^{2+}, and Ca^{2+}), is unaffected by Mg^{2+}, and is inhibited by Zn^{2+}, Cu^{2+}, and Fe^{3+} (4).

Tsp was initially suspected to use a mechanism different than classical serine, cysteine, aspartic, or metallo proteases because it was neither homologous to these enzymes nor inhibited by class-specific inhibitors of these proteases (4). To identify residues of Tsp required for proteolytic activity, 20 residues that were strongly or absolutely conserved among Tsp family members were individually changed by site-directed mutagenesis to alanine (20) (Table I). The resulting variants were then purified and assayed for proteolytic activity, structural integrity, and binding to a substrate, FITC-labeled insulin B chain. These studies identified two residues—Ser-430 and

TABLE I

EFFECTS OF MUTATIONS IN Tsp

Tsp variant[a]	Proteolytic activity	Defect
Wild type	+	
H19A	+	
H34A	+	
H60A	+	
H203A	+	
V229Q[b]	50%	Substrate binding
V229E[b]	40%	Substrate binding
D269A	+	
R371A	+	
S372A	+	
G375A	−	Structure
G376A	−	Structure
S428A	+	
S430A	−	Active site
S430C	+	
S432A	+	
E433A	−	Structure
D441A	−	?
D441N	+	
R444A	+	
E449A	+	
T452A	−	Structure
K455A	−	Active site
K455H	−	Active site
K455R	−	Structure
D505A	+	
H553A	+	
Δ1-205[b]	−	?
Δ335-660[b]	−	Active site

[a] Data from (20) unless otherwise noted.
[b] Data from (18).

Lys-455—with properties expected of active-site residues. Specifically, mutations at both positions eliminated proteolytic activity without significantly affecting Tsp structure or substrate binding. The corresponding residues in D1P (Ser-372 and Lys-397) are on the surface of the C-domain (Fig. 1A) and close enough to hydrogen bond after adjustments of their side-chain dihedral angles (3).

The properties of Ser430 substitution mutations in Tsp suggested that the wild-type side chain is likely to act as the nucleophile for peptide-bond attack. Namely, the Ala-430 enzyme was proteolytically inert, but the Cys-430 enzyme retained partial activity, which could now be inhibited by cysteine-modifying reagents such as iodoacetamide and *N*-ethylmaleimide (note that wild-type Tsp has no cysteines). In classical serine proteases such as subtilisin, cysteine has also been shown to be capable of replacing the active-site serine with retention of partial activity (*23, 24*). Other Ser and Thr residues are also conserved in the Tsp/D1P family, but mutations of these residues in Tsp either were silent or affected protein folding (Table I). Hence, both mutational studies and phylogenetic comparisons identify Ser-430 as the best candidate for the active-site nucleophile. Lys-455 in Tsp is also absolutely required for proteolytic activity (*20*). Variants with 16 different residues replacing Lys-455 were found to be inactive *in vivo*. In addition, purified Tsp variants containing Ala, Arg, or His at position 455 were inactive *in vitro*. The Ala-455 mutant had a circular dichroism spectrum and substrate-binding activity indistinguishable from those of wild-type Tsp, indicating that truncation of the Lys-455 side chain affects catalysis rather than structure or substrate recognition.

Although Tsp appears to use an active-site serine, it does not have a Ser-His-Asp catalytic triad like those in classical serine proteases such as chymotrypsin and subtilisin. First, each of the five histidines in Tsp was changed to alanine with no loss of proteolytic activity; by contrast, substitution of the active-site histidine in classical serine proteases abolishes activity (*25*). Second, there is no Tsp residue analogous to the triad Asp in classical serine proteases. Mutational studies show that Asp-441 plays some role in Tsp activity, but this residue can be functionally replaced by Asn and is a Glu in other family members (*20*). The active-site Asp in trypsin, however, cannot be replaced by Asn and is absolutely conserved (*26*). Finally, Lys-455 is critical for Tsp catalysis, whereas conventional serine proteases do not use an active-site lysine. Overall, these observations suggest that Tsp and D1P utilize a Ser-Lys catalytic dyad that may be similar to those used by the class A β-lactamases, LexA, UmuD, and the type I signal peptide peptidases (*27–33*). However, none of the latter enzymes show sequence or structural homology with the Tsp/D1P family (*3*). In conventional serine proteases, the serine hydroxyl in the Ser-His-Asp triad is activated for nucleophilic attack by transfer of its proton to the histidine, which in turn is stabilized by the aspartic acid. In Ser-Lys proteases, the serine is probably also the nucleophile, but the lysine presumably acts as the general base in place of histidine.

IV. Kinetic Constants

Kinetic parameters have been measured for Tsp degradation of the native Arc repressor protein, the BAS9 peptide *ac-NH*-WARAAAR-AAARBAAB-*COOH* (where B is aminobutyric acid), and the ssrA-tag peptide NH_2-AARAAK-(6-aminocaproyl)$_2$-ENYALAA-*COOH* with an

TABLE II

KINETIC CONSTANTS FOR CLEAVAGE OF PROTEIN AND PEPTIDE SUBSTRATES BY Tsp

Tsp variant	Substrate	C-terminal residues	$K_M(\mu M)$	K_{cat} (s^{-1})	K_{cat}/K_M ($\times 10^3 M^{-1}s^{-1}$)
wt	Arc[a]	GRIGA	50 +/− 36	0.19 +/− 0.05	3.8 +/− 2.9
	BAS9 peptide[a,b]	RBAAB	35 +/− 13	3.7 +/− 0.5	110 +/− 40
	Peptide 1[c,d]	YALAA	4.4 +/− 0.6	0.050 +/− 0.000	11
	Peptide 2[c,e]	YALAE			>0.058
V229Q	Peptide 1[c,d]	YALAA	9.5 +/− 1.3	0.050 +/− 0.010	5.3
V229E	Peptide 1[c,d]	YALAA	11.1 +/− 1.4	0.050 +/− 0.002	4.5

[a] Data from (8).

[b] BAS9 peptide is *ac-NH*-WARAAARAAARBAAB-*COOH* (where B is aminobutyric acid).

[c] Data from (22).

[d] Peptide 1 is NH_2-AARAAK-(6-aminocaproyl)$_2$-ENYALAA-*COOH* with an EDANS group attached to the N-terminal and a DABCYL group attached to the lysine side chain.

[e] Peptide 2 is identical to peptide 1 except for the last residue.

EDANS group attached to the N terminus and a DABCYL group attached to the lysine side chain (8, 22) (Table II). For Arc degradation, K_M was 50 μM, k_{cat} was 0.19 s^{-1}, and k_{cat}/K_M was $3.8 \times 10^3 M^{-1} s^{-1}$. For the BAS9 peptide, K_M was similar (35 μM), but k_{cat} and k_{cat}/K_M were 20- to 30-fold greater (3.7 s^{-1} and $1.1 \times 10^5 M^{-1} s^{-1}$, respectively). For the ssrA-tag peptide, K_M was 4.0 μM, k_{cat} was 0.086 s^{-1}, and k_{cat}/K_M was $2.2 \times 10^4 M^{-1} s^{-1}$. It was initially argued that k_{cat} for degradation of the BAS9 peptide was faster than that of Arc repressor because the latter protein is stably folded whereas the peptide is unstructured (8). This model, however, fails to explain why the ssrA-tag peptide, which is presumably also unstructured, has a k_{cat} value smaller than that for Arc degradation.

V. Substrate Specificity

What characteristics of a protein make it a substrate for Tsp? This question has been addressed by examining the susceptibility to Tsp cleavage of a variety of protein and peptide variants (8, 22). Four properties influence the degradation of proteins and peptides by Tsp: the identity of residues at the C terminus, the presence of a free α-carboxyl group, the thermodynamic stability of the native structure, and the presence of an appropriate cleavage site.

Tsp is unusual in that it recognizes substrate proteins in two distinct ways. The first recognition event depends on the C-terminal residues of the substrate. For example, Tsp was first purified based on its ability to degrade a variant of the N-terminal domain of λ repressor with the hydrophobic C-terminal sequence WVAAA, but not to degrade an otherwise identical variant with the polar C-terminal sequence RSEYE (4). Tsp's preference for nonpolar C-terminal residues has been confirmed for a variety of peptides and proteins (21, 22). More complete rules for substrate selectivity were deduced from randomization experiments in which the last three residues of a good substrate—cytochrome b_{562} with a WVAAA tail—were individually mutated and the resulting variants were assayed for degradation by Tsp (21) (Fig. 2). At the C-terminal residue, Tsp preferred small uncharged side chains in the order Ala > Cys > Val > Ser> Thr. The importance of an uncharged C-terminal residue is highlighted in the ssrA-tag peptide, where substitution of Glu for Ala at the C terminus reduced k_{cat}/K_M by at least 380-fold (22). At the penultimate position in cytochrome b_{562}-WVA$\underline{A}$A, there was less discrimination, but uncharged residues were again preferred in the order Ala, Tyr > Ile > Trp > Thr, Leu. At the antepenultimate or third position, the preference order was Ala > Leu > Val > Ile. In these studies, a strong correlation was also observed between the rates of substrate degradation *in vitro* and *in vivo,* indicating that Tsp does not require other cellular proteins for substrate discrimination (21). In addition to nonpolar C-terminal residues, efficient recognition by Tsp also appears to require a free α-COOH. For example, degradation of the BAS9 peptide was reduced by at least 10-fold when its C terminus was amidated (8).

In addition to the C-terminal tail sequence, the thermodynamic stability of native proteins affects their rate of cleavage by Tsp. For example, the Asn29→Lys variant of Arc repressor is much less stable than wild type and is degraded much faster by Tsp, whereas the Pro8 → Leu Arc variant is hyper stable and degraded much more slowly than wild type by Tsp (8). Because the mutations in these proteins do not alter the C-terminal tail sequence or sites of Tsp cleavage, the observed differences in degradation rates appear to be attributable to differences in protein stability (i.e., the fraction of time that a given molecule is denatured). This relationship between degradation rate and thermodynamic stability is consistent with a model in which bound substrates are only cleaved by Tsp in a transiently denatured form in which the peptide bonds of the substrate are available to the protease active site. By this model, local or global unfolding of a native substrate must precede cleavage.

Cleavage of appropriate substrates by Tsp depends not only on their C-terminal sequences but also on the presence of appropriate and accessible peptide-bond sequences. Tsp is capable of cleaving some substrates at numerous positions, and cleavage sites have been mapped for two substrates, λ repressor and Arc repressor (8). For both of these protein substrates, a

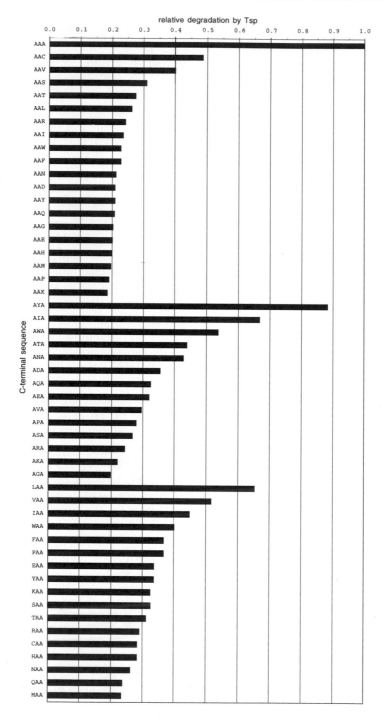

complex mixture of overlapping fragments was observed, even at early re-action times, suggesting that Tsp can make the first cleavage at any one of several different positions in individual substrate molecules. Tabulation of residues flanking the scissile bonds suggested a preference for small side chains such as Ala, Ser, and Val at the P1 position, with cleavage after Ile, Leu, Arg, and Lys being less common. At the P1′ position, 10 residues were found, including Ala, Ser, Val, Ile, Met, Leu, Tyr, Trp, Lys, and Arg. No pat-terns could be detected at positions farther from the scissile bond. These preferences are sufficiently broad that appropriate sites for Tsp cleavage are likely to be present in most proteins.

The observation that scissile bonds in substrate proteins were distinct from recognition determinants at the C terminus led to the hypothesis that Tsp contains distinct sites that mediate substrate binding and catalysis (8). Because PDZ domains usually mediate binding to specific C-terminal se-quences and interact directly with the α-COOH group, this domain of Tsp was an obvious candidate for mediating C-terminal recognition of sub-strates (9). Indeed, biochemical studies have confirmed that the isolated PDZ domain of Tsp can be expressed as a stable, soluble protein that medi-ates C-terminal recognition of substrates and functions independently of the active site (18). For example, intact Tsp binds the ssrA-tag peptide with a K_D of $1.8 \pm 0.3\ \mu M$, whereas the isolated PDZ fragment binds this peptide with a K_D of $1.9 \pm 0.3\ \mu M$. Moreover, an ssrA-tag peptide cross-links specifically to the PDZ domain, both in intact Tsp and in the isolated PDZ fragment (18). Further support comes from analysis of mutations in the PDZ domain. Val-229 is adjacent to the α-carboxylate-binding motif of the Tsp PDZ domain and is predicted to be involved in hydrophobic interactions with substrates. Mutation of Val-229 to Glu or Gln in intact Tsp resulted in a 2- to 3-fold increase in K_M for the ssrA-tag peptide but had no effect on k_{cat}, indicating that the affinity for the substrate has been reduced without affecting the catalytic step (18). Thus, the PDZ domain in Tsp is responsible for binding specificity by recognizing hydrophobic residues adjacent to the α-carboxyl group of good substrates.

Because the PDZ domain is distinct from the active site and can bind substrates independently, proteolysis is essentially a two-step process for Tsp. First the substrate is bound by the PDZ domain, and then proteolysis occurs when a peptide bond becomes available to the active site. This model is consistent with the importance of protein stability because binding of the C terminus of the substrate would tether it to Tsp until unfolding exposed

FIG. 2. Effects of the substrate C-terminal sequence on degradation by Tsp. Variants of cytochrome b_{562}-WVAAA that differ only in their three C-terminal residues were assayed for degradation by Tsp (21). The results were normalized relative to to the AAA variant.

a cleavage site. If the substrate was hyperstable, it could dissociate from Tsp before unfolding and cleavage occurred. This model can also explain the observation that short peptides can act as inhibitors of Tsp, but are not cleaved themselves (8, 18). The predominant cleavage sites in protein and peptide substrates of Tsp and D1P are generally 8 or more residues from the C terminus. Peptides of this general length or shorter could act as a competitive inhibitor by binding to the PDZ domain, but would not be cleaved efficiently because the peptide would be too short to reach the active site. Peptide-bond cleavage by Tsp as far as 99 residues from the C terminus has been observed in a 102-residue protein substrate (8). In a denatured protein tethered by its C terminus to the PDZ domain of Tsp, the chance of cleavage at any particular peptide bond should depend both on the flanking P1 and P1' side chains and on the probability of encounter of the peptide bond with the active site, which, in turn, will be a function of polypeptide length and flexibility.

The two-site model for Tsp activity raises some interesting but unanswered questions about the processivity of degradation. After an initial cleavage, does the C-terminal part of the substrate remain bound to the PDZ domain, allowing subsequent cleavage events to occur without dissociation and reassociation? Do the new C-terminal sequences created by cleavage after Ala, Ser, and Val residues create new sites for binding of released peptides to the PDZ domain, allowing these peptides to be rebound and cleaved again?

VI. Phylogenetic Distribution

Tsp and D1P family members have been identified in a wide array of organisms, including gram-negative and gram-positive bacteria, cyanobacteria, spirochetes, plants, and algae. Tsp is not, however, ubiquitous, even in bacteria. For example, there are no identifiable homologs in *Mycobacterium tuberculosis*, *Mycoplasma pneumoniae*, or *Mycoplasma genitalium* even though complete genome sequences are available for these species. Most bacteria have a single Tsp homolog, but *Bacillus subtilis* has two closely related homologs, and *Synechocystis* sp. PCC6803 has three. The regions of highest sequence conservation among Tsp and D1P family members correspond to the PDZ and active-site domains, with sequences adjacent to the active site being the most conserved (Fig. 3). It is not clear if the α-helical N-terminal domain of the *S. obliquus* D1P is conserved among Tsp family members. There is little sequence identity in this region, but it is predicted to be mostly α-helical in all members of the family. In addition, the two β strands that extend from the active-site domain into the N-terminal domain of *S. obliquus* D1P appear to be conserved. It is therefore possible that the structure of the N-terminal domain is similar in all Tsp family

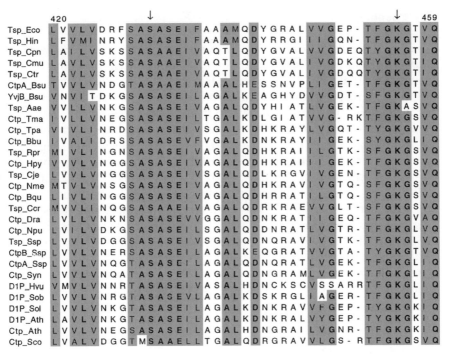

FIG. 3. Alignment of the active-site region of Tsp homologs. Similar residues are shown shaded, residues that are conserved in 90% of sequences are in bold, and the active site serine and lysine residues are marked by arrows. Residue numbers are shown for the first and last positions in the alignment using the *E. coli* Tsp nomenclature. Species and accession numbers are as follows: Tsp_Eco *Escherichia coli* P23865; Tsp_Hin *Haemophilus influenzae* P45306; Tsp_Cpn *Chlamydophila pneumoniae* F72063; Tsp_Cmu *Chlamydia muridarum* AAF39535; Tsp_Ctr *Chlamydia trachomatis* B71515; CtpA_Bsu *Bacillus subtilis* B69610; YvjB_Bsu E70042; Tsp_Aae *Aquifex aeolicus* F70369; Ctp_Tma *Thermotoga maritima* F72340; Ctp_Tpa *Treponema pallidum* H71342; Ctp_Bbu *Borrelia burgdorferi* F70144; Tsp_Rpr *Rickettsia prowazekii* A71677; Ctp_Hpy *Helicobacter pylori* D71827; Tsp_Cje *Campylobacter jejuni* CAB75148; Ctp_Nme *Neisseria meningitidis* AAF41707; Ctp_Bqu *Bartonella quintana* AAD04178; Tsp_Ccr *Caulobacter crescentus* communicated by The Institute for Genomic Research; Ctp_Dra *Deinococcus radiodurans* G75383; Ctp_Npu *Nostoc punctiforme* AAC45366; Tsp_Ssp *Synechocystis sp. PCC 6803* S77395; CtpB_Ssp S74579; CtpA_Ssp BAA10189; Ctp_Syn *Synechococcus* PCC7002 P42784; D1P_Hvu *Hordeum vulgare* (barley) T05975; D1P_Sob *Scenedesmus obliquus* T10500; D1P_Sol *Spinacia oleracea* (spinach) JH0263; D1P_Ath *Arabidopsis thaliana* CAA10694; Ctp_Ath T06752; Ctp_Sco *Streptomyces coelicolor* CAB72213.

members. Some Tsp homologs, such as *Streptomyces coelicolor*, contain little sequence outside the N-terminal, PDZ, and active-site domains, whereas those of *E. coli* and *Haemophilus influenzae* have large extensions before the N-terminal domain and after the active-site domain. It will be important to determine how sequence differences in Tsp family members affect their substrate specificity and biological function.

VII. Phenotypes

Bacteria with mutations in Tsp have several phenotypes. For example, *E. coli* with a null mutation in the *tsp* (or *prc*) gene exhibit a conditional lethal phenotype and die when grown in salt-free medium at temperatures above 40°C (*7*). Overexpression of Tsp in *E. coli* is lethal at elevated temperatures in LB medium, but there is no effect in hypotonic medium (*7, 34*). The lethal phenotype associated with *tsp* null mutations can be suppressed by overexpression of *degQ, degS, rlpA,* or *dksA* (*34*). DegQ and DegS are periplasmic proteases that also contain PDZ domains and probably have substrate specificities that partially overlap with Tsp (*34, 35*). The function of RlpA and DksA is unclear, but both induce a stress response, which may cause suppression of the Tsp phenotype (*34*). Strains with transposon insertions in *tsp* show increased susceptibility to several antibiotics (*36*) and also have a defect in long-chain fatty acid transport (*37*). The molecular basis for these *E. coli* phenotypes has not been established. Cyanobacteria and plants with a mutation in the *ctpA* gene that encodes D1P have an inactive photosystem II reaction center caused by failure to process the D1 protein (*38–40*).

VIII. Biological Functions

There appear to be two distinct classes of substrates for Tsp cleavage *in vivo:* proteins tagged at the C terminus by the SsrA system (*2*), and precursor proteins with C-terminal extensions (*1, 7, 38–41*). The SsrA quality-control pathway cotranslationally adds a peptide tag (*AANDENYALAA-COOH*) to the C terminus of proteins whose translation has stalled, which targets the tagged protein for degradation. In a strain lacking Tsp, the half-life of an SsrA-tagged periplasmic protein is significantly longer, indicating that Tsp is responsible for most SsrA-mediated degradation in the periplasm (*2*). Many of the characteristics of Tsp are ideally suited to this role of eliminating partially synthesized SsrA-tagged proteins. First, SsrA-tagged proteins should be efficiently bound by the PDZ domain of Tsp because they have an exposed, hydrophobic C-terminal sequence ending in LAA, which is one of the most preferred binding sequences. Second, once these tagged substrates are bound to Tsp, they should be efficiently degraded because they are unlikely to be stably folded.

Genetic evidence indicates that Tsp is also involved in the processing of C-terminal extensions from precursor proteins. Tsp mutants in *E. coli* are defective in the processing of an 11-residue C-terminal extension from the septum-forming enzyme penicillin-binding protein 3 (*1, 7, 41*). Likewise, mutations in the D1P enzyme in cyanobacteria and chloroplasts results in a defect in processing of the photosystem II protein D1 (*38–40*). Neither of these precursor proteins has a C-terminal sequence that is predicted to be a good substrate for Tsp: GRS is the sequence in the case of PBP3 and VNG in the case of D1. It is possible that there are two distinct sets of selectivity rules operating, one that permits recognition of the precursor substrates and one that recognizes SsrA-tagged proteins. Alternatively, recognition of precursors may involve additional factors such as membrane binding and/or the participation of other domains of the protease. For example, Liao *et al.* have suggested that a small cluster of Arg and Lys residues in the α-helical domain of D1P may serve as a phospholipid binding surface that increases the effective concentration of the protease near the precursor form of D1, which is an integral membrane protein in its mature form (*3*).

ACKNOWLEDGMENTS

We thank D. Liao, Q. Qian, D. Chisholm, D. Jordan, and B. Diner for communicating unpublished data about the D1P structure, and for providing Fig. 1A.

REFERENCES

1. Hara, H., Nishimura, Y., Kato, J., Suzuki, H., Nagasawa, H., Suzuki, A., and Hirota, Y. (1989). *J. Bacteriol.* **171,** 5882.
2. Keiler, K. C., Waller, P. R., and Sauer, R. T. (1996). *Science* **271,** 990.
3. Liao, D., Qian, Q., Chisholm, D. A., Jordan, D. B., and Diner, B. A. (2000). *Nat. Struct. Biol.* **7,** 749.
4. Silber, K. R., Keiler, K. C., and Sauer, R. T. (1992). *Proc. Natl. Acad. Sci. USA* **89,** 295.
5. Park, J. H., Lee, Y. S., Chung, C. H., and Goldberg, A. L. (1988). *J. Bacteriol.* **170,** 921.
6. Roseman, J. E., and Levine, R. L. (1987). *J. Biol. Chem.* **262,** 2101.
7. Hara, H., Yamamoto, Y., Higashitani, A., Suzuki, H., and Nishimura, Y. (1991). *J. Bacteriol.* **173,** 4799.
8. Keiler, K. C., Silber, K. R., Downard, K. M., Papayannopoulos, I. A., Biemann, K., and Sauer, R. T. (1995). *Protein Sci.* **4,** 1507.
9. Ponting, C. P. (1997). *Protein Sci.* **6,** 464.
10. Doyle, D. A., Lee, A., Lewis, J., Kim, E., Sheng, M., and MacKinnon, R. (1996). *Cell* **85,** 1067.
11. Morais Cabral, J. H., Petosa, C., Sutcliffe, M. J., Raza, S., Byron, O., Poy, F., Marfatia, S. M., Chishti, A. H., and Liddington, R. C. (1996). *Nature* **382,** 649.
12. Schultz, J., Hoffmuller, U., Krause, G., Ashurst, J., Macias, M. J., Schmieder, P., Schneider-Mergener, J., and Oschkinat, H. (1998). *Nat. Struct. Biol.* **5,** 19.
13. Kornau, H. C., Schenker, L. T., Kennedy, M. B., and Seeburg, P. H. (1995). *Science* **269,** 1737.
14. Kim, E., Niethammer, M., Rothschild, A., Jan, Y. N., and Sheng, M. (1995). *Nature* **378,** 85.
15. Songyang, Z., Fanning, A. S., Fu, C., Xu, J., Marfatia, S. M., Chishti, A. H., Crompton, A., Chan, A. C., Anderson, J. M., and Cantley, L. C. (1997). *Science* **275,** 73.

16. Stricker, N. L., Christopherson, K. S., Yi, B. A., Schatz, P. J., Raab, R. W., Dawes, G., Bassett, Jr., D. E., Bredt, D. S., and Li, M. (1997). *Nat. Biotechnol.* **15,** 336.
17. Schepens, J., Cuppen, E., Wieringa, B., and Hendriks, W. (1997). *FEBS Lett.* **409,** 53.
18. Beebe, K. D., Shin, J., Peng, J., Chaudhury, C., Khera, J., and Pei, D. (2000). *Biochemistry* **39,** 3149.
19. Liou, G. I., Geng, L., and Baehr, W. (1991). *Prog. Clin. Biol. Res.* **362,** 115.
20. Keiler, K. C., and Sauer, R. T. (1995). *J. Biol. Chem.* **270,** 28864.
21. Keiler, K. C., and Sauer, R. T. (1996). *J. Biol. Chem.* **271,** 2589.
22. Beebe, K. D., and Pei, D. (1998). *Anal. Biochem.* **263,** 51.
23. Nakatsuka, T., Sasaki, T., and Kaiser, E. T. (1987). *J. Am. Chem. Soc.* **109,** 3808.
24. Higaki, J. N., Gibson, B. W., and Craik, C. S. (1987). *Cold Spring Harb. Symp. Quant. Biol.* **52,** 615.
25. Carter, P., and Wells, J. A. (1988). *Nature* **332,** 564.
26. Craik, C. S., Roczniak, S., Largman, C., and Rutter, W. J. (1987). *Science* **237,** 909.
27. Slilaty, S. N., and Little, J. W. (1987). *Proc. Natl. Acad. Sci. USA* **84,** 3987.
28. Strynadka, N. C., Adachi, H., Jensen, S. E., Johns, K., Sielecki, A., Betzel, C., Sutoh, K., and James, M. N. (1992). *Nature* **359,** 700.
29. Little, J. W. (1993). *J. Bacteriol.* **175,** 4943.
30. Paetzel, M., Strynadka, N. C., Tschantz, W. R., Casareno, R., Bullinger, P. R., and Dalbey, R. E. (1997). *J. Biol. Chem.* **272,** 9994.
31. Paetzel, M., Dalbey, R. E., and Strynadka, N. C. (1998). *Nature* **396,** 186.
32. Tschantz, W. R., Sung, M., Delgado-Partin, V. M., and Dalbey, R. E. (1993). *J. Biol. Chem.* **268,** 27349.
33. Paetzel, M., and Dalbey, R. E. (1997). *Trends Biochem. Sci.* **22,** 28.
34. Bass, S., Gu, Q., and Christen, A. (1996). *J. Bacteriol.* **178,** 1154.
35. Waller, P. R., and Sauer, R. T. (1996). *J. Bacteriol.* **178,** 1146.
36. Seoane, A., Sabbaj, A., McMurry, L. M., and Levy, S. B. (1992). *J. Bacteriol.* **174,** 7844.
37. Azizan, A., and Black, P. N. (1994). *J. Bacteriol.* **176,** 6653.
38. Shestakov, S. V., Anbudurai, P. R., Stanbekova, G. E., Gadzhiev, A., Lind, L. K., and Pakrasi, H. B. (1994). *J. Biol. Chem.* **269,** 19354.
39. Anbudurai, P. R., Mor, T. S., Ohad, I., Shestakov, S. V., and Pakrasi, H. B. (1994). *Proc. Natl. Acad. Sci. USA* **91,** 8082.
40. Oelmuller, R., Herrmann, R. G., and Pakrasi, H. B. (1996). *J. Biol. Chem.* **271,** 21848.
41. Nagasawa, H., Sakagami, Y., Suzuki, A., Suzuki, H., Hara, H., and Hirota, Y. (1989). *J. Bacteriol.* **171,** 5890.

14

Co- and Posttranslational Processing: The Removal of Methionine

RALPH A. BRADSHAW • CHRISTOPHER J. HOPE •
ELIZABETH YI • KENNETH W. WALKER

Department of Physiology and Biophysics
College of Medicine, University of California
Irvine, California 92697

I. N-Terminal Processing: An Overview

It has been appreciated for over 30 years that protein synthesis in living organisms is initiated by the amino acid methionine (Met), which is coded

THE ENZYMES, Vol. XXII

by the trinucleotide AUG (*1, 2*). In eubacteria, the alpha amino group is formylated at the tRNA level, which does not occur in archaebacteria and eukaryotes (*3*). The singular and essentially universal use of an amino acid with only a single codon provides substantial opportunity for regulation, particularly when that amino acid is found in limiting quantities in virtually all organisms. However, if this residue were also retained on all of the various proteins synthesized, it would result in intracellular shortages of this amino acid and thus a possible inability of the host cell to meet the demands of protein synthesis as well as other metabolic functions (*4*). To avoid this problem, specific processing enzymes that allow the bulk of the initiator Met to be recovered, primarily as a cotranslational event, have evolved. In fact, the global distribution of this enzyme (in its various isoforms) with essentially invariant substrate specificity suggests that the recycling of initiator Met is a fundamental requirement of all living organisms.

The methionine aminopeptidases (MetAPs) are apparently dedicated to this task. Although there is a great deal of aminopeptidase activity present in both prokaryotic and eukaryotic cells (*5*), there are no other known enzymes, at present, with the same characteristics as this family. Generally, they are entirely specific for Met, they are not processive, and they have a selectivity that is defined by the size of the penultimate residue. These are properties that are not shared with other exopeptidases with amino-terminal specificity.

The profile of N-terminal processing, with respect to the removal of Met from protein substrates during translation, was initially defined from an examination of protein sequence data bases (*6, 7*) and the analysis of a number of mutant proteins, particularly of the cytochrome *c* family (*8*). The latter information in particular led to the correct prediction that the removal of initiator Met in both prokaryotes and eukaryotes was strongly influenced by the nature of the adjacent amino acid and that the principal property was the size of the side chain (*9*). Thus, it was suggested that proteins with any of the seven smallest amino acids (Gly, Ala, Ser, Thr, Pro, Cys, or Val) would be substrates for the enzymes removing the Met, whereas the remaining 13 amino acids would not. There are also suggestions that the antepenultimate residue can influence specificity. For example, the presence of proline (introduced by a point mutation) in this position can inhibit the normal removal of Met from a hemoglobin variant (*10, 11*). However, generally, the specificity appears to be overwhelmingly dependent only on the penultimate residue.

Site-directed mutagenesis studies, using a recombinant protein (thaumitin) expressed in yeast, confirmed this predicted profile and further defined the secondary modification of N^α-acetylation, unique to eukaryotes, that is closely connected to Met processing (Fig. 1) (*12*). The combination of these two enzymes, i.e., the aminopeptidases and acetyltransferases, define

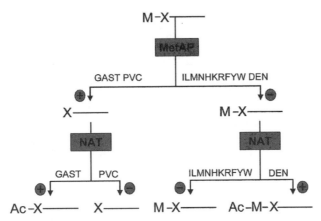

FIG. 1. Proposed pathway for the cotranslational modification of eukaryotic proteins by methionine aminopeptidase (MetAP) and N^{α}-acetyltransferase (NAT): M, initiator methionine; X, the penultimate residue; + and −, positive or negative enzyme actions. The categories of proteins produced when the indicated amino acid is the penultimate residue (X) are shown.

four classes of N-terminal structure found in intracellular eukaryotic proteins (13). These enzymes also act on extracellular, transmembrane, and imported mitochondrial proteins, but these modifications are lost following the removal of the various signal peptides involved in their translocation. The thaumitin study was subsequently extended by other studies with hemoglobin (14), iso-1-cytochrome c (15), and the proteins expressed in E. coli (16, 17). Importantly, the sequences of proteins found in data bases that have actually been subjected to direct analysis are entirely consistent with this profile, suggesting that (1) this substrate specificity is widespread, and (2) it is not particularly dependent on the substrate (although individual protein sequences may occasionally be expected to show variations, and this may ultimately prove to be important in certain instances, as described below). Most importantly, the predicted and demonstrated profiles from these experiments have been entirely substantiated by direct studies on the isolated enzymes from a number of species.

Our knowledge of the overall structural and functional features of the MetAP family and how it has probably evolved have developed rapidly in the last several years (18). There are, however, still substantial questions remaining with respect to some aspects of structure and function for which at present there is only speculation. This chapter provides a summary of the presently appreciated information regarding the MetAPs that are involved in cotranslational processing of the initiator Met with an indication of the questions that remain to be resolved.

II. Initiation of Protein Synthesis

A. PROKARYOTES

The universal usage of Met as the initiating amino acid in protein synthesis (*4*) is modified in eubacteria (and in the eukaryotic organelles thought to have arisen from eubacterial origins, i.e., mitochondria and chloroplasts) by the addition of a N-formyl group that is added to the charged tRNA (*3*). The origin of this group is 10-formyl tetrahydrofolate, and the reaction is catalyzed in eubacteria by the enzyme methionine-tRNA^{fMet} formyltransferase. The same one-carbon moiety is removed as a cotranslational event by a second enzyme, peptidyl deformylase [see (*19*)]. The latter enzyme has a nearly universal specificity, and virtually all proteins synthesized in *E. coli* (and presumably other bacterial as well) do not retain the N-formyl group. The genes for the transferase and deformylase (*fmt* and *def,* respectively) are cotranscribed in *E. coli* (*20*). Analysis of known mutations of these genes indicate that the peptidyl deformylase is essential while loss of the formyltransferase activity only results in severely impaired growth (*20*). Importantly, the double null-mutation acts similarly to the mutation in *fmt.* These results (*21*) indicate that the formylation of Met as a step in the initiation of protein synthesis is important for enhancing this essential activity but is not absolutely required. Solbiati *et al.* (*21*) have also demonstrated that the lethality associated with the null-mutation in the *def* gene is due not to the buildup of undeformylated proteins, but to the fact that these derivatives cannot be further modified by MetAP to remove the initiator residue.

As with MetAP, the peptidyl deformylase has been extensively studied over the past several years and considerable information is available regarding its structural and functional properties (*22–28*). Although the three-dimensional structures of the two enzyme types (peptidyl deformylase and MetAP) are not related (*23, 26, 29*) and they have somewhat different catalytic profiles (the deformylase appears to act on virtually all sequences, whereas the MetAP has a defined specificity), there are some interesting similarities, particularly with respect to catalytic mechanism. Both enzymes are monomeric and in eubacteria are presumably freely soluble. In addition, both are metal-dependent enzymes, and the identification of the active metal ion in each case has been somewhat controversial. Although originally identified as a zinc protein (*25*), it has now been suggested that the physiologically relevant metal ion of the peptidyl deformylase is ferrous iron (*24*), although cobaltous ion derivatives are highly active (*27*). By comparison, the zinc enzymes show only very weak activity. Interestingly, nickel can also be substituted for iron with virtually no effect on catalytic efficiency (*22, 24*). In contrast, the cobalt form of the enzyme is about an order of magnitude weaker in activity, but like the nickel enzyme is highly stable (*27*). The native ferrous ion form is extremely unstable, a property that inhibited study of this enzyme for many years (*30*). A similar story for MetAPs is described below. At present, there is no indication that the peptidyl deformylase and MetAP associate despite the fact that they probably both can act on the nascent polypeptide (although it is not clear when either acts during the translation process in eubacteria).

The role of N^α-formylation of the initiator Met as an enhancer, as opposed to a prerequisite for initiation, is consistent with the hypothesis that the lack of an active peptide deformylase is lethal, not because of the accumulation of formylated peptides, but because of the inhibition of subsequent MetAP action. Mazel *et al.* (*20*) suggested that there was weak MetAP activity against N^α-formylated Met substrates, but Solbiati *et al.* (*21*) established clearly that this does not occur *in vitro* or *in vivo*. In view of the need to recycle Met and the requirement of some enzymes for a free alpha amino group of the penultimate residue to function, it is likely that the development of MetAP activity preceded that of the formylation/deformylation modulation and probably represents one of the most primitive activities to be developed by living organisms.

B. EUKARYOTES

The initiation of eukaryotic protein synthesis is also through Met residues, and there is no corresponding formylation as is found in the eubacteria. Thus, the nascent polypeptide chains of all eukaryotic proteins bear unmodified Met and are therefore potentially substrates for direct processing by MetAP. The overall process regulating the various phases of initiation, elongation, and termination of protein synthesis are somewhat more complex in eukaryotes than prokaryotes and involve a greater number of factors (*31*). Curiously, one isoform of MetAP interacts with one important initiation factor (eIF2), an event not seen in eubacteria. This association is independent of its catalytic activity, and it likely represents another example of a protein with two independent functions, such as found in the coagulation protein, thrombin, which stimulates blood clot formation and induces cell mitogenicity (which presumably is important in subsequent wound healing events) as separate activities. As described below, the importance of these two functions associated with this MetAP isoform are not fully understood.

III. The Methionine Aminopeptidases

A. ISOLATION AND CHARACTERIZATION

The prediction (*9*) and subsequent determination (*12, 14, 15*) of the substrate profile of the enzymes involved in the removal of initiator Met provided a basis for their isolation and characterization. The strong specificity for N-terminal Met and the size restriction on the side chains of the penultimate residues were distinguishing characteristics from other aminopeptidases that had been previously identified (*32*). The susceptibility of Met-Pro substrates was particularly diagnostic. The first reports of the isolation of cotranslationally active MetAPs were from *E. coli* (*33*) and *Salmonella typhimurium* (*34*). Both preparations were monomeric proteins of approximately 30 kDa, and both were dependent on cobalt as a metal cofactor for activity. The *E. coli* enzyme was also strongly influenced by both K^+ and PO_4^{2-}. These enzyme preparations utilized an array of substrates ranging from tripeptides to full proteins, but were not active against dipeptides. However, later studies (*21*) have reported that dipeptides can be utilized. Ben-Bassat *et al.* (*33*) also reported the cloned gene sequence for the *E. coli* protein; the sequence for the *S. typhimurium* enzyme was reported subsequently (*35*). Comparison of these sequences established that these two proteins were greater than 85% identical. A third bacterial MetAP from *B. subtilis* has also been shown to be quite similar to the other two bacterial MetAPs (*36*).

The first isolation and characterization of a eukaryotic form of MetAP was reported by Chang *et al.* (*37*). Like the prokaryotic enzymes, it was monomeric and showed a strong requirement for divalent metals (particularly cobalt). It had an apparent molecular mass of approximately 34 kDa, similar to the bacterial enzymes, and the specificity expected for the cotranslationally active enzyme. Subsequently, yeast MetAP was cloned (*38*), which confirmed its sequential relatedness to the bacterial enzymes but also revealed some important differences. In the first place, the predicted molecular weight was found to be somewhat higher than that determined by SDS–PAGE, corresponding to a calculated value of 43,269. This was significantly larger than the molecular weight calculated from the sequences of the bacterial enzymes, which were 29,333 and 29,292, respectively, for the *E. coli* and *S. typhimurium* MetAPs (*33, 35*). Thus, the MetAPs from the two bacterial sources and yeast all showed aberrant behavior on SDS–PAGE, with the two prokaryotic forms giving somewhat higher values while the yeast enzyme showed an artificially low value. Nonetheless, a comparison of the yeast sequence to the bacterial MetAP sequences clearly indicated that the C-terminal portion of the yeast enzyme showed strong relatedness to the prokaryotic forms, confirming the presence of a homologous catalytic domain. The extra sequence (amounting to approximately 13 kDa) was found as an amino-terminal extension.

The isolation and characterization of porcine liver MetAP yielded an enzyme with structural properties that were quite distinct from both the prokaryote and yeast proteins (*39*). Its apparent molecular weight was found to be 67,000, which was considerably larger than any of the previously isolated enzymes, although it too behaved as a monomer in solution. It also apparently required cobalt ion for activity and showed the same substrate profile that have been established to be characteristic for cotranslational processing. It also was inhibited by added zinc ion, a feature observed for the yeast and bacterial enzymes.

Recombinant forms of several MetAPs including *E. coli*, yeast, human, and the archaebacteria (*P. furiosus*) have now been expressed and isolated (*29, 38, 40, 41*). These preparations have generally confirmed or extended the properties determined for the naturally occurring material. For example, the *P. furiosus* protein (*41*) has been shown to be stimulated primarily by cobalt and to show the same heat stability found in preparations of the native protein. Studies with recombinant proteins that deal with other important properties and characteristics of these enzymes will be described below.

B.   SEQUENCE/STRUCTURE STUDIES

Escherichia coli MetAP provided not only the first sequence for this family of proteins but also the first three-dimensional structure (*29*). The material used for this study was a recombinant protein prepared by the procedure described for the native enzyme (*33*). The main fold of the protein (Fig. 2) revealed that the bacterial MetAP had a symmetrical structure (described as a "pita bread" fold) that consisted of an essential antiparallel β sheet covered on one side by two α helices and by a C-terminal loop. The structural analysis also revealed the protein contained two cobalt ions that were coordinated by five protein ligands and were spaced 2.9 Å apart. The side chains

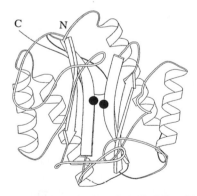

FIG. 2.   Ribbon diagram showing the "pita-bread" fold of *E. coli* MetAP. The view direction is essentially parallel to the local two-fold axis of symmetry, and the active site is marked by the two cobalt ions shown as solid circles. Reprinted with permission from Roderick and Matthews (*29*), copyright 1993, American Chemical Society.

of two aspartic acids, two glutamic acids, and a histidine residue provided the key elements of the chelate structure. Roderick and Matthews (*29*) concluded from this analysis that the MetAP protein possessed a unique three-dimensional fold that was unlike other known structures, particularly other aminopeptidases, in the protein data base.

By virtue of sequence similarity, the structure of the *E. coli* protein also provided a clear insight into the structure of the other bacterial MetAPs and the corresponding catalytic domain of the yeast protein described by Chang *et al.* (*37, 38*). However, it did not initially appear to readily accommodate the amino acid sequence of the rat protein (*42*). Partial sequence analysis of the porcine protein had indicated that it possessed the amino acid sequence of a protein already recorded in the data base (*43*). This protein, from rat liver, was identified from a biological activity distinct from that of an aminopeptidase, i.e., it was found to be an inhibitor of the phosphorylation of eIF2α (*44*). Previous studies with this protein had not revealed aminopeptidase activity or metal binding properties (*45, 46*).

Two studies utilizing sequence comparisons, structure predictions, and model building ultimately resolved this apparent ambiguity and revealed an interesting isoform distribution of the MetAPs in various living organisms (*18*). Bazan *et al.* (*47*) predicted that the apparently unique fold of the MetAP protein from *E. coli* had actually been presaged by the three-dimensional structure of creatine amidinohydrolase (creatinase) from *Pseudomonas putida* (*48*). In fact, not only did the MetAP of *E. coli* share a strikingly similar three-dimensional structure with this protein, but additional sequence comparisons suggested that aminopeptidase P (proline aminopeptidase) (*49*) and proline dipeptidase (or "prolidase") (*50*) also likely shared this basic three-dimensional structure. As shown in Fig. 3 (see color plate), subsequent structural analyses have confirmed that the aminopeptidase P does have the same fold as MetAP and creatinase (*51*). Bazan *et al.* (*47*) also identified a partial reading frame from an unknown sequence from the archaebacterium *Methanothermus fervidus* that was similar and therefore likely to be a member of the same family. Using a model building approach for both the rat and human MetAP sequences, Arfin *et al.* (*42*) concluded that the cobalt ligands of the *E. coli* enzyme were conserved in the higher eukaryotic MetAPs and that the rat (and human) MetAP could, in fact, be incorporated into the three-dimensional structure of the *E. coli* enzyme (although it required an additional surface loop in the catalytic domain that was not found in the *E. coli* and yeast enzymes). These workers also noted that the sequence of the rat (and human) MetAP was much more similar to another yeast open reading than the previously reported yeast MetAP sequence (*38*). They concluded that there were two classes of MetAPs present in eukaryotic cells, in contrast to eubacteria, which apparently contained only a single class. It was

(a)

(d)

(b)

(e)

(c)

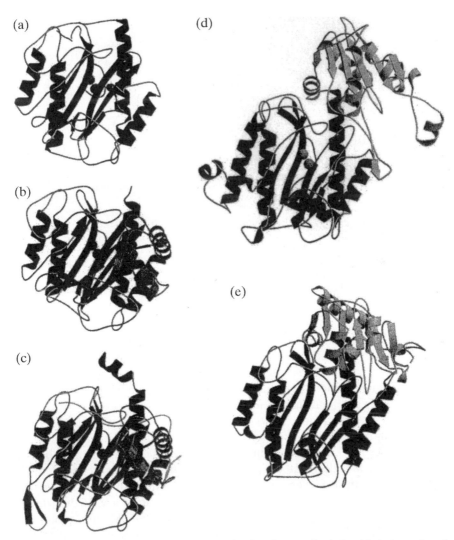

Fig. 3. (See color plate.) Ribbon drawings showing the overall relationship between type I and type II MetAPs and other known pita-bread enzymes. The pseudo twofold symmetry of the core domain is illustrated by red α helices and blue β strands. The α-helical subdomain insertion unique to the type II enzymes is shown in green. N-terminal domain extensions are shown in yellow. The metals of the active site, if present, are Co(II) (magenta) or Mn(II) (cyan). (a) *E. coli* MetAP, PDB accession code 2MAT (*59*). (b) *P. furiosis* MetAP-IIa, IXGS (*56*). (c) Human MetAP-IIb, 1BN5 (*57*). (d) Aminopeptidase P (AMPP), 1A16 (*71*). (e) Creatine amidinohydrolase (creatinase), 1CHM (*91*). The N-terminal domain extensions of creatinase and AMPP share the same fold, but are in a slightly different orientation in relation to the rest of the protein. The activity of creatinase is not metal dependent (*91*). Reprinted with permission from Lowther and Matthews (*51*).

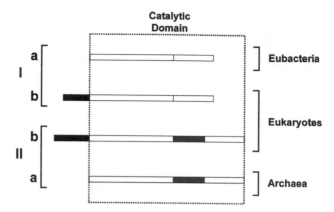

FIG. 4. (See color plate.) Domain organization of MetAPs. Type 1a enzymes are found in eubacteria, and type IIa enzymes are found in archaea. MetAPs of both types are present in eukaryotes and display additional N-terminal domains. The type Ib enzymes have zinc-finger domains (shown in blue), whereas the type Ib enzymes have stretches of polybasic and acidic residues (shown in red). The catalytic domain is shown in yellow and the inserted domain, characteristic of type II enzymes, is shown in orange. Reprinted by permission from Bradshaw *et al.* (*18*).

subsequently confirmed from a number of complete archaebacteria genome sequences that they also contained only a single form of MetAP, but that it was similar to the MetAP isoform originally isolated from porcine liver (*18, 39*). Using a genetic selection approach, Li and Chang (*52*) concluded independently that yeast contained two MetAP isoforms.

Figure 4 (see color plate) shows a schematic representation of the isoform distribution of MetAPs in prokaryotes and eukaryotes (*42*). Four principal enzyme types are known to occur. The eubacterial enzyme defined by the MetAP in *E. coli* is designated type I, whereas that found in archaebacteria, with the extra surface insertion in the catalytic domain, is denoted type II. The eukaryotic forms, with their N-terminal extensions, are further designated "b" types; the prokaryotic forms are labeled "a." All eukaryotes apparently contain both type I and type II enzymes, whereas the eubacteria and archaebacteria each contain only a single form.

A sequence comparison of the catalytic domains of representative members of each type is shown in Fig. 5 (see color plate). Panel A depicts an alignment that stresses the comparison of type I and II isoforms; Panel B compares the two prokaryotic and the two human sequences. It is readily seen that the type I proteins show 44% identity while the type II enzymes are more distant with only 24% identity. There is even more limited sequence identity in the prokaryotic and eukaryotic comparisons (14 and 16%). This in part reflects the very small number of residues conserved in all four isotypes. Most of these

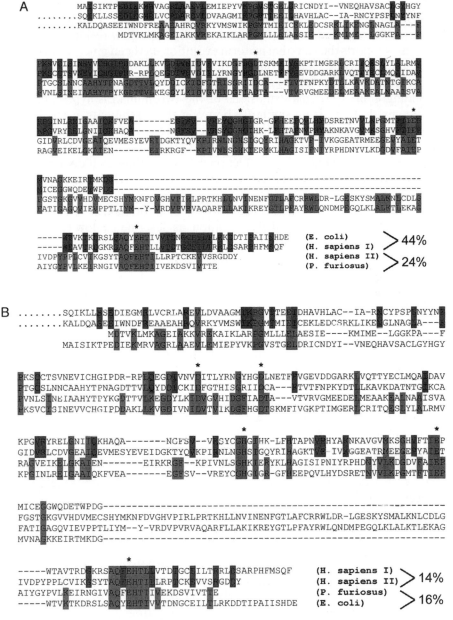

FIG. 5. (See color plate.) Alignment of the amino acid sequences of the catalytic domain for the MetAP of *E. coli*, *P. furiosus*, and human (types I and II). Residues conserved in all sequences are highlighted in red; residues present in each pair compared are marked in green (upper) or magenta (lower). (A) Comparison of type I enzymes vs type II enzymes; (B) comparison of both human enzymes vs the two prokaryotic enzymes.

represent metal binding or other catalytically important residues. Overall, this also supports the view that isoform evolution has been much more highly conserved than any species conservation.

Figure 6 (see color plate) shows the sequence comparison of the corresponding amino-terminal extensions of the two human isotypes (*42, 53*). Quite clearly there is no similarity between these segments at all. Type I amino terminal extensions (characteristic of the type Ib isoforms) contain two putative zinc fingers that are highlighted. It has been shown that this sequence can be deleted without effect on catalysis, although it does alter the growth properties of the host cell (yeast) (*54*). The type II sequence, which is abnormally rich in both acidic and basic residues, does not contain any recognized motif and there is no assigned function to this domain as yet. However, it may be the site of O-linked glycosylation, as described below (*45*).

Two additional observations, in addition to the sequence comparisons shown in Fig. 5, also support the view that both isoforms function as MetAPs and that they are indeed related in a fashion associated with proteins that have evolved from a common precursor. In the first place, Li and Chang (*55*) expressed the human cDNA corresponding to this protein in a baculovirus system and demonstrated unequivocally with a variety of protein substrates that it functioned as a MetAP with the characteristic sequence profile demonstrated for the porcine type II enzyme (*39*) and the yeast type I enzyme (*37*). They also demonstrated that the activity was dependent on bound colbalt ions. The second line of evidence was the determination of the three-dimensional structure of the *P. furiosus* enzyme (*56*). As with the *E. coli* structure, this analysis was also carried out with recombinant material and revealed a three-dimensional structure that was quite similar to that of *E. coli* protein, with the exception of the insertion in the catalytic domain, which was found to be composed primarily of three helices and is the key feature that distinguishes type I and type II isoforms (Fig. 3). At present, there is no information regarding the possible role of this insertion domain. It is unlikely that it plays a role in catalysis (considering that the rest of the domain so closely resembles that of the type I isoform), and therefore it is probably significant in some other physiological function of the type II isoform. It may be important for the alternative activity (inhibition of eIF2α phosphorylation) at least in eukaryotes. Constructs in which this region has been deleted may ultimately be informative in understanding the contribution it makes to the type II isoform activities.

The three-dimensional structure of the human type II enzyme has also been determined (*57*) (Fig. 3). Although the sequence similarity between the catalytic domains of the type II enzymes (see Fig. 5A) is only approximately 24%, the three-dimensional structures are virtually superimposable. Unfortunately, the amino-terminal domain was largely disordered in the structural analysis and, therefore, there is no information regarding the overall organization of this domain at present. The determination of the three-dimensional structure of the type II enzyme was also important in defining the modification of type II isozymes with a class of anti-angiogenic compounds based on fumagillin that are much less reactive toward the type I enzymes (see below).

The sequence predictions of Bazan *et al.* (*47*), which suggested that the MetAPs were members of a larger structurally related family, have been well substantiated. As shown in Fig. 3, two structures, in addition to the three MetAP structures, have been determined, which clearly show the "pita bread" fold found in *E. coli* MetAP (*51*). Aminopeptidase P is also a metal-dependent peptidase, but creatinase is not. However, there will also likely be additional members of this group. Given the universal distribution of the MetAP isozymes and the absolute requirement for this activity for all living organisms, these structures probably represent the most ancient members of this family. A proposed evolutionary scheme for the MetAP isoforms is given in the next section.

C. EVOLUTION

The distribution of isoforms of the MetAPs (Fig. 4) suggests that the two isoforms diverged relatively early and developed along separate lines of evolution, although they certainly arose from a common precursor (*18*). The facts that the archaebacteria contain only type II enzyme and that they are thought to be among the most primitive life forms on earth, suggest that this type probably more closely resembles the original precursor molecule. Activity would likely have appeared with the gene duplication event that produced the symmetrical bilobal structure that makes up the "pita bread" fold, and it seems probable that the separation that led to the two isoform lineages also must have happened very early. As indicated in the schematic diagram shown in Fig. 7, the opportunity for the primitive eukaryotic organisms to regain the well-defined and separately evolved prokaryotic form would have occurred during the endocytosis that led to the symbiotic formation of the mitochondria. One may presume that initially this premitochondrial organelle retained the MetAP to process proteins that were synthesized there. However, in due course, as the other mitochondrial genes were transferred to the nucleus of the developing eukaryotic species, the gene of the type I isoform would have also made this transition. As the eukaryotic cells underwent further development, both of the isoforms would then have obtained their additional structural features that are presumably important for their present day functions. Thus, the type I enzyme would have acquired the amino-terminal extension with the zinc figures, while the type II enzyme would have developed the amino terminal domain containing the polyacidic and basic sequences as well as the insertion in the catalytic domain.

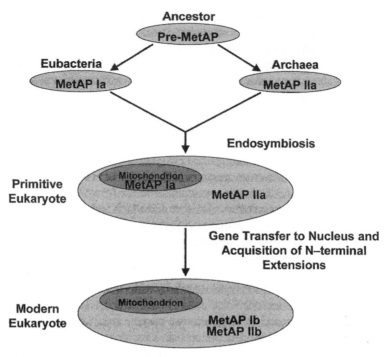

FIG. 7. Proposed evolution of the methionine aminopeptidases (MetAPs). In this scheme, the pre-MetAP is considered to be the progenitor of all MetAPs and is assumed to have the features of the type II enzyme. The type I enzymes are proposed to have been formed by mutational events during the development of bacteria and entered eukaryotes following endosymbiosis of mitochondria. Both type I and type II enzymes subsequently acquired the features of the larger isozymes (Ib and IIb). Reprinted by permission from Bradshaw *et al.* (*18*).

The question of why both enzymes with similar catalytic specificities were retained through eukaryotic development cannot be entirely answered by the acquisition of other functions by the type II isoform. The overall catalytic specificity has been retained (with only relatively minor variations in kinetic constants), suggesting that there must be independent functions associated with N-terminal processing itself. The main activity is the routine cleavage of Met that occurs at the surface of the ribosome as the nascent polypeptide chain emerges during translation, and one of the two isoforms is likely to play this role. The presence of the putative zinc fingers make the type I enzyme an attractive candidate for this function (*54*). The type II isoform may also be tethered to one or more organelles or molecules in the cell, but at the moment there is no evidence as to what these might be. It is clear that it must interact with eIF2, which is part of the initiation complex, and it is entirely possible that the MetAP II isoform may also become associated with ribosomes in some fashion during the formation of the initiation complex. One possible role that has been suggested (*18*) is that it may take part in the processing of upstream open reading frames (uORF), particularly with the genes of proteins involved in cell cycle regulation and related activities. These relative short sequences must be translated and processed before the ribosomes can move to the mature open reading frame of the principal protein encoded by that mRNA (*31*), and such a role for MetAP II would be consistent with its apparent involvement in the proliferation of lymphocytes and endothelial cells (see below).

D.   CATALYTIC PROPERTIES

1.   *Metal Cofactors*

a.   *Identification.* Although there is clear agreement that all of the isoforms of MetAP require a divalent metal ion for catalytic activity, there are substantive differences with respect to which metal ions can be utilized and, therefore, are the physiologically relevant ones. As summarized below, it is possible, even likely, that there are species variations in the metal ions used. It may even be the case that different metals are utilized within the same species (*58*).

The first preparation of substantially homogeneous MetAP to be obtained was from a hyperproducing strain of *E. coli* (*33*). The activity of this enzyme was sensitive to EDTA and it was only restored by 0.2 mM CoCl$_2$. Zn^{2+}, Mn^{2+}, and Mg^{2+} salts at 1 mM were not effective. Tris buffer also caused loss of activity that was restored with sodium or potassium phosphate. This last observation is consistent with the report of Lowther *et al.* (*59*) that defined a sodium (or potassium) specific binding site in the *E. coli* enzyme.

The fundamental observation that a preparation of apo MetAP, usually generated by exposure to EDTA, can be fully reactivated by the addition of Co^{2+} has since been observed for all of the isoforms of MetAP. This has naturally led to the general classification of this enzyme family as cobalt-dependent (*18, 51*). However, identification of the "native" metal ion for any source or isoform has been considerably more challenging. It is well known that cobalt ions can substitute for other metals, particularly zinc ions, in reconstitution experiments, usually with the retention of substantive if not full activity (*60*). This has, in fact, been highly valuable as an experimental approach because of useful properties of Co^{2+} not shared by Zn^{2+}. However, the natural utilization of non-corrin cobalt in enzyme systems appears to be very restricted (*61*), thus making the MetAP family a part of an apparently quite small group. A brief summary of the metal ion requirements of each isoform is given below.

Although the original preparation of *E. coli* MetAP was shown to be stimulated by Co^{2+}, the isolated protein was not analyzed for metal ion content. Wingfield *et al.* (*35*) subsequently did analyze preparations of MetAP from both *E. coli* and *S. typhimurium* by X-ray fluorescence and found only substoichiometric amounts of zinc and iron, but no cobalt. They concluded that the catalytically critical Co^{2+} is loosely bound and lost during purification. D'souza and Holz (*62*) also addressed the issue of the native metal ion for *E. coli* MetAP and concluded that it was most likely to be Fe^{2+}. They showed that there was a substantive increase in iron associated with the increase in recombinant MetAP production in whole cell measurements and found, in the strict absence of oxygen, that MetAP could be reconstituted to yield a better catalyst with ferrous iron than either cobalt or zinc, as judged by the determination of kinetic constants with Met-Gly-Met-Met as substrate. However, the specific activity of the Fe^{2+} enzyme was only 80% that of the Co^{2+} enzyme. All of the reconstitution experiments were done at near equimolar ratios of metal to protein. They also pointed out that the peptide deformylase of *E. coli* has been shown to be an iron enzyme (*23*) and suggested that iron may be a common regulator of N-terminal processing in some bacteria. Lowther *et al.* (*59*) found Zn^{2+} to be equivalent to Co^{2+} if the enzyme was treated with EDTA immediately prior to the assay. They also found that higher concentrations of Zn^{2+} were inhibitory (see below).

The eukaryotic type I MetAP has been studied in detail with the recombinant yeast enzyme. The initial characterization of the yeast MetAP I with respect to metal ion usage showed inhibition by chelators and excess zinc and restoration of catalytic activity of apoenzyme with only Co^{2+} (*37*). In a subsequent, more extensive study (*63*), it was shown that at lower concentrations, zinc was as effective as cobalt in restoring activity, and that manganese and nickel showed only minimal ability to do so (Fig. 8). Furthermore, in the presence of 5 m*M* reduced glutathione, the cobalt-dependent activity was lost while the zinc-dependent activity was actually enhanced. Interestingly, D'souza and Holz (*62*) observed no effect of GSH on the activity of reconstituted *E. coli* MetAP with any of the metal ions they studied. The results of Walker and Bradshaw (*63*) strongly supported the view that zinc ion was likely to be the cofactor utilized by the native enzyme in yeast.

The type II enzyme from either archaebacteria or eukaryotes has been studied in less detail with respect to metal dependency. Tsunasawa *et al.* (*41*) have reported the isolation and characterization of recombinant MetAP from *P. furiosus* and found that the enzyme was indeed inactivated by chelators and that only Co^{2+} and Mg^{2+} restored activity. The latter was about one-third as effective. Other metal ions, including Zn^{2+}, did not restore any activity. The concentrations utilized were not reported. Studies with the porcine enzyme (*39*) showed a similar apparent preference for Co^{2+}, but the

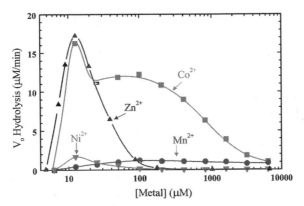

FIG. 8. Activity of 10 nM apo-rMetAP recharged with CoCl$_2$ (■) and Co-rMetAP in increasing ZnCl$_2$ (▲), MnCl$_2$ (●), or NiCl$_2$ (▼). The initial velocity was determined by measuring the hydrolysis of 60 μM MSSHRWDW over a 90 s interval. Reprinted by permission from Walker and Bradshaw (63).

possible role of other metals was not carefully assessed. A recombinant form of human MetAP II (55) has been isolated from a baculovirus expression system. It was sensitive (inhibited) to chelation and to high zinc ion concentrations. The apo enzyme was converted to an active holo form by Co^{2+} and Mn^{2+}, with the former about twice as effective.

In summary, the type I enzyme in yeast (and possibly other eukaryotes as well) most likely utilizes zinc ions. It is less clear what the eubacterial forms use as Co^{2+}, Fe^{2+}, and Zn^{2+} all form active complexes. The type II isozymes have not been as carefully studied. It is noteworthy that non-corrin cobalt-dependent enzymes occur, as has been determined, only in eubacteria and archaebacteria with the apparent exception of the MetAPs (61). If the type Ib enzyme is truly Zn^{2+} dependent, that suggests that the eukaryotic type IIb protein may be as well, based on the arguments as to the unlikely existence of non-corrin cobalt-dependent enzymes in eukaryotes (63). That would leave the type IIa form of archaebacteria (and possibly the type Ia MetAP of eubacteria) as the only true Co^{2+} enzyme. Interestingly, it suggests that despite the clear homology seen in the sequence comparisons, the well-conserved 3-D structure, and the similarities in substrate specificities, there are likely real differences in the metal cofactors used by various MetAP isoforms.

b. *Coordination.* The initial structural analysis by Roderick and Mathews (29) of *E. coli* MetAP defined the cobalt ligation, showing two atoms of Co^{2+} bound per enzyme monomer. Appropriately, they were careful to point out that the *in vivo* metal had not been determined. The metal ions were located in a pocket of the central β sheet and marked the active site. The intra-Co^{2+} distance was first reported as 2.9 Å, but a new structure (at 1.9 Å resolution) has extended that distance to 3.2 Å (*59*). As shown in Fig. 9, there are five protein ligands (His-171, Glu-204, Asp-97, Asp-108, and Glu-235 in *E. coli* numbering) and these are preserved in identity and position in all of the MetAPs [although the numbering is altered in the eukaryote forms because of the N-terminal extensions and also in the type II isoforms because of the insert in the catalytic domain (Table I)]. His-171 and Glu-204 are monodentate while the others are bidentate (Asp-97 to only Co-2; the other two have oxygen ligands to each Co). In addition, there are two water molecules (not shown), one suspended between the two metal centers and one that serves as a terminal ligand to Co-2. The geometry of the complex to Co-1 is basically a trigonal bipyramid and that of Co-2, a distorted octahedron. The proposed role of the metal centers (and their ligands) in the mechanism is described below.

2. Substrate Specificity

The binding of substrate to MetAP has been defined to a reasonable degree by the structure of the *E. coli* MetAP-inhibitor complex determined by Lowther *et al.* (*59*). The inhibitor utilized contained a bestatin derivative where a norleucine side chain was built onto a 2-hydroxy-3-amino acid base to produce 3(*R*)-amino-2(*S*)-hydroxyheptanoic acid (AHHpA). This was attached to the α-amino group of Ala-Leu-Val-Phe-OMe. As expected, the AHHpA acted as an analog of Met with the hydroxy and amino groups both chelating to the cobalt ions as well as the oxygen (O-1) of the carboxyl of the amide linkage joining AHHpA to the Ala. The inhibitor displaced the solvent molecules seen in the uncomplexed protein. The AHHpA side chain, mimicking that of the obligatory Met, occupies a hydrophobic pocket (P_1) formed by His-79, Cys-59, Cys-70, Tyr-62, Tyr-65, Phe-177, and Trp-221 (Table I). The tapering geometry of this binding site supports the view that only a Met side chain (or an isosteric homolog such as norleucine) would be recognized. Consistent with this, Ben-Bassat *et al.* (*33*) reported that the sulfoxide derivative of Met was not cleaved. However, Solbiati *et al.* (*21*) found both sulfoxide and sulfone derivatives to be quite good substrates. This discrepancy remains to be resolved. The adjacent Ala (the defining residue position in determining MetAP substrate specificity) is found in a shallower pocket (S_1') made up primarily of Glu-204, Gln-233, Met-206, and Tyr-168. The peptide backbone of the inhibitor forms H-bonds with the protein and the conserved His-79 and -178 participate in these interactions. Site-directed mutagenesis studies in which each was converted to Ala gave derivatives that were substantially reduced (H79A, $\sim 10^5$-fold, H178A

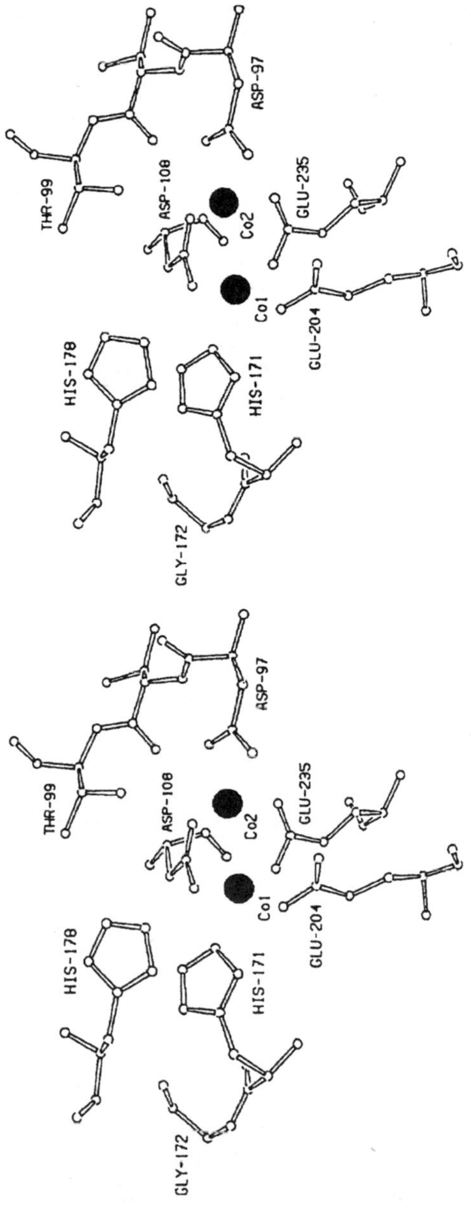

FIG. 9. Stereoview of the active site of *E. coli* MetAP including cobalt ions (●) and protein ligands. Reprinted with permission from Roderick and Matthews (29), copyright 1993, American Chemical Society.

TABLE I

FUNCTIONALLY SIGNIFICANT RESIDUES IN METAP ISOTYPES

Function	*E. coli*[a]	Sc1[a]	Pf[a]	hII[a]
S_1	C59[b]	P182	F50	P220
S_1	Y62[b]	Y185	[P234]	[M384]
S_1	Y65[b]	F188	[I205]	[H382]
S_1	C70[b]	C193	N53	G222
Fum	H79[b]	H202	H62	H231[b]
Me	D97[b]	D219[b]	D82	D215
Me	D108[b]	D230	D93	D260
S_1'	Y168	Y291	L150	L328
Me	H171[b]	H294	H153	H331
S_1'	F177[b]	F300	L160	I338
Cat	H178[b]	H301	H161[b]	H339[b]
Cat	—	—	H173[b]	—
Me/S_1'	E204[b]	E327	E187	E364
S_1	M206	M329[b]	F189	F360
S_1	W221[b]	W344	Y265	Y444
S_1'	Q233	Q356[b]	Q278	Q457
Me	E235[b]	E358	E280	E459
S_1'	—	—	L268	L447

[a] Residues in each row occupy related positions in their 3-D structures (unless enclosed by brackets) for the *E. coli, P. furiosus* (Pf), and human type II (hII) enzymes. The assignments for *S. cerevisiae* (Sc1) assume occupancy similar to that of the *E. coli* enzyme.

[b] Derivatives produced by site-directed mutagenesis of this site have been characterized. See text for details. Abbreviations: S_1 binding site for Met; S_1', binding site for penultimate residue; Me, metal ligand; Fum, site of fumagillin modification; Cat, residue implicated in catalysis.

~50-fold, as compared to wild-type) indicating their important role in catalysis. His-79 is also the site of fumagillin modification (*64*) and corresponds to His-231 in the human type II enzyme, also the site of covalent modification by fumagillin and related analogs (*65*) (see below).

Chiu *et al.* (*66*) have also carried out site-directed mutagenesis on *E. coli* MetAP. In this study, all of the metal ligands (see Table I) and six residues located in the putative binding site (S_1) that recognizes the Met side chain (see also above) were altered. The substitutions chosen were intended to be chemically conservative. When each of the metal ligands was substituted, it resulted in a nearly complete loss in enzymatic activity. There was a concomitant loss in Co^{2+} binding. Thus, the disruption of just one ligand results in the loss of both metal ions. In contrast, substitution of the residues in the

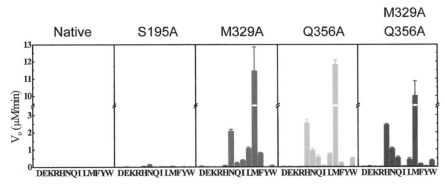

FIG. 10. Initial velocity of *S. cerevisiae* MetAP I on nonnormal substrates: native; S195A; M329A; Q356A; and M329A/Q356A. Assays were conducted using substrates of family MXSHRWDW, where the penultimate (X) amino acid is indicated below each bar. Initial velocities were determined by linear regression analysis. Three initial velocities were determined for each sample, and the error bars were calculated as ± S.E. Note that the initial velocities for the MMSHRWDW peptide represent the lower limit of the velocity, since two products are formed during the reaction (MSHRWDW and SHRWDW). Reprinted by permission from Walker and Bradshaw (*68*).

S_1 site was much less deleterious. Only two residues, Cys-70 and Trp-221, when substituted with Ser and Leu, respectively, caused a significant loss in activity. C70S showed 54% activity and W221L, 27%. The double mutant (C70S, W221L) was only 6% active. Another double mutant, C59S/C70S, showed about the same level of activity as C70S alone, and modification of His-63 to Leu, when combined with the Cys-59 mutation, was without effect. Kinetic evaluation of the C70S and W221L mutants showed increased K_m and decreased k_{cat} values.

The substitution of Asp97, a metal ligand, was much more destructive in *E. coli* MetAP than when the corresponding residue in yeast MetAP I (Asp-291) was modified (*67*). In that study, only about 50% of the cobalt ion was lost with a fivefold decrease in enzymatic activity. This residue is a bidentate chelator but both ligands are to Co-2 (*29*). However, Klinkenberg *et al.* (*67*) substituted Asn in contrast to Chiu *et al.* (*66*), who used Ala. *B. subtilis* MetAP has a Gln residue corresponding to Glu-235, suggesting that such an amide–acid substitution may be better tolerated than the complete elimination of a group capable of acting as a metal ligand.

The contribution of the putative S_1' residues to the specificity of type I MetAP activity was examined with recombinant yeast enzyme (*68*) (Fig. 10). Two of the four residues identified from the structure of the *E. coli* enzyme (*29*) to potentially line this pocket (Met-329 and Gln-356, equivalent to Met-206 and Gln-233 in *E. coli*) were changed to Ala by site-directed mutagenesis.

The M329A derivative had an average catalytic efficiency 1.5-fold higher with normal substrates and, in addition, cleaved substrates with Asn, Gln, Ile, Leu, Met, and Phe in the penultimate position of the substrate. The Q233A derivative was only about a third as efficient as a catalyst as native (or recombinant) enzyme with normal substrates, but also showed an expanded repertoire that included substrates with His, Asn, Gln, Leu, Met, Phe, Trp in the position adjacent to the N-terminal Met. A Ser-195 to Ala mutation was without effect. Thus, only substrates with residues with a formal charged (Asp, Glu, Lys, Arg) or Tyr in position 2, the penultimate position, were uncleaved. Interestingly, this study also showed that Met-Asn substrates were cleaved at a significant rate by native enzyme. This study thus confirms the prediction of the location of the S_1' site.

Although 3-D structures of type II MetAP from *P. furiosus* (56) and human (57) are known, there is little direct information about residues involved in activity. However, there are likely inferences that can be drawn. In the first place, as described by Tahirov *et al.* (56), there are several rearrangements in the structure that distinguish the prokaryotic type I and II proteins. There are two short deletions in the N and C termini and a 62-residue insertion between Gly-220 and Trp-221 of the *E. coli* protein; there are also corresponding changes to accommodate the large insertion and to contribute to its stabilization. In addition, there are some small variations in overall organization, particularly as associated with the helical structures. Second, although the metal ligand structure is similar, there are some differences in the position and interaction of other residues. For example, the H-bonding pattern of His-153 (His-171 in *E. coli*) is different: in the *E. coli* protein the imidazole ND1 enjoys a split interaction between the main chain oxygen atoms of Phe-177 and Gly-172; in *P. furiosus*, the ND1 H-bond is to the Leu-160 oxygen. Overall, it was concluded that there were two principal structural features that were likely to be characteristic differences between the two isotypes. The first is the substitution of Lys-154 (Val, Thr, and Ser in other type II proteins) with Gly in all type I isotypes, and the other is the insertion of a Trp or Tyr at position 158 (*P. furiosus* numbering) in the type II enzymes.

The S_1 pocket of the *P. furiosus* MetAPII is comprised of Phe-50, Pro-51, Leu-160, Ile-205, Met-207, Pro-234, and Tyr-265 (Table I). This structure is quite different then the type I S_1 site (see Table I). Interestingly, most of these residues are located in the regions that are most different in the two enzyme classes. For example, Tyr-62 and -65 of *E. coli* MetAP are located in the insertion sequence that divides Phe-50 and Pro-51 of the archaebacterial MetAP. However, these and other differences that are likely characteristic of the two isotypes do not alter the selectivity for Met that is bound in these pockets.

As would be expected, the *P. furiosus* enzymes shows a higher heat stability than the *E. coli* enzyme (*41*). An analysis of the potential contributing factors showed that ion pairs occur in significantly higher amounts in the archaebacterial protein and that these stabilize helices by the formation of intrahelical pairs (especially with helix D) as well as strengthening the whole tertiary structure by cross-linking distant segments (*69*). Although the number of H bonds was found to be the same, there was a significant difference in the nature of these bonds in the *P. furiosus* enzyme relative to the *E. coli* protein, which, along with internal water and the location of Pro residues, were also thought to contribute to the thermal stability. It has also been reported that the MetAP of *P. furiosus* forms amyloid-like fibrils when treated with guanidine-HCl (*70*). These structures form by the conversion of the protein to β structures during denaturation, a phenomenon observed with other proteins as well.

Table I summarizes the residues that are involved in metal binding, substrate interactions and catalysis. Some of the assignments are by inference to studies with other family members, where more detailed studies utilizing site-directed mutagenesis or known 3-D structures have been reported.

3. *Reaction Mechanism*

The improved resolution of the *E. coli* MetAP structure, including the determination of the binding site of the AHHpA-peptide inhibitor and the possible roles of His-79 and -178, yielded two potential catalytic mechanisms (*59*). These differed primarily as to whether the substrate carboxyl oxygen interacted with Glu-204 or Co-1 and His-178. Utilizing various derivatives of Met, including phosphorus-containing transition state analogs, Lowther *et al.* (*58*) proposed a refined mechanistic model based on four assumptions: (1) the substrate α-amino group binds to Co2 (with the displacement of a solvent molecule); (2) a noncovalent *gem*-diolate tetrahedral intermediate is formed after attack by the metal-bridging water or hydroxide ion; (3) the carboxyl oxygen of the scissile bond binds to Co-2 and His-178; and (4) His-79 forms an H bond with the nitrogen of the scissile bond.

The overall reaction mechanism proposed is shown in Fig. 11. In this scheme, the substrate binds to both metal ions, which changes the coordination sphere of Co-1 from 5 to 6, and likely activates the nucleophilic hydroxide ion, bound between the two metals, for attack on the carbonyl carbon of the scissile bond. Proton transfer to Glu-204 (from the hydroxide ion) would follow. The tetrahedral intermediate formed (as found for all proteases) is stabilized by interactions with both histidines (79 and 178) and by interaction with Co-1. The reaction would be completed by donation of the Glu-204 proton to the newly formed α-amino group of the substrate penultimate residue. The return of Co-1 to 5-fold coordination would accompany

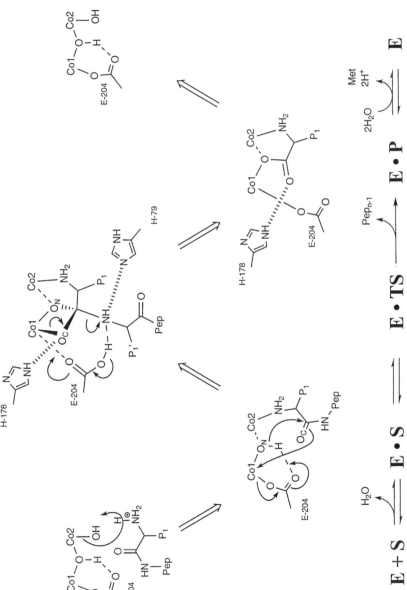

FIG. 11. Proposed reaction mechanism for cMetAP. The alteration in the electronic nature of the metal center on substrate binding is presumed to activate the nucleophile (O_N) and facilitate proton transfer to Glu-204. The carboxy anion of the resulting tetrahedral intermediate, originally from the oxygen atom of the scissile bond (O_C), is stabilized by the expanded coordination sphere of Co-1 and interactions with His-79 and His-178. Breakdown of the intermediate to products returns the coordination number of Co-1 to 5 while retaining the metal-bridging and His-178 interactions. Regeneration of the active site occurs with the release of Met and the deprotonation of solvent molecules. (See Ref. 51.)

the breakdown of the transition-state intermediate followed by the release of free Met still coordinated to the metal center and H-bonded to His-178.

This mechanism is similar to that proposed for prolidase (aminopeptidase P) (71) and has analogous features to other proteases and similar enzymes. The role of Glu-204 as a proton shuttle is particularly exemplary of this, as is the proposed use of histidine. The model also clearly provides an excellent rationale for the observations that His-79 is crucial for activity (as well as being the site of modification for fumagillin), whereas His-178 is dispensable. The interaction of His-79 with the substrate backbone clearly is important, perhaps because the substrate makes no contact with the enzyme beyond the penultimate residue (explaining the overwhelming influence of the penultimate residue on specificity).

E. FUNCTIONAL CONSIDERATIONS

1. *Eubacteria*

The function of MetAP in eubacteria, as presently understood, is to remove N-terminal initiator Met residues, following deformylation, for compatible substrates. It is not entirely clear whether this is a co- or post-translational process. These enzymes lack the N-terminal extensions of the eukaryote type I isoforms and, therefore, probably do not bind to ribosomes, although this has not been demonstrated directly. As shown by Chang *et al.* (72), the *E. coli map* gene, located next to the *rpsB* gene at min 4, is essential for cell growth.

2. *Yeast*

Although a simple eukaryote, yeast contains both MetAP isoforms (in the b or extended form). As judged by a null mutation, it was initially concluded that MetAP I (coded by *map1*) was not essential (38). However, the identification of the second gene (*map2*) suggested the possibility of functional redundancy. Indeed, a null mutation in this gene also was not lethal, although it produced a slower growth rate as well; however, the double null was lethal (52). This confirmed that MetAP activity is essential for viability in yeast, and it is likely that this is true for all life forms. What is less clear is whether there are unique functions for the two enzymes in yeast, as can clearly be attributed to the isoforms in higher eukaryotes (see below). Klinkenberg *et al.* (67) have described a MetAP I mutant (D219N) that, they concluded on the basis of growth experiments, acts as a dominant negative inhibitor of MetAP I and that can also interfere with the function of MetAP II. More detailed molecular characterization will be required to determine the basis of this observation and to how MetAP I and II may interact *in vivo*.

3. Higher Eukaryotes

a. *MetAP II as an Inhibitor of Protein Synthesis.* The MetAP isoforms of higher eukaryotes, unlike yeast, display significant functional diversity. This is particularly true of the type II enzyme. Although both are likely still involved in some fashion in the processing of initiator Met from nascent chains, MetAP II has a well-defined second activity, namely an interaction with a eukaryotic initiation factor, eIF2, that blocks phosphorylation of the α subunit and thus promotes translation (*44, 73*). This identification actually provided the first characterization of MetAP II from a eukaryote (*43, 46*).

In a detailed series of studies, these workers established that the eIF2 associated activity was due to a protein of apparent molecular mass of 67 kDa [and they subsequently designated the protein p67 accordingly (*44*)], that it was mono O-linked glycosylated at multiple sites, (*45*), and that it bound to eIF2γ (*46*). They also showed that deglycosylation eliminated the inhibitory activity and proposed that the carbohydrate interacted with the α subunit to sterically block the heme-regulated inhibitor (HRI) and double-stranded RNA-activated inhibitor (dsI), both of which are kinases that modify eIF2α (*44*). Subsequently, Wu *et al.* (*43*) cloned the cDNA for this protein from rat liver, which predicted a 480 amino acid sequence. [This sequence probably contained a minor error near the C terminus that became evident, because of the very high degree of identity, when the human MetAP II sequence was deduced (*42, 53*)]. Following the isolation and characterization of porcine MetAP (*39*), the sequence analysis of isolated peptides revealed the apparent coidentity of MetAP II and p67, which was presaged by the sequence comparisons of Bazan *et al.* (*47*) and confirmed by the sequence analysis of the human cDNA (*42, 53*). This was established by direct assays of recombinant human MetAP II by Griffith *et al.* (*74*).

There are a number of intriguing issues raised by this dual functionality that are related both to structure and physiological role. First, the O-linked glycosylation that is apparently essential for the inhibitory activity does not seem to be involved in the catalytic functions of the protein. Indeed, the 3-D structure of human MetAP II (*57*), which was basically limited to the catalytic domain, showed no carbohydrate modifications. This suggests that this modification is restricted to the N-terminal extension. Yeast MetAP I is fully active catalytically without this domain (*54*), but a similar determination has not yet been made for the type II isozyme. Second, the binding of MetAP II (p67) to the eIF2γ subunit is, as would be expected, a reversible phenomenon (*75*). However, it is unclear what controls this process. There is no evidence to suggest that p67 remains associated with eIF2 when it forms the 43S ribosomal initiation complex. Is this displacement associated with other events during the complex formation, such as the binding of the charged tRNA? This also raises the question of whether there is any connection between

the two MetAP activities. Finally, there is the question of cellular localization. It has been suggested that MetAP I may be tethered to the ribosome
through the zinc finger motifs of its N-terminal extension (38). If MetAP II
also functions as the eIF2α phosphorylation inhibitor, such a localization
would be apparently contradictory. Therefore, it must either be associated
with another structure (or molecule) other than eIF2α or be free in solution.
If the latter is the case, its role as an N-terminal processing enzyme is likely
to be secondary, at least temporally, to that of MetAP I (but would allow
redundancy of this essential physiological process, as seen in yeast).

   b.   *MetAP II as a Mediator of Angiogenesis.* One of the more interesting
developments in studies on N-terminal processing and the role of MetAPs
has been the identification of MetAP II as the principal target of a class of
antiangiogenesis/immunosuppressive drugs (40, 74). The parent compound,
fumagillin (Fig. 12), has been known for 50 years (76) and was first introduced
as an antibiotic to treat amoebiasis. In 1990, Ingber *et al.* (77) determined

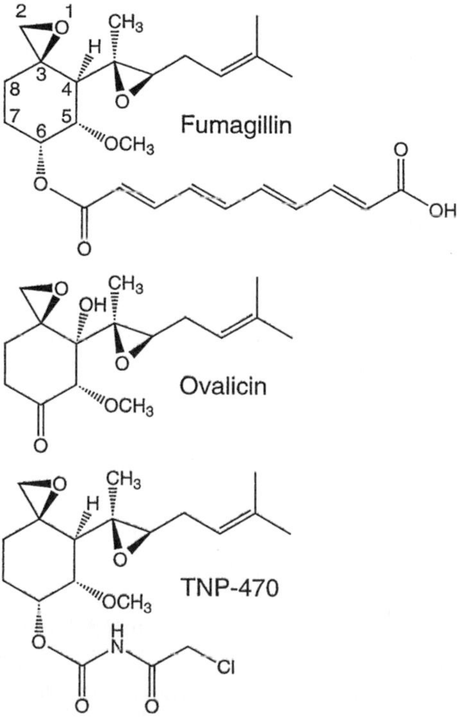

FIG. 12. Antiangiogenesis compounds that covalently inhibit MetAPs. Numbering scheme
taken from Griffith *et al.* (74).

from an opportunistic contamination that this compound had potent endothelial cell-rounding activity and was able to inhibit angiogenesis. Subsequently, several analogs were developed with superior properties, and one of these, AGM-1470 (also known as TNP-470), is now under clinical investigation as an antitumor drug (78). Independently, ovalicin, isolated from the culture medium of the ascomycete *P. eurotium ovalis* (79), was shown to be a strong immunosuppressive by blocking the proliferation of lymphocytes (80). Although there are some differences, the two activities (antiangiogenesis and immunosuppression) have now been shown to be shared by both compounds (81).

In 1997, two groups (40, 74) showed independently that MetAP II is the target for these drugs and that the observed cellular responses result from an irreversible covalent modification that leads to the inactivation of the processing capability. They do not, however, block the eIF2-related activity (74). Subsequent studies demonstrated that the site of modification in human MetAP II was His-231 (equivalent to His-79 in *E. coli* MetAP I) and that the ring epoxide, as opposed the side chain epoxide, was responsible for the reaction (Fig. 12) (65). This was confirmed by the structural analysis of the inhibited form of human MetAP II (57). The clear specificity for the type II isoform prompted Liu *et al.* (57) to speculate on the structural variations between the two isoforms that might account for this specificity and to suggest that MetAP I could not be equivalently modified. However, Lowther *et al.* (64) showed that both *E. coli* and yeast type I enzymes were inhibited, albeit at 10^3 greater concentrations of reagent, with ovalicin. The *E. coli* enzyme was also inhibited by fumagillin, but the yeast enzyme was not. Modification was limited to His-79 (in the *E. coli* enzyme) and occurred via the ring epoxide. The inhibition clearly results from the modification of a catalytically essential residue (see above).

A number of studies have focused on the molecular basis for the suppression of endothelial cell (and lymphocyte) division by the MetAP II inhibition. The fact that the inhibition produces arrest, rather than cytotoxicity, has naturally led to the suggestion that the cell cycle, and the molecules that control it, are the most likely targets. Consistent with this, Hori *et al.* (82) reported the suppression of cyclin D1 expression in mid G_1 phase, and Abe *et al.* (83) found that AGM 1470 inhibited the phosphorylation of the retinoblastoma gene product (Rb). They also showed that candidate Rb kinases, cdc2 and cdk2, were inhibited, although the effect was not direct. The inhibitor also blocked growth factor-induced mRNA expression of cdc2 and cyclinA, but not that of cdk2, cdk4, and cyclin D1. Zhang *et al.* (84) have shown the AGM-1470 treatment of endothelial cells leads to the activation of the p53 pathway, resulting in the accumulation of the cyclin-dependent kinase inhibitor p21[WAF1/CIP1]. However, the direct role of MetAP II remains unclear. Turk *et al.* (*85*) have shown that inhibition of MetAP II does not lead to a change in N^α-myristoylation, but they did find differences with respect to the N-terminal processing of select proteins. The initiator Met of glyceraldehyde-3-phosphate dehydrogenase and cyclophilin A was only partially processed in MetAP II inhibited cells. A similar observation has been made for endothelial nitric oxide synthase (eNOS) (*86*). Model peptides from these proteins, along with controls, clearly established that these proteins were processed substantially better by MetAP II than by MetAP I, despite the fact that the N-terminal sequences (M-G, M-P, and M-V) are all readily processed by both enzymes in *in vitro* and *in vivo* assays in the context of other substrates or different overall N-terminal sequences. This observation suggests that the seemingly similar specificity of the two isotypes of MetAP maybe affected by cellular factors/context that have not yet been identified. It is also interesting that certain cell types are clearly much more sensitive to MetAP II inhibition than others. The basis for this response also remains to be elucidated. It has also been reported that mutations in the putative MetAP II gene in *Drosphila* also leads to growth and developmental defects (*87*). Thus, the involvement of N-terminal processing in controlling cell division may be a fairly widespread phenomenon.

   c. *Protein Turnover: The N-End Rule.* In 1986, Bachmair *et al.* (*88*) showed, using a fusion protein of ubiquitin in frame with a β-galactosidase derivative, that the N-terminal residue of a protein strongly affects its turnover. This classic experiment is shown in schematic form in Fig. 13. When the fusion proteins were expressed in yeast, they were readily cleaved by ubiquitinase to yield the β-galactosidase protein (minus any initiator Met). By substituting the potential N-terminal residue with different amino acids, it was shown that those with small side chains produced long-lived derivatives, whereas those with larger side chains were turned over quite rapidly. Extensive additional experimentation has defined this process (termed the N-end rule) in great detail (*89*). The clear correlation of the specificity of the MetAP family to this profile suggested that the processing of N termini and subsequent intracellular stability were linked (*13*). Initiator Met retention clearly protects proteins against premature degradation while still allowing the bulk of proteins (~50% in yeast) to be processed and thereby return substantial Met to the free amino acid pool. However, what is less clear is whether the "protected" proteins—those that retain initiator Met—can eventually be converted to an unstable form (and be subsequently degraded) by downstream removal of the Met (by non-MetAP enzymes). The evidence that this is so is quite limited and suggests that the N-end rule is reserved for quite special functions, such as has been suggested for the degradation of skeletal muscle proteins (*90*).

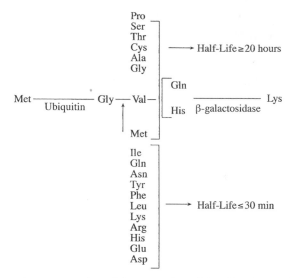

FIG. 13. Schematic representation of the expression of the ubiquitin–β-galactosidase constructs in yeast; sequences were modified to alter the N-terminal residue of the β-galactosidase portion as indicated. When expressed, the ubiquitin was immediately removed by ubiquitinase (by cleavage of the scissile bond, indicated by the vertical arrow) to produce each β-galactosidase product (except for the Pro derivative, which was poorly processed by the ubiquitinase). The two classes of proteins, based on stability, are indicated. [After A. Bachmeier *et al.* (*88*).]

IV. Future Directions

Although there has been substantial progress in elucidating the events/agents surrounding the removal of the initiator Met, there remain many unresolved issues, some of which have already been described. Central to all of these, at least in eukaryotic cells, is the role of each MetAP isotype. However, there are also structural questions such as the metal ions used *in vivo* (is there more than one in some cases?), the importance of posttranslational modifications, in particular glycosylation, and the role of the distinguishing domains. Cellular localization and *in vivo* specificity are equally important issues for future study.

Although it has not been stressed, the control of N-terminal structures is important in the expression of recombinant proteins, both in terms of investigative studies and in the development of therapeutics. Engineered proteins can bear different N termini than found in their native counterparts, particularly if a protein that is normally found in the extracellular compartment is made in a nonsecretory vector. The development of a MetAP with an expanded specificity is an example of a reagent that may be useful for *in vitro* processing (*68*). Additional MetAP derivatives may expand this range further and improve their usefulness.

There are also wider questions about N-terminal processing that have not been addressed here in any detail. The relationship of eubacteria MetAP to the peptide deformylase is unclear and, in eukaryotes, there are quite a large number of additional modifications, e.g., myristoylation and acetylation, that are also important for structure/function relationships in mature proteins. Collectively they represent the first co/posttranslational modifications that intracellular proteins undergo and they play a major role in controlling activity, stability, and translocation. The elucidation of how these are synchronized will be important in the developing field of proteomics.

Acknowledgments

The authors thank Dr. Darren Tyson for help in preparing some of the figures and Ms. Anita Amabile for preparing the manuscript. Studies emanating from the Bradshaw Laboratory were supported by NIH research grant DK 32465.

References

1. Jackson, R., and Hunter, T. (1970). *Nature* **227,** 672.
2. Wilson, D. B., and Dintzis, H. M. (1970). *Proc. Natl. Acad. Sci. USA* **66,** 1282.
3. Marcker, K., and Sanger, F. (1964). *J. Mol. Biol.* **8,** 835.
4. Meinnel, T., Mechulam, Y., and Blanquet, S. (1993). *Biochimie* **75,** 1061.
5. Rawlings, N. D., and Barrett, A. J. (1993). *Biochem. J.* **290(Pt 1),** 205.
6. Persson, B., Flinta, C., von Heijne, G., and Jornvall, H. (1985). *Eur. J. Biochem.* **152,** 523.
7. Flinta, C., Persson, B., Jornvall, H., and von Heijne, G. (1986). *Eur. J. Biochem.* **154,** 193.
8. Tsunasawa, S., Stewart, J. W., and Sherman, F. (1985). *J. Biol. Chem.* **260,** 5382.
9. Sherman, F., Stewart, J. W., and Tsunasawa, S. (1985). *Bioessays* **3,** 27.
10. Barwick, R. C., Jones, R. T., Head, C. G., Shih, M. F., Prchal, J. T., and Shih, D. T. (1985). *Proc. Natl. Acad. Sci. USA* **82,** 4602.
11. Prchal, J. T., Cashman, D. P., and Kan, Y. W. (1986). *Proc. Natl. Acad. Sci. USA* **83,** 24.
12. Huang, S., Elliott, R. C., Liu, P. S., Koduri, R. K., Weickmann, J. L., Lee, J. H., Blair, L.C., Ghosh-Dastidar, P., Bradshaw, R. A., Bryan, K. M., Einarson, B., Kendall, R. L., Kolacz, K. H., and Saito, K. (1987). *Biochemistry* **26,** 8242.
13. Arfin, S. M., and Bradshaw, R. A. (1988). *Biochemistry* **27,** 7979.
14. Boissel, J. P., Kasper, T. J., and Bunn, H. F. (1988). *J. Biol. Chem.* **263,** 8443.
15. Moerschell, R. P., Hosokawa, Y., Tsunasawa, S., and Sherman, F. (1990). *J. Biol. Chem.* **265,** 19638.
16. Dalboge, H., Bayne, S., and Pedersen, J. (1990). *FEBS Lett.* **266,** 1.
17. Hirel, P. H., Schmitter, M. J., Dessen, P., Fayat, G., and Blanquet, S. (1989). *Proc. Natl. Acad. Sci. USA* **86,** 8247.
18. Bradshaw, R. A., Brickey, W. W., and Walker, K. W. (1998). *Trends Biochem. Sci.* **23,** 263.

19. Schmitt, E., Guillon, J. M., Meinnel, T., Mechulam, Y., Dardel, F., and Blanquet, S. (1996). *Biochimie* **78,** 543.
20. Mazel, D., Pochet, S., and Marliere, P. (1994). *EMBO J.* **13,** 914.
21. Solbiati, J., Chapman-Smith, A., Miller, J. L., Miller, C. G., and Cronan, J. E. J. (1999). *J. Mol. Biol.* **290,** 607.
22. Becker, A., Schlichting, I., Kabsch, W., Schultz, S., and Wagner, A. F. (1998). *J. Biol. Chem.* **273,** 11413.
23. Becker, A., Schlichting, I., Kabsch, W., Groche, D., Schultz, S., and Wagner, A. F. (1998). *Nat. Struct. Biol.* **5,** 1053.
24. Groche, D., Becker, A., Schlichting, I., Kabsch, W., Schultz, S., and Wagner, A. F. (1998). *Biochem. Biophys. Res. Commun.* **246,** 342.
25. Meinnel, T., and Blanquet, S. (1995). *J. Bacteriol.* **177,** 1883.
26. Meinnel, T., Blanquet, S., and Dardel, F. (1996). *J. Mol. Biol.* **262,** 375.
27. Rajagopalan, P. T. R., Grimme, S., and Pei, D. (2000). *Biochemistry* **39,** 779.
28. Ragusa, S., Blanquet, S., and Meinnel, T. (1998). *J. Mol. Biol.* **280,** 515.
29. Roderick, S. L., and Matthews, B. W. (1993). *Biochemistry* **32,** 3907.
30. Adams, J. M. (1968). *J. Mol. Biol.* **33,** 571.
31. Kozak, M. (1999). *Gene* **234,** 187.
32. Taylor, A., ed. (1996). "Aminopeptidases." Molecular Biology Intelligence Unit, R. G. Landes Company, Austin, TX.
33. Ben-Bassat, A., Bauer, K., Chang, S. Y., Myambo, K., Boosman, A., and Chang, S. (1987). *J. Bacteriol.* **169,** 751.
34. Miller, C. G., Strauch, K. L., Kukral, A. M., Miller, J. L., Wingfield, P. T., Mazzei, G. J., Werlen, R. C., Graber, P., and Movva, N. R. (1987). *Proc. Natl. Acad. Sci. USA* **84,** 2718.
35. Wingfield, P., Graber, P., Turcatti, G., Movva, N. R., Pelletier, M., Craig, S., Rose, K., and Miller, C. G. (1989). *Eur. J. Biochem.* **180,** 23.
36. Suh, J. W., Boylan, S. A., Oh, S. H., and Price, C. W. (1996). *Gene* **169,** 17.
37. Chang, Y. H., Teichert, U., and Smith, J. A. (1990). *J. Biol. Chem.* **265,** 19892.
38. Chang, Y. H., Teichert, U., and Smith, J. A. (1992). *J. Biol. Chem.* **267,** 8007.
39. Kendall, R. L., and Bradshaw, R. A. (1992). *J. Biol. Chem.* **267,** 20667.
40. Sin, N., Meng, L., Wang, M. Q., Wen, J. J., Bornmann, W. G., and Crews, C. M. (1997). *Proc. Natl. Acad. Sci. USA* **94,** 2362.
41. Tsunasawa, S., Izu, Y., Miyagi, M., and Kato, I. (1997). *J. Biochem. (Tokyo)* **122,** 843.
42. Arfin, S. M., Kendall, R. L., Hall, L., Weaver, L. H., Stewart, A. E., Matthews, B. W., and Bradshaw, R. A. (1995). *Biochemistry* **92,** 7714.
43. Wu, S., Gupta, S., Chatterjee, N., Hileman, R. E., Kinzy, T. G., Denslow, N. D., Merrick, W. C., Osterman, J. C., and Gupta, N. K. (1993). *J. Biol. Chem.* **268,** 10796.
44. Datta, B., Chakrabarti, D., Roy, A. L., and Gupta, N. K. (1988). *Proc. Natl. Acad. Sci. USA* **85,** 3324.
45. Datta, B., Ray, M. K., Chakrabarti, D., Wylie, D. E., and Gupta, N. K. (1989). *J. Biol. Chem.* **264,** 20620.
46. Ray, M. K., Chakraborty, A., Datta, B., Chattopadhyay, A., Saha, D., Bose, A., Kinzy, T. G., Wu, S., Hileman, R. E., Merrick, W. C., and Gupta, N. K. (1993). *Biochemistry* **32,** 5151.
47. Bazan, J. F., Weaver, L. H., Roderick, S. L., Huber, R., and Matthews, B. W. (1994). *Proc. Natl. Acad. Sci. USA* **91,** 2473.
48. Hoeffken, H. W., Knof, S. H., Bartlett, P. A., Huber, R., Moellering, H., and Schumacher, G. (1988). *J. Mol. Biol.* **204,** 417.
49. Henrich, B., Monnerjahn, U., and Plapp, R. (1990). *J. Bacteriol.* **172,** 4641.
50. Haas, E. S., Daniels, C. J., and Reeve, J. N. (1989). *Gene* **77,** 253.
51. Lowther, W. T., and Matthews, B. W. (2000). *Biochim. Biophys. Acta.* **1477,** 157.

52. Li, X., and Chang, Y. H. (1995). *Proc. Natl. Acad. Sci. USA* **92,** 12357.
53. Li, X., and Chang, Y. H. (1995). *Biochim. Biophys. Acta* **1260,** 333.
54. Zuo, S., Guo, Q., Ling, C., and Chang, Y.-H. (1995). **246,** 247.
55. Li, X., and Chang, Y. H. (1996). *Biochem. Biophys. Res. Commun.* **227,** 152.
56. Tahirov, T. H., Oki, H., Tsukihara, T., Ogasahara, K., Yutani, K., Ogata, K., Izu, Y., Tsunasawa, S., and Kato, I. (1998). *J. Mol. Biol.* **284,** 101.
57. Liu, S., Widom, J., Kemp, C. W., Crews, C. M., and Clardy, J. (1998). *Science* **282,** 1324.
58. Lowther, W. T., Zhang, Y., Sampson, P. B., Honek, J. F., and Matthews, B. W. (1999). *Biochemistry* **38,** 14810.
59. Lowther, W. T., Orville, A. M., Madden, D. T., Lim, S., Rich, D. H., and Matthews, B. W. (1999). *Biochemistry* **38,** 7678.
60. Vallee, B. L., and Holmquist, B. (1980). In "Methods for Determining Metal Ion Environment in Proteins" (D. W. Darnall and R. G. Wilkins, eds.), Vol. 2, pp. 27–74. Elsevier, Amsterdam.
61. Kobayashi, M., and Shimizu, S. (1999). *Eur. J. Biochem.* **261,** 1.
62. D'souza, V. M., and Holz, R. C. (1999). *Biochemistry* **38,** 11079.
63. Walker, K. W., and Bradshaw, R. A. (1998). *Protein Sci.* **7,** 2684.
64. Lowther, W. T., McMillen, D. A., Orville, A. M., and Matthews, B. W. (1998). *Proc. Natl. Acad. Sci. USA* **95,** 12153.
65. Griffith, E. C., Su, Z., Niwayama, S., Ramsay, C. A., Chang, Y. H., and Liu, J. O. (1998). *Proc. Natl. Acad. Sci. USA* **95,** 15183.
66. Chiu, C. H., Lee, C. Z., Lin, K. S., Tam, M. F., and Lin, L. Y. (1999). *J. Bacteriol.* **181,** 4686.
67. Klinkenberg, M., Ling, C., and Chang, Y. H. (1997). *Arch. Biochem. Biophys.* **347,** 193.
68. Walker, K. W., and Bradshaw, R. A. (1999). *J. Biol. Chem.* **274,** 13403.
69. Ogasahara, K., Lapshina, E. A., Sakai, M., Izu, Y., Tsunasawa, S., Kato, I., and Yutani, K. (1998). *Biochemistry* **37,** 5939.
70. Yutani, K., Takayama, G., Goda, S., Yamagata, Y., Maki, S., Namba, K., Tsunasawa, S., and Ogasahara, K. (2000). *Biochemistry* **39,** 2769.
71. Wilce, M. C., Bond, C. S., Dixon, N. E., Freeman, H. C., Guss, J. M., Lilley, P. E., and Wilce, J. A. (1998). *Proc. Natl. Acad. Sci. USA* **95,** 3472.
72. Chang, S. Y., McGary, E. C., and Chang, S. (1989). *J. Bacteriol.* **171,** 4071.
73. Chakraborty, A., Saha, D., Bose, A., Chatterjee, M., and Gupta, N. K. (1994). *Biochemistry* **33,** 6700.
74. Griffith, E. C., Su, Z., Turk, B. E., Chen, S., Chang, Y. H., Wu, Z., Biemann, K., and Liu, J. O. (1997). *Chem. Biol.* **4,** 461.
75. Chakraborty, A., Saha, D., Bose, A., Hileman, R. E., Chatterjee, M., and Gupta, N. K. (1994). *Indian J. Biochem. Biophys.* **31,** 236.
76. McCowen, M., Callender, M., and Lawlis, J. (1951). *Science* **113,** 202.
77. Ingber, D., Fujita, T., Kishimoto, S., Sudo, K., Kanamaru, T., Brem, H., and Folkman, J. (1990). *Nature* **348,** 555.
78. Castronovo, V., and Belotti, D. (1996). *Eur. J. Cancer* **32A,** 2520.
79. Sigg, H. P., and Weber, H. P. (1968). *Helv. Chim. Acta* **51,** 1395.
80. Hartmann, G. R., Richter, H., Weiner, E. M., and Zimmermann, W. (1978). *Planta Med.* **34,** 231.
81. Turk, B. E., Su, Z., and Liu, J. O. (1998). *Bioorg. Med. Chem.* **6,** 1163.
82. Hori, A., Ikeyama, S., and Sudo, K. (1994). *Biochem. Biophys. Res. Commun.* **204,** 1067.
83. Abe, J., Zhou, W., Takuwa, N., Taguchi, J., Kurokawa, K., Kumada, M., and Takuwa, Y. (1994). *Cancer Res.* **54,** 3407.
84. Zhang, Y., Griffith, E. C., Sage, J., Jacks, T., and Liu, J. O. (2000). *Proc. Natl. Acad. Sci. USA* **97,** 6427.

85. Turk, B. E., Griffith, E. C., Wolf, S., Biemann, K., Chang, Y. H., and Liu, J. O. (1999). *Chem. Biol.* **6,** 823.
86. Yoshida, T., Kaneko, Y., Tsukamoto, A., Han, K., Ichinose, M., and Kimura, S. (1998). *Cancer Res.* **58,** 3751.
87. Cutforth, T., and Gaul, U. (1999). *Mech. Dev.* **82,** 23.
88. Bachmair, A., Finley, D., and Varshavsky, A. (1986). *Science* **234,** 179.
89. Varshavsky, A. (1996). *Proc. Natl. Acad. Sci. USA* **93,** 12142.
90. Solomon, V., Baracos, V., Sarraf, P., and Goldberg, A. L. (1998). *Proc. Natl. Acad. Sci. USA* **95,** 12602.
91. Coll, M., Knof, S. H., Ohga, Y., Messerschmidt, A., Huber, R., Moellering, H., Russmann, L., and Schumacher, G. (1990). *J. Mol. Biol.* **214,** 597.

15

Carboxypeptidases E and D

LLOYD D. FRICKER

Department of Molecular Pharmacology
Albert Einstein College of Medicine
Bronx, New York 10461

I. Introduction

The role of carboxypeptidases in the generation of biologically active peptides was predicted in the late 1960s based on the sequence of proinsulin (*1*). In this precursor, the biologically active domains are separated by pairs of basic amino acids from the inactive "C" domain. It was proposed that a trypsin-like endopeptidase would initially cleave the proinsulin at the

THE ENZYMES, Vol. XXII

pairs of basic residues, generating intermediates containing C-terminal basic amino acids. Then, a carboxypeptidase would remove these basic amino acids. Other prohormones were also found to be produced from larger precursors, and in most cases the biologically active peptides were separated from the intervening sequences by pairs of basic amino acids (1). Thus, the proposed processing route for proinsulin was assumed to function for most neuroendocrine peptides, although at the time it was not clear if there would be enzymes specific for each neuroendocrine peptide precursor or general enzymes that could cleave most precursors.

A search for neuropeptide-cleaving carboxypeptidases led to the discovery of carboxypeptidase E (CPE) in the early 1980s. CPE is widely distributed in the neuroendocrine system and has a broad substrate specificity (providing the C terminus is a basic amino acid). Because no additional secretory pathway carboxypeptidase was detected in any of the early studies, it was falsely assumed that CPE was the only mammalian peptide processing carboxypeptidase. However, when mice that lacked CPE activity (because of a point mutation in the CPE gene) were found to correctly process small amounts of neuroendocrine peptides, the possibility of additional peptide processing carboxypeptidases was considered. Two other CPE-related enzymes had already been identified (carboxypeptidases N and M), but these were not predicted to function in the intracellular processing of peptides because of their distribution and enzymatic properties. A search for novel members of the CPE-gene family revealed five additional proteins with approximately 40–50% amino acid sequence identity with CPE. Of these other members of the CPE gene family only one, named carboxypeptidase D (CPD), is likely to participate in the intracellular biosynthesis of neuroendocrine peptides. The potential physiological functions of the other members of the family are briefly described below, following a detailed description of CPE and CPD.

II. Carboxypeptidase E

A. DISCOVERY

The first studies examining prohormone processing carboxypeptidases were done by Zuhlke and Steiner et al. (2) and resulted in the detection of a metallocarboxypeptidase with a pH optimum around 5.5. This is close to the predicted pH at which peptide processing occurs, based on the results of pulse-chase analysis, which indicated that prohormone processing occurred following packaging of the prohormone into post-Golgi secretory vesicles (1). The intracellular pH of the Golgi is near neutral, the *trans*-Golgi network is slightly acidic, and then mature secretory vesicles have an intravesicular pH of 5–5.5 (1, 3, 4). The finding by Zuhlke and Steiner that the insulin-processing carboxypeptidase was inhibited by chelating agents indicated that it was a metallopeptidase and thus distinct from the serine and cysteine carboxypeptidases of the lysosomes (2).

A search for an enkephalin-processing carboxypeptidase in bovine adrenal medulla chromaffin vesicles (i.e., the peptide-containing secretory vesicles of the adrenal) also turned up a metallocarboxypeptidase with an acidic pH optimum (5). Initially named "enkephalin convertase," this enzyme was purified and extensively characterized (5–7). This carboxypeptidase was able to sequentially remove basic residues from the C terminus of enkephalin peptides containing Arg-Arg extensions, which resembles the endogenous processing intermediate (6, 7). Although initially thought to be specific for the production of enkephalin, further studies revealed that this enzyme was present in all bovine and rat peptide-processing neuroendocrine tissues examined (6, 8). In some of these tissues, enkephalin convertase was further localized to the peptide-containing secretory vesicles (5, 9–11). Biochemical studies revealed that this enzyme was not specific for enkephalin precursors but was able to cleave a variety of peptides with C-terminal Lys and/or Arg extensions (6, 9, 10). Based on this broader endocrine role, enkephalin convertase was renamed carboxypeptidase E (for "endocrine"). Subsequently, this enzyme was also named carboxypeptidase H and given the designation E.C. 3.4.17.10 (12). Although the names "CPE" and "CPH" were used with equal popularity for several years, in the past few years virtually all of the publications on this enzyme have used the name "CPE."

In addition to being discovered as a peptide processing enzyme, CPE has also been "rediscovered" multiple times. CPE was found to be "glycoproteins J and K" in a survey of membrane-bound glycoproteins of the bovine adrenal medulla (13). CPE was also discovered as a developmentally regulated rat brain cDNA (14), as an autoantigen associated with type 1 diabetes (15), as a cDNA enriched in human ocular ciliary epithelium (16), and as a "prohormone sorting receptor" (17). In addition, CPE has been detected in screens of libraries using the yeast two-hybrid approach with a variety of ligands. However, many of the proteins used for "bait" in the two-hybrid screen are cytosolic, and since CPE is thought to be entirely within the lumenal side of the secretory pathway it is unlikely that the binding of CPE to various "bait" proteins is physiologically relevant; it more likely represents the nonspecific interaction of a sticky protein.

B.   PROPERTIES AND DISTRIBUTION OF CARBOXYPEPTIDASE E

The nucleotide sequence of CPE has been determined for a wide variety of species, including human (14), bovine (18), rat (19, 20), mouse (GenBank U23184), anglerfish (21), and mollusk (Aplysia) (22). The deduced amino acid sequence is highly conserved among species, with greater than 90% amino acid sequence identity among all the mammalian species of CPE and 80% amino acid sequence identity between fish and mammalian CPE. Although *Aplysia* CPE has only 45% amino acid sequence identity with mammalian CPE, the enzymatic properties are remarkably conserved (*23*). Mammalian CPE contains a fairly long 27-residue signal peptide, a short 15-residue pro peptide, a 330-residue carboxypeptidase domain with approximately 20% amino acid sequence identity with mammalian carboxypeptidase A (CPA) and carboxypeptidase B (CPB), an additional 75-residue region that is conserved in CPD but is not present in CPA or CPB (and which has structural homology to transthyretin), and then a short 30-residue C-terminal region that is distinct from other proteins in the database (Fig. 1).

Although multiple forms of CPE protein are found, there is no evidence for alternative RNA splicing (*18–20*). Different forms of CPE mRNA are detected in some tissues but these forms either arise through alternative polyadenylation (*18, 24*) or transcription initiation sites (*25*). The heterogeneity in CPE mRNA is entirely within the 5′ or 3′ untranslated regions and the protein coding region is not affected. The various forms of CPE protein arise through differential posttranslational processing. For example, the cleavage of the 15-residue Pro domain is not complete in bovine pituitary, and proCPE can be detected in this tissue. Interestingly, proCPE is enzymatically active with properties identical to those of the mature form (*26*), unlike most proenzymes that have greatly reduced catalytic activity relative to the mature form. Removal of the N-terminal pro domain is catalyzed by furin at the RRRRR sequence immediately adjacent to the N terminus of the mature form (*27*). This cleavage occurs in immature secretory vesicles or another post-Golgi compartment (*27*). Additional cleavages occur within

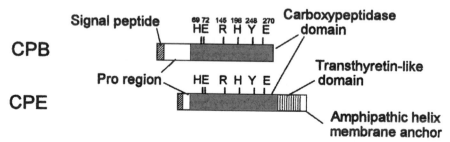

Fig. 1. Comparison of the domain organization of mammalian CPB and CPE. The key active site residues are indicated, using the numbering system of mature CPA and CPB. His-69, Glu-72, and His-196 are Zn^{2+} binding ligands. Arg-145 binds to the C terminus of the substrate. Tyr-248 binds substrate and contributes to the K_m. Glu-270 is involved in the deprotonation of water and/or the protonation of the amide leaving group following hydrolysis of the peptide bond.

the C-terminal region of CPE that remove 1–2 kDa of the protein (28, 29). Although the exact cleavage site has not been determined, this cleavage alters the membrane-binding ability of CPE (28, 29). C-terminally truncated deletion mutants of CPE show extremely low membrane binding but otherwise have similar enzymatic properties (30). The C terminus of CPE is predicted to form an amphipathic α helix that binds peripherally to membranes (28). This binding occurs at the acidic pH values of mature secretory vesicles (5–5.5) but not at neutral pH (31). Peptides corresponding to the C-terminal region of CPE are able to mimic this pH-dependent membrane binding (28). The pH-dependent membrane binding of CPE was proposed to function in the sorting of this protein into the secretory pathway (32), although this hypothesis is not supported by the finding that the sorting domain is distinct from the membrane-binding domain (30).

The distribution of CPE mRNA has been determined in rat and other species using Northern blots (18–20) and in situ hybridization analysis (33–35). The distribution of CPE protein has been determined in a variety of mammalian species using enzyme assays (5, 6, 8), inhibitor binding autoradiography (36–39), and immunohistochemistry (40). All of these techniques revealed a neuroendocrine distribution for CPE. In rat, levels of CPE are highest in pituitary, brain, endocrine pancreas, and adrenal medulla. CPE is also present in heart, lung, intestine, and other tissues where peptide processing occurs. Within the brain, CPE is broadly distributed, with highest levels in the hippocampus and lowest levels in the cerebellum (8). This distribution generally fits with the distribution of known neuropeptides; although some peptides are detected in the cerebellum, the levels of peptides in this brain region are substantially lower than in other brain regions. In brain, CPE is generally localized to neurons, although expression is also detected in a small subset of nonneuronal cells (33–35, 41). CPE is also present in cultured astrocytes (42–44), and although this cell type is not traditionally thought to be a peptide producing cell, several studies have found astrocytes to synthesize enkephalin and other neuropeptides (45, 46). In addition, CPE is detected in brain ependyma and choroid plexus epithelium (47) and in adipose tissue (48). The role of CPE in these cell types is not known; it is possible that unknown peptides are present that require CPE for their biosynthesis. CPE is not detected in adult rat liver, muscle, or most other nonneuroendocrine tissues (18–20). This broad neuroendocrine distribution is generally consistent with a role for CPE in the processing of a large number of bioactive peptides.

The subcellular distribution of CPE also fits with the predictions that neuroendocrine peptide processing occurs in the late secretory pathway, beginning in the trans-Golgi network and continuing in secretory vesicles as they mature (1). CPE is present in immature and mature secretory vesicles in endocrine (*11*) and neuroendocrine cell types (*49*). Also, CPE is cosecreted from neuroendocrine cells along with the peptide hormones via the regulated pathway (*50*).

The substrate specificity of CPE has been examined with a variety of peptides that correspond to processing intermediates following the action of the endopeptidases on the prohormone precursors. CPE cleaves C-terminal basic amino acids (Lys, Arg, and His) from peptides but does not have detectable activity toward any other residues (*6, 51*). Of these basic residues, CPE has the highest k_{cat}/K_m values toward peptides with Arg and slightly lower values with C-terminal Lys (*6*). Peptides with C-terminal His are slowly cleaved by CPE, but this cleavage is not common within the secretory pathway as the majority of the endopeptidase cleavage sites are Lys-Arg or Arg-Arg (*51*). Aside from the high specificity toward peptides with C-terminal basic residues, CPE generally has less specificity for amino acids in other positions (*6*). Except for the Pro-Lys bond of α-neoendorphin, which is cleaved extremely slowly (*51*), all other peptides with C-terminal Lys or Arg are generally good substrates for CPE (*6*). Using mass spectrometry to detect cleavage, CPE was tested with a mixture of 12 peptides of the sequence Tyr-Glu-Pro-Gly-Ala-Pro-Ala-Ala-Gly-Xaa-Arg, where Xaa = either Gly, Ala, Ser, Pro, Thr, Leu, Asp, Gln, Met, Phe, Tyr, or Trp (*52*). The peptide with Met in the penultimate position was cleaved slightly faster than the average peptide, those with Gly or Trp were cleaved slightly more slowly than the average peptide, and the peptide with Pro in this position was cleaved much more slowly than the average peptide (*52*). Although Pro is not favored in the penultimate position, small substrates with the sequence dansyl-Pro-Ala-Arg are cleaved with a rate similar to that for dansyl-Phe-Ala-Arg (*6*).

Although the substrate specificity for CPE differs slightly from that of CPD (described in the next section), this difference is not sufficient for a selective assay for CPE. Earlier reports of specific or selective CPE assays (*53, 54*) were published prior to the discovery of CPD, and thus all of the literature describing the measurement of CPE in crude tissues is subject to reinterpretation. In tissues with a high ratio of CPE to CPD, the reported activity presumably represented CPE, especially if only the soluble fraction of the tissue was used (CPD is primarily membrane-bound, as described below). Studies that examined "CPE" activity in membrane fractions of tissues with a lower ratio of CPE to CPD presumably represented a mixture of the two activities. Until specific substrates and/or inhibitors of either CPE or CPD are developed, the only method to specifically measure CPE activity is to physically separate it from CPD prior to assay. A simple method to separate these activities is affinity chromatography on a *p*-aminobenzoyl-Arg

Sepharose resin (55). Both enzymes bind to this resin, but CPE elutes from the column when the pH is raised to 8.0 whereas CPD remains bound (55).

CPE is inhibited by metal chelating agents such as 1,10-phenanthroline and EDTA, as expected for a metalloprotease (5, 6). Although the active-site metal has not been determined, it is most likely a Zn^{2+} based on the presence of this metal in the active site of the distantly related CPA and CPB (56, 57). CPE has been found to bind Ca^{2+}; this binding was predicted based on conservation of the Ca^{2+}-binding region of the bacterial carboxypeptidase T (CPT) in CPE (58). However, the presence or absence of Ca^{2+} does not have a large effect on the enzymatic activity of CPE (discussed in Section II,C), unlike the presence of the active-site metal, which is critical for enzyme activity. Replacement of the active site metal with Co^{2+} activates CPE approximately 5- to 10-fold, depending on the substrate, although this is not likely to be physiological (5, 6). Other metallocarboxypeptidases are also activated *in vitro* by Co^{2+} (57, 59). Although the extent of the activation of CPE by Co^{2+} initially seemed to be much greater than the activation of other metallocarboxypeptidases by this metal and was proposed to be useful for the selective assay of CPE in crude tissue homogenates (5, 6, 8), this activation is not a distinguishing feature of CPE. For example, the Co^{2+}-induced activation of carboxypeptidase M is only 40% at pH 7.5 (the optimum of carboxypeptidase M) but becomes much larger (500%) at acidic pH values (59). Similar results were found with carboxypeptidase N (Lloyd Fricker, unpublished). Thus, even though a Co^{2+}-activated enzyme is detected in serum when measured at pH 5.5, this activity is largely due to carboxypeptidase N and not to CPE.

In contrast to Co^{2+}, CPE is inhibited by other metals such as Cd^{2+}, Hg^{2+}, and Cu^{2+} (5, 7). The effect of mercuric salts may reflect interference with cysteine residues rather than direct binding of the metal to the active site region. The powerful cysteine-modifying reagent *p*-chloromercuriphenyl sulfonate potently inhibits CPE (7). Other compounds that attack sulfhydryl groups (*N*-ethylmaleimide) do not substantially inhibit CPE (7).

Several compounds initially developed as active site-directed inhibitors of other carboxypeptidases are more effective inhibitors of CPE (60), although this may be due in part to the fact that CPE is assayed at acidic pH and these inhibitors may be more potent at low pH (59). Guanidinoethylmercapto-succinic acid (GEMSA) was developed as an inhibitor of carboxypeptidase B (CPB), which it inhibits with a low micromolar K_i (61, 62). This compound has a side chain guanidino group and a carboxylate that resembles the C terminus of arginine, but in place of the amide linkage found in peptides containing a C-terminal arginine, GEMSA has a second carboxylate group that was predicted to bind to the active-site metal. GEMSA inhibits CPE with a K_i of approximately 10 nM (60). Although GEMSA was initially considered to be selective for CPE, this compound inhibits CPD with a similar K_i (discussed below). In addition to GEMSA, CPE and CPD are potently inhibited by 2-mercaptomethyl-3-guanidinothiopropanoic acid, a derivative of GEMSA that contains a thiol group in place of the metal-binding carboxylate group of GEMSA (60).

C. Regulation of Carboxypeptidase E

Typically, when several enzymes function together in the same pathway, the rate-limiting enzyme is highly regulated while the remaining enzymes are less regulated. This is the case for CPE, and for peptide biosynthesis in general. The first and rate-limiting step in the production of bioactive peptides is the endopeptidase cleavage by prohormone convertases 1 and 2, which are regulated by a number of stimuli (discussed elsewhere in this volume). In contrast, the carboxypeptidase cleavage step is not rate limiting for the production of most neuroendocrine peptides, exceptions being the conversion of α-neoendorphin into β-neoendorphin, which requires cleavage of a Pro-Lys bond, and the generation of rat β-endorphin 1-26, which requires cleavage of a C-terminal His-27 residue; both of these cleavages are catalyzed extremely slowly by CPE (51).

Most proteases, and many peptidases, are activated only within certain environments, thus preventing inappropriate cleavage. As discussed above, CPE is not activated by cleavage of a zymogen; both proCPE and CPE have comparable enzyme activities (26). However, CPE activity is tightly regulated by pH (63). CPE is essentially inactive at the neutral pH of the endoplasmic reticulum and Golgi apparatus and is maximally active at the intravesicular pH of mature secretory granules, pH 5–5.5. Kinetic analysis revealed that two protons are required to activate CPE between pH values of 7 and 5.5 (63). Thus, rather than increase 10-fold with each unit drop in pH (which represents a 10-fold increase in proton concentration), CPE is activated 100-fold between 7.0 and 6.0 (63). Similarly, on secretion of CPE into the plasma or into the synaptic cleft of the nervous system, CPE is greatly deactivated by the increase in pH to neutral levels.

In addition to the pH gradient within the secretory pathway, there is also a Ca^{2+} gradient with total levels of this ion (bound plus free) in the millimolar range within the mature secretory vesicles (64). Whereas the secretory vesicle prohormone convertases are substantially activated by both decreasing pH and increasing Ca^{2+}, CPE is primarily activated by pH and not by Ca^{2+}. As discussed above, CPE binds Ca^{2+}, presumably within a Ca^{2+}-binding loop conserved among selected members of the metallocarboxypeptidase gene family (58). Interestingly, this Ca^{2+}-binding loop is not present in several of the mammalian carboxypeptidases but is present in CPE, one of the domains of CPD (discussed below), and in bacterial carboxypeptidases such as CPT from *Thermoactinomyces vulgaris* (58, 65, 66). Divergence of eukaryotes and prokaryotes occurred roughly 2 billion years ago, and so any conserved features would be expected to play a critical role in the function. Despite this conservation, CPE activity is only slightly affected by Ca^{2+} (58), although this is difficult to study because strong chelators inhibit enzyme activity by removal of the active site metal (presumably zinc). Low concentrations (micromolar) of calcium chelators activate CPE approximately 50–100% but higher concentrations (millimolar) inhibit enzyme activity (58). Although low concentrations of Ca^{2+} do not affect CPE activity or stability, concentrations of 1 mM or more decrease the stability of CPE at elevated temperature, suggesting conformational changes (58). The lack of a substantial effect of Ca^{2+} on CPE activity raises the question as to the function of this binding. Ca^{2+} has been found to facilitate the aggregation of CPE at slightly acidic pH values, and this may contribute to the sorting of CPE into secretory vesicles (67).

CPE has been proposed to be inhibited by its reaction products (68, 69). However, the physiological significance of this mechanism of inhibition is unclear. Although the levels of peptide products in the secretory vesicle may reach the 5–10 millimolar levels required to observe a small degree of product inhibition (68), these levels will only be achieved once the enzyme has finished its task. Because of the specificity of CPE for basic C-terminal amino acids, it does not seem necessary to inactivate this enzyme within the secretory vesicle to prevent nonspecific degradation of the peptides.

Even though the carboxypeptidase cleavage is not a highly regulated step in the production of most neuroendocrine peptides, there are some situations where CPE mRNA and/or proteins levels are regulated. Many studies have investigated the regulation of CPE in a variety of cell lines. In the PC12 rat adrenal medulla cell line, CPE mRNA is elevated about 50–100% on treatment with phorbol esters for 1–6 days or on treatment with forskolin for 6 days (70). Depolarization of PC12 cells by extracellular K^+ causes the stimulation of CPE secretion (71). Prolonged stimulation causes a significant drop in cellular CPE levels and a compensatory increase in CPE mRNA (71). However, these effects are modest, with the maximal stimulation of mRNA approximately 65% above the control level (71). Similarly, the induction of secretory vesicles in the rat pituitary GH_4C_1 cell line leads to a modest 80% increase in CPE activity (72). CPE mRNA is not regulated by this treatment of GH_4C_1 cells (72) or by treatment of AtT-20 cells with a variety of secretagogues (73). The largest change in CPE mRNA levels in a cell line was found with the ODM-2 ciliary epithelial cell line; treatment of these cells with either veratridine or phorbol esters produces a 2- to 5-fold increase in CPE mRNA (74).

CPE regulation has also been studied in primary cultures of mammalian cells. Treatment of cultured astrocytes with a phorbol ester produces a 50–70% increase in levels of CPE mRNA (75). Treatment of either cultured astrocytes or the C6 glioma cell line with nitric oxide-producing agents decreases levels of CPE-like activity by 60–70% (76) although the assay used detects both CPE and CPD. Since there was no change in CPE mRNA levels in this study, the effect is either due to translational/posttranslational effects on CPE protein, or effects on CPD protein and/or mRNA. Cultured bovine adrenal medulla chromaffin cells have also been used to study the regulation of CPE, but using a nonspecific assay that detects other carboxypeptidases in addition to CPE and CPD (77). Thus, the finding of kinetic changes in carboxypeptidase activity in response to reserpine treatments may be due to an effect on CPE or another enzyme present (77).

The regulation of CPE in intact animals has also been studied. In rats, hypoglycemic shock and the subsequent reflex splanchnic nerve stimulation of the adrenal medulla leads to a massive secretion of catecholamines, enkephalin peptides, and all other soluble contents of the secretory vesicle including CPE (78). In response to this significant depletion of adrenal medullary CPE and other secretory vesicle proteins, there is an up regulation in the mRNAs of these proteins. Adrenal medullary CPE mRNA is elevated 55–70% 6 hr to 2 days after the hypoglycemic shock and returns to normal by 7 days (78). Treatment of rats with haloperidol, a dopamine antagonist, leads to a 100% increase in levels of CPE mRNA in the intermediate pituitary after 2–3 weeks (79). Levels of CPE mRNA in the anterior pituitary or in a variety of brain regions are not altered by this treatment (79). Similarly, reserpine treatment of rats does not influence the level of CPE mRNA in rat brain paraventricular nucleus (80) or rat adrenal medulla ganglion cells (81) but does transiently increase the level of this mRNA in rat adrenal medulla (81). Denervation of rat adrenal decreases the CPE-like enzyme activity, although the assay detects both CPE and CPD in this tissue (82). Kainic acid-induced seizures result in a 150% increase in CPE mRNA in hippocampus 1 day following the seizures, and this returns to normal by 10 days (83). CPE mRNA levels vary approximately 2- to 3-fold in the rat retina throughout the light–dark cycle, with peak levels occurring soon after the onset of either the light or the dark cycle (i.e., a change in the lighting condition) (84).

The rat CPE gene was isolated, the gene structure determined, and the promoter region identified (85, 86). The CPE promoter region lacks a conventional TATA box and appears to rely on an SP1 sequence and an element with similarity to the "initiator" element (87). Additional elements are present in the 5' flanking region that influence transcriptional activity, but most of these elements contribute only modestly (86). Despite the relatively low magnitude of regulation of CPE mRNA, gene expression is highly regulated in a tissue-specific fashion (18–20, 86). As described above, CPE is abundant in the pituitary and other neuroendocrine tissues and is not detectable in liver, muscle, and many other nonendocrine tissues (18–20). The elements responsible for this dramatic tissue-dependent expression of CPE have not been identified and do not appear to be within 10 kilobases of the 5' flanking region based on studies expressing reporter constructs in neuroendocrine and nonneuroendocrine cell lines (86).

D. Naturally Occurring Mutations of Carboxypeptidase E

Quite often, identification of the genes responsible for naturally occurring mutant phenotypes leads to major advances in biology. This was the case for the *fat* mutation, which spontaneously occurred in an inbred mouse at The Jackson Laboratory (88). This is a recessive mutation; mice with a single *fat* allele have the same phenotype as wild-type mice. The *fat/fat* mice develop obesity starting around 10–14 weeks of age, eventually becoming approximately twice the mass of wild-type littermates (88, 89). Because the *fat/fat* mice do not eat much more food than controls, the defect is thought to be due to nutrient partitioning (89). In addition to being overweight, mice recessive for the *fat* mutation are sterile (both males and females). Furthermore, when transferred from the HRS/J strain to the C57BLKS/J strain, male *fat/fat* mice develop hyperglycemia by about 14 weeks, which lasts until the mice are about 30–40 weeks and then undergoes spontaneous remission (89). Interestingly, female mice are not as severely affected by this mutation and have blood sugar levels that are slightly elevated but not significantly different from those of controls (89).

After many years of work, Jurgen Naggert, Ed Leiter, and colleagues at The Jackson Laboratory mapped the *fat* mutation to within 1 centimorgan of the CPE locus on mouse chromosome 8 (90). However, CPE was not thought to be a likely candidate gene because the phenotype of the mutant was limited to sterility, obesity, and hyperglycemia (the last being limited to male mice of a certain strain), whereas a defect in CPE was expected to have broader neuroendocrine consequences (possibly even embryonic lethality). In addition, studies on proinsulin processing in mice recessive for the *fat* mutation showed a defect in the endopeptidase processing step as well as the carboxypeptidase processing step (90). For these reasons, CPE was not considered a prime candidate. Initial studies on CPE activity in mice recessive for the *fat* mutation showed a severe defect in levels of active enzyme in all tissues examined (*90*). Although some enzyme activity was detected with an assay assumed to be specific for CPE (*90*), this residual activity was subsequently found to be due to CPD (described below). Thus, the *fat* mutation completely eliminates CPE activity (*91*). Further analysis revealed a point mutation within the coding region of the CPE gene that substituted a Pro for Ser in position 202 (numbering from the beginning of the mature form, and not from the initiation Met). When this point mutation was recreated in rat CPE and the protein expressed in the baculovirus system (which produced high levels of wild-type CPE activity) the resulting protein was completely inactive (*90*). Furthermore, this mutant was not secreted from either insect or mammalian cell lines but was degraded prior to transport to the Golgi apparatus (*92, 93*). Computer modeling using the distantly related CPA and CPB structures suggested that this Ser in position 202 was present in a β sheet and that substitution of the Pro would severely perturb the structure. Further genetic analysis revealed that the *fat* mutation was inseparable from the point mutation in the CPE gene, and so this mutation was renamed Cpe^{fat} (*90*).

As expected from the broad neuroendocrine distribution of CPE, the processing of many peptides is affected in Cpe^{fat}/Cpe^{fat} mice. Peptides such as insulin, Leu-enkephalin, neurotensin, neuromedin N, neuropeptide-Glu-Ile, gastrin, and cholecystokinin contain C-terminal basic residues in Cpe^{fat}/Cpe^{fat} mice (*90, 91, 94–97*). Unexpectedly, small amounts of the fully processed peptides are detectable in mice lacking CPE activity; these peptides are present in levels ranging from 10 to 50% of the levels of these peptides in nonmutant littermate control mice (*90, 91, 94–97*). A likely interpretation of this observation is that another carboxypeptidase contributes to the processing of neuroendocrine peptides; a search for novel enzymes resulted in the identification of CPD as well as several other candidates (discussed below).

The Cpe^{fat}/Cpe^{fat} mice have been used for the discovery of new neuroendocrine peptides (*98*). This technique makes use of the fact that peptides with C-terminal basic amino acid extensions accumulate in Cpe^{fat}/Cpe^{fat} mice, but these peptides are not detectable in wild-type mice (*98*). Affinity purification of these C-terminally extended peptides is readily accomplished on a resin containing anhydrotrypsin, which binds peptides with C-terminal basic residues (*98*). Alternatively, a mutant form of CPE that lacks active site residues but still binds substrate can be used, but the commercially available anhydrotrypsin resin is adequate for the affinity purification. The eluate from this column contains all peptides that are normally processed by CPE. Using mass spectrometry to identify the peptides present in this complex mixture, a large number of known peptides have been found in this affinity column eluate from Cpe^{fat}/Cpe^{fat} mouse brain and pituitary (Lloyd Fricker,

unpublished). In addition to these known peptides, five previously unidentified mouse peptides were discovered using this technique (98). Two of
these peptides correspond to peptides previously detected in bovine adrenal
medulla, for which partial sequence information was reported (99). All five of
these mouse peptides arise from a single novel precursor, named proSAAS
(98). The function of the proSAAS-derived peptides is not clear, but the
proSAAS precursor as well as partially processed intermediates containing the C-terminal region of proSAAS are potent inhibitors of prohormone
convertase 1 (98, 100, 101). It is likely that additional novel peptides can be
discovered using the Cpe^{fat}/Cpe^{fat} mice.

Because of many similarities between the mechanisms of body weight
regulation in mice and humans, it is possible that mutations in human CPE
contribute to obesity. Several variants of CPE have been found (discussed
below). However, none of these mutations causes CPE to be completely
inactive, as is the case for the mouse *fat* mutation. Furthermore, the frequency
of the variants in the human population is so low that they are unlikely to
account for the vast majority of human obesity.

Human CPE maps to chromosome 4 q28 (102). Like the rat CPE gene (85),
the human gene contains nine exons and spans approximately 60 kilobases
(103). A study of approximately 300 obese and/or diabetic Japanese men
and women found three variants within a small number of patients (103).
Two of these variants are within the noncoding 5′ flanking region and the
third is a silent mutation within the coding region (103). However, there is no
association of any of these three variants and either type 2 diabetes or obesity
(103). Additional variants within the CPE gene have been found in other
populations. A study of obese people in West Virginia revealed a nucleotide
change that converts Glu-359 to Lys (David Nevin, unpublished). A study of
obese New Yorkers revealed a nucleotide change converting Glu-465 to Gly
(Rudy Leibel, unpublished). Because Glu-359 and Glu-465 are conserved
in all known mammalian forms of CPE, and in anglerfish these residues
are either Glu or Asp, the Glu-359Lys and Glu-465Gly variants of human
CPE were expressed in the baculovirus system and the proteins purified and
characterized. Both of these variants have catalytic properties and substrate
specificities that are identical to those of wild-type CPE (Yimei Qian and
Lloyd Fricker, unpublished). Also, genetic analysis indicated that neither
variant correlates with obesity or hyperglycemia in family members of the
patients in which the original variants were detected. Thus, it is unlikely that
either of these variants is responsible for human obesity.

A study of 945 individuals of Ashkenazi Jewish origin revealed several
additional variants within the CPE gene (52). Two of these variants are
within the noncoding regions of the gene, one is within the coding region
but does not alter the amino acid sequence, and two alter the amino acid

sequence: Arg-283 to Trp and Ala-423 to Val (52). The change of Ala-423 to Val in human CPE is not expected to produce a large change in the enzyme activity because some species of CPE (such as *Xenopus*) have a Val in this position, and other related carboxypeptidases generally have either an Ala or a Val in this position. However, Arg-283 is highly conserved among mammalian CPE, and a Lys residue is found in anglerfish and *Aplysia* CPE. Furthermore, most known metallocarboxypeptidases have a Lys or Arg in the comparable position, and none have a bulky Trp residue. When the Arg-283Trp variant was expressed in the baculovirus system and tested, it was found that the enzymatic properties differed from wild-type CPE expressed in the same system (52). Specifically, the maximum rate of CPE activity is approximately fivefold lower for the Arg-283Trp variant than for wild-type human CPE. In addition, the Arg-283Trp CPE has a sharper pH optimum than wild-type CPE. Thus, the variant CPE is substantially less active at intermediate pH values of 5.5 to 6.5, values that are found within the immature secretory vesicles. The Arg-283Trp variant has the same substrate specificity as wild-type CPE, indicating that the mutation is selective for the catalytic activity and not substrate binding. However, Arg-283 is not predicted to be a critical residue for catalytic activity, and the effect is likely to be secondary to structural changes.

It is not clear if the Arg-283Trp variant is responsible for human obesity or diabetes (52). The frequency of the variant is approximately 1% of the Ashkenazi population and does not correlate with obesity or type 2 diabetes (52). A similar frequency was found in a cross-section of people living in the United States (52). However, in the Ashkenazi population, the presence of the variant strongly correlated with the age of onset of type 2 diabetes (52). Thus, it is possible that the CPE variant contributes to diabetes along with other genes. Also, with a frequency of 1%, approximately 1 in 10,000 people would be expected to be homozygous for the Arg-283Trp variant, and this would be expected to have a greater effect, as is the case for mice with the *fat* mutation. However, it is clear from the various genetic analyses that mutations in CPE do not play a significant role in either human obesity or diabetes.

E.   PHYSIOLOGICAL FUNCTIONS OF CARBOXYPEPTIDASE E

CPE functions in the processing of neuroendocrine peptides, based on the distribution, enzyme properties, and consequences of the lack of CPE in the Cpe^{fat}/Cpe^{fat} mice (discussed above). Although CPE is not the only secretory pathway carboxypeptidase, the altered processing of a large number of neuroendocrine peptides in the Cpe^{fat}/Cpe^{fat} mice implies that CPE is the major route by which C-terminal Lys and/or Arg residues are removed from peptide processing intermediates (following the endopepti-dase step). An unexpected finding is that levels of the substrates for the endoproteases are also elevated in the Cpe^{fat}/Cpe^{fat} mice (*90, 91, 94–97*); this implies that CPE is required for the full activation of the endopepti-dases. Recent studies have shown that the endopeptidases are inhibited by peptides containing C-terminal basic residues; these peptides can either be normal prohormone processing intermediates (*104*), the Pro domain of the endopeptidases themselves (*105*), or specific inhibitors of the endopepti-dases (*98, 106*). The specific inhibitors, designated proSAAS and 7B2, are slowly cleaved by prohormone convertase 1 and 2 (respectively), but even after cleavage the products remain as potent inhibitors of the endopepti-dase until a carboxypeptidase removes the C-terminal basic residues (*101, 106*). Thus, the full activation of the peptide processing endopeptidases re-quires the presence of an active carboxypeptidase, and presumably CPE contributes to this activation.

In addition to its enzymatic role, it is possible that CPE performs other functions within the cell or after secretion. As described above, CPE has been discovered several times on searches for proteins with a variety of functions. In many cases, CPE was found based on its ability to bind to another protein. When the CPE-binding protein is nonneuroendocrine and/or not present in the secretory pathway (such as cytosolic proteins), this interaction is likely to be an artifact. CPE is a very sticky protein and readily forms aggregates with itself (*67*) and with other proteins (*107*) under some conditions. Because CPE is restricted to the secretory pathway in neuroendocrine cells, only interactions that occur in this compartment are possibly relevant.

One alternative function that has received considerable attention is the potential role of CPE in the sorting of neuroendocrine peptides into se-cretory vesicles. The mechanism by which neuroendocrine peptides are tar-geted to the regulated secretory pathway has been a major question for many years. CPE was identified during a search for proteins that bind to an N-terminal fragment of proopiomelanocortin (*17*). The binding of this peptide fragment to membranes, which was interpreted as representing the binding to CPE, was competed by full-length proopiomelanocortin, proenkephalin, proinsulin, and chromogranin A (*17*). The secretion of proopiomelanocortin-derived peptides from Cpe^{fat}/Cpe^{fat} mouse pituitaries was largely constitutive, indicating a defect in the sorting to the regulated pathway (*17, 108*). In addition, a cell line expressing antisense CPE also showed a large reduction in the regulated secretion of proopiomelanocortin, proinsulin, and proenkephalin, but not chromogranin A (*17, 109*). Peng Loh and colleagues interpreted these results to indicate that CPE is the sorting receptor for the entry of prohormones into the regulated pathway. However, there are several major problems with this hypothesis. First, the stoichiometry of prohormone to CPE is not even close to 1 : 1; estimates from the AtT-20 cell line have indicated a ratio of CPE:proopiomelanocortin of 1 : 50 to 1 : 100. Thus, it is not clear how one molecule of CPE binding one molecule of proopiomelanocortin could contribute to the sorting of the bulk of the prohormone. Second, CPE is not retrieved from the immature vesicles to the *trans*-Golgi network (unlike CPD and other resident *trans*-Golgi network proteins). Thus, CPE cannot be a sorting receptor in the same fashion as the KDEL receptor, which cycles between the endoplasmic reticulum and the *cis*-Golgi (*110*). Third, most cell lines expressing antisense CPE mRNA do not show an altered secretion of prohormones; the two lines reported by Dr. Loh and colleagues required screening of a large number of cell lines and may therefore represent a clonal variation unrelated to the reduction of CPE levels. Fourth, when tested using equilibrium dialysis, purified CPE is unable to bind to the same iodinated proopiomelanocortin peptides found to bind to crude secretory vesicle membranes by Dr. Loh (our unpublished observation), raising the possibility that the binding in the crude vesicle membranes is not due to CPE. Fifth, and most importantly, both primary cultures of Cpe^{fat}/Cpe^{fat} mouse pancreatic islets (*111*) and beta-cell lines derived from the Cpe^{fat}/Cpe^{fat} mice (*93*) secrete proinsulin in a regulated manner despite the complete absence of functional CPE. Thus, it is clear that CPE is not required for the sorting of proinsulin, and further studies are needed to elucidate whether CPE plays a role in the sorting of any prohormone.

III. Carboxypeptidase D

A. DISCOVERY

The finding that a small amount of normal peptide processing occurs in the Cpe^{fat}/Cpe^{fat} mice (discussed above) implies that another carboxypeptidase functions in the secretory pathway and that this second carboxypeptidase has a partially overlapping role with CPE. Furthermore, the Cpe^{fat}/Cpe^{fat} mice have low but detectable levels of CPE-like activity in brain, pituitary, and pancreatic islets despite the fact that the $Ser^{202}Pro$ mutation completely eliminates CPE activity (*90*). This apparent paradox is best explained by the presence of a second CPE-like enzyme. Upon further studies, it was found that the residual CPE-like activity from the Cpe^{fat}/Cpe^{fat} mice was largely membrane-bound and that in contrast to CPE, this second activity could not be released from membranes by neutral pH (*55*). Also, this CPE-like enzyme was retained on a *p*-aminobenzoyl-Arg affinity column, as was CPE, but whereas pH 8.0 released CPE the novel activity remained bound (*55*). This allowed the purification of the novel enzyme away from CPE.

Both CPE and the novel activity were detected in bovine, rat, and wild-type mouse tissues whereas only the second activity was found in Cpe^{fat}/Cpe^{fat} mouse tissues (91).

Purification of the novel carboxypeptidase to homogeneity revealed a protein with an apparent migration on denaturing polyacrylamide gels of 180 kDa (55). The N-terminal partial amino acid sequence was unique and so this novel activity was named metallocarboxypeptidase D (CPD; EC 3.4.17.22). As a side point, the name carboxypeptidase D has also been recently used to refer to a class of serine carboxypeptidases from various species; these enzymes have no similarity to metalloCPD.

Independent of the studies on carboxypeptidases in Cpe^{fat}/Cpe^{fat} mice, CPD was simultaneously discovered by two other groups working in non-mammalian systems. Kazuyuki Kuroki, Don Ganem, and colleagues found a 180 kDa protein in duck liver that bound to viral particles of duck hepatitis B (112). Specifically, the preS protein of these viral particles was thought to be essential for the initial binding of the particles to a cellular protein (113). Cloning of the cDNA encoding the 180 kDa preS binding protein revealed a novel member of the metallocarboxypeptidase family that contains three CPE-like domains followed by a potential transmembrane region and then a 58 residue cytosolic tail (Fig. 2) (114). Subsequently, human, rat, and mouse carboxypeptidase D were also found to have a domain structure identical to the duck 180 kDa protein (65, 115, 116). The duck 180 kDa protein was shown to encode an enzyme with similar properties to mammalian CPD (117).

The other nonmammalian system in which CPD was codiscovered is the fruit fly, *Drosophila melanogaster* (118). Stephen Settle, Kenneth Burtis, and colleagues isolated the gene responsible for the *silver* mutation and found that it encoded a novel protein with approximately 35% amino acid sequence identity to mammalian CPE (118). The *silver* mutation exists in several forms; the most severe is embryonic lethal, but it is not clear if this is due to the loss of CPD or that of neighboring genes (118). Two mutants of the *silver* gene that arise from the insertion of P-elements within the middle of the CPD gene are viable but have silvery cuticles and altered wing shape (118). The

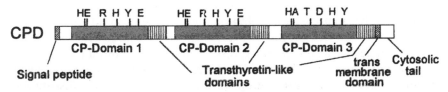

FIG. 2. Domain organization of mammalian and duck CPD. The indicated residues are those in positions comparable to the active-site residues of CPB shown in Fig. 1.

silver gene was initially considered to be a CPE homolog because CPD had not been discovered at the time. However, the amino acid sequence identity between the *silver* gene product and mammalian CPD (41%) is higher than with mammalian CPE (35%). Also, the *silver* gene product contains multiple carboxypeptidase domains (*118*), as does mammalian CPD (*65, 115, 116*). In the initial report, two CPE-like domains were found, followed by a domain that had weak similarity to CPE but only extended through half of the carboxypeptidase domain (*118*). Analysis of the recently reported *Drosophila* genome reveals that this partial domain represents a splice variant and that the full third carboxypeptidase domain is present in the genome (Lloyd Fricker, unpublished analysis). Also, this third domain is followed in the genome by a potential transmembrane domain and a cytosolic tail of about 60 amino acids (Lloyd Fricker, unpublished analysis). It is likely that this form exists; searches of *Drosophila* expressed sequence tag cDNA sequences revealed several clones corresponding to this longer unspliced form.

CPD homologs have also been detected in other species. *Aplysia californica,* a marine mollusk, has a CPD-like enzyme with four carboxypeptidase-like domains, although only the first two have high amino acid sequence identity to mammalian CPD (*119*). The enzymatic properties of *Aplysia* CPD are similar to those of mammalian CPD (discussed below). The nematode *Caenorhabditis elegans* has 10 genes for metallocarboxypeptidase-like proteins, of which three are most similar to the CPE-subfamily, and one of these has a predicted transmembrane domain and cytosolic tail. Amino acid comparisons of this predicted *C. elegans* gene product (from the F59A3 gene fragment) show the highest amino acid sequence similarity to CPD. Although there is only a single carboxypeptidase domain, there are approximately 400 amino acids between this domain and the putative transmembrane domain. Thus, the overall linear structure of this predicted *C. elegans* gene product is more similar to mammalian CPD than to any other mammalian carboxypeptidase, raising the possibility that the key features of CPD have been conserved through evolution.

B. PROPERTIES AND DISTRIBUTION OF CARBOXYPEPTIDASE D

Unlike CPE, which is primarily restricted to neuroendocrine tissues, CPD has a much broader tissue distribution and is present in all tissues examined (*41, 65, 112, 115, 120*). However, CPD is not expressed in all tissues at the same levels, with highest expression in brain and pituitary and much lower expression in muscle (*65, 112, 115, 120*). Within brain, immunohistochemistry and *in situ* hybridization were used to examine the distribution of CPD protein and mRNA (*41*). Both are present in neurons as well as glia, whereas CPE is largely restricted to neurons and a subpopulation of nonneuronal cells (41). CPD distribution has also been examined in placenta using immunohistochemistry (121). In this tissue, expression of CPD is detected in trophoblasts and chorionic villous endothelial cells (121). Many cells do not have detectable levels of CPD protein, indicating that this expression is not ubiquitous (41, 121).

The intracellular distribution of CPD has been examined in cell lines and in tissue (41, 49, 122). In AtT-20 cells, CPD is enriched in the trans-Golgi network and immature secretory vesicles (49, 122). However, most mature secretory vesicles do not have detectable levels of CPD protein, suggesting that although this protein is initially sorted into immature secretory vesicles, it is largely retrieved back to the trans-Golgi network rather than allowed to remain in the maturing vesicles (49). CPD is also detected on the cell surface, although this is only transient as the protein is rapidly internalized to an intracellular compartment with the same distribution as the trans-Golgi network (122). Further discussion of the intracellular trafficking of CPD is provided in a later section of this review.

The topographical orientation of CPD has been examined using antisera to either the three carboxypeptidase domains or the predicted cytosolic tail domain. As predicted by the presence of a signal peptide on the N terminus of the deduced protein sequence, the carboxypeptidase domains are lumenal and are transiently exposed on the cell surface following exocytosis (122). Antisera to the cytosolic tail do not react with cell surface CPD unless the cells are permeabilized with detergents (122). Thus, CPD is a type I membrane protein with the enzymatically active domains inside the trans-Golgi network and immature secretory vesicles.

The enzyme properties of CPD are consistent with a metallocarboxypeptidases that functions within the late Golgi, the trans-Golgi network, and/or the immature secretory vesicles. As discussed above, there is a pH gradient within these compartments, ranging from neutral within the Golgi to around 6 in the immature vesicles and 5–5.5 in the mature vesicles (3, 4). Unlike CPE, which is activated as the pH drops from 7 to 5.5, CPD is maximally active throughout the range of 5.5 to 7 (55). This broad pH optimum is due to overlapping pH optima of the first and second carboxypeptidase domains, the first showing an optimum closer to neutral pH while the second is maximally active in the acidic range (117, 123). Interestingly, the third domain is inactive toward substrates containing C-terminal Lys or Arg at any pH value examined (123). This observation fits with the absence of critical active site residues in this third domain (discussed below in Section III,D on crystal structure). Although the properties of the individual domains of CPD have been examined only for duck CPD, the high degree of amino acid sequence identity within the first and second domains among species implies that the properties have been conserved through evolution. In addition, CPD from

Aplysia, duck, and mammals have broad pH optima consistent with a role in the Golgi and secretory vesicles (*55, 117, 119*).

The substrate specificity of CPD has been examined with a variety of small synthetic peptides (*55, 117*). In addition to studies using full-length CPD, the individually expressed first and/or second domains of duck CPD have been examined (*117*). However, kinetic analysis of these individually expressed domains suggested that the large deletions altered the structure, and so a series of point mutations were made in a critical active-site residue within each of the domains. The mutated residue corresponds to Glu-270 in CPA and CPB; this residue is involved in general acid/base catalysis (*56, 57*). A previous study had shown that mutation of the Glu-270 equivalent in CPE to Gln caused a complete loss of activity without affecting substrate binding (*124*). The corresponding residues in the first domain of CPD, the second domain of CPD, or both the first and second domains of CPD were mutated to Gln and the enzymatic properties examined after expression and purification (*123*). The double mutant was completely inactive (*123*), as expected from the deletion analysis which indicated that the third domain does not have carboxypeptidase activity (*117*). Kinetic analysis of several substrates revealed that the first domain has a higher K_{cat} for substrates containing C-terminal Arg vs Lys, while the converse is true of the second domain (*123*); it prefer Lys over Arg. Mass spectrometric analysis of a mixture of peptides containing the sequence Tyr-Glu-Pro-Gly-Ala-Pro-Ala-Ala-Gly-Xaa-Arg, where Xaa = 12 different amino acids (described above for CPE), revealed that both the first and second domains had generally similar preferences for the penultimate amino acid (*123*). Except for peptides with X = Pro, which were poorly cleaved by both domains, CPD cleaved the various peptides in the mixture with only small differences in the rates (*123*). Thus, the major differences between the first and second domains of CPD are in their pH optima and their preferences for C-terminal Lys vs Arg residues.

Both the first and second domains are inhibited by chelating agents such as 1,10-phenanthroline and EDTA (*117, 123*). These two domains are also inhibited by reagents that affect sulfhydryl groups, although with different sensitivities; the first domain is inhibited by lower concentrations of these reagents, compared to the second domain. Active site-directed inhibitors such as GEMSA and MGTA (discussed above for CPE) are approximately fivefold more potent toward the first domain than the second domain (*123*).

The physiological importance of the various domains of CPD has not been directly examined, although some information can be gleaned from the various *silver* mutations. As discussed above, three mutations of the *silver* gene have been characterized; one is a large chromosomal deletion, and two others result from insertional interruption of the CPD gene (*118*). Although the exact site of insertion is not yet known, Southern blot analysis (*118*) together with an understanding of the gene structure based on the *Drosophila* genome sequence suggests that both of these insertions are within the region encoding the second domain of *Drosophila* CPD. It is possible that these mutants produce functional CPD containing an intact first domain. Further analysis to identify the exact forms of CPD produced in the various *silver* mutants are needed to pinpoint the defect and provide a better understanding of the physiological function of the various domains.

C. Intracellular Routing of Carboxypeptidase D

Although primarily found in the TGN, CPD cycles between this compartment and the cell surface (*122*). Like other TGN proteins that also cycle to the cell surface, the cytosolic tail is essential for the trafficking of the protein. Studies on the routing of CPD have explored three areas: the compartments in which CPD is present, the elements within CPD that control the routing, and the cellular proteins that interact with CPD to influence routing. In addition to being present in the TGN, CPD also is detected in large vesicles that resemble the immature secretory vesicles within AtT-20 cells (*49*). However, nearly all of the mature dense-core secretory vesicles are devoid of CPD, indicating that this protein is removed from the immature vesicles as they mature (*49*). Based on these observations, and on the results of deletion analysis discussed below, it appears that the trafficking of CPD is similar to that of furin and other TGN proteins; these proteins are enriched in the TGN either by a retention mechanism or by retrieval from post-TGN vesicles. The retrieval pathway provides a local recycling loop for CPD within the TGN. Another local recycling loop is thought to be present at the cell surface (*125*). Following uptake of CPD from the cell surface, the protein can return to the cell surface, return to the TGN, or enter lysosomes. Movement between the various compartments may be controlled by phosphorylation/dephosphorylation, as proposed for the trafficking of furin (*125*).

Studies examining the regions of CPD that affect transport have identified several important elements (*126*). Removal of 49 of the 58 residues of the CPD cytosolic tail causes the protein to accumulate on the cell surface or within the cell surface recycling loop (*126*). Attachment of the transmembrane domain and cytosolic tail of CPD to albumin, which is normally secreted via the constitutive secretory pathway, causes the albumin to accumulate in the TGN and to cycle to and from the cell surface (*127*). Deletion of the C-terminal 10 amino acids greatly increases the rate of entry of CPD into nascent secretory vesicles that bud from the TGN, suggesting that this region contains a TGN-retention element (Oleg Varlamov and Lloyd Fricker, unpublished). The half-life of the C-terminally truncated protein is also shorter than that of wild-type CPD (*126*). Thus, decreasing the retention of CPD within the TGN leads to a greater flux of protein to the cell surface and subsequently to lysosomes. Point mutations have revealed that the phenylalanine within a FHRL sequence is important for the return of CPD from the cell surface to the TGN (*126*); this FHRL resembles a tyrosine-containing motif found in a variety of other TGN proteins (*126*).

Several proteins or protein complexes have been found to bind to the cytosolic tail of CPD, including adaptor protein 1 (AP1), adaptor protein 2 (AP2), a protein designated PACS1, and protein phosphatase 2A (PP2A) (*127*) (and Oleg Varlamov, Gary Thomas, and Lloyd Fricker, unpublished). All of these proteins or protein complexes have been implicated in the trafficking of other TGN proteins and are presumably involved in the intracellular movement of CPD. The analysis of deletion and point mutations suggests that AP1 and AP2 bind to a FHRL sequence in the CPD tail (Oleg Varlamov and Lloyd Fricker, unpublished), which is similar to the YxxL motifs known to bind to this protein complex (*128*). PACS1 binds to the CPD tail only after phosphorylation by casein kinase II (Oleg Varlamov, Lloyd Fricker, and Gary Thomas, unpublished), as is the case for binding of PACS1 to furin (*129*). PACS1 is thought to mediate the retrieval of phosphorylated TGN proteins to the TGN and/or the cell surface via a local recycling loop. In this model, a dephosphorylation step is required for the proteins to escape the local recycling loop (*125, 129, 130*). PP2A was implicated as the phosphatase that mediates this step for the trafficking of furin (*125*). The recent finding that PP2A binds to the cytosolic tail of CPD (*127*) is consistent with this model.

D.   Crystal Structure of Carboxypeptidase D Domain 2

Based on the similar crystal structures of the pancreatic carboxypeptidases (CPA, CPB) with the distantly related bacterial CPT (*131*), it was predicted that CPE, CPD, and other metallocarboxypeptidases would also have a generally similar structure. However, alignment of the amino acid sequences of the various carboxypeptidases showed a number of additional inserts within CPE and D that were not present in CPA or CPB; these inserts were predicted to correspond to extra loops within these proteins. Also, CPD and all other members of the CPE/D subfamily (see Section IV) contain an extra region of approximately 75 amino acids located C-terminal to the 330-residue region that has sequence similarity to the pancreatic and bacterial carboxypeptidases. This extra 75-residue domain is generally conserved to the same degree as the carboxypeptidase domain within members of the CPE/D subfamily, but does not have amino acid sequence similarity to other proteins in the database. Thus, the structure of this C-terminal addition and the shorter internal loops could not be accurately predicted, and so the crystal structure of a member of the CPE/D subfamily was needed. Many attempts to crystallize CPE failed, possibly because of the tendency for this protein to aggregate at high concentrations (67). Fortunately, the second carboxypeptidase domain of CPD (which includes the extra region present in all members of the CPE/D subfamily but absent from pancreatic or bacterial carboxypeptidase) could be expressed in high levels using the *Pichia* yeast expression system, and the purified protein did not aggregate at high concentration. Crystallization of the protein was performed using the single drop vapor diffusion method with an ammonium sulfate buffer, and the structure was then determined of the single mercury derivative (132).

As predicted, the general structure of the carboxypeptidase domain of CPD domain 2 (hereafter referred to as CPD-2) is similar to the structures of the pancreatic and bacterial carboxypeptidases (132). The CPD-2 carboxypeptidase subdomain shares the central twisted eight-stranded β-sheet structure that is flanked by α helices, which is common to proteases displaying the alpha/beta hydrolase fold (132). All of the catalytically essential residues found in the pancreatic carboxypeptidases are present in CPD-2 in a spatially similar orientation within the active site. Although most of these catalytically essential residues are the same and are located in a similar position in the linear sequence, an exception is Arg-71 in CPA and CPT, which interacts with the P1 carbonyl oxygen of the substrate. In CPD-2, this residue is functionally replaced by Lys-277 within the catalytic site (132). Another difference is within the residues that bind to the substrate side chain and provide specificity for aromatic (CPA) or basic residues (CPD, CPB). In the pancreatic carboxypeptidases, this specificity is imparted by residue 255, which is an Asp in CPB and a Leu or Ile in CPA (56, 57). However, the amino acid in a linearly similar position in CPD-2 (as well as in most other members of the regulatory carboxypeptidase subfamily) is Gln-257 (132). Instead, the role of providing specificity for substrates with basic C-terminal residues in CPD-2 is carried out by Asp-192 which is in a position similar to Ser-207 in CPB (132). Thus, both CPD-2 and CPB contain a polar residue (Gln-257 in CPD-2, Ser-207 in CPB) and an acidic residue (Asp-192 in CPD-2, Asp-255 in CPB) in the side-chain binding pocket, thus conferring specificity for a positively charged amino acid in the C-terminal position.

One major difference between the structure of CPD-2 and the pancreatic/ bacterial carboxypeptidases is in the loops between some of the α helices and the β sheets. Most importantly, there are several differences in the chain segments that form the funnel-like access to the active site cleft. CPD-2 contains a 16-residue loop (Tyr-225 His-241) that is not present in the pancreatic carboxypeptidases and that is located at the beginning of strand VII at the funnel border (132). Another segment of CPD-2, Ser-124–Val-133, folds inwards and partially covers access to the active site (132). On the other side of the active site rim, a 14-residue loop is missing in CPD-2, further altering the geometry of the region responsible for access to the active-site cleft (132). These differences account for the inability of potato carboxypeptidase inhibitor to bind CPD-2; this potato protein is a potent inhibitor of pancreatic carboxypeptidases but is completely inactive toward CPD and other related carboxypeptidases (unpublished). Differences in the loops surrounding the active site funnel region may also partially explain why CPD-2 does not rapidly cleave short dipeptide substrates of CPB (such as benzoyl-Gly-Arg) but prefers substrates with at least three amino acids (55).

The crystal structure of CPD-2 revealed that the 75-residue C-terminal region that is absent from the pancreatic carboxypeptidases (but that is conserved in other members of the CPE/D subfamily) forms a β-barrel structure with seven strands, each connected by short loops (132). This structure shares topological similarity with transthyretin, a serum protein that transports thyroid hormones throughout plasma (132). Transthyretin forms homotetramers and also forms a complex with retinol-binding proteins (133). Several other proteins also have subdomains with a similar topological structure, including cyclodextrin glucanotransferase, glucoamylase, and protocatechuate 3,4-dioxygenase (132). The function of this transthyretin-like beta-barrel subdomain in CPD-2 is not known. It is possible that the transthyretin-like subdomain is involved in the transport of small molecules, as is transthyretin itself. However, the transthyretin-like domain is present in all members of the CPE/D subfamily, and it seems unlikely that every member of this diverse group of proteins would have retained this function during evolution. For example, genes from diverse species such as nematodes (C. elegans) and plants (Arabidopsis thaliana) are predicted to encode proteins that contain a CPE/D-like carboxypeptidase domain and then a transthyretin-like subdomain. Thus, the transthyretin-like subdomain appears to be essential for protein function, possibly playing a role in folding. In support of this, deletion of even a small portion of the C-terminal of the transthyretin-like subdomain eliminates activity of CPE and causes the protein to be degraded intracellularily, which is a hallmark of unfolded proteins (30). Thus, the transthyretin-like domain in CPE/D-like carboxypeptidases may be important for protein folding.

E.   Physiological Functions of CPD

Based on the broad tissue distribution, the intracellular localization within the secretory pathway, and the broad substrate specificity (discussed in the preceding sections), the major function of CPD appears to be in the processing of proteins and peptides that transit the secretory pathway. Although this role includes in part the processing of neuroendocrine peptides, this is not likely to be the major role of CPD, since this enzyme cannot completely compensate for the absence of CPE in Cpe^{fat}/Cpe^{fat} mice (discussed in Section II). Instead, the major role of CPD is presumably the processing of intermediates following the action of furin and related endoproteases within the *trans*-Golgi network. These substrates include growth factors, growth factor receptors, and other proteins destined for secretion via the constitutive pathway (*134*). This role is supported by the results of disruption of the CPD gene in *Drosophila* (discussed above). It is likely that CPD also plays a role in the processing of proteins within the endocytic pathway, following internalization from the cell surface. Furin and related endopeptidases have been implicated in the processing of bacterial and viral toxins on endocytosis (*134*). With the exception of *Pseudomonas* exotoxin, which requires C-terminal processing of basic residues to liberate a KDEL-like endoplasmic reticulum–targeting sequence (*135*), it is not known whether the C-terminal processing step is vital for the activation of proteins.

The role of the third carboxypeptidase-like domain in CPD from various species remains a mystery. This domain does not contain several of the residues known to be within the active site that are essential for catalytic activity and/or substrate binding. Also, as mentioned above, the third domain has no detectable carboxypeptidase activity toward a number of peptide substrates. Despite the absence of carboxypeptidase activity, the high degree of conservation of this domain among duck and mammalian species and the presence of a related domain even in *Aplysia* and *Drosophila* strongly suggest an essential function for this domain. Studies on duck CPD have shown that this third domain is the region responsible for the binding to duck hepatitis B virus particles (*117*). It is conceivable that this third domain functions as a binding site for an endogenous protein and that the virus mimics this natural interaction. Although CPD is unlikely to act as a conventional receptor because of the lack of the appropriate intracellular domains, this protein may function in the transport of other molecules either through the secretory pathway (as proposed for CPE) or within the endocytic pathway. However, searches for endogenous proteins that can compete for the binding of duck hepatitis B preS protein to duck CPD have not been successful (unpublished). It is also possible that the third domain of CPD has hydrolase activity towards a nonpeptide substrate, or even carboxypeptidase activity with the right substrate under very specific conditions. Although the absence of several critical active-site residues argues against this possibility, other residues could substitute for the missing residues and so this possibility cannot be discounted. Further studies examining the effect of disruption of CPD, and specifically of the third domain of CPD, will be important to address the question of function.

IV. Other Members of the Carboxypeptidase E/D Subfamily

Although this review is focussed only on CPE and CPD, some of the other members of the metallocarboxypeptidase family also function in neuropeptide processing even though they are not present in the regulated secretory pathway. These other family members include carboxypeptidase N (CPN), carboxypeptidase M (CPM), and carboxypeptidase Z (CPZ), as well as proteins designated CPX-1, CPX-2, and AEBP1 (which is also known as aortic carboxypeptidase-like protein, ACLP). These family members all have a single carboxypeptidase and transthyretin-like domain with approximately 40–50% amino acid sequence identity with CPE (Fig. 3). In addition to these domains, CPZ has an N-terminal domain with approximately 20–30% amino acid sequence identity with the cysteine-rich domain of the frizzled receptors, and CPX-1, CPX-2, and AEBP1/ACLP have N-terminal domains with approximately 20–30% amino acid sequence identity with discoidin, a slime-mold lectin involved in cell-cell interactions (Fig. 3).

Of these family members, only CPM, CPN, and CPZ have been proposed to play a role in peptide processing, but they all are thought to function outside of the cell. CPM is attached to the surface of a variety of cell types via a phosphatidylinositol glycan linkage (*136–138*), CPN circulates in plasma as a complex with a binding protein (*139*), and CPZ is present in the extracellular matrix of several cell types (*140*). All three of these enzymes are able to remove C-terminal basic residues and are thus capable of completing the processing of any peptides that are not properly processed within the cell.

CPX-1 (*141*), CPX-2 (*142*), and AEBP1/ACLP (*143–145*) are not predicted to encode active carboxypeptidases because of the absence of key active-site residues, as with the third domain of CPD (discussed above). Although AEBP1 has been reported to be enzymatically active (*143, 146*), our own laboratory has not been able to replicate this finding using the same protein and procedures (Lloyd Fricker, unpublished). CPX-1 and CPX-2 were also unable to cleave a variety of standard carboxypeptidase substrates under a range of assay conditions (*141, 142*). Interestingly, all of these "inactive" members of the CPE subfamily are highly conserved between human and rodent, including the "missing" active-site residues, which suggests that these proteins perform an important function, perhaps as binding proteins rather than active enzymes. Further studies are needed to identify the function of these proteins.

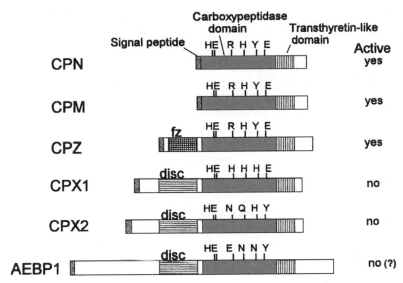

FIG. 3. Comparison of the domain organization of various mammalian members of the carboxypeptidase E/D subfamily. All members contain a signal peptide, a carboxypeptidase-like domain, a domain with structural homology to transthyretin, and then a C-terminal region of variable length. In addition, CPZ contains a Cys-rich domain with homology to the Wnt-binding region of the frizzled receptors ("fz"), and CPX-1, CPX-2, and AEBP1 contain a region with homology to discoidin ("disc"), a slime-mold protein involved in cell–cell aggregation. The indicated residues are those in positions comparable to the active site residues of CPB shown in Fig. 1. Also indicated is whether the expressed protein is active toward standard carboxypeptidase substrates.

V. Concluding Remarks

In summary, CPE is the primary carboxypeptidase involved with the processing of neuroendocrine peptides. Although the carboxypeptidase step proceeds after the initial endopeptidase step, the activity of the carboxypeptidase is required for full activity of the endopeptidases. In the absence of CPE activity, only small amounts of neuroendocrine peptides are correctly processed. It is likely that CPD is the enzyme that performs these cleavages in the absence of CPE. CPD is broadly distributed throughout many cell types but is primarily present in the TGN, whereas the bulk of peptide processing occurs in later compartments. Thus, CPD is not able to completely correct for the deficiency of CPE in Cpe^{fat}/Cpe^{fat} mice.

Even after 20 years of research on CPE, there are many interesting areas for further studies on this enzyme and on other carboxypeptidases. A key question is the function of the "inactive" members of the CPE gene family, including the third domain of CPD and proteins designated CPX-1, CPX-2, and AEBP-1. Another question concerns the role of the transthyretin-like subdomain that is present in all members of the CPE subfamily of metal-locarboxypeptidases. Understanding how CPE and CPD are sorted to their different intracellular sites is another area of investigation. Further studies are also needed to explore alternative roles for CPE and CPD, especially to test the hypothesis that CPE functions as a "sorting receptor" for prohormones. Finally, it is relatively easy to isolate Lys- and Arg-extended peptides from animals lacking carboxypeptidases, providing an efficient tool for identifying endogenous substrates of these enzymes.

ACKNOWLEDGMENTS

The writing of this manuscript was supported by Research Scientist Development Award DA-00194 from the United States National Institutes of Health. Special thanks to Dr. Lakshmi Devi for critical reading of the manuscript.

REFERENCES

1. Steiner, D. F. (1991). *In* "Peptide Biosynthesis and Processing" (L. D. Fricker, ed.), p. 1. CRC Press, Boca Raton.
2. Zuhlke, H., Kohnert, K. D., Jahr, H., Schmidt, S., Kirscke, H., and Steiner, D. F. (1977). *Acta. Biol. Med. Germ.* **36,** 1695.
3. Russell, J. T. (1984). *J. Biol. Chem.* **259,** 9496.
4. Johnson, R. G., and Scarpa, A. (1976). *J. Biol. Chem.* **251,** 2189.
5. Fricker, L. D., and Snyder, S. H. (1982). *Proc. Natl. Acad. Sci. USA* **79,** 3886.
6. Fricker, L. D., and Snyder, S. H. (1983). *J. Biol. Chem.* **258,** 10950.
7. Supattapone, S., Fricker, L. D., and Snyder, S. H. (1984). *J. Neurochem.* **42,** 1017.
8. Fricker, L. D., Supattapone, S., and Snyder, S. H. (1982). *Life Sci.* **31,** 1841.
9. Docherty, K., and Hutton, J. C. (1983). *FEBS Lett.* **162,** 137.
10. Hook, V. Y. H., and Loh, Y. P. (1984). *Proc. Natl. Acad. Sci. USA* **81,** 2776.
11. Guest, P. C., Ravazzola, M., Davidson, H. W., Orci, L., and Hutton, J. C. (1991). *Endocrinol.* **129,** 734.
12. Webb, E. C. (1986). *Eur. J. Biochem.* **157,** 1.
13. Laslop, A., Fischer-Colbrie, R., Hook, V., Obendorf, D., and Winkler, H. (1986). *Neurosci. Lett.* **72,** 300.
14. Manser, E., Fernandez, D., Loo, L., Goh, P. Y., Monfries, C., Hall, C., and Lim, L. (1990). *Biochem. J.* **267,** 517.
15. Castano, L., Russo, E., Zhou, L., Lipes, M. A., and Eisenbarth, G. S. (1991). *J. Clin. Endocrinol. Metab.* **73,** 1197.
16. Escribano, J., Ortego, J., and Coca-Prados, M. (1995). *J. Biochem.* **118,** 921.
17. Cool, D. R., Normant, E., Shen, F., Chen, H., Pannell, L., Zhang, Y., and Loh, Y. P. (1997). *Cell* **88,** 73.
18. Fricker, L. D., Evans, C. J., Esch, F. S., and Herbert, E. (1986). *Nature (Lond.).* **323,** 461.

19. Fricker, L. D., Adelman, J. P., Douglass, J., Thompson, R. C., von Strandmann, R. P., and Hutton, J. (1989). *Mol. Endocrinol.* **3,** 666.
20. Rodriguez, C., Brayton, K. A., Brownstein, M., and Dixon, J. E. (1989). *J. Biol. Chem.* **264,** 5988.
21. Roth, W. W., Mackin, R. B., Spiess, J., Goodman, R. H., and Noe, B. D. (1991). *Mol. Cell. Endocrinol.* **78,** 171.
22. Fan, X., and Nagle, G. T. (1996). *DNA Cell Biol.* **15,** 937.
23. Juvvadi, S., Fan, X., Nagle, G. T., and Fricker, L. D. (1997). *FEBS Lett.* **408,** 195.
24. Fricker, L. D. (1988). *In* "Molecular Biology of Brain and Endocrine Peptidergic Systems" (M. Chretien and K. W. McKerns, eds.), p. 189. Plenum Press, New York.
25. Smith, D. R., Pallen, C. J., Murphy, D., and Lim, L. (1992). *Mol. Endocrinol.* **6,** 713.
26. Parkinson, D. (1990). *J. Biol. Chem.* **265,** 17101.
27. Song, L., and Fricker, L. D. (1995). *J. Neurochem.* **65,** 444.
28. Fricker, L. D., Das, B., and Angeletti, R. H. (1990). *J. Biol. Chem.* **265,** 2476.
29. Fricker, L. D., and Devi, L. (1993). *J. Neurochem.* **61,** 1404.
30. Varlamov, O., and Fricker, L. D. (1996). *J. Biol. Chem.* **271,** 6077.
31. Fricker, L. D. (1988). *J. Cell. Biochem.* **38,** 279.
32. Mitra, A., Song, L., and Fricker, L. D. (1994). *J. Biol. Chem.* **269,** 19876.
33. MacCumber, M. W., Snyder, S. H., and Ross, C. A. (1990). *J. Neurosci.* **10,** 2850.
34. Birch, N. P., Rodriguez, C., Dixon, J. E., and Mezey, E. (1990). *Mol. Brain Res.* **7,** 53.
35. Schafer, M. K.-H., Day, R., Cullinan, W. E., Chretien, M., Seidah, N. G., and Watson, S. J. (1993). *J. Neurosci.* **13,** 1258.
36. Lynch, D. R., Strittmatter, S. M., and Snyder, S. H. (1984). *Proc. Natl. Acad. Sci. USA* **81,** 6543.
37. Strittmatter, S. M., Lynch, D. R., De Souza, E. B., and Snyder, S. H. (1985). *Endocrinol.* **117,** 1667.
38. Lynch, D. R., Strittmatter, S. M., Venable, J. C., and Snyder, S. H. (1986). *J. Neurosci.* **6,** 1662.
39. Lynch, D. R., Strittmatter, S. M., Venable, J. C., and Snyder, S. H. (1987). *Endocrinol.* **121,** 116.
40. Lynch, D. R., Braas, K. M., Hutton, J. C., and Snyder, S. H. (1990). *J. Neurosci.* **10,** 1592.
41. Dong, W., Fricker, L. D., and Day, R. (1999). *Neurosci.* **89,** 1301.
42. Vilijn, M. H., Das, B., Kessler, J. A., and Fricker, L. D. (1989). *J. Neurochem.* **53,** 1487.
43. Klein, R. S., Das, B., and Fricker, L. D. (1992). *J. Neurochem.* **58,** 2011.
44. Klein, R. S., and Fricker, L. D. (1992). *Brain Res.* **569,** 300.
45. Vilijn, M. H., Vaysse, P. J. J., Zukin, R. S., and Kessler, J. A. (1988). *Proc. Natl. Acad. Sci. USA* **85,** 6551.
46. Shinoda, H., Marini, A. M., Cosi, C., and Schwartz, J. P. (1989). *Science* **245,** 415.
47. Gee, P., Rhodes, C. H., Fricker, L. D., and Angeletti, R. H. (1993). *Brain Res.* **617,** 238.
48. Maeda, K., Okubo, K., Shimomura, S., Mizuno, K., Matsuzawa, Y., and Matsubara, K. (1997). *Gene* **190,** 227.
49. Varlamov, O., Eng, F. J., Novikova, E. G., and Fricker, L. D. (1999). *J. Biol. Chem.* **274,** 14759.
50. Mains, R. E., and Eipper, B. A. (1984). *Endocrinology* **115,** 1683.
51. Smyth, D. G., Maruthainar, K., Darby, N. J., and Fricker, L. D. (1989). *J. Neurochem.* **53,** 489.
52. Chen, H., Jawahar, S., Qian, Y., Duong, Q., Chan, G., Parker, A., Meyer, J. M., Moore, K. J., Chayen, S., Gross, D. J., Glasser, B., Permutt, M. A., and Fricker, L. D. (2001). *Hum. Mutat.*, in press.
53. Fricker, L. D. (1994). *Neuroprotocols* **5,** 151.

54. Fricker, L. D. (1995). *Meth. Neurosci.* **23,** 237.
55. Song, L., and Fricker, L. D. (1995). *J. Biol. Chem.* **270,** 25007.
56. Hartsuck, J. A., and Lipscomb, W. N. (1971). *In* "The Enzymes" (P. D. Boyer, ed.), p. 1. Academic Press, New York.
57. Folk, J. E. (1971). *In* "The Enzymes" (P. D. Boyer, ed.), p. 57. Academic Press, New York.
58. Nalamachu, S. R., Song, L., and Fricker, L. D. (1994). *J. Biol. Chem.* **269,** 11192.
59. Skidgel, R. A. (1991). *Meth. Neurosci.* **6,** 373.
60. Fricker, L. D., Plummer, T. H. J., and Snyder, S. H. (1983). *Biochem. Biophys. Res. Commun.* **111,** 994.
61. McKay, T. J., and Plummer, T. H. J. (1978). *Biochem.* **17,** 401.
62. Plummer, T. H. J., and Ryan, T. J. (1981). *Biochem. Biophys. Res. Commun.* **98,** 448.
63. Greene, D., Das, B., and Fricker, L. D. (1992). *Biochem. J.* **285,** 613.
64. Chanat, E., and Huttner, W. B. (1991). *J. Cell Biol.* **115,** 1505.
65. Xin, X., Varlamov, O., Day, R., Dong, W., Bridgett, M. M., Leiter, E. H., and Fricker, L. D. (1997). *DNA Cell Biol.* **16,** 897.
66. Smulevitch, S. V., Osterman, A. L., Galperina, O. V., Matz, M. V., Zagnitko, O. P., Kadyrov, R. M., Tsaplina, I. A., Grishin, N. V., Chestukhina, G. G., and Stepanov, V. M. (1991). *FEBS Lett.* **291,** 75.
67. Song, L., and Fricker, L. D. (1995). *J. Biol. Chem.* **270,** 7963.
68. Hook, V. Y. H., and LaGamma, E. F. (1987). *J. Biol. Chem.* **262,** 12583.
69. Hook, V. Y. (1990). *Life Sci.* **47,** 1135.
70. Laslop, A., Tschernitz, C., and Eiter, C. (1994). *Neurosci.* **59,** 477.
71. Das, B., Sabban, E. L., Kilbourne, E. J., and Fricker, L. D. (1992). *J. Neurochem.* **59,** 2263.
72. Fricker, L. D., Reaves, B. J., Das, B., and Dannies, P. S. (1990). *Neuroendocrinol.* **51,** 658.
73. Thiele, E. A., and Eipper, B. A. (1990). *Endocrinology* **126,** 809.
74. Ortego, J., Escribano, J., Crabb, J., and Coca-Prados, M. (1996). *J. Neurochem.* **66,** 787.
75. Klein, R. S., and Fricker, L. D. (1993). *J. Neurochem.* **60,** 1615.
76. Devi, L., Petanceska, S., Liu, R., Arbabha, B., Bansinath, M., and Garg, U. (1994). *J. Neurochem.* **62,** 2387.
77. Hook, V. Y. H., Eiden, L. E., and Pruss, R. M. (1985). *J. Biol. Chem.* **260,** 5991.
78. Fricker, L. D., Rigual, R. J., Diliberto, E. J. J., and Viveros, O. H. (1990). *J. Neurochem.* **55,** 461.
79. Grigoriants, O., Devi, L., and Fricker, L. D. (1993). *Mol. Brain Res.* **19,** 161.
80. Mahata, S. K., Mahata, M., Fischer-Colbrie, R., and Winkler, H. (1993). *Mol. Brain Res.* **19,** 83.
81. Laslop, A., Mahata, S. K., Wolkersdorfer, M., Mahata, M., Srivastava, M., Seidah, N. G., Fischer-Colbrie, R., and Winkler, H. (1994). *J. Neurochem.* **62,** 2448.
82. McMillian, M. K., Hudson, P. M., Lee, D. Y., Thai, L., Hung, G. H., and Hong, J. S. (1993). *Dev. Brain Res.* **71,** 75.
83. Mahata, S. K., Gruber, B., Mahata, M., Roder, C., Fischer-Colbrie, R., and Sperk, G. (1993). *Acta Neuropathol.* **86,** 590.
84. Schlamp, C. L., and Nickells, R. W. (1996). *J. Neurosci.* **16,** 2164.
85. Jung, Y. K., Kunczt, C. J., Pearson, R. K., Dixon, J. E., and Fricker, L. D. (1991). *Mol. Endocrinol.* **5,** 1257.
86. Jung, Y. K., Kunczt, C. J., Pearson, R. K., Fricker, L. D., and Dixon, J. E. (1992). *Mol. Endocrinol.* **6,** 2027.
87. Jung, Y. K., and Fricker, L. D. (1994). *Biochimie* **76,** 336.
88. Coleman, D. L., and Eicher, E. M. (1990). *J. Hered.* **81,** 424.
89. Leiter, E. H., Kintner, J., Flurkey, K., Beamer, W. G., and Naggert, J. K. (1999). *Endocrine* **10,** 57.

90. Naggert, J. K., Fricker, L. D., Varlamov, O., Nishina, P. M., Rouille, Y., Steiner, D. F., Carroll, R. J., Paigen, B. J., and Leiter, E. H. (1995). *Nature Genet.* **10,** 135.
91. Fricker, L. D., Berman, Y. L., Leiter, E. H., and Devi, L. A. (1996). *J. Biol. Chem.* **271,** 30619.
92. Varlamov, O., Leiter, E. H., and Fricker, L. D. (1996). *J. Biol. Chem.* **271,** 13981.
93. Varlamov, O., Fricker, L. D., Furukawa, H., Steiner, D. F., Langley, S. H., and Leiter, E. H. (1997). *Endocrinology* **138,** 4883.
94. Rovere, C., Viale, A., Nahon, J., and Kitabgi, P. (1996). *Endocrinology* **137,** 2954.
95. Udupi, V., Gomez, P., Song, L., Varlamov, O., Reed, J. T., Leiter, E. H., Fricker, L. D., and Greeley, G. H. J. (1997). *Endocrinology* **138,** 1959.
96. Lacourse, K. A., Friis-Hansen, L., Rehfeld, J. F., and Samuelson, L. C. (1997). *FEBS Lett.* **416,** 45.
97. Cain, B. M., Wang, W., and Beinfeld, M. C. (1997). *Endocrinology* **138,** 4034.
98. Fricker, L. D., McKinzie, A. A., Sun, J., Curran, E., Qian, Y., Yan, L., Patterson, S. D., Courchesne, P. L., Richards, B., Levin, N., Mzhavia, N., Devi, L. A., and Douglass, J. (2000). *J. Neurosci.* **20,** 639.
99. Sigafoos, J., Chestnut, W. G., Merrill, B. M., Taylor, L. C. E., Diliberto, E. J., and Viveros, O. H. (1993). *J. Anat.* **183,** 253.
100. Qian, Y., Devi, L. A., Mzhavia, N., Munzer, S., Seidah, N. G., and Fricker, L. D. (2000). *J. Biol. Chem.* **275,** 23596.
101. Cameron, A., Fortenberry, Y., and Lindberg, I. (2000). *FEBS Lett.* **473,** 135.
102. Hall, C., Manser, E., Spurr, N. K., and Lim, L. (1993). *Genomics* **15,** 461.
103. Utsunomiya, N., Ohagi, S., Sanke, T., Tatsuta, H., Hanabusa, T., and Nanjo, K. (1998). *Diabetologia* **41,** 701.
104. Day, R., Lazure, C., Basak, A., Boudreault, A., Limperis, P., Dong, W., and Lindberg, I. (1998). *J. Biol. Chem.* **273,** 829.
105. Zhong, M., Munzer, J. S., Basak, A., Benjannet, S., Mowla, S. J., Decroly, E., Chretien, M., and Seidah, N. G. (1999). *J. Biol. Chem.* **274,** 33913.
106. Zhu, X., Rouille, Y., Lamango, N. S., Steiner, D. F., and Lindberg, I. (1996). *Proc. Natl. Acad. Sci. USA* **93,** 4919.
107. Rindler, M. J. (1998). *J. Biol. Chem.* **273,** 31180.
108. Shen, F., and Loh, Y. P. (1997). *Proc. Natl. Acad. Sci. USA* **94,** 5314.
109. Normant, E., and Loh, Y. P. (1998). *Endocrinology* **139,** 2137.
110. Munro, S., and Pelham, H. R. B. (1987). *Cell* **48,** 899.
111. Irminger, J., Verchere, B., Meyer, K., and Halban, P. A. (1997). *J. Biol. Chem.* **272,** 27532.
112. Kuroki, K., Cheung, R., Marion, P. L., and Ganem, D. (1994). *J. Virol.* **68,** 2091.
113. Ishikawa, T., Kuroki, K., Lenhoff, R., Summers, J., and Ganem, D. (1994). *Virology* **202,** 1061.
114. Kuroki, K., Eng, F., Ishikawa, T., Turck, C., Harada, F., and Ganem, D. (1995). *J. Biol. Chem.* **270,** 15022.
115. Tan, F., Rehli, M., Krause, S. W., and Skidgel, R. A. (1997). *Biochem. J.* **327,** 81.
116. Ishikawa, T., Murakami, K., Kido, Y., Ohnishi, S., Yazaki, Y., Harada, F., and Kuroki, K. (1998). *Gene* **215,** 361.
117. Eng, F. J., Novikova, E. G., Kuroki, K., Ganem, D., and Fricker, L. D. (1998). *J. Biol. Chem.* **273,** 8382.
118. Settle, S. H. J., Green, M. M., and Burtis, K. C. (1995). *Proc. Natl. Acad. Sci. USA* **92,** 9470.
119. Fan, X., Qian, Y., Fricker, L. D., Akalal, D. B., and Nagle, G. T. (1999). *DNA Cell Biol.* **18,** 121.
120. Song, L., and Fricker, L. D. (1996). *J. Biol. Chem.* **271,** 28884.

121. Reznik, S. E., Salafia, C. M., Lage, J. M., and Fricker, L. D. (1998). *J. Histochem. Cytochem.* **46,** 1359.
122. Varlamov, O., and Fricker, L. D. (1998). *J. Cell Sci.* **111,** 877.
123. Novikova, E. G., Eng, F. J., Yan, L., Qian, Y., and Fricker, L. D. (1999). *J. Biol. Chem.* **274,** 28887.
124. Qian, Y., Varlamov, O., and Fricker, L. D. (1999). *J. Biol. Chem.* **274,** 11582.
125. Molloy, S. S., Anderson, E. D., Jean, F., and Thomas, G. (1999). *Trends Cell Biol.* **9,** 28.
126. Eng, F. J., Varlamov, O., and Fricker, L. D. (1999). *Mol. Biol. Cell* **10,** 35.
127. Varlamov, O., Kalinina, E., Che, F., and Fricker, L. D. (2001). *J. Cell Sci.* **114,** 311.
128. Marks, M. S., Ohno, H., Kirchhausen, T., and Bonifacino, J. S. (1997). *Trends Cell Biol.* **7,** 124.
129. Wan, L., Molloy, S. S., Thomas, L., Liu, G., Xiang, Y., Rybak, S. L., and Thomas, G. (1998). *Cell* **94,** 205.
130. Molloy, S. S., Thomas, L., Kamibayashi, C., Mumby, M. C., and Thomas, G. (1998). *J. Cell Biol.* **142,** 1399.
131. Teplyakov, A., Polyakov, K., Obmolova, G., Strokopytov, B., Kuranova, I., Osterman, A., Grishin, N., Smulevitch, S., Zagnitko, O., Galperina, O., matz, M., and Stepanov, V. (1992). *Eur. J. Biochem.* **208,** 281.
132. Gomis-Ruth, F. X., Companys, V., Qian, Y., Fricker, L. D., Vendrell, J., Aviles, F. X., and Coll, M. (1999). *EMBO J.* **18,** 5817.
133. Schreiber, G., and Richardson, S. J. (1997). *Comp. Biochem. Physiol. B. Biochem. Mol. Biol.* **116,** 137.
134. Nakayama, K. (1997). *Biochem. J.* **327,** 625.
135. Hessler, J. L., and Kreitman, R. J. (1997). *Biochem.* **36,** 14577.
136. Skidgel, R. A., Davis, R. M., and Tan, F. (1989). *J. Biol. Chem.* **264,** 2236.
137. Tan, F., Chan, S. J., Steiner, D. F., Schilling, J. W., and Skidgel, R. A. (1989). *J. Biol. Chem.* **264,** 13165.
138. Skidgel, R. A., Tan, F., Deddish, P. A., and Li, X. (1991). *Biomed. Biochim. Acta* **50,** 815.
139. Plummer, T. H. J., and Erdos, E. G. (1981). *Meth. Enzymol.* **80,** 442.
140. Novikova, E. G., Reznik, S. E., Varlamov, O., and Fricker, L. D. (2000). *J. Biol. Chem.* **275,** 4865.
141. Lei, Y., Xin, X., Morgan, D., Pintar, J. E., and Fricker, L. D. (1999). *DNA Cell Biol.* **18,** 175.
142. Xin, X., Day, R., Dong, W., Lei, Y., and Fricker, L. D. (1998). *DNA Cell Biol.* **17,** 897.
143. He, G. P., Muise, A., Li, A. W., and Ro, H. S. (1995). *Nature* **378,** 92.
144. Ohno, I., Hashimoto, J., Shimizu, K., Takeoka, K., Ochi, T., Matsubara, K., and Okubo, K. (1996). *Biochem. Biophys. Res. Commun.* **228,** 411.
145. Layne, M. D., Endege, W. O., Jain, M. K., Yet, S., Hsieh, C., Chin, M. T., Perrella, M. A., Blanar, M. A., Haber, E., and Lee, M. (1998). *J. Biol. Chem.* **273,** 15654.
146. Muise, A. M., and Ro, H. S. (1999). *Biochem. J.* **343,** 341.

Author Index

Numbers in regular font are reference numbers and indicate that an author's work is referred to although the name is not cited in the text. Numbers in italics refer to the page numbers on which the complete reference appears.

A

Abe, J., 414, *419*
Abrahmsen, L., 68, *74*
Abrami, L., 227, *235*
Accarino, M., 191, 192, *197*
Achstetter, T., 262, *286*
Acquati, F., 191, 192, *197*
Adachi, H., 45, 51, *55,* 122, *125,* 377, *386*
Adamec, J. (Chapter Author), 77, 83, 84, 85, 88, *98, 99*
Adams, J. M., 391, *418*
Adams, M. D., 4, 7, *23, 24,* 36, *54,* 122, *125*
Adelman, J. P., 423, 424, 425, 431, *449*
Aebi, U., 351, *369*
Affholter, J. A., 88, *99*
Agabian, N., 283, *289*
Agard, D. A., 360, 361, 364, 365, *370*
Aghion, J., 136, *158*
Ahn, K., 351, *369*
Aimoto, S., 361, *370*
Aitken, J. R., 268, *287*
Aizawa, T., 192, *197*
Akalal, D. B., 438, 440, *451*
Akamatsu, T., 249, 250, *258*
Akey, C. W., 59, *73*
Akil, H., 301, 303, 307, *330*
Akita, M., 29, *53*
Akopian, T. N., 345, 360, *369, 370*
Akrim, M., 134, 136, *158*
Albano, M., 134, *158*
Albert, M. J., 138, *158*
Albertson, N. H., 136, *158*

Alconda, A., 79, *97*
Alden, R. A., 264, *287*
Aldrich, C., 345, *369*
Aldrich, H. C., 339, 340, 342, 345, 347, 350, 353, 361, 363, *367, 368*
Alexander, P. A., 264, *287*
Alexciev, K., 116, *124*
Allen, R. G., 200, *230,* 266, *287*
Allison, D. S., 79, 80, *96, 97*
Allsop, A. E., 41, *55,* 104, *123*
Alm, R. A., 131, 132, *157, 158*
Alsmark, U. C. M., 7, *25*
Alsobrook, J. P. II, 95, *100*
Altschul, S. F., 5, 12, 15, *24,* 32, *54*
Altuvia, Y., 345, *369*
Amarante, P., 254, *258*
Amaya, Y., 83, 84, *98*
Amemiya, A., 307, *330*
Amherdt, M., 169, *194*
Amsterdam, A., 345, *369*
Anbudurai, P. R., 384, 385, *386*
Andersen, J. B., 20, *26*
Anderson, C. M., 103, *123*
Anderson, E. D., 183, *196,* 203, 205, 210, 211, 215, 221, 223, 227, *232, 233, 234,* 238, 240, 243, 244, 254, *256, 257, 258,* 264, 283, *287, 289,* 298, 315, *329,* 441, 442, *452*
Anderson, J. M., 375, *385*
Anderson, J. P., 254, *258*
Anderson, S., 221, *234,* 324, *332*
Andersson, J. A., 7, *25*
Andersson, S. G. E., 7, *25*
André, D., 110, 118, 119, 120, 121, *123*

AUTHOR INDEX

Casjens, S., 7, *25*
Castano, J. G., 352, 358, *370*
Castano, L., 423, *448*
Castner, B. J., 254, *258*
Castronovo, V., 414, *419*
Catlin, G. H., 58, *73*
Cavadini, P., 83, *98*
Cavener, D., 110, *123*
Cawley, N. X., 191, *197*, 302, *332*
Cejka, Z., 336, 339, 340, 345, 353, *367, 368*
Cereghino, J. L., 269, *288*
Cerretti, D. P., 254, *258*
Cerundolo, V., 363, 364, *370*
Chaal, B. K., 105, 106, 107, 109, 110, *123*
Chabin, R., 45, *55*
Chaddock, A. M., 39, *53*
Chai, T., 15, *26*
Chaidez, J., 68, *74*
Chakrabarti, D., 394, 398, 412, *418*
Chakraborty, A., 394, 412, *418, 419*
Chambers, T. M., 220, 227, *234*
Chambre, I., 84, 85, 88, *98*
Chan, A. C., 375, *385*
Chan, G., 426, 433, 434, *449*
Chan, S. J., 190, *197*, 200, 202, *230, 232*, 238, 240, 244, 250, *256, 258*, 264, 285, *286*, 325, *332*, 446, *452*
Chanal, A., 28, *53*
Chanat, E., 428, *450*
Chance, R., 171, *194*
Chance, R. E., 199, *230*
Chang, C. N., 31, *54*, 58, *73*
Chang, E. Y., 303, *332*
Chang, S. Y., 7, 8, 10, *24*, 392, 393, 401, 404, 411, *418, 419*
Chang, Y.-H., 392, 393, 394, 396, 398, 400, 402, 403, 406, 407, 411, 412, 413, 414, 415, *418, 419, 420*
Chapman, C., 254, *258*
Chapman, R. E., 61, *73*, 204, 212, *232*
Chapman-Smith, A., 390, 391, 392, 404, *418*
Chatterjee, A. K., 136, *158*
Chatterjee, M., 412, *419*
Chatterjee, N., 394, 412, *418*
Chatterjee, S., 38, *55*
Chattopadhyay, A., 394, 412, *418*
Chaudhury, C., 375, 376, 381, *386*
Chayen, S., 426, 433, 434, *449*
Che, F., 441, 442, *452*
Checler, F., 192, *197*, 253, 254, *258*

Chemelli, R. M., 307, *330*
Chen, A., 202, *231*
Chen, C., 128, 151, *157*
Chen, H., 423, 426, 433, 434, 435, *448, 449*
Chen, J. S., 202, *231*
Chen, K. C. S., 129, 132, *157*
Chen, L., 28, *53*
Chen, M., 28, *53*
Chen, P., 342, 343, 347, 351, 353, 357, 362, 365, *368*
Chen, S., 412, 413, 414, *419*
Chen, S. J., 86, *99*
Chen, X., 33, 36, 44, *54*, 61, 62, 63, 65, 67, 71, 72, *74*, 120, *124*
Chen, Z., 86, *99*
Cheng, D., 240, 242, 243, 244, *256, 257*, 297, *329*
Cheng, M. Y., 86, *99*
Cheng, S. Y., 31, 39, *53*
Chestnut, W. G., 433, *451*
Chestukhina, G. G., 429, *450*
Cheung, R., 437, 438, *451*
Chew, A., 89, 92, 93, 94, 95, 96, *100*
Chi, Y. I., 89, *100*
Chiang, Y., 221, *234*, 324, *332*
Chillingworth, T., 7, *25*
Chin, M. T., 446, *452*
Chinzei, T., 61, *73*
Chisholm, D. A., 373, 376, 377, 385, *385*
Chishti, A. H., 375, *385*
Chiu, C. H., 406, 407, *419*
Cho, S., 172, 173, 174, 176, *194*
Chretien, M., 164, 177, *193, 195*, 199, 200, 201, 202, 203, 204, 206, 218, 220, 221, 222, 223, *230, 231, 232, 234*, 238, 240, 242, 243, 244, 250, 254, *256, 257, 258*, 269, 276, 285, *286*, 293, 294, 295, 296, 297, 300, 301, 302, 303, 304, 305, 307, 308, 309, 310, 314, 315, 316, 321, 323, *328, 329, 330, 331, 332*, 425, 435, *449, 451*
Christen, A., 384, *386*
Christian, J. L., 225, 226, *235*
Christie, G., 254, *258*
Christopherson, K. S., 375, *386*
Chu, L. L., 176, *194*
Chun, J. Y., 202, *231*
Chung, C. H., 340, *368*, 374, *385*
Chung, Y. S., 134, 136, *158*
Churcher, C., 7, *24, 25*
Ciechanover, A., 336, *367*

Subject Index

K

L

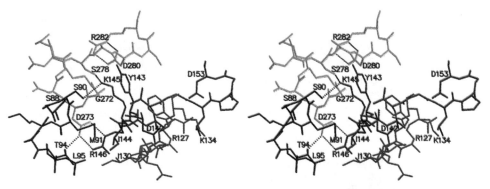

CHAPTER 2—FIGURE 6 The residues conserved among type I SPases (boxes B-E) make up the active site region of *E. coli* SPase I. A ball and stick stereographic representation of the residues that make up box B (88–95, black), box C (127–134, red), box D (142–153, blue), and box E (272–282, green). This figure was made with the program MOLSCRIPT (*104*). (See text page 43.)

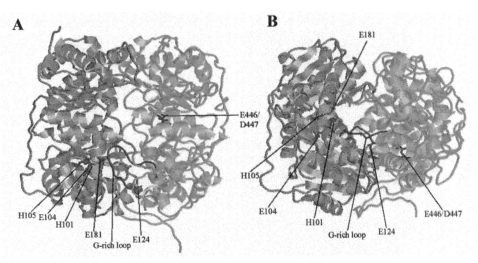

CHAPTER 4—FIGURE 5 Simulation model of rat MPP. A model of MPP was built using the bovine cytochrome bc_1 complex core proteins as reference proteins, using the Insight II/Homology Software Package. (*A*) Parallel view of the two subunits (*B*) View rotated 90° around the horizontal axis relative to A. α-MPP and β-MPP are shown in green and blue, respectively. The amino acids are numbered counting from the first amino acid of the precursor protein, as in Fig. 4. Modified from (*108*).

TGN / PLASMA MEMBRANE

budding

furin

clathrin

adaptor

PACS-1

retrieval

ENDOSOMAL COMPARTMENTS

B

furin-cd

budding

PACS-1/CK2-mediated retrieval

.........SYKGLPPEAWQEECPS$_{773}$DS$_{775}$EEDE.............
          P   P

Y   tyrosine motif

AC   phos. acidic cluster

CHAPTER 8—FIGURE 5 **Mechanism of cd-directed furin trafficking.** The cellular sorting machinery (A) and cytoplasmic domain sequences (B) utilized in directing the efficient budding of furin from the TGN/plasma membrane and retrieval from endosomes. See text for details.

Proprotein Convertase Family

Enzyme		Amino Acids
Subtilisin	DH N S	382
ykexin	DH N S	814
mPC2	DH D S	637
rPC4	DH N S	654
mPC1	DH N S	753
rPC7	DH N S	783
hfurin	DH NS	794
rPC5-A	DH N S	915
rPC5-B		1877
hPACE4-A	DH N S	969
hPACE4-E	DH N S	975

- Signal peptide
- Pro-segment
- Catalytic domain
- RGD(S) sequence
- P domain
- Ser/Thr-rich domain
- Cysteine-rich domain
- Amphipathic region
- Transmembrane domain
- Cytoplasmic domain
- N-glycosylation site

CHAPTER 9—FIGURE 1 **Diagrammatic representation of the structural motifs in mammalian proprotein convertases.** Comparison with bacterial subtilisin BPN' and yeast kexin. The C terminus of PACE4-E contains a hydrophobic sequence that could serve as a signal for GPI anchoring.

A
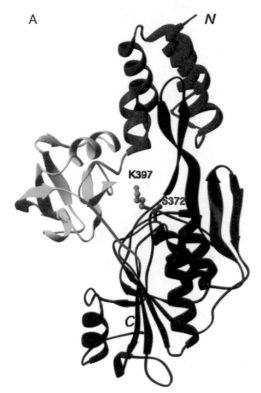

CHAPTER 13—FIGURE 1 Structure of D1P and Tsp homologs. (A) Crystal structure of *S. obliquus* D1P. The PDZ domain is shown in yellow, the active-site domain is blue, and the N-terminal domain is red. The active-site serine (S372 in the D1P numbering) is shown in red, and the active-site lysine (K379) is shown in cyan. This figure was kindly provided by D. Liao, Q. Qian, D. Chisholm, D. Jordan, and B. Diner. (B) Domain structure of Tsp and D1P homologs. The N-terminal domain (NTD), PDZ domain, and active-site domain for each homolog was predicted from alignments with *S. obliquus* D1P. The species and accession numbers are as follows: D1P_Sob *Scenedesmus obliquus* T10500; D1P_Ath *Arabidopsis thaliana* CAA10694; D1P__Sol *Spinacia oleracea* (spinach) JH0263; CtpA_Ssp *Synechocystis* sp. PCC 6803 BAA10189; Tsp_Ssp S77395; Ctp_Sco *Streptomyces coelicolor* CAB72213; Ctp_Dra *Deinococcus radiodurans* G75383; Ctp_Bbu *Borrelia burgdorferi* F70144; CtpA_Bsu *Bacillus subtilis* B69610; Tsp_Aae *Aquifex aeolicus* F70369; Tsp_Cpn *Chlamydophila pneumoniae* F720634; Tsp_Hin *Haemophilus influenzae* P45306; Tsp_Eco *Escherichia coli* P23865.

B
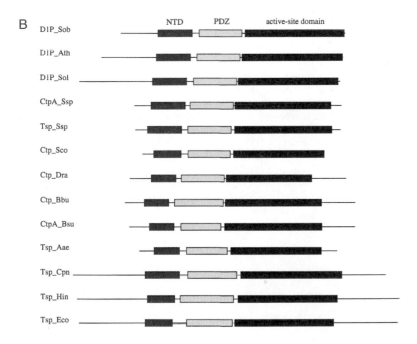

CHAPTER 14—FIGURE 3 Ribbon drawings showing the overall relationship between type I and type II MetAPs and other known pita-bread enzymes. The pseudo two fold symmetry of the core domain is illustrated by red α helices and blue β strands. The α-helical subdomain insertion unique to the type II enzymes is shown in green. N-terminal domain extensions are shown in yellow. The metals of the active site, if present, are Co(II) (magenta) or Mn(II) (cyan). (a) *E. coli* MetAP, PDB accession code 2MAT (*59*). (b) *P. furiosis* MetAP-IIa, IXGS (*56*). (c) Human MetAP-IIb, IBN5 (*57*). (d) Aminopeptidase P (AMPP), 1A16 (*71*). (e) Creatine amidinohydrolase (creatinase), 1CHM (*91*). The N-terminal domain extensions of creatinase and AMPP share the same fold, but are in a slightly different orientation in relation to the rest of the protein. The activity of creatinase is not metal dependent (*91*). Reported with permission from Lowther and Matthews (*51*). (See text page 395.)

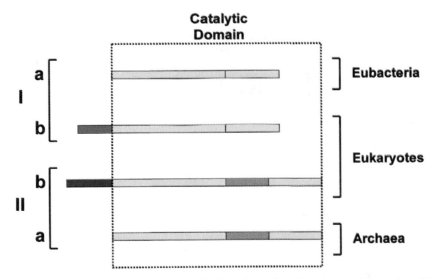

CHAPTER 14—FIGURE 4 Domain organization of MetAPs. Type Ia enzymes are found in eubacteria, and type IIa enzymes are found in archaea. MetAPs of both types are present in eukaryotes and display additional N-terminal domains. The type Ib enzymes have zinc-finger domains (shown in blue), whereas the type Ib enzymes have stretches of polybasic and acidic residues (shown in red). The catalytic domain is shown in yellow and the inserted domain, characteristic of type II enzymes, is shown in orange. Reprinted by permission from Bradshaw *et al.* (*18*). (See text page 396.)

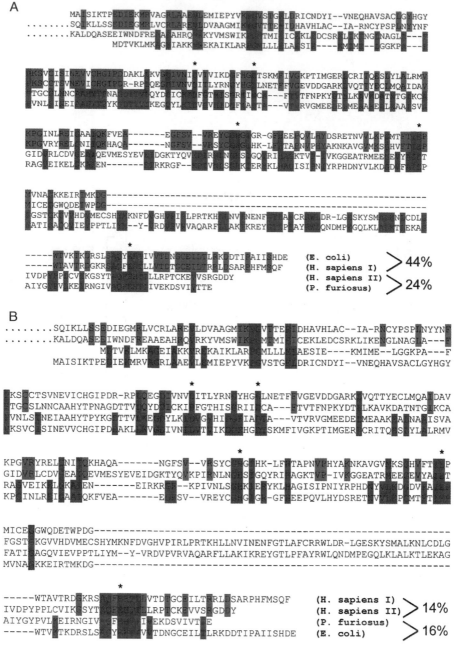

Chapter 14—Figure 5 Alignment of the amino acid sequences of the catalytic domain for the MetAP of *E. coli, P. furiosus,* and human (types I and II). Residues conserved in all sequences are highlighted in red; residues present in each pair compared are marked in green (upper) or magenta (lower). (A) Comparison of type I enzymes vs type II enzymes; (B) comparison of both human enzymes vs the two prokaryotic enzymes. (See text page 397.)

MALFQRAGSMAAVETRVCETDGCSSEAKLQCPTCIQLGIQGS
YFCSQECFKGSWATHKLLHKKAKDEKAKREVSSWTVEGDINT
DPWAGYRYTGKLRPHYPLMPTRPVPSYIQRPDYADHPLGMSE
MAISIKTPEDIEKMRVAGRLAAEVLEMIEPYVKPGVSTGELD
. . . . . .

MAGVEEVAASGSHLNGDLDPDDREEGAASPAEEAAKKKRRKKKK
SKGPSAAGEQEPDKESGASVDEVARQLERSALEDKERDEDDEDG
DGDGDGATGKKKKKKKKKRGPKVGTDPPSVPICDLYPNGVFPKG
QECEYPPTQDGRTAAWRTTSEEKKALDQASE

CHAPTER 14—FIGURE 6 Amino acid sequences of the N-terminal domains of the human MetAP isotypes. Upper panel: type I; lower panel: type II. The putative zinc finger motifs of the type I sequence are highlighted in red (with the actual proposed Zn^{2+} ligands given in white) and acidic/basic-rich sequences of the type II protein shown in red and green.

ISBN 0-12-122723-5

90051